Escherichia coli:
Virulence Mechanisms of a Versatile Pathogen

Escherichia coli

Virulence Mechanisms of a Versatile Pathogen

Edited by

Michael S. Donnenberg
Division of Infectious Diseases
Department of Medicine
University of Maryland School of Medicine
Baltimore, Maryland

ACADEMIC PRESS

An imprint of Elsevier Science

Amsterdam Boston London New York Oxford Paris San Diego
San Francisco Singapore Sydney Tokyo

This book is printed on acid-free paper. 

Academic Press
An imprint of Elsevier Science
525 B Street, Suite 1900, San Diego, California 92101-4495, USA
http://www.academicpress.com

Academic Press
An imprint of Elsevier Science
84 Theobald's Road, London WC1X 8RR, UK
http://www.academicpress.com

Library of Congress Control Number: 2002104261

International Standard Book Number: 0-12-220751-3

PRINTED IN THE UNITED STATES OF AMERICA
02 03 04 05 06 MB 9 8 7 6 5 4 3 2 1

To Berni for her support and understanding,
to the past and present members of my laboratory
for their dedication and hard work,
and to the authors for sharing their knowledge.

Contents

PART I

Escherichia coli The Organism

1 *The Genomes of Escherichia coli K-12 and Pathogenic E. coli*

Nicole T. Perna, Jeremy D. Glasner, Valerie Burland, and Guy Plunkett III

PART II

Escherichia coli Pathotypes

3 *Enteropathogenic Escherichia coli*

T. Eric Blank, Jean-Philippe Nougayrède, and Michael S. Donnenberg

4 *Enterohemorrhagic and Other Shiga Toxin–Producing Escherichia coli*

Cheleste M. Thorpe, Jennifer M. Ritchie, and David W. K. Acheson

9 *Meningitis-Associated Escherichia coli*

Kwang Sik Kim

PART III

Escherichia coli Virulence Factors

10 *Adhesive Pili of the Chaperone-Usher Family*

James G. Bann, Karen W. Dodson, Carl Frieden, and Scott J. Hultgren

11 *Type IV Pili*

W. Schreiber and Michael S. Donnenberg

12 *The LEE-Encoded Type III Secretion System in EPEC and EHEC: Assembly, Function, and Regulation*

J. Adam Crawford, T. Eric Blank, and James B. Kaper

Contributors

Numbers in parentheses indicate the pages on which the authors' contributions begin.

David W. K. Acheson (119)
Department of Epidemiology and Preventive Medicine
University of Maryland School of Medicine
Baltimore, Maryland 21201

Farah Bahrani-Mougeot (239)
Division of Infectious Diseases
Department of Medicine
University of Maryland School of Medicine
Baltimore, Maryland 21201

James G. Bann (289)
Department of Biochemistry and Molecular Biophysics
Washington University School of Medicine
St. Louis, Missouri 63110-1093

T. Eric Blank (81, 337)
Division of Infectious Diseases
Department of Medicine
University of Maryland School of Medicine
Baltimore, Maryland 21201

Valerie Burland (3)
Department of Genetics
University of Wisconsin–Madison
Madison, Wisconsin 53706

J. Adam Crawford (337)
Center for Vaccine Development
Department of Microbiology and Immunology
University of Maryland School of Medicine
Baltimore, Maryland 21201

William A. Day (209)
Department of Microbiology and Immunology
Uniformed Services University of the Health Sciences
F. Edward Hebert School of Medicine
Bethesda, Maryland 20814-4799

Karen W. Dodson (289)
Department of Molecular Microbiology and Immunology
Washington University School of Medicine
St. Louis, Missouri 63110-1093

Michael S. Donnenberg (81, 239, 307)
Division of Infectious Diseases
Department of Medicine
University of Maryland School of Medicine
Baltimore, Maryland 21201

Eric A. Elsinghorst (155)
Department of Clinical Laboratory Sciences
University of Kansas Medical Center
Kansas City, Kansas 66160

Carl Frieden (289)
Department of Biochemistry and Molecular Biophysics
Washington University School of Medicine
St. Louis, Missouri 63110-1093

Jeremy D. Glasner (3)
Department of Genetics
University of Wisconsin–Madison
Madison, Wisconsin 53706

Nereus W. Gunther IV (239)
Department of Microbiology and Immunology
University of Maryland School of Medicine
Baltimore, Maryland 21201

Colin Hughes (361)
Department of Pathology
Cambridge University
Cambridge CB2 1QP United Kingdom

Scott J. Hultgren (289)
Department of Molecular Microbiology and Immunology
Washington University School of Medicine
St. Louis, Missouri 63110-1093

James R. Johnson (55)
Veterans Affairs Medical Center
Department of Medicine
University of Minnesota
Minneapolis, Minnesota 55417

James B. Kaper (337)
Center for Vaccine Development
Department of Microbiology and Immunology
University of Maryland School of Medicine
Baltimore, Maryland 21201

Kwang Sik Kim (269)
Pediatric Infectious Diseases Division
Johns Hopkins University School of Medicine
Baltimore, Maryland 21136

Vassilis Koronakis (361)
Department of Pathology
Cambridge University
Cambridge CB2 1QP United Kingdom

Anthony T. Maurelli (209)
Department of Microbiology and Immunology
Uniformed Services University of the Health Sciences
F. Edward Hebert School of Medicine
Bethesda, Maryland 20814-4799

Harry L. T. Mobley (239)
Department of Microbiology and Immunology
University of Maryland School of Medicine
Baltimore, Maryland 21201

James P. Nataro (189)
Center for Vaccine Development
University of Maryland School of Medicine
Baltimore, Maryland 21201

Jean-Philippe Nougayrède (81)
Division of Infectious Diseases
Department of Microbiology and Immunology
University of Maryland School of Medicine
Baltimore, Maryland 21201

Nicole T. Perna (3)
Department of Animal Health and Biomedical Sciences

University of Wisconsin–Madison
Madison, Wisconsin 53706

Guy Plunkett III (3)
Department of Genetics
University of Wisconsin–Madison
Madison, Wisconsin 53706

Jennifer M. Ritchie (119)
Division of Geographic Medicine and Infectious Diseases
Tufts-New England Medical Center
Boston, Massachusetts 02111

Thomas A. Russo (379)
Department of Medicine and Microbiology
The Center for Microbial Pathogenesis
Veteran's Administration Medical Center
SUNY Buffalo
Buffalo, New York 10214

W. Schreiber (307)
Division of Infectious Diseases
Department of Medicine
University of Maryland School of Medicine
Baltimore, Maryland 212-1

Theodore Steiner (189)
Division of Infectious Diseases
University of British Columbia
Vancouver, British Columbia, Canada

Cheleste M. Thorpe (119)
Division of Geographic Medicine and Infectious Diseases
Tufts-New England Medical Center
Boston, Massachusetts 02111

Introduction

Michael S. Donnenberg
University of Maryland School of Medicine, Baltimore, Maryland

In 1885, the German pediatrician Theodore Escherich first described the organism that bears his name. Since that time, *Escherichia coli* may have become the most thoroughly understood organism on earth. Indeed, we owe much of our knowledge of intermediary metabolism, genetic recombination, DNA replication, RNA transcription, and protein synthesis, sorting, and secretion to studies carried out in *E. coli* [1].

Escherich discovered during his studies of the fecal flora of neonates and infants that this organism was universally present in healthy individuals (cited in [2]). In fact, *E. coli* is the most abundant facultative anaerobe present in the intestine of humans and many other warm-blooded species. As Escherich noted, *E. coli* colonizes neonates shortly after birth, an event that probably occurs during delivery because these initial strains usually are identical serologically to those found in the mother [3]. *E. coli* bacteria remain with us throughout life, although particular strains come and go over time. Most of these strains are nonpathogenic, coexisting in harmony with their hosts. Indeed, the relationship may be symbiotic, in that the bacteria, in addition to benefiting from the host, may synthesize cofactors and contribute to colonization resistance against pathogenic organisms.

This benign view of *E. coli* belies the fact that this species also can be regarded as the quintessential pleuripotential pathogen capable of causing a wide variety of illnesses in a broad array of species. Susceptible hosts include birds, pigs, cattle, rabbits, sheep, and humans. The gastrointestinal tract, the meninges, and the kidneys are among the target organs affected by *E. coli*. Diseases resulting from *E. coli* infections include diarrhea, dysentery, overwhelming sepsis, and the hemolytic-uremic syndrome. Thus this Jeckel and Hyde species can both coexist peacefully with its host and cause devastating illness. What is responsible for this paradox?

The answer lies in the existence of different strains of *E. coli* with differing pathogenic potential. Indeed, as early as 1897, Lesage postulated this point of view (cited in [2]). This concept gained gradual scientific support and ultimately achieved general acceptance when Bray established that strains that we now term *enteropathogenic E. coli* (EPEC) were the cause of devastating outbreaks of neonatal diarrhea [4]. Since then, a plethora of pathogenic varieties of this organism has been described.

The differences in the ability of stains to cause disease and the diverse syndromes caused by different strains can be attributed to the existence in some strains of specific genes encoding virulence factors and to the capacity of *E. coli* for genetic exchange. Even the genome of a nonpathogenic *E. coli* K-12 strain shows evidence of extensive lateral transfer and plasticity [5]. Indeed, Tatum and Lederberg first described bacterial recombination in *E. coli* [6], and genetic transfer via conjugation remains an important mode by which the species acquires new genes, as evidenced by the large number of pathogenic factors present in various strains of *E. coli* on plasmids. Other critical virulence determinants are encoded within the genomes of bacteriophages or within pathogenicity islands, mysterious blocks of foreign DNA encoding virulence factors, the precise origins of which remain obscure. Indeed, a pathogenic serotype O157:H7 *E. coli* strain has hundreds of blocks of DNA that are lacking in a nonpathogenic *E. coli* K-12 strain, and the K-12 strain has many blocks missing from the pathogenic strain [7]. Both strains also have many integrated bacteriophage genomes. Thus the pathogenic potential of a particular *E. coli* strain depends on the repertoire of specific virulence genes present within its genome. Particular virulence gene combinations define specific pathotypes of *E. coli*, and each pathotype has a propensity to cause a limited variety of clinical syndromes.

The first part of this book therefore is devoted to the general themes of how pathogenic *E. coli* strains differ from nonpathogenic strains and of the relationships among different lineages of pathogenic and nonpathogenic strains. Chapter 1 describes in more detail than was possible in the original publications the genomes of two strains of *E. coli*, one highly virulent and one harmless, and highlights the similarities and differences between these strains. Chapter 2 provides insight into the evolution of pathogenic *E. coli* strains (including *Shigella* spp., which properly should be regarded as strains of *E. coli*) by analyzing the genetic relationships among extant strains. This chapter for the first time provides emphasis on extraintestinal as well as intestinal pathogenic strains.

The relatively large number of *E. coli* pathotypes, the similarity in the names given to different pathotypes, inconsistencies in usage in the literature, advances in our understanding of pathogenesis, and the emergence of new pathotypes have all contributed to the complexity of the nomenclature for pathogenic *E. coli*. This nomenclature may be viewed as existing in a state of continuous flux as new strains are described and the relationships among previously described pathotypes are clarified. Figure 1 represents an attempt to illustrate these complex relation-

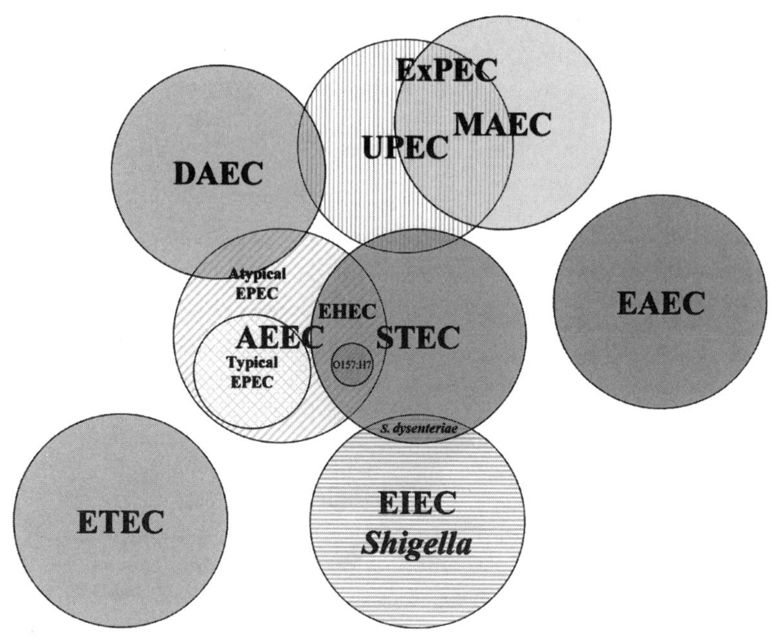

FIGURE 1 Venn diagram illustrating the complex relationships among different pathotypes of *E. coli* that cause disease in humans. Extraintestinal pathogenic *E. coli* (ExPEC) strains include meningitis-associated *E. coli* (MAEC, *yellow*) and uropathogenic *E. coli* (UPEC, *vertical stripes*). These strains share many virulence factors, and it is clear that single clones can cause both types of infections [8]. It is less clear whether or not strains exist that are capable of causing one syndrome and not the other. Among the UPEC, some strains exhibit diffuse adherence to tissue culture cells and share with diffuse adhering *E. coli* (DAEC, *orange*) the same adhesins. DAEC is a heterogeneous pathotype that has been epidemiologically linked to diarrhea. There are reports of DAEC strains recovered from individuals with both urinary tract infections (UTIs) and diarrhea [9]. There are also reports of Shiga toxin–producing *E. coli* (STEC, *green*) strains causing UTI [10]. STEC are defined by production of Shiga toxins, usually encoded by bacteriophages. Among STEC, some strains are also capable of attaching intimately to epithelial cells, effacing microvilli, and eliciting the formation of adhesion pedestals composed of cytoskeletal proteins, a property that defines the attaching and effacing *E. coli* (AEEC, *diagonal stripes*). Such strains, which are both STEC and AEEC, are known as enterohemorrhagic *E. coli* (EHEC). The most important serotype found within the EHEC pathotype is O157:H7. AEEC strains that do not produce Shiga toxins are referred to as enteropathogenic *E. coli* (EPEC). Among EPEC, many strains produce a bundle-forming pilus and attach to tissue culture cells in a localized adherence pattern. These are referred to as typical EPEC (*checkered*), whereas those which do not are known as atypical EPEC. There are some strains of atypical EPEC that exhibit diffuse adherence. Enteroinvasive *E. coli* (EIEC, *horizontal stripes*) invade tissue culture cells with high efficiency, multiply in the cytoplasm, and spread from cell to cell. These strains include the organisms commonly classified in the genus *Shigella*, which in fact is a subset of the species *E. coli*. Strains classified as *S. dysenteriae* produce Shiga toxins and therefore are members of both the EIEC and STEC pathotypes. Enteroaggregative *E. coli* (EAEC, *blue*) is a heterogeneous pathotype that causes acute and persistent diarrhea and is defined by its pattern of adherence. Enterotoxigenic *E. coli* (ETEC, *violet*) strains cause acute diarrhea and are defined by production of heat-labile and/or heat-stable enterotoxins. See Color Plate 1.

ships. It is useful to view pathogenic strains as belonging to two groups: those which cause gastrointestinal illness and those which cause extraintestinal infections. However, there may be strains with virulence potential that bridges these boundaries. Among the extraintestinal strains, it seems likely that most, if not all, strains capable of causing neonatal meningitis also can cause urinary tract infections, although the converse may not be true. Among the gastrointestinal pathotypes, the situation is even more complex, especially given the overlap in attributes of EPEC and Shiga toxin–producing *E. coli* (STEC). However, a precise lexicon remains possible within the classification scheme presented. Nevertheless, it is likely that new strains will emerge that have traits currently attributed to more that one pathotype to further complicate the picture.

The second part of this book contains chapters detailing the molecular pathogenesis of infections due to each of the *E. coli* pathotypes that cause human disease. It should be recognized that additional pathogenic varieties exist that cause disease exclusively in nonhuman species. These chapters provide an in-depth profile of each of these categories of organisms. A remarkable feature of this section is the shear number of distinct molecular pathways to human disease used by these different classes of the same species. Many of these pathogenic strategies are employed by other species that cause disease in animals and humans. Thus *E. coli* can serve as a model organism for the study of bacterial pathogenesis as well as intermediary metabolism.

Despite our attempts to categorize strains of *E. coli*, there is much overlap in the mechanisms of pathogenesis for various pathotypes. Similar virulence pathways may be pursued by more than one type of strain. For example, pili of the chaperone-usher family are ubiquitous among pathogenic and nonpathogenic *E. coli* strains. Type III secretion systems play an important role in the pathogenesis of EPEC, enterohemorrhagic *E. coli* (EHEC), and enteroinvasive *E. coli* infections. Type IV pili are produced by EPEC and some strains of enterotoxigenic *E. coli* (ETEC). Hemolysins of the RTX family are produced by many strains of *E. coli* that cause extraintestinal infections, by EHEC, and occasionally by other strains associated with intestinal infections. Many strains of pathogenic *E. coli*, especially those which cause extraintestinal infections, elaborate polysaccharide capsules, and all strains make lipopolysaccharide. To allow these critical virulence factors to be explored in more detail than is possible in the second section, the final part of this book contains chapters devoted to virulence systems that are common to more than one pathotype.

The interrelationships among various pathogenic and nonpathogenic *E. coli* strains, the complexities of the disease pathways navigated by each pathotype, and the overlap in virulence mechanisms employed by different types reveal an intricate web of information about the organism. Yet there is still much to learn. Despite our advances in the cellular and molecular details of the interactions between these organisms and host cells, we remain ignorant of the mechanisms by which most strains of *E. coli* actually cause disease. For some pathotypes, such as diffuse adhering and enteroaggregative *E. coli*, the virulence mechanisms that define the group remain largely mysterious. For other pathotypes that we thought we understood, such as ETEC, further research has revealed new surprises, the significance of which has yet to be fully realized. The entire genomic sequence of a uropathogenic *E. coli* strain is completed and should be available soon. This and the genomes of other prototypic strains likely will reveal unanticipated genes that may help to unravel disease mechanisms, clarify relationships among pathotypes, and provide insight into the evolution of the species. It is also likely that additional pathotypes of *E. coli* lurk unrecognized, awaiting characterization until new assays are applied to strains isolated from patients and controls. *E. coli* has been subject to intensive scrutiny for more than a century and will continue to be regarded with interest for a long time to come.

REFERENCES

1. Neidhardt, F. C. (1996). *Escherichia coli and Salmonella: Cellular and Molecular Biology*, 2d ed. ASM Press, Washington.
2. Robins-Browne, R. M. (1987). Traditional enteropathogenic *Escherichia coli* of infantile diarrhea. *Rev. Infect. Dis.* 9:28–53.
3. Bettelheim, K. A., Breadon, A., Faiers, M. C., *et al.* (1974). The origin of O serotypes of *Escherichia coli* in babies after normal delivery. *J. Hyg. (Lond)* 72:67–70.
4. Bray, J. (1945). Isolation of antigenically homogeneous strains of *Bact. coli neapolitanum* from summer diarrhoea of infants. *J. Pathol. Bacteriol.* 57:239–247.
5. Lawrence, J. G., and Ochman, H. (1998). Molecular archaeology of the *Escherichia coli* genome. *Proc. Natl. Acad. Sci. USA* 95:9413–9417.
6. Tatum, E. L., and Lederberg, J. (1947). Gene recombination in the bacterium *Escherichia coli*. *J. Bacteriol.* 53:673–684.
7. Perna, N. T., Plunkett, G., III, Burland, V., *et al.* (2001). Genome sequence of enterohaemorrhagic *Escherichia coli* O157:H7. *Nature* 409:529–533.
8. Russo, T. A., and Johnson, J. R. (2000). Proposal for a new inclusive designation for extraintestinal pathogenic isolates of *Escherichia coli* (ExPEC). *J Infect. Dis.* 181:1753–1754.
9. Germani, Y., Bégaud, E., Duval, P., and Le Bouguénec, C. (1997). An *Escherichia coli* clone carrying the adhesin-encoding *afa* operon is involved in both diarrhoea and cystitis in twins. *Trans. R. Soc. Trop. Med. Hyg.* 91:573–573.
10. Tarr, P. I., Fouser, L. S., Stapleton, A. E., *et al.* (1996). Hemolytic-uremic syndrome in a six-year-old girl after a urinary tract infection with Shiga-toxin-producing *Escherichia coli* O103:H2. *New Engl. J. Med.* 335:635–638.

PART I

Escherichia coli The Organism

CHAPTER 1

The Genomes of Escherichia coli K-12 and Pathogenic E. coli

Nicole T. Perna
Department of Animal Health and Biomedical Sciences, University of Wisconsin–Madison, Madison, Wisconsin

Jeremy D. Glasner
Department of Genetics, University of Wisconsin–Madison, Madison, Wisconsin

Valerie Burland
Department of Genetics, University of Wisconsin–Madison, Madison, Wisconsin

Guy Plunkett III
Department of Genetics, University of Wisconsin–Madison, Madison, Wisconsin

INTRODUCTION

Escherichia coli K-12 holds a special place in the hearts and laboratories of experimental molecular biologists and microbiologists. It is perhaps not surprising that this model organism–laboratory reagent–industrial workhorse was among the first microorganisms targeted for genome sequencing. Given the sheer volume of biologic research targeted at or using *E. coli* K-12, what is perhaps more surprising is the amount learned through relatively simple analysis of genome sequence and the number of questions generated by the evidence that so much remains unknown about what features are encoded within this genome, how networks of regulation orchestrate the genes to produce a dynamic living organism, and the relationship between the *E. coli* K-12 genome and those of the phenotypically diverse group of organisms we know as *E. coli*. Fortunately, many of these questions are already under investigation using genome-scale approaches to ascertaining function, comparative genomics to reveal the evolutionary and population history, and in laboratories worldwide now using the genome sequence as a resource to accelerate progress on more directed projects. Here we attempt to provide an overview of the history, current state, and future of *E. coli* genomics.

Escherichia coli: Virulence Mechanisms of a Versatile Pathogen
ISBN 0-12-220751-3

HISTORY OF *E. COLI* GENOMICS

Currently, there are published and publicly available *E. coli* genome sequences for *E. coli* K-12 [1], a benign laboratory model organism, and *E. coli* O157:H7 [2,3], an enterohemorrhagic pathogen. Additional genome sequencing projects are ongoing at the University of Wisconsin addressing *E. coli* strains that exhibit distinct phenotypes, at least with respect to their potential to cause human diseases. A genome sequence of an *E. coli* strain associated with urinary tract infections, CFT073, is nearly complete, and data have been available via a Web server for over 1 year (*www.genome.wisc.edu*). Data collection is nearing completion for RS218, a strain associated with neonatal sepsis and meningitis. Other closely related genomes also are being explored, ranging from *Shigella flexneri*, the causative agent of dysentery, to *Yersinia pestis*, the causative agent of the black plague, a more distantly related member of the family Enterobacteriaceae. Other enterobacterial genome sequencing projects either have been completed or are underway elsewhere, including several *Salmonella* isolates (Washington University–St. Louis, University of Illinois, and Sanger Centre), another *Y. pestis* strain (Sanger Centre), *Y. pseudotuberculosis* (Lawrence Livermore National Laboratory/Institute Pasteur), and *Klebsiella pneumoniae* (Washington University–St. Louis). For a list of completed and ongoing microbial genome projects, see the TIGR Web site at *www.tigr.org*. In all probability, this group will remain one of the best-sampled clades of closely-related organisms.

The *E. coli* K-12 Strain MG1655 Genome-Sequencing Project at the University of Wisconsin–Madison

A systematic *E. coli* K-12 genome-sequencing project began with genome mapping in the late 1980s in the laboratory of Dr. Frederick R. Blattner at the University of Wisconsin–Madison. The 4,639,221-basepair (bp) chromosome of strain MG1655 published by this group in 1997 represented the largest genome sequence completed in a single laboratory at that point in time [1]. The fully annotated sequence is deposited in GenBank (U00096) and is available as a single sequence through the Entrez Genomes Division; it is split into 400 records of approximately 11,500 bp each through the Entrez Nucleotide Database (Accession Numbers AE000111–AE000500). These sequence data are also available online directly from the Wisconsin group at *www.genome.wisc.edu*.

The original *E. coli* K-12 strain was isolated by Lederberg in 1922 from a convalescent patient suffering from an unrelated pathology. As an experimental model organism, derivatives of this strain underwent extensive handling in countless laboratories. *E. coli* K-12 strains have been subjected to repeated experimental mutagenic strategies, including treatment with ultraviolet light, EMS, x-rays, and acridine dyes to cure them of the F-plasmid, excise phage lambda, and procure variants with phenotypes of value in the laboratory environment [4]. After examining the records of Barbara Bachmann and the *E. coli* Genetic Stock Center at Yale University, MG1655 was selected as the available strain most similar to the original isolate that had undergone a minimum of mutagenic assaults [4].

The basic strategy employed in this project varied over its history as sequencing technologies developed. Approximately one-third of the genome was determined from a series of mapped overlapping lambda clones (~10 kbp each) proceeding counterclockwise from the 0/100 minute region of the chromosome. Analyses of subsections of completed genome sequence were released in a series of publications beginning in 1992 [5–10] and deposited into GenBank (Accession Numbers U00039, L10328, M87049, L19201, U00006, U14003, and U18997). The remaining two-thirds of the genome sequence was determined from random-shotgun libraries of larger fragments (100–200 kbp) recovered by pulsed-field gel electrophoresis (PFGE) of restriction endonuclease–digested genomic DNA from MG1655-derived strains constructed by introducing novel rare restriction sites into the chromosome using a mini-Tn10-derivative transposon vector [11]. The PFGE fragment isolation was subject to approximately 15% contamination from other areas of the genome, and sequences collected from these extraneous DNA templates were used to bolster coverage genome-wide.

Other *E. coli* K-12 Genome-Sequencing Projects

Even prior to any concerted efforts to complete a genome sequence of *E. coli* K-12, a substantial amount of sequence data was already available. By 1989, about 20% of the total chromosome sequence was known from the piecemeal contributions from different independent laboratories [12]. In contrast to the several large-scale sequencing efforts initiated after this point, these sequences are derived from a number of different K-12 strains and are distributed throughout the chromosome according to the research interests of each individual group.

An evolving consortium of Japanese laboratories has collected most of the genome sequence of a second *E. coli* K-12 strain, W3110 [13–22]. The bulk of this sequence was determined from the Kohara lambda clone library of *E. coli* K-12 strain W3110 [23]. This group began sequencing mapped clones beginning from the 0/100 minute mark and moving clockwise around the chromosome, focusing on filling gaps in the already available sequence data. Analysis of these data has been the subject of a series of publications, and sequences are available through GenBank as large, contiguous segments (Accession Numbers D10483, D26562, and D83536) and as individual sequences corresponding to each Kohara phage clone (Accession Numbers D90699–D90892, but note that some numbers in the sequence are not used). The sequence data and additional information also are available online at *http://ecoli.aist-nara.ac.jp/*. This project is ongoing and is expected to result in a complete W3110 chromosome sequence.

Two additional groups also have published large, contiguous *E. coli* K-12 sequences. A team at Harvard University, led by G. Church, reported two large, contiguous sequences from strain BHB2600 as part of a program to establish new methods in sequence determination (Accession Numbers U00007 and U00008). A group at Stanford, headed by R. W. Davis, has deposited four overlapping entries from strain MG1655 that total 526,200 bp in length (Accession Numbers U70214, U73857, U82664, and U82598). Each group has annotated its sequences independently, leading to multiple interpretations of the same feature. Coupled with true redundancy within the genome, in the form of multigene families, the redundancy in the GenBank database resulting from overlap between K-12 genome sequencing projects can lead to confusion in interpreting BLAST search results. Readers are cautioned to research the origin and history of each individual sequence and annotation carefully when drawing conclusions from the deposited genome data.

E. coli O157:H7 Genome-Sequencing Projects

Just as 1997 was a busy year for the genomics of *E. coli* K-12, 2001 saw the publication of two genomes from enterohemorrhagic *E. coli* O157:H7 strains [2,3,24]. Again, a group from the University of Wisconsin and a Japanese group from the Kazusa Institute worked independently toward genome sequences of closely related *E. coli*, but this time the strain selection was of a decidedly regional bent. The Wisconsin group chose O157:H7 strain EDL933, an isolate from Michigan ground beef linked to a multistate U.S. outbreak in 1982 in which *E. coli* O157:H7 was first associated with human disease [25]. The Japanese group chose O157:H7 Sakai, an isolate associated with the largest known outbreak of O157:H7 worldwide—a 1996 incident involving over 6000 children infected through the school lunch system in Sakai, Japan [26].

The Wisconsin group employed a whole-genome random shotgun cloning approach with the assistance of an optical restriction map [2,27], whereas the Japanese group used a strategy of shotgun cloning with several different vectors holding diverse-sized inserts that were end-sequenced [3]. Unlike the "manhandled" K-12 strains, both natural O157:H7 isolates contained plasmids, and these were sequenced and published by both groups [28,29], as were the Shiga toxin–bearing phage BP-933W [30] and VT-Sakai 2a [31,32], discussed in more detail later. Comparable results were obtained with both sequencing strategies, although the unanticipated plethora of prophage-related sequences proved a formidable obstacle to assembly of shotgun sequencing data for the Wisconsin group, and at the time of this writing, directed efforts to close the final gap in the chromosome sequence are still underway.

Independent analyses of the O157:H7 genomes collected by each group have yielded notably similar results. Observations made from comparisons of O157:H7 strain genomes with K-12 strain genomes have revealed remarkable genetic diversity within the *E. coli*.

ANALYSIS AND COMPARISON OF *E. COLI* GENOMES

Physical/Structural Characteristics of the *E. coli* Chromosome

E. coli have a circular chromosome. The complete K-12 MG1655 chromosome is 4,639,221 bp in length [1]. The O157:H7 Sakai strain has a 5,498,450-bp chromosome [3]. Bergthosson and Ochman [33] conducted a study of *E. coli* chromosome length using PFGE of rare-cutting restriction endonuclease–digested (I-*Ceu*I) genomic DNAs from 35 strains from an *E. coli* reference collection (ECOR) [34]. Their results suggest that *E. coli* chromosomes range from 4.500 to 5.520 million basepairs (Mbp). Thus K-12 and O157:H7 strains are close to the two extremes. Fine-scale dissection of the causes and consequences of differences between the K-12 and O157:H7 chromosomes is the subject of much of this chapter. Here we focus on global characteristics of *E. coli* genomes. A fully extended *E. coli* chromosome is roughly 1000 times longer than the cell itself [35]. *In vivo*, the chromosome exists as a looped and folded nucleoid structure of 50 to 100 individual supercoiled domains (reviewed in [36]). The genome sequences themselves have yielded no clues to the outstanding question of what maintains this structure.

Replication of the *E. coli* chromosome begins from a single origin (termed *oriC*) and proceeds bidirectionally [37]. Termination occurs approximately opposite the origin somewhere within a region bounded by several *Ter* sites [38,39]. The origin and the seven *Ter* sites known from the K-12 chromosome are all conserved in the O157:H7 EDL933 genome. The two halves of the *E. coli* genome defined by the origin and terminus of replication are referred to as *replichores* [1]. The two replichores of K-12 MG1655 are close to equal in length, as are the replichores of O157:H7 EDL933. In fact, there is a marked symmetry of replichore length throughout the natural isolates of *E. coli* used in the study of chromosome length discussed earlier [33]. Chromosomal inversions that include the origin or terminus may be subject to stabilizing selection to equalize the replichores in *Salmonella* [40–42]. Interestingly, the O157 Sakai genome is somewhat unbalanced in comparison with EDL933. The two replichores of O157 Sakai differ by about 290 kbp. EDL933 differs from both O157 Sakai and K-12 MG1655 by a relatively large (~426 kbp) inversion that includes the terminus (see below). In part due to this inversion, the two replichores of EDL933 are more equal in length.

On average, the *E. coli* genome is about 52% G + C. Local variations in G + C content are observed throughout the chromosome and often are associated with lineage-specific elements (see below for more detail). Like many prokaryote genomes [43], there is strand-specific bias in nucleotide use known as *skew*. This skew is not limited to coding features and is independent of a bias in orientation of genes [1]. Rather, the skew tracks well with the replicon structure, with the leading strand of each replichore tending to exhibit 1–2% more Gs than Cs. For this reason, skew appears to be related to an inherent bias in replication, but the issue of whether it is driven by a mutation bias or by natural selection remains to be determined (reviewed in [44]). Higher-order compositional biases also can be observed in the *E. coli* genomes [1]. In particular, an octamer (8-bp sequence) known as *Chi* and associated with recombination exhibits a strand-specific skew that follows the replichore organization. Additional octamers related to Chi also exhibit this bias. A common subsection of these skewed octamers includes a sequence used in the initiation of primer synthesis for discontinuous replication of the lagging strand, providing one possible link between a compositional skew and the molecular mechanism of replication. These skewed oligomers have been exploited to design strand-biased primers for a genome-scale polymerase chain reaction (PCR)–based survey of natural variation in the *E. coli* population [45].

Backbone and Islands: The Mosaic *E. coli* Chromosomes

From our current vantage point, it is difficult to discuss the content of any of the *E. coli* genomes without first considering what we know from the comparison of the genomes of *E. coli* K-12 and *E. coli* O157:H7. The two groups of strains share a core 4.1-Mbp basically colinear chromosomal frame that is punctuated by lineage-specific elements that cumulatively account for the size difference and many of the phenotypic differences between the strains. Figure 1 illustrates the segmented structure of the *E. coli* genomes. We refer to the shared regions of chromosome as *backbone* and the lineage-specific elements as *islands* (*O-islands* are found in O157:H7 EDL933, and *K-islands* are specific to K-12 MG1655) [2]. In total, there are 177 O-islands greater than 50 bp in length present in O157:H7 strain EDL933 but not K-12 MG1655, and there are 234 K-islands greater than 50 bp in MG1655. O- and K-islands are found at the same relative position of the chromosome in 146 cases. Although the thresholds and nomenclature differ, the same basic conclusion is drawn from comparison of O157 Sakai with MG1655. Hayashi *et al.* [3] report 296 strain-specific *S-loops*, or regions greater than 19 bp, in O157 Sakai and 325 *K-loops* in MG1655, of which 203 loops are found at analagous sites of the two chromosomes. The

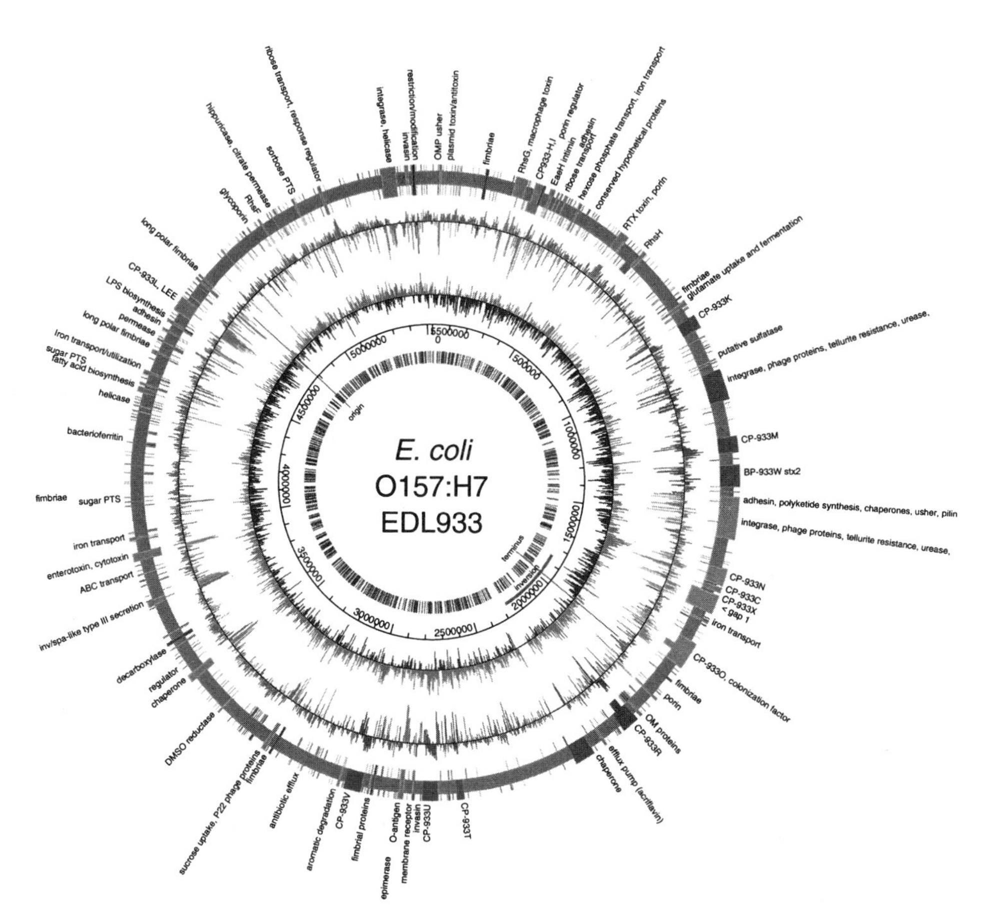

FIGURE 1 Circular map comparing the EDL933 and MG1655 genomes. Outer circle shows the distribution of islands: shared colinear backbone (*blue*); positions of EDL933-specific sequences (O-islands) (*red*); MG1655-specific sequences (K-islands) (*green*); O-islands and K-islands at the same locations in the backbone (*tan*); hypervariable (*purple*). Second circle: G + C content calculated for each gene >100 aa, plotted around the mean value for the whole genome, color coded like outer circle. Third circle: GC skew for codon third position, calculated for each gene >100 aa; positive values (*lime*), negative (*dark green*). Fourth circle: scale in bp. Fifth circle: distribution of the highly skewed octamer Chi (GCTGGTGG), *bright blue* and *purple* indicating the two DNA strands. The origin and terminus of replication and the chromosomal inversion are shown. The gap at ~2,300,000 is now closed, and the GenBank file and Web site are updated accordingly. [Reprinted with permission from Perna et al., 2001, *Nature* 409(6819):529–533; map figures created by Genvision (DNASTAR).] See Color Plate 2.

coding potential and evolutionary dynamics underlying this mosaic chromosome are the subject of the remainder of this section.

Evolution of Island/Strain-Specific Loops and Backbone

K-12 and O157:H7 strains last shared a common ancestor about 4.5 million years ago. This time point corresponds to approximately the midpoint in the 10-million-year-long radiation of *E. coli* strains [46]. As K-12 and O157:H7 diverged, regions of backbone from the ancestral chromosome have been retained and have accumulated base substitutions and small insertions or deletions, whereas lineage-specific islands have been introduced into the chromosome by horizontal transfer. While this simple model seems likely to hold for the majority of backbone and island segments, we anticipate that more complex evolutionary histories will be revealed for some segments of backbone and that some of the islands identified in the genome comparison will prove to be ancestral and deleted from one strain rather than acquired by the other. Establishing the polarity of insertion/deletion events will be greatly aided by the availability of a third *E. coli* genome sequence. In particular, the uropathogenic strain CFT073 is expected to cluster (unpublished data) with a clade of strains that form an outgroup to all other known *E. coli* [46]. Islands common to either K-12 or O157:H7 and CFT073 could be explained most parsimoniously as ancestral. Here we note that regions of the genome that are ancestral may still represent horizontal transfers into the *E. coli* population; they simply predate strain divergence.

This view of an *E. coli* genome as a composite of regions evolving clonally, periodically disrupted by replacement of blocks of existing alleles by homologous recombination, and introduction of new segments over time by horizontal transfer nicely fits the model of clonal frames of various ages posited by Milkman and colleagues in an in-depth series of studies of the *trp* region [47–50]. It also reconciles countless observations of strong linkage disequilibrium among alleles of housekeeping genes distributed throughout the chromosome with the diversity of *E. coli* genome sizes (for a review, see [51]).

Backbone and Divergence of Shared Regions

Backbone regions of the chromosome defined in the comparison of *E. coli* K-12 MG1655 and O157:H7 EDL933 are very similar. The average nucleotide identity between strains is 98.5%. A comparable value of 98.31% is reported for the comparison of O157 Sakai and MG1655. The backbone, although discontinuous, is colinear in all the sequenced *E. coli* genomes, with the exception of a large inversion spanning the terminus of replication that distinguishes EDL933 from both O157 Sakai and K-12 MG1655. The precise end points of the inverted segment are not entirely clear. Originally, from comparison of EDL933 and MG1655, the end points were assessed as a 121-bp repeat at the boundaries of a 426-kbp region of the EDL933 chromosome. The inverted repeat occurs directly adjacent to a junction between backbone and an O-island containing a cryptic phage in EDL933, whereas the 121-bp sequence occurs only once in MG1655. On publication of the O157 Sakai sequence, we initiated a comparison of the O157:H7 isolates and discovered that similarity between the O157:H7 isolates extends beyond the boundary defined by the 121-bp inverted repeat, suggesting that the end points of the inversion lie within two O-islands (S-loops) that both correspond to lambdoid prophage (see below). No other clear intragenomic rearrangements are observed in the backbone regions of the chromosomes, although interpretation of the distribution of repetitive elements and homologous subsections of the numerous cryptic lambdoid prophages (see below) suggests some internal juggling of information content.

Despite the strong conservation of nucleotide sequence, the net effect of base substitution should not be overlooked. Across the entire 4.1-Mbp backbone we have defined close to 70,000 single nucleotide polymorphisms that distinguish between K-12 MG1655 and O157:H7 EDL933 [2]. Although the majority of nucleotide substitutions lie at synonymous third-codon positions, this rate of divergence is sufficient to render 80% of the homologous protein products different by at least one amino acid replacement. Small insertion/deletion differences occur within coding regions common to the strains, resulting in fusions of genes in one strain that are distinct in the other strain and creating frame-shift mutations in otherwise shared coding regions. It is interesting that two such cases figure prominently in the detection of O157:H7. Assays for growth on sorbitol and β-glucuronidase activity are standard diagnostic tests for O157:H7. EDL933, typical of O157:H7 isolates, is negative in both. In the EDL933 backbone, the PTS components IIB and IIC for sorbitol/glucitol (*srlA* and *srlE* in K-12) are truncated by frame shifts. The β-glucuronidase protein UidA in K-12 is truncated from 604 to 237 aa by a frame shift in the backbone of EDL933.

Still other shared coding features exhibit signs of a more complex divergence pattern. Some genes span one or more junctions between backbone and lineage-specific regions. Other features fall into or contain regions that we have classified as *hypervariable* to distinguish them from surrounding regions of the backbone [2]. These hypervariable segments show much lower similarity between MG1655 and EDL933 than the regions classified as backbone yet encode related features at the same relative position in the two chromosomes. In the most extreme example, the proteins from MG1655 and EDL933 share only 34% amino acid identity. This pattern is reminiscent of elevated levels of divergence observed in *E. coli* sequences that are believed to be under selection for diversity, like genes involved with DNA restriction/modification systems [52], and for fimbriae [53] and flagella [54], where surface expression necessitates evasion of the host immune system. A similar observation was made from comparison of two homologous pathogenicity islands, the locus of enterocyte effacement (LEE) of O157:H7 EDL933 and the LEE of an enteropathogenic *E. coli* (EPEC) strain,

E2348/69. Excess polymorphism is found in discontinuously encoded gene products that interact directly with the host, whereas levels of similarity typical of clonal divergence are observed in genes that encode components of a type III secretory apparatus encoded adjacent to and interdigitated among the hypervariable patches [55]. In most of these cases, the rates of divergence at both synonymous and nonsynonymous sites are elevated in these regions. This is important because although selection alone might account for elevated rates of polymorphism in amino acid sequences of proteins, elevated rates at synonymous, or silent, sites are indicative of higher rates of mutation in the hypervariable regions than in the surrounding backbone. Studies that closely examine the shifting phylogenetic signal across homologous sequences can clearly define the boundaries of intralocus recombination events (e.g., see [52–54]). Homologous recombination among divergent alleles coupled with selection to differentially preserve the relatively rare recombinants seems most consistent with the observations. The particularly high levels of divergence among the LEE elements described earlier and among several of the hypervariable regions identified in the genome-wide comparison show that the divergence level between donor and recipient sequences for these recombination events could greatly exceed that of the divergence between *E. coli* and *Salmonella* [56].

Islands Acquired by Horizontal Transfer and/or Differential Gene Loss

The genome comparisons described here highlight the need for a reassessment of the island concept. The classic pathogenicity island consists of a large genome region with distinct boundaries delineating it from adjacent backbone, G + C content different from the genome average, insertion at a tRNA gene, and presence of an integrase, transposase, or other potential recombinase that could have mobilized the segment [57]. These features are interpreted as evidence of acquisition by horizontal transfer. Several large islands having the classic form have now been described in various species and reviewed comprehensively by Hacker and Kaper [58]. The islands described here (see Fig. 1) were identified by alignment of the EDL933 and K-12 genomes using a suffix-array algorithm to locate maximal exact matches and thus were based entirely on differences between the genomic DNA sequences, without reference to the genes encoded or the presence of other specific features. From these unbiased comparisons, it is clear that the classic large islands are a small subset of the lineage-specific segments of EDL933, all of which could qualify as islands whose origins are outside the O157:H7 serogroup and even outside *E. coli*.

In some cases, a potential relationship between genes from an island and pathogenic functions seems obvious. In other cases, islands encode products of a discernible function, but if they are related to pathogenicity, exactly how is unclear. Still other EDL933 islands encode only open reading frames (ORFs) for which no function can yet be postulated. More information about the structure of islands and functions encoded by islands is found below. An equally difficult problem for the island paradigm is the observation that the benign *E. coli* K-12 has a genome that is populated extensively with islands, including at least one that meets all other criteria for a classic pathogenicity island. Furthermore, many of the islands in both strains show no characteristics of autonomous mobility yet clearly exhibit a base composition or codon usage atypical for *E. coli* backbone, suggestive of horizontal transfer.

The occurrence of islands of similar size but entirely different content in each strain at the same site relative to the backbone may explain why methods based on restriction digestion lead to the conclusion that *E. coli* genomes differ by a relatively small number of large elements [59]. This type of methodology has been applied to compare a uropathogenic *E. coli* (strain J96) with K-12 and has successfully identified four large J96-specific elements and two deletions relative to K-12 [60].

Evidence of extensive horizontal transfer in the genome sequences was not entirely unexpected. Prior studies of the MG1655 sequence alone gave clues to the mosaic structure. On the basis of a correspondence analysis of codon usage prior to completion of the MG1655 genome sequence, Medigue *et al.* [61] predicted that up to 10% of the *E. coli* K-12 chromosome was derived from horizontal transfer. Lawrence and Ochman [62] analyzed the complete genome using codon-position-specific nucleotide composition to estimate the time of acquisition of putative horizontal transfers. The comparison of MG1655 and the O157:H7 genomes confirms that many of the predicted horizontal transfers do show a lineage-specific distribution. Now it is possible to collect genome-scale presence and absence data by hybridizing genomic DNA to spotted arrays of all the genes of the K-12 genome. Ochman and Jones [63] used whole-genome spotted arrays to detect differences in genome content among four *E. coli* strains, including K-12 W3110. Their analysis indicates that a minimum of 37 insertion and 30 deletion events are necessary to account for the gene content of MG1655, given the distribution of genes among the other four strains and the known phylogenetic relationship of clonally evolving genes of these strains. The biggest limitation of the hybridization-based approach is that a gene in the *E. coli* population will be detected only if it has been included already on the array. We are now in a position to create composite MG1655 and EDL933 arrays. Analysis of random amplified polymorphic DNA in the ECOR collection revealed a large pool of strain-specific elements in the *E. coli* population [64]. Any additional lineage-specific elements identified in ongoing *E. coli* genome projects or from studies of individual loci should be added to arrays.

Gene Content of the *E. coli* Genomes

The density of coding regions is very high in microbial genomes. In MG1655, 88% of the genome encodes known or putative proteins, and 1% encodes stable RNAs [1]. Comparable numbers are obtained from both O157:H7 genomes. In this discussion of the coding capacity of the *E. coli* genomes, we have opted to focus on MG1655 and EDL933. The fact that both these genomes were annotated by a single group using the same basic methodology facilitates comparison across strains. In particular, MG1655 and EDL933 ORFs were defined using GeneMark [65,66], and putative functions were attributed using the same classification system [67,68]. Table I enumerates the functional classes in this system and the number of genes in each class observed in MG1655 and EDL933. Genes from each strain are subdivided according to whether they occur within backbone, in lineage-specific segments, or exhibit a more complex relationship between the two strains. Clearly, many of the lineage-specific gene products in both strains are of undetermined function. These fall into one of two functional categories (0 and 60.95) depending on whether any inference can be made about the general physiologic role of the protein. For example, a putative protein that exhibits similarity to other transport proteins but for which no substrate is indicated would be "unclassified" (60.95) rather than "unknown" (0). Some broad categories have very few lineage-specific genes, including "central intermediary metabolism" and "amino acid biosynthesis." A very large difference between strains is seen in the "phage- and prophage-related" categories. This difference is largely due to a revision of the classification system. At the time MG1655 was annotated, ORFs that fell within phage boundaries were assigned to a variety of categories, especially "unknown." For EDL933, we expanded the classification system to include subcategories within the general designation and assigned even unknown ORFs within phage boundaries to one of the "phage- and prophage-related" subcategories.

Although the two O157:H7 genomes were published too recently to allow time for a rigorous comparison, it is clear that there is a general concordance of observations about gene functions between the two research groups. Features that clearly differ between the O157:H7 isolates include a large island present once in O157 Sakai but twice in EDL933 (see below) and a cryptic prophage specific to O157 Sakai (see below) (unpublished observations).

EDL933 Genes, Operons, and Islands

There are 177 O-islands of EDL933 ranging from 87 kbp to less than 1 kbp, each designated *OI#*. ORFs have been assigned unique identifiers of the form *Z0000*. Sixteen islands are phages or phage remnants, which are treated as a separate class in this analysis (see below). The largest two islands, OI#43 and OI#48, are identical, suggesting a recent duplication event. Five islands are of the classic pattern, associated with a tRNA and containing integrase and transposase genes that could account for acquisition. Table II lists the islands larger than 5 kbp, and Fig. 2 shows the 12 largest. Most of these contain genes with potential relevance to

TABLE I Functional Attributions for the ORFs Annotated in *E. coli* K-12 Strain MG1655 and O157:H7 Strain EDL933 Classified by Whether the Genes Lie Entirely within Backbone Regions of the Chromosome, Lie Entirely within Lineage-Specific Regions of the Chromosome, or Exhibit a More Complex Relationship between Strains (Includes Both Hypervariable Genes and Genes That Span the Junction between Backbone and Lineage-Specific Segments)

Functional group number	Functional class description (general)	Functional class description (intermediate)	MG1655 backbone	MG1655 specific	MG1655 complex	EDL933 backbone	EDL933 specific	EDL933 complex
0	Unknown proteins	Unknown function	1070	191	102	1124	401	91
1.1	Degradation of small molecules	Carbon compounds	97	14	2	97	12	2
1.2	Degradation of small molecules	Amino acids	17	1	0	17	0	0
1.25	Degradation of small molecules	Amines	7	1	1	8	0	0
1.3	Degradation of small molecules	Fatty acids	5	5	0	5	0	0
1.4	Central intermediary metabolism	Phosphorus compounds	17	0	0	18	0	0
1.6	Degradation of small molecules	Other	0	0	0	0	14	0
2.1	Energy metabolism, carbon	Glycolysis	17	0	0	16	0	0
2.2	Energy metabolism, carbon	Pyruvate dehydrogenase	5	0	0	4	0	1
2.3	Energy metabolism, carbon	TCA cycle	15	0	2	15	1	1
2.51	Energy metabolism, carbon	Oxidative branch, pentose pathway	2	0	0	2	0	0
2.52	Central intermediary metabolism	Nonoxidative branch, pentose pathway	6	2	0	6	0	0
2.71	Energy metabolism, carbon	Aerobic respiration	32	0	0	32	0	0
2.72	Energy metabolism, carbon	Anaerobic respiration	79	0	1	79	0	1
2.8	Energy metabolism, carbon	Electron transport	25	0	0	24	0	1

(continues)

TABLE I (*Continued*)

Functional group number	Functional class description (general)	Functional class description (intermediate)	MG1655 backbone	MG1655 specific	MG1655 complex	EDL933 backbone	EDL933 specific	EDL933 complex
2.81	Energy metabolism, carbon	Fermentation	19	0	1	19	0	1
3	Central intermediary metabolism	Pool, multipurpose conversions of intermediary metabolism	57	5	1	57	1	1
3.05	Central intermediary metabolism	Glyoxylate bypass	4	1	0	4	0	0
3.06	Central intermediary metabolism	Entner-Douderoff	3	0	0	3	0	0
3.07	Central intermediary metabolism	Polyamine biosynthesis	7	0	1	8	0	0
3.11	Central intermediary metabolism	Gluconeogenesis	4	0	0	4	0	0
3.12	Central intermediary metabolism	Nucleotide hydrolysis	2	0	0	2	0	0
3.14	Central intermediary metabolism	Sugar-nucleotide biosynthesis, conversions	14	4	0	14	0	0
3.15	Central intermediary metabolism	Nucleotide interconversions	10	2	0	10	0	0
3.16	Central intermediary metabolism	Amino sugars	9	0	1	9	0	0
3.18	Central intermediary metabolism	Not defined	3	0	0	3	0	0
3.19	Central intermediary metabolism	Miscellaneous glucose metabolism	3	0	0	3	0	0
3.2	Central intermediary metabolism	Sulfur metabolism	10	0	0	10	0	0
5	Energy transfer, ATP-proton motive force	ATP-proton motive force interconversion	9	0	0	9	0	0
6	Global regulatory functions	Global regulatory functions	46	3	3	45	0	2

6.1	Nucleoid-related functions	Nucleoid-related functions	0	0	0	1	0	0
9.11	Amino acid biosynthesis	Glutamate	1	0	0	1	0	0
9.12	Amino acid biosynthesis	Glutamine	5	0	0	5	0	0
9.13	Amino acid biosynthesis	Arginine	11	1	0	11	0	0
9.14	Amino acid biosynthesis	Proline	3	0	0	3	0	0
9.21	Amino acid biosynthesis	Aspartate	1	0	0	1	0	0
9.22	Amino acid biosynthesis	Asparagine	3	0	0	3	0	0
9.23	Amino acid biosynthesis	Lysine	9	0	0	9	0	0
9.24	Amino acid biosynthesis	Threonine	3	0	1	3	0	1
9.26	Amino acid biosynthesis	Methionine	8	0	0	8	0	0
9.31	Amino acid biosynthesis	Glycine	1	0	0	1	0	0
9.32	Amino acid biosynthesis	Serine	3	0	0	3	0	0
9.33	Amino acid biosynthesis	Cysteine	5	0	0	5	0	0
9.51	Amino acid biosynthesis	Phenylalanine	3	0	0	2	0	0
9.52	Amino acid biosynthesis	Tyrosine	3	0	0	3	0	0
9.53	Amino acid biosynthesis	Tryptophan	8	0	1	10	0	0
9.54	Amino acid biosynthesis	Chorismate	8	0	0	8	0	0
9.6	Amino acid biosynthesis	Histidine	9	0	0	9	0	0
9.71	Amino acid biosynthesis	Alanine	2	0	0	2	0	0
9.81	Amino acid biosynthesis	Isoleucine	13	0	1	13	0	0
9.82	Amino acid biosynthesis	Leucine	5	0	1	5	0	1
16	Nucleotide biosynthesis	Purine ribonucleotide biosynthesis	22	0	0	22	0	0
17	Nucleotide biosynthesis	Pyrimidine ribonucleotide biosynthesis	10	0	0	10	0	0
18	Central intermediary metabolism	2′-Deoxyribonucleotide metabolism	8	0	1	8	0	1
19	Central intermediary metabolism	Salvage of nucleosides and nucleotides	17	0	1	17	0	1
22.05	Biosynthesis of cofactors, carriers	Biotin carboxyl carrier protein (BCCP)	1	0	0	1	0	0
22.1	Biosynthesis of cofactors, carriers	Biotin	9	0	0	9	0	0

(continues)

TABLE I (*Continued*)

Functional group number	Functional class description (general)	Functional class description (intermediate)	MG1655 backbone	MG1655 specific	MG1655 complex	EDL933 backbone	EDL933 specific	EDL933 complex
22.2	Biosynthesis of cofactors, carriers	Folic acid	8	0	1	8	0	1
22.3	Biosynthesis of cofactors, carriers	Lipoate	2	0	0	2	0	0
22.4	Biosynthesis of cofactors, carriers	Molybdopterin	10	0	2	9	2	2
22.5	Biosynthesis of cofactors, carriers	Pantothenate	3	0	1	4	0	0
22.6	Biosynthesis of cofactors, carriers	Pyridoxine	6	0	0	6	0	0
22.7	Biosynthesis of cofactors, carriers	Pyridine nucleotide	6	0	0	6	0	0
22.8	Biosynthesis of cofactors, carriers	Thiamine	8	0	0	9	0	0
22.9	Biosynthesis of cofactors, carriers	Riboflavin	5	0	0	7	0	0
22.91	Biosynthesis of cofactors, carriers	Thioredoxin, glutaredoxin, glutathione	10	0	0	10	0	0
22.92	Biosynthesis of cofactors, carriers	Menaquinone, ubiquinone	13	0	1	13	0	1
22.93	Biosynthesis of cofactors, carriers	Heme, porphyrin	14	0	1	14	1	1
22.98	Biosynthesis of cofactors, carriers	Cobalamin	6	0	0	6	1	0
22.99	Biosynthesis of cofactors, carriers	Enterochelin	6	0	0	4	0	0
23	Fatty acid biosynthesis	Fatty acid and phosphatidic acid biosynthesis	23	0	0	23	11	1

40.05	Macromolecule synthesis, modification	Ribosomal and stable RNAs	24	0	1	24	0	1
40.1	Macromolecule synthesis, modification	Ribosomal proteins: synthesis, modification	56	0	0	56	0	0
40.2	Macromolecule synthesis, modification	Ribosomes: maturation and modification	5	0	0	4	0	0
40.31	tRNAs	tRNA	73	4	9	82	13	3
40.32	aa-tRNAs	Amino acyl tRNA synthesis; tRNA modification	40	0	0	40	0	0
40.4	Basic proteins	Basic proteins: synthesis, modification	7	0	0	5	0	0
40.5	Macromolecule synthesis, modification	DNA: replication, repair, restriction, modification	75	1	7	74	7	8
40.6	Macromolecule synthesis, modification	Proteins: translation and modification	33	1	0	33	0	0
40.7	Macromolecule synthesis, modification	RNA synthesis, modification, DNA transcription	22	0	0	20	0	0
40.8	Macromolecule synthesis, modification	Polysaccharides (cytoplasmic)	5	0	1	5	0	1
40.91	Macromolecule synthesis, modification	Lipoprotein	9	0	1	9	0	1
40.93	Macromolecules	Lipopolysaccharide	3	6	4	3	0	2
40.96	Macromolecule synthesis, modification	Phospholipids	10	0	1	9	2	1
41.1	Macromolecule degradation	Degradation of RNA	11	0	0	11	0	0
41.2	Macromolecule degradation	Degradation of DNA	18	5	2	18	1	0
41.3	Macromolecule degradation	Degradation of proteins, peptides, glycoprotein	28	1	0	28	5	0
41.5	Macromolecule degradation	Degradation of polysaccharides	3	0	0	3	0	0
50.1	Cell envelope	Outer membrane constituents	11	2	3	11	4	4
50.15	Cell envelope	Inner membrane	4	0	0	3	0	0

(continues)

TABLE I (*Continued*)

Functional group number	Functional class description (general)	Functional class description (intermediate)	MG1655 backbone	MG1655 specific	MG1655 complex	EDL933 backbone	EDL933 specific	EDL93 complex
50.2	Cell exterior constituents	Surface polysaccharides and antigens	15	4	0	16	16	3
50.3	Cell exterior constituents	Surface structures	51	0	6	53	17	16
50.5	Cell envelope	Murein sacculus, peptidoglycan	31	1	1	32	0	0
51	Transport of small molecules	Amino acids, amines	48	0	8	56	0	1
52	Transport of small molecules	Cations	42	6	3	43	15	2
53	Transport of small molecules	Carbohydrates, organic acids, alcohols	80	2	2	81	21	4
54	Transport of small molecules	Nucleosides, purines, pyrimidines	6	0	0	6	0	0
56	Transport of small molecules	Anions	21	0	1	21	0	0
58	Transport of small molecules	Other	12	0	0	12	0	0
58.1	Folding and ushering proteins	Chaperones	7	0	0	7	1	2
58.2	Cell division	Cell division	28	4	1	28	0	1
58.3	Chemotaxis, motility	Chemotaxis and mobility	12	0	0	12	0	0
58.4	Transport of large molecules	Protein, peptide secretion	24	0	0	25	30	1
58.45	Extracellular functions	Secreted proteins	0	0	0	0	20	2
58.5	Adaptation	Osmotic adaptation	13	0	1	13	0	1
58.6	Protection responses	Detoxification	9	0	2	10	0	1
58.7	Protection responses	Cell killing	2	1	0	4	6	0
60.2*	Phage- and prophage-related functions	Other or unknown	11	15	3	5	390	38
60.21*	Phage- and prophage-related functions	DNA packaging, phage assembly	0	0	0	0	46	9

60.22*	Phage- and prophage-related functions	Replication	0	0	0	0	9	0
60.23*	Phage- and prophage-related functions	Regulation	0	0	0	0	32	2
60.24*	Phage- and prophage-related functions	Integration, recombination	0	0	0	0	42	4
60.25*	Phage- and prophage-related functions	Lysis	0	0	0	0	36	3
60.26*	Phage- and prophage-related functions	Structural component	0	0	0	0	77	10
60.3	Laterally acquired elements	Colicin-related functions	4	0	1	4	0	1
60.4	Laterally acquired elements	Plasmid-related functions	1	0	0	1	0	0
60.41	Plasmid-related functions	Replication and maintenance	0	0	0	0	2	0
60.42	Plasmid-related functions	Plasmid transfer	0	0	0	0	2	0
60.5	Protection responses	Drug/analogue sensitivity	37	1	2	37	3	4
60.8	Adaptation	Adaptations, atypical conditions	14	1	1	20	16	1
60.9	Laterally acquired elements	Unknown function	3	52	3	2	80	1
60.91*	IS- and transposon-related functions	Transposases	0	0	0	0	37	1
60.95	Not classified	Not classified	882	127	75	872	75	51
Subtotals			3675	464	266	3739	1449	291
Strain total					4405			5479

Note: For some categories (marked with an asterisk), the classification system was expanded after the annotation of MG1655; direct comparisons across strains should be exercised with caution for these designations. Totals for EDL933 reflect annotation updates since the initial publication [2].

TABLE II O-Islands Larger than 5 kb

OI#	Size (bp)	Description
43	87,563	Large cluster inserted at *serW* tRNA includes many ORFs similar to known genes: integrase, urease, and tellurite resistance clusters; complement resistance protein; diacylglycerol kinase; colicin immunity protein; secreted proteins; outer membrane receptor; DNA repair protein; several ORFs similar to CP-933L and other cryptic P4-like prophages, IS elements, and IS complex clusters
48	87,549	Duplicate of OI#43 inserted at *serX*
172	44,434	Inserted at *leuX*; integrase, unknowns, putative resolvase, putative helicases, IS fragment
7	35,915	Many unknowns, putative macrophage toxin, ClpB paralogue, RhsG
47	31,727	Large cluster including many unknowns, a pilin-like structural subunit and chaperone, putative secreted protein, HecB-like protein, and several genes encoding components of a fatty acid or polyketide biosynthesis system
28	25,165	Putative novel RTX-like exoprotein and transport system
122	23,455	Inserted at *pheV*, putative P4-type integrase, IS elements and partials, putative enterotoxin, cytotoxin, and PagC-like virulence factors
115	16,948	Large cluster encoding a putative type III secretion system, response regulator for two-component system, InvF-like and putative secreted proteins, putative lipoprotein
138	15,393	Unknowns, putative enzymes of fatty acid biosynthesis (acyl carriers, synthases, and reductase), unclassified putative enzymes
84	14,188	O157 O-antigen biosynthesis, complete cluster
35	13,966	Putative glutamate uptake and fermentation system plus several unknowns
102	12,807	Inserted at *argW*, integrase, phage P22 gp7 and 14, unknowns, putative resolvase; sucrose permease, hydrolase, and regulator, fructokinase
30	11,614	RhsH, unknowns
167	9,307	Putative response regulator, putative aldolase, putative ABC transport system for ribose, putative histidine kinase, unknown
140	9,057	Iron transport system homologous to the Chu system of *S. dysenteriae*
20	8,114	Ferric iron uptake system similar to *afuABC* of *Actinobacillus* and two-component regulatory system similar to *uhpABC*, hexose phosphate transport; REP region
70	7,888	Putative drug-resistance efflux pump
153	7,425	Unknowns, REP region
123	7,304	Cluster encoding a putative iron transport system, one unknown
154	6,943	Fimbrial subunits, usher and chaperones
55	6,737	TonB-dependent outer membrane receptor, molybdenum and/or iron transport proteins, and several unknowns
139	6,564	Unknowns, putative PTS subunits, putative xylulose kinase, phosphocarrier, aldolase, and regulator

(*continued*)

TABLE II (*Continued*)

OI#	Size (bp)	Description
1	6,506	Putative type I fimbrial protein, usher and chaperone, unknowns
166	6,318	Putative PTS for sorbose, REP region
95	6,185	Aromatic degradation cluster, LysR-type regulator, partial BoxC repeat
9	6,133	Unknowns including three ORFs similar to cluster implicated in regulation of a *P. aeruginosa* porin for imipenem/basic amino acid uptake, begins within B0282-*yagP*, putative regulator
144	6,091	Unknowns, putative adhesin
141	5,904	Putative fimbrial subunits, usher and chaperone
173	5,646	Putative invasin, putative transport protein, unknowns, REP region
162	5,521	Rhs core, unknowns, part of IS element
145	5,407	Putative LPS biosynthesis enzymes
14	5,279	Unknowns
17	5,258	Putative sugar (possibly ribose) transport system
90	5,131	Putative regulator of molybdate metabolism

pathogenic functions. Table III lists 19 of the smaller islands (0.5–5 kbp), also revealing many genes of potential importance to the lifestyle and virulence of EDL933. Many of the smaller islands contain genes of unknown function. A further 46 islands are smaller than 500 bp, and island borders often occur within genes. Below we discuss some of most interesting functional units observed in the EDL933 islands and their potential relevance to pathogenesis of disease.

Adhesins and Toxins

The genes of the LEE, the Shiga toxins, and the (plasmid-encoded) hemolysin have been studied extensively. The large plasmid of EDL933 encodes a huge protein bounded by IS elements with interesting regions of similarity to the clostridial ToxAB proteins [29]. Two ORFs (Z4332-3) in the island OI#122 at tRNA *pheV* also showed similarity to the N-terminal region of ToxA. These two ORFs also match adjacent regions in the N-terminus of a 3233-aa ORF identified in EPEC 2348/69 and EHEC O111:H- that are responsible for inhibition of lymphocyte activation and adhesion to CHO cells, respectively [69,70]. The two short ORFs in EDL933 are immediately followed by an IS element.

The *pheV* island OI#122 also contains a homologue (Z4326) of the enterotoxin ShET2 of *Shigella* and an ORF (Z4321) with 45% identity to PagC of *S. enterica*, which is similar to the invasion protein Ail of *Yersinia* spp. and to the Lom serum resistance protein of phage 933W [30,71]. Sequences similar to the EAST-1 enterotoxin gene [72] were found in two partial IS1414/285-like regions

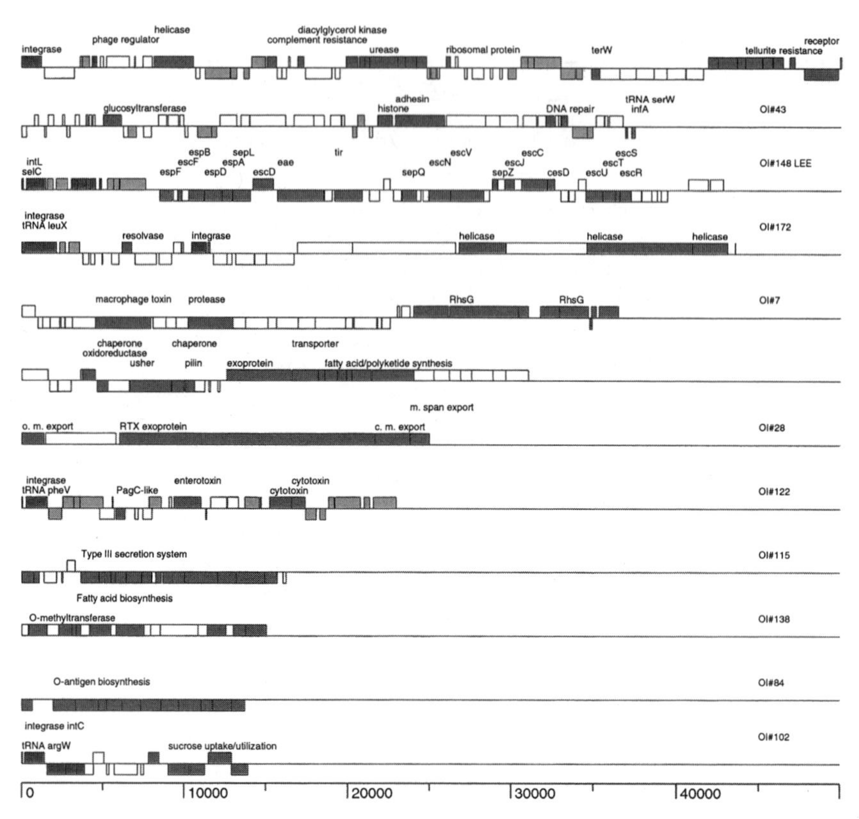

FIGURE 2 Twelve largest O-islands. Boxes represent ORFs and their orientation by position above or below the axis line. O-island identification numbers are shown at the right of each map. *Red* ORFs, function or putative function assigned; *open boxes*, unknown function; *turquoise*, tRNA; *green*, phage; *yellow*, IS elements. OI#43 is 87 kb and occupies the first two panels. OI#48 (not shown) is identical to OI#43. Abbreviations: m, membrane; o. m., outer membrane; c. m., cytoplasmic membrane. Labels are aligned with the left border of the corresponding ORF. See Color Plate 3.

as in enterotoxigenic and enteroaggregative *E. coli* (ETEC and EAEC) [73,74], but neither could be expressed due to mutations at the 5′(5′) end.

In OI#28 a putative new RTX determinant [75] was found consisting of a 4700-residue ORF (Z0615) with B- and D-like transport genes downstream (Z0634-5) and a TolC-like exoprotein secretory gene upstream (Z0608). The 9-aa repeat signature of the exported protein is absent, but the N-terminal 100-residue repeats are present. A candidate virulence gene (Z0250) in OI#7 is 20.6% similar to 949 aa of *Legionella pneumophila* IcmF, linked to survival in macrophages [76].

Nine adhesin candidates were located by homology to EPEC AIDA-1 and examined for features of autotransporters [77–81]. Identical candidates (Z1211

TABLE III Selected Small O-Islands

OI#	Size (bp)	Description
106	4,859	DMSO reductase subunits
11	4,769	Several oxidoreductases, unknowns, and a transcriptional regulator
15	4,643	Putative secreted protein similar to AIDA-I (diffuse adherence in DAEC 2787)
71	4,141	IS element, putative chaperone
134	3,508	DNA processing protein, helicase (three ORFs)
61	3,280	Putative fimbrial genes including an usher, chaperone, and structural subunit
82	3,275	IS element, putative membrane receptor
119	3,267	Unknowns, putative ATP-binding components of ABC transport system, part of a putative kinase
110	2,936	Putative decarboxylase and regulator, unknowns, REP region
163	2,691	Putative hippuricase, putative citrate permease
126	2,326	PTS subunits and putative deacetylase
109	2,077	Putative regulator, unknown
97	2,006	Putative antibiotic efflux protein, N-terminus of putative regulator
159	1,751	Putative glycoporin
62	1,548	Porin, unknown
85	1,376	Putative UDP-gal-4-epimerase
130	734	Putative leader peptidase, putative bacterioferritin, unknown
13	725	Putative dehydrogenase
2	635	Topoisomerase toxin-antitoxin system

and 1651) were found in the duplicate islands OI#43 and OI#48. These are large ORFs with several characteristics of a β-barrel outer membrane pore [78,80], 68% identical to the EPEC AIDA-1, and similar to virulence factors from several species. Another compelling candidate, Z0402 in OI#15, is 55% identical over 477 aa to AIDA-1 and 31% identical over 1063 aa to VirG/IcsA, an *S. flexneri* virulence factor [82]. All three candidates have signatures associated with autocatalytic cleavage of passenger and effector segments and show other common adhesin features of unknown function.

Other candidates include Z1542 in OI#47, similar to filamentous hemagglutinin of *B. pertussis* [83], and HecA, a putative virulence factor of *Erwinia chrysanthemi* [84], AIDA-1, and other surface or virulence proteins. Z5029 in OI#148 is similar to Hsf and Hia, surface fibrils of *Haemophilus influenzae* [85]; to UspA1, a *Moraxella catarrhalis* adhesin [86]; and part of YadA, a *Y. enterocolitica* invasin. Autotransporter signatures also were observed. Z5932 in OI#173 is intimin-like; Z1200 and Z1640 in OI#43 and OI#48 are putative membrane proteins. Ten candidates were located in backbone regions or were related to K-12 ORFs:

Z2691, Z2196, Z3135, Z3948, Z3449, Z2322, Z0469, Z2658, Z3487, and finally Z0375 encoding a full EaeH intimin.

Iron Uptake

In the human host, iron is intracellular or else largely sequestered in molecules such as hemoglobin. Iron required by infecting bacteria must be taken up actively from the host environment by ABC transport systems using siderophores enterobactin, aerobactin, or ferrichrome. O157:H7 uses both free heme and hemoglobin directly as iron sources [87]. Transport is regulated by TonB and Fur. Iron uptake loci and islands are shown in Fig. 3*A*.

The *fep* locus (Z0724-38), using enterobactin, and the *fhu* locus (Z0161-4, Z5968, and Z1741), using ferrichrome and other ferric hydroxamates, are located in backbone (see Fig. 3A). EDL933 can use ferric citrate [87] but not by the *fec* locus, which is absent. In K-12, the *fec* operon is located on a 39-kbp island of classic form at tRNA *leuX*, including a P4 phage integrase and a sugar PTS as well as *fec* (Fig. 4). In EDL933, a locus designated *chu* (OI#140, Z4910-4919) is homologous to *shu* in *S. dysenteriae* [88,89]. A similar locus exists in many pathogenic *E. coli* [88].

Three other iron transport systems were identified (see Fig. 3*A*). OI#20 contains genes (Z0458-60) homologous to the *Actinobacillus pleuropneumoniae afuABC* locus, encoding a putative siderophore-independent iron transport system. The second system (OI#123, Z4381-6) includes a putative siderophore receptor and associated proteins with similarity to ferric citrate and ferric hydroxamate uptake systems in *Synechocystis* and *Bacillus subtilis*. A third locus, OI#55, includes a TonB-dependent outer membrane receptor (*prrA*, Z1961), ABC transporter components, and several ORFs of unknown function.

Fimbriae

These filamentous structures on the bacterial surface attach to host cells during infection. Ten loci in EDL933 encode fimbrial proteins, all members of the type I and pap family with a common secretion mechanism [90]. Three of the fimbrial loci are in the K-12 backbone; *sfmACDHF-fimZ* (Z0686-693) and *fimBEAICDFGH* (Z5910-5918) are typical orthologues. The third locus (Z1288-92) contains divergent paralogues of *fimH*, *fimG*, *fimF*, an intact *fimD*, *fimC*, and *fimA*. This locus is missing *fimB*-, *fimE*-, and *fimI*-like genes and the invertible regulatory region.

OI#141 (Z4965-8) and OI#154 (Z5220-5) contain only fimbrial genes, and their base composition suggests acquisition by horizontal transfer. Both loci are similar to the *lpfABCDE* locus of *Salmonella* [91] (see Fig. 3*B*) encoding the proteins of long polar fimbriae that are involved in adhesion to murine Peyer's

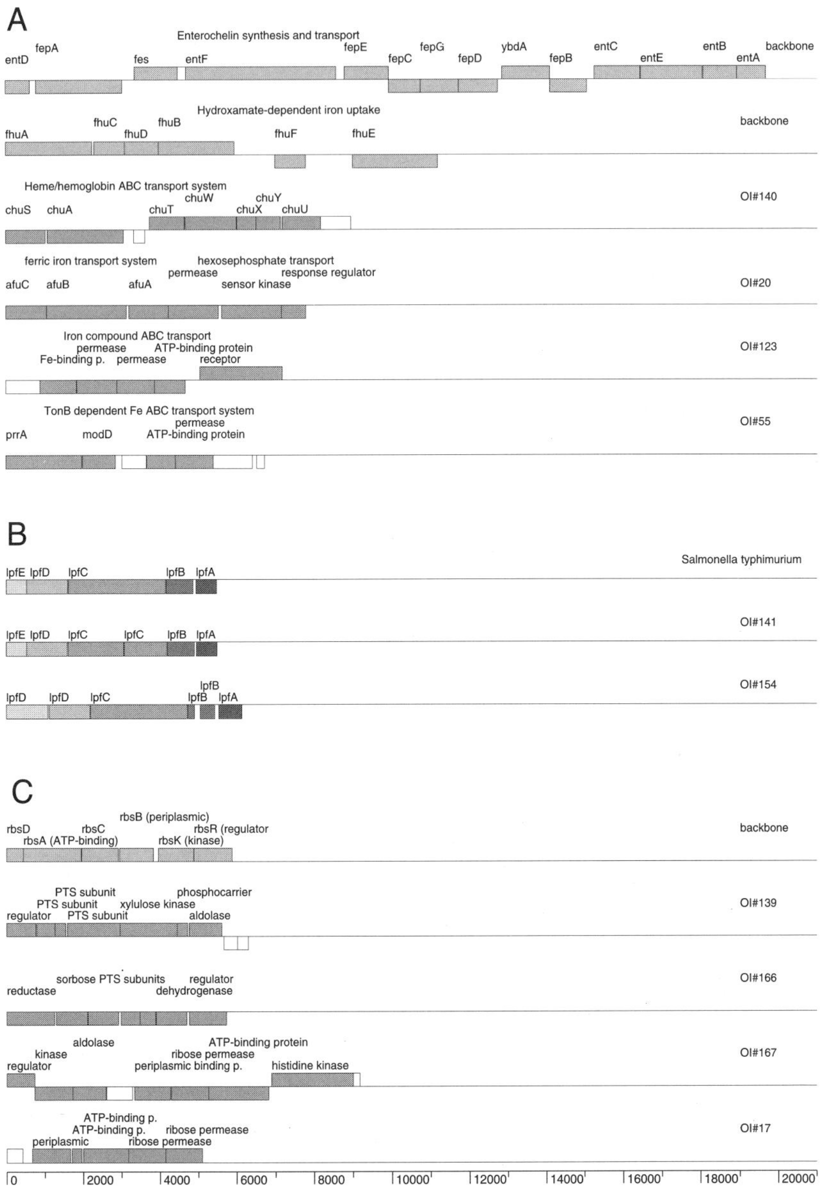

FIGURE 3 Maps of EDL933 islands and backbone segments encoding specific functions. *Filled boxes*, ORFs with function or putative function assigned; *open boxes*, ORFs of unknown function. *A*, *C*, backbone ORFs *light gray;* O-island ORFs *dark gray. A*. Systems for iron uptake and utilisation. *B*. Comparison of the *lpf* loci found in EDL933 with that of *S. typhimurium*. Homologous ORFs are the same shade of gray. *C*. Sugar transport loci of EDL933.

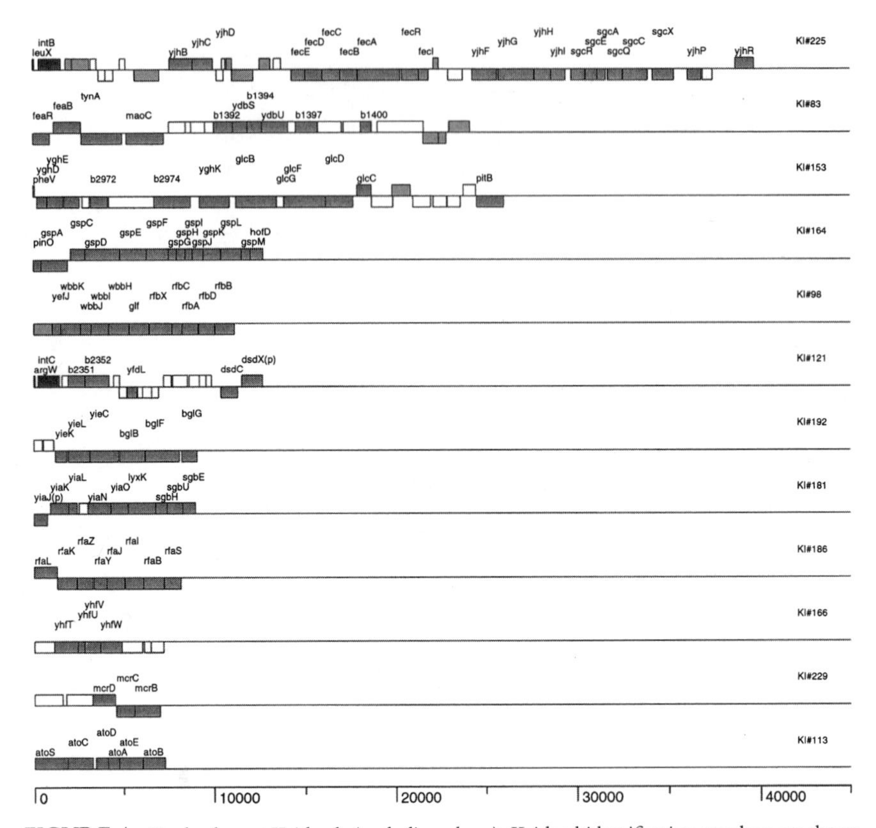

FIGURE 4 Twelve largest K-islands (excluding phage). K-island identification numbers are shown at the right. *Mauve* ORFs, function or putative function assigned; *open boxes*, unknown function; *turquoise*, tRNA; *green*, phage; *yellow*, IS elements. (p) indicates that the ORF spans the island boundary. See Color Plate 4.

patches. In OI#141, the *lpfC* homologue is disrupted, as is *lpfB* in OI#154. This island has no homologue of *lpfE*, the minor structural subunit, but *lpfD* is duplicated. Preliminary evidence confirms that the OI#141 locus contributes to EDL933 adherence (A. Torres and J. Kaper, personal communication).

OI#1 (Z0022-24) is related to putative fimbrial components YehBCD in K-12, although not part of the backbone. They are also similar to components of other type I fimbrial systems and are putatively identified as an usher, a periplasmic chaperone, and the major fimbrial subunit. The remaining four fimbrial biosynthesis loci (Z3596-3601, Z3276-80, Z0146-52, and Z0868-72) are hypervariable regions, i.e., genes homologous to K-12 in the same order at the same relative chromosomal position, but the level of similarity is far lower than normal for clonally diverging orthologs (see above).

Acid Resistance

Intestinal pathogens and commensals can survive the extremely low pH of the human stomach and the organic acids in the intestine. In *E. coli*, three systems contribute to acid resistance [92]: a glucose-repressible oxidative system [93] and antiporters for arginine and glutamate [94]. All the genes for these systems are found in the EDL933 backbone, which is shared with K-12. Urease also may be important for local pH control in several pathogens [95–98]. Both OI#43 and OI#48 (see Fig. 2) contain complete urease operons (Z1142-8 and Z1581-7) [99], very similar to *ureA-G* genes from *Klebsiella aerogenes* [100]. Although the proteins contain the key residues for activity, urease function has yet to be demonstrated in EDL933.

Metabolic Capacity

Lineage-specific regions of EDL933 encode various putative metabolic or catabolic functions not found in K-12. A few examples are described here. EDL933 has a backbone set of fatty acid biosynthesis genes and also two unique groups, in OI#138 (Z4850-9) and OI#47 (Z1544-9). The first group may have the capacity for modular synthetic assembly of molecules such as polyketides. They include novel acyl carrier proteins (ACP), β-ketoacyl synthases, ACP dehydratase, CoA transacylase, and a putative enzyme similar to a part of a polyketide synthase from *Streptomyces*, most conserved in the area of the active site. These ORFs are similar to proteins from *Rhizobium, Bacillus, Streptomyces, Haemophilus*, and others but not *E. coli*.

Two DMSO reductase operons are in shared backbone, and a third is unique in EDL933 (OI#106, Z3783-5). DMSO is an alternative electron acceptor for anaerobic respiration. The extra system may provide a growth advantage in the anaerobic conditions of the mammalian intestine.

Six EDL933-specific loci encode sugar transport systems, including both phosphotransferase (PTS) and ATP-binding systems (see Fig. 3*C* and examples in Figs. 2 and 3*A*). The high-affinity ribose transport system *rbsABCKR* is intact in EDL933 backbone. Two islands, OI#17 (Z0414-9) and OI#167 (Z5684-93), encode two more putative ribose high-affinity ABC transport systems. Other islands containing sugar PTS genes are OI#139 (Z4874-83) and OI#166 (Z5613-9) with specificity for sorbose. A putative hexose phosphate transport system in OI#20 (Z0461-3) provides an alternative to the backbone *uhpATC* system. In OI#102, Z3623-6 encode proteins for sucrose utilization; EDL933 can ferment sucrose (unpublished).

ORFs Z3390-5 in OI#95 are similar to enzymes of the aerobic degradation pathways of aromatic compounds in *Sphingomonas* and *Pseudomonas*, including putative salicylate hydroxylase, glutathione-*S*-transferase, gentisate dioxygenase, and isomerase-decarboxylase enzymes [101–103]. A putative hydroxybenzoate transporter [104] and a LysR-type regulator are also present. Salicylate and gentisate are key intermediates in the catabolism of napthalene.

Other Loci of Interest

The LIM island of enteropathogenic *E. coli* encoding TrcA, a chaperone-like putative virulence factor, was located at *potB* in EPEC and an O157:H7 strain [105]. In EDL933, LIM is at a different site flanked by IS629 and then prophage CP-933Y. OI#108 includes the LIM ORFs Z3918-21, the IS, and CP-933Y. It is possible that the LIM ORFs are part of this phage. A different prophage, CP-933N, is located at *potB* in EDL933.

Tellurite resistance is also used as a diagnostic assay for O157:H7 [106]. Islands OI#43 and OI#48 both encode *terZABCDEF* (Z1171-7 and Z1610-6) and *terW* (Z1164 and Z1603), homologous to the *ter* locus on the plasmid R478 of *Serratia marcesans* [107,108]. OI#2 (Z0056-7) encodes a toxin-antitoxin system similar to *letAB* of "the F-plasmid" [109] that kills plasmid-free cells. Other plasmid systems are found in the backbone and in cryptic prophage. The roles of these chromosomally-encoded proteins are unknown. OI#43 and OI#48 encode homologues of TraT, a lipoprotein of many enterobacterial plasmids with the role of preventing conjugation between bacteria carrying like plasmids. Since serum resistance also has been associated with TraT [110], it could be considered a possible virulence factor.

O-Antigen Biosynthesis

Both K-12 and O157:H7 encode specific O-antigens and core oligosaccharides, and lineage-specific genes encoding the enzymes for their synthesis. Six of the 12 O-antigen genes (OI#84, Z3192-3204) are highly specific for O157 [111]. A putative UDP-galactose-4-epimerase (Z3206) similar to WfbT in *Vibrio cholerae* [112] is separated from the group by a small backbone segment. The biosynthetic enzymes of the core polysaccharide of EDL933 are encoded by the *waa* genes (formerly *rfa*; OI#145 and OI#146 and hypervariable segments). The six proteins WaaCLDJYI (Z3137-42) are lineage-specific. The K-12-specific genes *rfbABCDX*, *wbbHIJK*, and *glf* encode the O16 antigen of K-12, although it is not expressed due to interruption by an IS element. Many of the genes responsible for synthesis of the core oligosaccharide are also in a K-12-specific segment, *rfaLKZYJIBSPQ*, also called *waaQORBULPYSZ* [113,114].

K-Islands

The K-12-specific islands also present an interesting array of genes either known or likely to confer discernible phenotypes. Table IV lists the largest K-islands, some of which are also illustrated in Fig. 4. Genetic features in K-12-specific

TABLE IV K-Islands Larger than 5 kb (End Points Are MG1655 Coordinates)

KI#	Start	End	Size (bp)	Description
225	4494059	4534178	40,120	*leuX* island, integrase, IS elements, unknowns, citrate-dependent iron uptake system, sugar transport system
71	1195515	1222786	27,272	Prophage e14, unknowns, putative transport system
83	1470184	1444234	25,951	*fea* operon (aromatic aldehyde oxidation), IS elements, unknowns
153	3108459	3133150	24,692	*pheV* island, putative general secretion pathway proteins, glycolate oxidase, *pitB*, IS element, unknowns
164	3451092	3463961	12,870	General secretion pathway proteins GspA–M, calcium-binding replication initiation protein, bacterioferrin (part)
98	2099311	2111401	12,091	O-antigen: *wbbH-K, glf, rfbA-D,X*
121	2465721	2476986	11,266	*argW* island, integrase, putative glycan biosynthesis enzyme, RNA polymerase beta, ligase, *dsdDX*
192	3895068	3904493	9,426	*bglBFG*, putative enzymes, unknowns
181	3739443	3748430	8,988	Putative sugar uptake and utilization
186	3794508	3802883	8,376	Core lipopolysaccharide biosynthesis: *rfaLKZYJIBS*
166	3502537	3510270	7,734	Putative permease, hydrolase, and mutase; unknowns
229	4569489	4577072	7,584	*mcrBCE*, unknowns
113	2318020	2325385	7,366	*ato* operon: short-chain fatty acid utilization
161	3358254	3365333	7,080	*glt* operon regulator, putative chaperone and outer membrane protein, IS element, putative regulator
218	4304398	4311465	7,068	Putative ABC transport system, possibly ribose
173	3616619	3623309	6,691	*RhsB*
157	3183086	3189723	6,638	Putative fimbrial and membrane proteins, IS element, unknown
96	2068493	2074799	6,307	Part of prophage CP4-44, outer membrane fluffing protein, putative histone and DNA repair proteins
142	2858439	2864535	6,097	Putative transport system
49	764376	770302	5,927	Transcriptional regulator of succinylCoA synthetase operon, sugar hydrolase, protein modification enzyme HrsA
25	379208	384417	5,210	Putative enzymes, IS element, unknowns
224	4487671	4492757	5,087	Gluconate transport and utilization

islands include several metabolic/catabolic capacities, e.g., *feaBR* and the *ato* and *glc* operons. FeaB, also called *PadA*, oxidizes a specific class of toxic aromatic aldehydes in the environment. The *ato* and *glc* operons encode proteins that enable short-chain fatty acids to be used, feeding them into the TCA cycle and into glycolysis via the glyoxalate bypass [115]. The *glc* operon permits the metabolism of glycolate, a two-carbon product of plant and algal metabolism that can be used as a carbon source via the glycerate pathway [116].

Phages, Cryptic Prophages, and Phage Remnants in *E. coli* Genomes

Phages and Evolution

Phages have played a seminal role in the development of modern biology [117]. More important, along with plasmids, bacteriophages have had a significant impact on the evolution of bacteria. In addition to the ongoing infection and response cycles that are thought to have influenced the evolution of restriction/modification systems and cell surface variations, they have played a more direct role as mediators of genetic exchange [118]. Finally, temperate phages and their cryptic remnants actually can evolve as part of the chromosome, with functions possibly being coopted by their hosts for new uses [119].

The phages detected in the completed *E. coli* genomes are summarized in Table V. Given the examples of cointegrated phages, different phages using the same attachment site, and ongoing variation by recombination, swapping, and acquisition of modules, it seems clear that the mosaic evolutionary structure of phage genomes first described by Campbell and Botstein [120] and subsequently elaborated on by others [121–123] represents a microcosm of what is seen in the bacterial genome as a whole, at least within the Enterobacteriacea.

E. coli K-12 Phages

The original *E. coli* K-12 strain isolated in 1922 carried bacteriophage lambda as a prophage. Although the sequenced *E. coli* K-12 strains MG1655 and W3110 were cured of lambda, this bacteriophage has served as a paradigm for studies of phage biology and the ability of phages to alter their hosts [124–126].

Earlier studies had indicated that *E. coli* carried several cryptic lambdoid prophages that were incapable of productive growth but carried genes capable of complementing and/or rescuing mutants of viable phages on infection [119]. Completion of its genome sequence revealed that *E. coli* K-12 MG1655 carried at least eight recognizable phages: DLP12, e14, Rac, Qin, CP4-6, CP4-44, CP4-57, and Eut. The W3110 strain shares most of these but lacks the Eut phage [1]. Several other regions in the K-12 sequence were termed *phage remnants*, including integrase genes (*intB* and *intC*) that seemed to lack association with other phage-like sequences. In light of comparisons with the O157:H7 genome, some of these can now be recognized as islands (see above and Fig. 4 and Table IV).

TABLE V Prophages in the Sequenced Genomes of *E. coli* K-12 MG1655 and *E. coli* O157:H7 EDL933

Strain: phage name	Length (bp)	Integration site	Direct repeat	Comments
MG1655: CP4-6	34,308	tRNA *thrW* (b0244)	60 bp	CP4 element; contains *argF*
MG1655: DLP12	21,302	tRNA *argU* (b0536)	47 bp	Lambdoid phage; contains *nmpC*, *emrE*, *appY*, *ompT*
MG1655: e14	15,204	*icdA* (b1136)	11 bp	Contains *pin*, *lit*, *mcrA*
MG1655: Rac	23,060	*ydaO* (b1344)	26 bp	Lambdoid phage; contains *trkG*, *sbcA*, *recEFT*
MG1655: Qin	20,377	Between *ydfJ* (b1543) & *rspB* (b1580)	ND	Lambdoid phage; contains *cspBFI*, *relBF*, *dicBFAC*
MG1655: CP4-44	12,873	Between *cobU* (b1993) & *yeeX* (b2007)	ND	CP4 element; contains *flu*
MG1655: Eut	6,790	Between *eutB* (b2441) & *eutA* (b2451)	8 bp	(Not present in W3110)
MG1655: CP4-57	22,031	tmRNA *ssrA* (b2621)	8 bp	CP4 element
(Lambda)	48,502	Between *ybhC* (b0772) & *ybhB* (b0773)	7 bp	(Excised in MG1655 and W3110; contains *lom*, *bor*)
EDL933: CP-933H	10,586	tRNA *thrW* (Z0305)	47 bp	Lambdoid phage, similar to Sp1 of Sakai, integrated in tandem with CP-933I; replaces CP4-6 of K-12
EDL933: CP-933I	12,895	tRNA *thrW* (Z0305)	16 bp	P4-like phage, similar to Sp2 of Sakai, integrated in tandem with CP-933H; replaces CP4-6 of K-12
EDL933: CP-933K	38,588	Between *ybhC* (Z0943) & *ybhB* (Z0992)	13 bp	Lambdoid phage, similar to Sp3 of Sakai; replaces lambda of original K-12; contains *lom*
EDL933: CP-933M	45,244	tRNA *serT* (Z1388)	60 bp	Lambdoid phage, similar to Sp4 of Sakai; contains *lom*
EDL933: BP-933W	61,663	*wrbA* (Z1423)	7 bp	Viable lambdoid phage, similar to Sp5 of Sakai; contains Stx2, *stk*, *bor*, *lom*, 3 tRNAs

(continues)

TABLE V *(continued)*

Strain: phage name	Length (bp)	Integration site	Direct repeat	Comments
EDL933: CP-933N	~47,315	*potB* (Z1830)	ND	Lambdoid phage, similar to Sp6 of Sakai; contains 2 tRNAs
EDL933: CP-933C	15,227	Between *ycfD* (Z1833) & *phoQ* (Z1858)	21 bp	Similar to "genetic island" of *H. influenzae*, Sp7 of Sakai
EDL933: CP-933X	~54,000	*icdA* (Z1865)	ND	Lambdoid phage, similar to Sp8 of Sakai; replaces e14 of K-12, with adjacent deletion; locus spans remaining gap; contains *lom*, *bor*, catalase
EDL933: CP-933O	~80,826	*yciD* (Z2034)	ND	Lambdoid phages (2–4 phages), similar to Atlas of ECOR, Qin of K-12, Sp9, Sp11, Sp12 of Sakai; involved in EDL933 inversion; contains *lom*, 2 tRNAs
EDL933: CP-933R	49,797	*ydaO* (Z2416)	26 bp	Lambdoid phage, similar to Rac of K-12, Sp10 of Sakai; contains Cu/Zn superoxide dismutase, *lom*, 2 tRNAs
EDL933: CP-933P	~57,984	Before Z2569 (b1582 of K-12; *rspBA* deleted)	ND	Lambdoid phage, similar to Qin of K-12, Atlas of ECOR, Sp12, Sp9 of Sakai; involved in EDL933 inversion; contains *lom*, 3 tRNAs
EDL933: CP-933T	21,120	tRNA *leuZ* (Z2997)	121 bp	P2-like phage, similar to Sp13 of Sakai
EDL933: CP-933U	45,177	tRNA *serU* (Z3131)	61 bp	Lambdoid phage, similar to Sp14 of Sakai; contains *lom*, 3 tRNAs
EDL933: CP-933V	48,916	*yehV* (Z3376)	21 bp	Lambdoid phage, similar to Sp15 of Sakai; contains Stx1, Cu/Zn superoxide dismutase, *lom*
EDL933: CP-933Y	~21,681	tmRNA *ssrA* (Z3914)	ND	Lambdoid phage, similar to Sp17 of Sakai; replaces CP4-57 of K-12

E. coli O157:H7 Phages

The Shiga toxins encoded by *E. coli* O157:H7 (and other EHEC strains) were known to be encoded by bacteriophages in at least some cases [127–130], and both the sequenced strains carry Stx2-encoding prophages capable of lytic growth and production of infectious particles. BP-933W, the Stx2 phage of EDL933 [30], and VT2-Sa (aka Sp5), the Stx2 phage of the Sakai strain [31,32], are nearly identical over most of their genomes and share an integration site in the *wrbA* gene of their respective hosts. However, the early regulation and replication regions of these two phages are markedly different, suggesting that the two phages are derived from a common ancestor but have undergone several rounds of subsequent genetic exchange.

Although the strain carries no other viable phages, the phage load of EDL933 is almost twice that of MG1655, with 16 multigenic regions of the chromosome related to previously described bacteriophages and with some loci containing more than one distinct element. The EDL933 prophages are designated *cryptic prophages* (CPs) to indicate their lack of a full complement of functional phage genes. These elements vary considerably in length and exhibit a mosaic of patchy similarities to each other, bacteriophage lambda, the CPs of K-12, other bacteriophage sequences, and nothing at all.

Thirteen of the EDL933 CPs have a lambda-like organization and a gene content composed of recognizable homologues of members of the extended lambdoid family of prophages. All the lambdoid phages are represented by similar entities in the Sakai strain, with common integration sites and gene organizations. However, as with the viable Stx2 phages, the CPs exhibit evidence of continued evolution involving recombinational exchange—not suprising given that lambda RFLPs have been suggested as a means of subtyping O157 strains [131].

The most complicated instance involves the locus designated CP-933O. The arrangement of genes in this region indicates that it contains the remnants of at least two phages (e.g., there are two distinct integrases) and possibly as many as four. As noted earlier, this region also was involved in the inversion found in EDL933. The 121-bp inverted repeats initially posited as the end points of the inversion originally were a direct repeat flanking the prophage at the present site of CP-933O. This repeat sequence and location, along with the adjacent integrase sequence, identify the original prophage at that site as being a member of the Atlas family [47,48] identified in the ECOR collection of natural *E. coli* strains [34]. The other end of the inversion involved CP-933P, at the site occupied by Qin in K-12; a second partial integrase in that region also may reflect multiple phage elements. Although the order of events that lead to the present state in EDL933 remains unclear, it must have involved additional recombinational exchanges before and/or after the inversion itself.

In addition to the large family of lambdoid phages, both O157:H7 genomes carry a P4-like phage (CP-933I, Sp2) integrated in tandem with a lambdoid phage (CP-933H, Sp1) at the site occupied by CP4-6 in K-12. CP-933T (Sp13) is related to bacteriophage P2 and phi-186, and CP-933C (Sp7) has similarities to a phage-like "genetic island" in *Haemophilus influenzae* [132]. Other regions contain phage-like sequences, blurring the distinction between prophage and island (see above): The sucrose-utilization island OI#102 contains P22-like genes, and both the urease-tellurite islands (OI#43 and OI#48) and the previously described LEE (OI#148) [55] contain similarities to the CP4-family prophages of K-12. CP4-44 of K-12 is also present in EDL933 but contains internal substitutions. In a situation analogous to the Eut phage in the two K-12 genomes, the Sakai O157:H7 strain carries a Mu-like phage (Sp18) entirely absent from EDL933 [3,24].

Phages and Pathogenesis

The existence of pathogenicity factors encoded by lysogenic phages is a widespread phenomenon, including phage-encoded toxins, phage-mediated serotype conversion, and enhanced bacterial defenses against macrophages [133–141]. In O157:H7, both Stx1 and Stx2 toxin-converting phages encode the Shiga toxin subunits within an apparent Q antiterminator-dependent transcript such that maximal toxin production would be coupled with host lysis and phage release. It has been demonstrated that phage induction in response to treatment of O157:H7 with some antibiotics does lead to increased expression of the Shiga toxins [142–144]. The location of *stx* genes within lambdoid phage late regions is preserved in all Stx-encoding loci, although in some cases no Q-homologue is observed [143,145–148]. Other potential pathogenicity determinants also are encoded by prophages in the O157:H7 genome, including nine *lom/pagC/rck*-like genes, two copies of *bor*, two copper/zinc superoxide dismutases, a catalase, and the previously described putative serine/threonine kinase of BP-933W (absent in the Sakai strain).

As additional genomic sequences become available (see above), the involvement of bacteriophage elements will continue to be of interest. At least four lambdoid-like regions are present in the unfinished CFT073 genome (unpublished), and the murine toxin plasmid pMT1 of *Y. pestis* contains a partial lambda-like region [149,150].

Plasmids

A large single-copy plasmid known as *F* was present in early isolates of K-12. Studies of this plasmid have elucidated details of replication, partition, incompatibility, and transfer mechanisms. The 99-kbp F plasmid recently has been fully sequenced by Sampei and Mizobuchi (GenBank Accession Number AP001918).

Plasmids are found frequently in clinical or environmental *E. coli* samples, and plasmid profiling is now used commonly to investigate the genetic structure of bacterial populations [51,151,152]. Large F-like plasmids exist in *E. coli* pathogens, where they often carry genes involved in or essential for pathogenicity.

Two F-like virulence plasmids are fully sequenced, the 92-kbp pO157 from EDL933 and O157 Sakai [28,29] and the EAF plasmid pB171 (68.8 kbp) in EPEC [153]. Plasmids of *S. flexneri* [154,155], *Y. pestis* [149,150,156], and *Salmonella typhi* [157] also have been sequenced completely. Several loci that may contribute to the infectious mechanism are encoded on pO157 [158], including EHEC-Hly (hemolysin). The *bfp* operon encoding a bundle-forming pilus is found on pB171. Other large plasmids, although not sequenced fully, are known in ETEC, encoding a colonization factor (CFA/1) and enterotoxin on a 92-kbp plasmid [159] and AAF fimbriae on pAA2 (~100 kbp) in EAEC [160]. Large plasmids also have been detected in some uropathogenic *E. coli* strains and extraintestinal *E. coli* such as K1 [161]. To date, no characterization or sequence is available for UPEC or K1 plasmids, but plasmid-specific sequences have been detected in a K1 genome project in progress at the University of Wisconsin.

A number of small plasmids also exist in *E. coli* strains, and it is not unusual to find three or more different-sized plasmids in a single host. In addition to pO157, EDL933 has a 3.3-kbp plasmid encoding a mobilization protein that is identical to that in O157 Sakai [28] and nearly identical to a third version sequenced in Europe [162]. Many plasmids are capable of transfer among and even outside related species and, with transposons, have been implicated widely in the alarming spread of resistance to antibiotic drugs [163]. One example among many documented cases is the plasmid-mediated multidrug-resistant *Salmonella typhi* that is responsible for typhoid fever outbreaks that were difficult to treat in many countries [164,165].

FUNCTIONAL ANALYSES OF *E. COLI* GENES

Functional Characterization prior to Large-Scale Sequencing

Genetic and Biochemical Analyses of Gene Functions

It has been more than 50 years since Gray and Tatum published the first use of x-rays to induce nutritional requirements in K-12 [166]. Until recently, their basic approach, random mutagenesis followed by screens or selection for altered phenotypic properties, was the most common means for the identification of genes that are involved in a particular biologic process. Genes were discovered one at a time, and the ultimate identification of their DNA sequence was a long and complex process. In other cases, biochemical methods have been employed to purify components involved in a process and in some cases have lead to discovery of the gene(s) encoding these products. Again, these methods tended to focus on one protein at a time or analysis of total cellular protein. Completion of the genome sequence allows a retrospective examination of the effectiveness of classic genetic and biochemical methods in the discovery of gene functions. Of the predicted 4405 genes in *E. coli*, fewer than half are represented by mutant strains [167]. Biochemical analyses have led to the identification of many genes that encode highly abundant products in the cell (e.g., translational machinery). Genetic methods have identified genes that encode a diversity of functions, but it is clear that a large fraction of predicted genes in the chromosome never was isolated in genetic screens or biochemical purification, most likely due to biases in screening procedures and great disparities in relative abundance of different RNAs and proteins. Perhaps many genes are expressed only under conditions that are not achieved typically in laboratory settings but are encountered often by *E. coli* in its natural environment. The sequencing of the genome and development of new tools that exploit this information for genome-scale analyses of gene activity promise to provide insight into the workings of genes that were untouched by prior approaches.

Biochemical Characterization of the Proteome

In 1975, the method of two-dimensional (2D) gel electrophoresis was introduced [168], and it allowed separation of complex mixtures of proteins into distinct polypeptide species. In a sense, this was the first genome-scale analysis tool because in principle it allows detection of most of the proteome in each experiment. In a practical sense, the method cannot detect all proteins in the cell due to limitations imposed by protein insolubility, extreme isoelectric points, and low-abundance proteins. In addition, the proteins visualized by 2D gel electrophoresis usually are anonymous until further (less high-throughput) experiments reveal their identity. Since gel-running conditions vary between applications and are difficult to standardize between laboratories, the determination of spot identities by downstream analyses (such as mass spectrometry or protein sequencing) limits the amount of data that can be correlated with the activity of particular genes. Despite these challenges, 2D gel electrophoresis has been used widely to analyze the *E. coli* proteome and has lead to the most complete descriptions of quantitative protein expression profiles (see [169]). To date, there exists no alternative technology for comprehensive analysis of the proteome, although the success of miniaturized arrays for analysis of the transcriptome is driving the development of equivalent systems for protein analysis.

Analysis of Transcriptional Regulation

In the mid-1960s, Jacob *et al.* [170] described fusions of the *lacY* gene to the *purE* gene to study gene activities. Further developments in the method led to its use as a general tool for studying gene transcription [171]. Reporter gene fusions are relatively easy to construct but do not allow simultaneous monitoring of multiple transcripts in a cell.

In 1987, Kohara, Akiyama, and Isono described the construction of a mapped collection of lambda clones covering the complete K-12 genome [23]. DNA from a minimal overlapping set of about 400 lambda clones was spotted on nylon filters and used to measure global changes in gene expression by serial hybridization of labeled RNA samples [172]. These early whole-genome arrays were used to detect changes in transcript levels due to heat shock, osmotic shock, nutrient starvation, anaerobiosis, growth in the mouse gut, and as a result of mutation in global regulatory genes. Many changes in expression were detected by this method, but because each lambda clone contains, on average, 10 genes, the identification of specific affected genes requires additional experimentation. The possibility that a single lambda clone might harbor genes that are both induced and repressed suggests that not all significant changes were detected by this approach.

Functional Characterization in the Postgenomic Era

Analysis of the Transcriptome

Using the complete nucleotide sequence, a set of reporter constructs that cover approximately 30% of the *E. coli* transcriptional units was generated [173]. These fusions were made by ligating random genomic DNA fragments to the *lux-CDEBA* operon encoding the bioluminescent protein luciferase. To map the random fusions, the constructs were end-sequenced to reveal the genomic location of the cloned fragment. Several fusions were tested and used to confirm predicted changes in gene expression from microarray experiments. Experiments with these fusions are still limited to single genes but may provide a generally useful means for independent confirmation of results obtained using array-based methods.

With completion of the MG1655 sequence [1], it became possible to use hybridization approaches that allow single-gene resolution. Based on the annotated ends of the genes, PCR primers were designed to precisely amplify each ORF from K-12 as an individual fragment [174]. These DNAs were spotted on nylon membranes as well as glass microscope slides and used to monitor changes in expression following heat shock or treatment with IPTG. In general, both methods were successful in identifying altered transcripts; however, the simultaneous hybridization of control and experimental samples labeled with fluorophors with different emission spectra permits more rapid and controlled experimentation. Distribution of complete sets of *E. coli* genome primers and nylon membranes spotted with each ORF by Sigma-Genosys has facilitated the use of these whole-genome arrays by additional laboratories.

Recently, Selinger *et al.* [175] described the use of oligonucleotide arrays for whole-genome expression profiling in *E. coli*. These high-density oligonucleotide arrays, manufactured by Affymetrix, contain, on average, one 25-nucleotide-long probe for every 30 bp of the genome. On average, there is one probe pair (one probe is identical to the genome sequence, and its mate contains a single-nucleotide difference) for every 6 bp of intergenic sequence and one probe pair for every 60 bp of coding sequences. The authors report the comparison of RNA prepared from cells grown to log phase with cells at stationary phase. Analysis of the hybridization data suggested that transcripts could be detected for the majority of genes at levels below one copy per cell. Interestingly, the authors also observed hybridization to the antisense probes for much of the genome. This method is appealing because it may allow identification of precise transcriptional start and stop sites that are often difficult to predict by computational methods. The comparison of two phases in the growth of the culture revealed a number of genes previously described to be affected under these conditions as well as several genes that were not known to be altered prior to these experiments.

Ultimately, the strategies for functional characterization of genes will mimic the history of DNA sequencing, following a path of single-gene analysis to the systematic determination of genome content. Table VI describes the use of DNA arrays to study genome functions in *E. coli*. Although the approach of microarray hybridization is systematic in the sense that it can assay the complete transcriptome, implementation of the technology has been far from systematic. This is not due to deliberate decisions by investigators but because this is a young technology still undergoing development. Table VI contains four distinct approaches to whole-genome hybridization and numerous differences in culture conditions, RNA preparation, and labeling, rendering comparisons between data sets difficult. The lack of a commonly used database for deposition of these experiments also limits their accessibility. Despite the shortcomings, these early successes with array-based experiments lead to the inevitable conclusion that these methods will become a routine part of the arsenal of techniques available to the modern molecular biologist.

Analyzing the Proteome

The most obvious advantage of having a complete nucleotide sequence for *E. coli* was that it provided a means for prediction of the complete proteome. Blattner *et al.* [1] predict 4290 protein-coding genes from the MG1655 sequence. It is far harder to predict how many of these proteins undergo posttranslational modification and how many forms of each protein there may be. Currently, 2D gels are used widely for experimental analysis of cellular proteins.

A good example of the power of combining 2D gel technology with genome information is the identification of *in vivo* substrates for the chaperonin GroEL

TABLE VI Whole-Genome DNA Array Experiments with *E. coli*

Experimental conditions assayed	Method employed	Reference
Constitutive *marA* expression versus *marA*Δ	Panorama nylon arrays	[187]
sdiA overexpression versus wild type	Glass microarrays	[188]
IPTG treatment, purification of *E. coli* messages by polyadenylation	Glass microarrays	[189]
Acetate-induced acid tolerance response of O157:H7	Panorama nylon arrays	[190]
Integration host factor mutant (*himA*Δ) versus wild type	Panorama nylon arrays	[191]
Stationary versus log-phase growth	Affymetrix oligonucleotide arrays	[175]
Replication fork progression	Glass microarrays	[192]
IPTG induction, heat shock, osmotic shock, nutrient starvation, stationary-phase growth, anaerobic growth, growth in a gnotobiotic mouse gut, several pleiotropic mutants	Nylon membranes of lambda clones	[172]
Heat shock	RT-PCR and hybridization of nylon membranes of lambda clones	[193]
Heat shock, IPTG treatment	Nylon membranes, glass microarrays	[174]
Minimal versus rich medium	Panorama nylon membranes	[194]
IPTG treatment, minimal versus rich medium, exponential versus stationary phase	Glass microarrays	[195]
Metal ion tolerance	Panorama nylon membranes	[196]
Tryptophan metabolism	Glass microarrays	[197]

[176]. The authors used immunoprecipitation with anti-GroEL antibodies to purify substrates complexed with the chaperonin. These peptides were separated using 2D gels, and the identity of 110 different spots was determined by trypsin digestion followed by peptide-mass fingerprinting using matrix-assisted laser desorption/ionization time-of-flight mass spectrometry. The correlation of masses of trypsin-digested peptides with masses of proteins predicted from the genome sequence provides the mechanism for linking particular spots with specific genes.

Reverse Genetics

The accumulation of DNA sequence data of unknown function has driven the development of so-called reverse genetic methods that allow site-directed mutagenesis of the bacterial chromosome. In recent years, *E. coli* researchers have developed several technologies that promise to permit the high-throughput generation of mutant strains for analysis of gene functions. In particular, the use of phage lambda–encoded recombination functions that elevate the rates of homologous recombination of ectopic linear DNA fragments is likely to stimulate mutant construction efforts [177–179].

Systematic Approaches to Functional Genomics

It is often assumed that identical (or extremely similar) sequences obtained from the same species have the same function. While it may be safe to assume that if both genes were expressed (a phenotype not usually predictable from sequence analysis alone), they would encode products with similar activities, the function of the products often will depend on the activities of numerous other products in the cell. Since the function of any gene product may depend on the genotype at other loci, it is preferable to perform systematic functional genomic experiments using the same strains that were used to determine the reference genome sequence. With the ultimate goal of understanding the precise relationship between genotype and phenotype, we cannot discount the importance of any polymorphism in altering phenotype. We are engaged in the systematic functional characterization of the MG1655 strain by mutagenesis and gene expression studies. Another group is systematically studying gene functions from the W3110 K-12 strain [180]. Multiple groups using different protocols to address similar problems will play an important role in addressing the robustness of experiments defining *E. coli* K-12 genome functions.

In Silico Analyses of Genome Functions

Nucleotide and protein sequence alignments have been the workhorse for researchers annotating functional features from complete genome sequences. Similarity of sequences often provides clues regarding functional roles but is seldom definitive.

The complete genome sequence has spurred the development of increasingly more complicated and systematic mathematical modeling of cellular processes. Edwards and Palsson [181–183] assembled gene functional information and biochemical literature to create a genome-scale description of the metabolic capabilities of *E. coli* MG1655. The method of flux-balance analysis uses this "metabolic genotype" to determine the constraints operating on cell behavior and ultimately allows prediction of growth characteristics under different environmental conditions. The model was used to predict the optimal performance of the metabolic network of MG1655 under different growth conditions [184]. The analysis defines a constrained space of allowable growth outcomes. A set of growth experiments was conducted to measure growth rates under a set of different conditions. The results of the experiment suggest that the *E. coli* metabolic network is optimized to achieve maximal growth under the test conditions. As the body of functional information about genomes grows, mathematical models that predict the outcome of cellular behavior sometimes may replace current methods of analysis that rely heavily on laboratory experimentation.

The rapid accumulation of complete microbial DNA sequences and the promise of many microarray experiments in the future have driven the development of computational tools that allow the prediction of regulatory sequences. AlignACE, a sequence-alignment program based on the Gibbs sampling algorithm, facilitates the identification of significant alignments from a set of input sequences [185]. The set of input sequences could be, for example, a set of sequences upstream of genes that were observed to be induced in a particular expression experiment or a set of genes that are known to function in a metabolic pathway. Using this approach, McGuire *et al.* [186] identified sets of predicted regulatory sequences from 17 complete microbial genomes. As additional data accumulate, tools that integrate functional information with sequence information will become increasingly important.

THE FUTURE OF *E. COLI* GENOMICS

The immediate future of genome-scale studies of *E. coli* seems self-evident. Ongoing genome projects will reveal the generality of observations made from the limited comparisons to date. One open question is how many entirely novel lineage-specific elements are present in the *E. coli* population. Another is what proportion of the same chromosomal sites occupied by lineage-specific elements in O157:H7 or K-12 contains completely different elements in other strains. A limited number of *E. coli* genomes will be sequenced to completion barring a radical reduction in sequencing costs. The implication is that if we want to understand the distribution of genes, operons, and islands in a more complete sample of the population, indirect methods must be employed. Genome-scale PCR- and microarray-based approaches are showing promise, but efforts must be scaled up to address such a diverse species with such a dynamic and mosaic genome structure effectively. This diversity also will figure prominently in efforts to unravel the complex relationships between *E. coli* genotypes and *E. coli* phenotypes. Thus far very little research has been devoted to whole-genome expression profiling and proteomics of *E. coli* strains other than K-12. This will change as additional genome sequences become available and protocols are optimized for cross-strain microarray hybridizations and high-throughput proteomics technologies develop. We look forward to the integration of all this information to generate interactive *in silico* models of the *E. coli* cell. As all these technologies converge, new approaches to the detection of *E. coli* pathogens and prevention and treatment of disease caused by *E. coli* will become available. Finally, as we combat known pathogenic *E. coli*, we should expect new variants to emerge that are resistant to our treatment strategies or evade standard detection procedures. Knowing this, we must strive to develop genomic assays that are both rapid and rapidly adaptable.

ACKNOWLEDGMENTS

We acknowledge the vision and leadership of Frederick R. Blattner, in whose laboratory much of this work was carried out. We also acknowledge our colleague Bob Mau for genome-comparison software, the sequencing team of the Genome Center, and many other colleagues and collaborators for their valuable contributions.

REFERENCES

1. Blattner, F. R., Plunkett, G., III, Bloch, C. A., *et al.* (1997). The complete genome sequence of *Escherichia coli* K-12. *Science* 277:1453–1474.
2. Perna, N. T., Plunkett, G., III, Burland, V., *et al.* (2001). Genome sequence of enterohaemorrhagic *Escherichia coli* O157:H7. *Nature* 409:529–533.
3. Hayashi, T., Makino, K., Ohnishi, M., *et al.* (2001). Complete genome sequence of enterohemorrhagic *Escherichia coli* O157:H7 and genomic comparison with a laboratory strain K-12. *DNA Res.* 8:11–22.
4. Bachmann, B. J. (1987). Derivations and genotypes of some mutant derivatives of *Escherichia coli* K-12. In *Escherichia coli and Salmonella typhimurium: Cellular and Molecular Biology* (F. C. Neidhardt, J. L. Ingraham, K. B. Low, *et al.*, eds.), Vol. 2, pp. 1190–1219. American Society for Microbiology, Washington.
5. Daniels, D. L., Plunkett, G., III, Burland, V., *et al.* (1992). Analysis of the *Escherichia coli* genome: DNA sequence of the region from 84.5 to 86.5 minutes. *Science* 257:771–778.
6. Burland, V., Plunkett, G., III, Daniels, D. L., *et al.* (1993). DNA sequence and analysis of 136 kilobases of the *Escherichia coli* genome: Organizational symmetry around the origin of replication. *Genomics* 16:551–561.
7. Plunkett, G., III, Burland, V., Daniels, D. L., *et al.* (1993). Analysis of the *Escherichia coli* genome: III. DNA sequence of the region from 87.2 to 89.2 minutes. *Nucleic Acids Res.* 21:3391–3398.
8. Blattner, F. R., Burland, V., Plunkett, G., III, *et al.* (1993). Analysis of the *Escherichia coli* genome: IV. DNA sequence of the region from 89.2 to 92.8 minutes. *Nucleic Acids Res.* 21:5408–5417.
9. Sofia, H. J., Burland, V., Daniels, D. L., *et al.* (1994). Analysis of the *Escherichia coli* genome: V. DNA sequence of the region from 76.0 to 81.5 minutes. *Nucleic Acids Res.* 22:2576–2586.
10. Burland, V., Plunkett, G., III, Sofia, H. J., *et al.* (1995). Analysis of the *Escherichia coli* genome: VI: DNA sequence of the region from 92.8 through 100 minutes. *Nucleic Acids Res.* 23:2105–2119.
11. Mahillon, J., Kirkpatrick, H. A., Kijenski, H. L., *et al.* (1998). Subdivision of the *Escherichia coli* K-12 genome for sequencing: manipulation and DNA sequence of transposable elements introducing unique restriction sites. *Gene* 223:47–54.
12. Masters, M., and Plunkett, G., III (1998). Bacterial genomes: the *Escherichia coli* model. In *ICRF Handbook of Genome Analysis* (N. K. Spurr, B. D. Young, and S. P. Bryant, eds.), Vol. 2, pp. 715–744. Blackwell Science, Oxford, U.K.
13. Yura, T., Mori, H., Nagai, H., *et al.* (1992). Systematic sequencing of the *Escherichia coli* genome: Analysis of the 0–2.4 min region. *Nucleic Acids Res.* 20:3305–3308.
14. Fujita, N., Mori, H., Yura, T., *et al.* (1994). Systematic sequencing of the *Escherichia coli* genome: Analysis of the 2.4–4.1 min (110,917–193,643 bp) region. *Nucleic Acids Res.* 22:1637–1639.

15. Oshima, T., Aiba, H., Baba, T., *et al.* (1996). A 718-kb DNA sequence of the *Escherichia coli* K-12 genome corresponding to the 12.7–28.0 min region on the linkage map. *DNA Res.* 3:137–155.
16. Oshima, T., Aiba, H., Baba, T., *et al.* (1996). A 718-kb DNA sequence of the *Escherichia coli* K-12 genome corresponding to the 12.7–28.0 min region on the linkage map (supplement). *DNA Res.* 3:211–223.
17. Aiba, H., Baba, T., Hayashi, K., *et al.* (1996). A 570-kb DNA sequence of the *Escherichia coli* K-12 genome corresponding to the 28.0–40.1 min region on the linkage map (supplement). *DNA Res.* 3:435–440.
18. Aiba, H., Baba, T., Hayashi, K., *et al.* (1996). A 570-kb DNA sequence of the *Escherichia coli* K-12 genome corresponding to the 28.0–40.1 min region on the linkage map. *DNA Res.* 3:363–377.
19. Itoh, T., Aiba, H., Baba, T., *et al.* (1996). A 460-kb DNA sequence of the *Escherichia coli* K-12 genome corresponding to the 40.1–50.0 min region on the linkage map. *DNA Res.* 3:379–392.
20. Itoh, T., Aiba, H., Baba, T., *et al.* (1996). A 460-kb DNA sequence of the *Escherichia coli* K-12 genome corresponding to the 40.1–50.0 min region on the linkage map (supplement). *DNA Res.* 3:441–445.
21. Yamamoto, Y., Aiba, H., Baba, T., *et al.* (1997). Construction of a contiguous 874-kb sequence of the *Escherichia coli*-K12 genome corresponding to 50.0–68.8 min on the linkage map and analysis of its sequence features. *DNA Res.* 4:91–113.
22. Yamamoto, Y., Aiba, H., Baba, T., *et al.* (1997). Construction of a contiguous 874-kb sequence of the *Escherichia coli* K-12 genome corresponding to 50.0–68.8 min on the linkage map and analysis of its sequence features (supplement). *DNA Res.* 4:169–178.
23. Kohara, Y., Akiyama, K., and Isono, K. (1987). The physical map of the whole *E. coli* chromosome: Application of a new strategy for rapid analysis and sorting of a large genomic library. *Cell* 50:495–508.
24. Hayashi, T., Makino, K., Ohnishi, M., *et al.* (2001). Complete genome sequence of enterohemorrhagic *Escherichia coli* O157:H7 and genomic comparison with a laboratory strain K-12 (supplement). *DNA Res.* 8:47–52.
25. Riley, L. W., Remis, R. S., Helgerson, S. D., *et al.* (1983). Hemorrhagic colitis associated with a rare *Escherichia coli* serotype. *New Engl. J. Med.* 308:681–685.
26. Michino, H., Araki, K., Minami, S., *et al.* (1999). Massive outbreak of *Escherichia coli* O157:H7 infection in school children in Sakai City, Japan, associated with consumption of white radish sprouts. *Am. J. Epidemiol.* 150:787–796.
27. Lim, A., Dimalanta, E. T., Potamousis, K. D., *et al.* (2001). Shotgun optical maps of the whole *Escherichia coli* O157:H7 genome. *Genome Res.* 11:1584–1593.
28. Makino, K., Ishii, K., Yasunaga, T., *et al.* (1998). Complete nucleotide sequences of 93-kb and 3.3-kb plasmids of an enterohemorrhagic *Escherichia coli* O157:H7 derived from Sakai outbreak. *DNA Res.* 5:1–9.
29. Burland, V., Shao, Y., Perna, N. T., *et al.* (1998). The complete DNA sequence and analysis of the large virulence plasmid of *Escherichia coli* O157:H7. *Nucleic Acids Res.* 26:4196–4204.
30. Plunkett, G., III, Rose, D. J., Durfee, T. J., *et al.* (1999). Sequence of Shiga toxin 2 phage 933W from *Escherichia coli* O157:H7: Shiga toxin as a phage late-gene product. *J. Bacteriol.* 181:1767–1778.
31. Miyamoto, H., Nakai, W., Yajima, N., *et al.* (1999). Sequence analysis of Stx2-converting phage VT2-Sa shows a great divergence in early regulation and replication regions. *DNA Res.* 6:235–240.
32. Makino, K., Yokoyama, K., Kubota, Y., *et al.* (1999). Complete nucleotide sequence of the prophage VT2-Sakai carrying the verotoxin 2 genes of the enterohemorrhagic *Escherichia coli* O157:H7 derived from the Sakai outbreak. *Genes Genet. Syst.* 74:227–239.
33. Bergthorsson, U., and Ochman, H. (1995). Heterogeneity of genome sizes among natural isolates of *Escherichia coli*. *J. Bacteriol.* 177:5784–5789.

34. Ochman, H., and Selander, R. K. (1984). Standard reference strains of *Escherichia coli* from natural populations. *J. Bacteriol.* 157:690–693.
35. Cairns, J. (1963). The chromosome of *Escherichia coli. Cold Spring Harbor Symp. Quant. Biol.* 28:43–46.
36. Pettijohn, D. E. (1996). The nucleoid. In *Escherichia coli and Salmonella: Cellular and Molecular Biology* (F. C. Neidhardt, R. Curtis, III, J. L. Ingraham, *et al.*, eds.), Vol. 1, pp. 158–166. ASM Press, Washington.
37. Masters, M., and Broda, P. (1971). Evidence for the bidirectional replication of the *Escherichia coli* chromosome. *Nature New Biol.* 232:137–140.
38. Kuempel, P. L., Pelletier, A. J., and Hill, T. M. (1989). Tus and the terminators: The arrest of replication in prokaryotes. *Cell* 59:581–583.
39. Hill, T. M. (1992). Arrest of bacterial DNA replication. *Annu. Rev. Microbiol.* 46:603–633.
40. Sanderson, K. E., and Liu, S. L. (1998). Chromosomal rearrangements in enteric bacteria. *Electrophoresis* 19:569–572.
41. Liu, S. L., and Sanderson, K. E. (1995). Rearrangements in the genome of the bacterium *Salmonella typhi. Proc. Natl. Acad. Sci. USA* 92:1018–1022.
42. Liu, S. L., and Sanderson, K. E. (1996). Highly plastic chromosomal organization in *Salmonella typhi. Proc. Natl. Acad. Sci. USA* 93:10303–10308.
43. Lobry, J. R. (1996). Asymmetric substitution patterns in the two DNA strands of bacteria. *Mol. Biol. Evol.* 13:660–665.
44. Frank, A. C., and Lobry, J. R. (1999). Asymmetric substitution patterns: A review of possible underlying mutational or selective mechanisms. *Gene* 238:65–77.
45. Kim, J., Nietfeldt, J., and Benson, A. K. (1999). Octamer-based genome scanning distinguishes a unique subpopulation of *Escherichia coli* O157:H7 strains in cattle. *Proc. Natl. Acad. Sci. USA* 96:13288–13293.
46. Reid, S. D., Herbelin, C. J., Bumbaugh, A. C., *et al.* (2000). Parallel evolution of virulence in pathogenic *Escherichia coli. Nature* 406:64–67.
47. Milkman, R., and Bridges, M. M. (1993). Molecular evolution of the *Escherichia coli* chromosome: IV. Sequence comparisons. *Genetics* 133:455–468.
48. Milkman, R., and Bridges, M. M. (1990). Molecular evolution of the *Escherichia coli* chromosome: III. Clonal frames. *Genetics* 126:505–517.
49. Milkman, R., and Stoltzfus, A. (1988). Molecular evolution of the *Escherichia coli* chromosome: II. Clonal segments. *Genetics* 120:359–366.
50. Stoltzfus, A., Leslie, J. F., and Milkman, R. (1988). Molecular evolution of the *Escherichia coli* chromosome: I. Analysis of structure and natural variation in a previously uncharacterized region between *trp* and *tonB. Genetics* 120:345–358.
51. Selander, R. K., Caugant, D. A., and Whittam, T. S. (1987). Genetic structure and variation in natural populations of *Escherichia coli.* In *Escherichia coli and Salmonella typhimurium: Cellular and Molecular Biology* (F. C. Neidhardt, J. L. Ingraham, K. B. Low, *et al.*, eds.), Vol. 2, pp. 1625–1648. American Society for Microbiology, Washington.
52. Sharp, P. M., Kelleher, J. E., Daniel, A. S., *et al.* (1992). Roles of selection and recombination in the evolution of type I restriction-modification systems in enterobacteria. *Proc. Natl. Acad. Sci. USA* 89:9836–9840.
53. Boyd, E. F., and Hartl, D. L. (1998). Diversifying selection governs sequence polymorphism in the major adhesin proteins *fimA*, *papA*, and *sfaA* of *Escherichia coli. J. Mol. Evol.* 47:258–267.
54. Reid, S. D., Selander, R. K., and Whittam, T. S. (1999). Sequence diversity of flagellin (*fliC*) alleles in pathogenic *Escherichia coli. J. Bacteriol.* 181:153–160.
55. Perna, N. T., Mayhew, G. F., Posfai, G., *et al.* (1998). Molecular evolution of a pathogenicity island from enterohemorrhagic *Escherichia coli* O157:H7. *Infect. Immun.* 66:3810–3817.
56. Ochman, H., and Wilson, A. C. (1987). Evolutionary history of enteric bacteria. In *Escherichia coli and Salmonella typhimurium: Cellular and Molecular Biology* (F. C. Neidhardt, J. L.

Ingraham, K. B. Low, *et al.*, eds.), Vol. 2, pp. 1649–1654. American Society for Microbiology, Washington.

57. Hacker, J., Bender, L., Ott, M., *et al.* (1990). Deletions of chromosomal regions coding for fimbriae and hemolysins occur *in vitro* and *in vivo* in various extraintestinal *Escherichia coli* isolates. *Microb. Pathog.* 8:213–225.
58. Hacker, J., and Kaper, J. B. (2000). Pathogenicity islands and the evolution of microbes. *Annu. Rev. Microbiol.* 54:641–679.
59. Rode, C. K., Melkerson-Watson, L. J., Johnson, A. T., *et al.* (1999). Type-specific contributions to chromosome size differences in *Escherichia coli*. *Infect. Immun.* 67:230–236.
60. Melkerson-Watson, L. J., Rode, C. K., Zhang, L., *et al.* (2000). Integrated genomic map from uropathogenic *Escherichia coli* J96. *Infect. Immun.* 68:5933–5942.
61. Medigue, C., Rouxel, T., Vigier, P., *et al.* (1991). Evidence for horizontal gene transfer in *Escherichia coli* speciation. *J. Mol. Biol.* 222:851–856.
62. Lawrence, J. G., and Ochman, H. (1998). Molecular archaeology of the *Escherichia coli* genome. *Proc. Natl. Acad. Sci. USA* 95:9413–9417.
63. Ochman, H., and Jones, I. B. (2000). Evolutionary dynamics of full genome content in *Escherichia coli*. *EMBO J.* 19:6637–6643.
64. Hurtado, A., and Rodriguez-Valera, F. (1999). Accessory DNA in the genomes of representatives of the *Escherichia coli* reference collection. *J. Bacteriol.* 181:2548–2554.
65. Lukashin, A. V., and Borodovsky, M. (1998). GeneMark.hmm: New solutions for gene finding. *Nucleic Acids Res.* 26:1107–1115.
66. Borodovsky, M., McIninch, J. D., Koonin, E. V., *et al.* (1995). Detection of new genes in a bacterial genome using Markov models for three gene classes. *Nucleic Acids Res.* 23:3554–3562.
67. Riley, M. (1993). Functions of the gene products of *Escherichia coli*. *Microbiol. Rev.* 57:862–952.
68. Riley, M. (1998). Systems for categorizing functions of gene products. *Curr. Opin. Struct. Biol.* 8:388–392.
69. Klapproth, J. M., Scaletsky, I. C., McNamara, B. P., *et al.* (2000). A large toxin from pathogenic *Escherichia coli* strains that inhibits lymphocyte activation. *Infect. Immun.* 68:2148–2155.
70. Nicholls, L., Grant, T. H., and Robins-Browne, R. M. (2000). Identification of a novel genetic locus that is required for *in vitro* adhesion of a clinical isolate of enterohaemorrhagic *Escherichia coli* to epithelial cells. *Mol. Microbiol.* 35:275–288.
71. Pulkkinen, W. S., and Miller, S. I. (1991). A *Salmonella typhimurium* virulence protein is similar to a *Yersinia enterocolitica* invasion protein and a bacteriophage lambda outer membrane protein. *J. Bacteriol.* 173:86–93.
72. Savarino, S. J., McVeigh, A., Watson, J., *et al.* (1996). Enteroaggregative *Escherichia coli* heat-stable enterotoxin is not restricted to enteroaggregative *E. coli*. *J. Infect. Dis.* 173:1019–1022.
73. Elias, W. P., Jr., Czeczulin, J. R., Henderson, I. R., *et al.* (1999). Organization of biogenesis genes for aggregative adherence fimbria II defines a virulence gene cluster in enteroaggregative *Escherichia coli*. *J. Bacteriol.* 181:1779–1785.
74. McVeigh, A., Fasano, A., Scott, D. A., *et al.* (2000). IS1414, an *Escherichia coli* insertion sequence with a heat-stable enterotoxin gene embedded in a transposase-like gene. *Infect. Immun.* 68:5710–5715.
75. Welch, R. A. (1991). Pore-forming cytolysins of gram-negative bacteria. *Mol. Microbiol.* 5:521–528.
76. Segal, G., Purcell, M., and Shuman, H. A. (1998). Host cell killing and bacterial conjugation require overlapping sets of genes within a 22-kb region of the *Legionella pneumophila* genome. *Proc. Natl. Acad. Sci. USA* 95:1669–1674.
77. Benz, I., and Schmidt, M. A. (1992). AIDA-I, the adhesin involved in diffuse adherence of the diarrhoeagenic *Escherichia coli* strain 2787 (O126:H27), is synthesized via a precursor molecule. *Mol. Microbiol.* 6:1539–1546.

78. Loveless, B. J., and Saier, M. H., Jr. (1997). A novel family of channel-forming, autotransporting, bacterial virulence factors. *Mol. Membr. Biol.* 14:113–123.
79. Maurer, J., Jose, J., and Meyer, T. F. (1999). Characterization of the essential transport function of the AIDA-I autotransporter and evidence supporting structural predictions. *J. Bacteriol.* 181:7014–7020.
80. Henderson, I. R., Navarro-Garcia, F., and Nataro, J. P. (1998). The great escape: structure and function of the autotransporter proteins. *Trends Microbiol.* 6:370–378.
81. Suhr, M., Benz, I., and Schmidt, M. A. (1996). Processing of the AIDA-I precursor: removal of AIDAc and evidence for the outer membrane anchoring as a beta-barrel structure. *Mol. Microbiol.* 22:31–42.
82. Goldberg, M. B., Barzu, O., Parsot, C., *et al.* (1993). Unipolar localization and ATPase activity of IcsA, a *Shigella flexneri* protein involved in intracellular movement. *J. Bacteriol.* 175:2189–2196.
83. Relman, D. A., Domenighini, M., Tuomanen, E., *et al.* (1989). Filamentous hemagglutinin of *Bordetella pertussis*: nucleotide sequence and crucial role in adherence. *Proc. Natl. Acad. Sci. USA* 86:2637–2641.
84. Kim, J. F., Ham, J. H., Bauer, D. W., *et al.* (1998). The *hrpC* and *hrpN* operons of *Erwinia chrysanthemi* EC16 are flanked by *plcA* and homologues of hemolysin/adhesin genes and accompanying activator/transporter genes. *Mol. Plant Microbe Interact.* 11:563–567.
85. St. Geme, J. W., III, Cutter, D., and Barenkamp, S. J. (1996). Characterization of the genetic locus encoding *Haemophilus influenzae* type b surface fibrils. *J. Bacteriol.* 178:6281–6287.
86. Cope, L. D., Lafontaine, E. R., Slaughter, C. A., *et al.* (1999). Characterization of the *Moraxella catarrhalis uspA1* and *uspA2* genes and their encoded products. *J. Bacteriol.* 181:4026–4034.
87. Torres, A. G., and Payne, S. M. (1997). Haem iron-transport system in enterohaemorrhagic *Escherichia coli* O157:H7. *Mol. Microbiol.* 23:825–833.
88. Wyckoff, E. E., Duncan, D., Torres, A. G., *et al.* (1998). Structure of the *Shigella dysenteriae* haem transport locus and its phylogenetic distribution in enteric bacteria. *Mol. Microbiol.* 28: 1139–1152.
89. Mills, M., and Payne, S. M. (1995). Genetics and regulation of heme iron transport in *Shigella dysenteriae* and detection of an analogous system in *Escherichia coli* O157:H7. *J. Bacteriol.* 177:3004–3009.
90. Sauer, F. G., Mulvey, M. A., Schilling, J. D., *et al.* (2000). Bacterial pili: molecular mechanisms of pathogenesis. *Curr. Opin. Microbiol.* 3:65–72.
91. Baumler, A. J., and Heffron, F. (1995). Identification and sequence analysis of *lpfABCDE*, a putative fimbrial operon of *Salmonella typhimurium*. *J. Bacteriol.* 177:2087–2097.
92. Lin, J., Smith, M. P., Chapin, K. C., *et al.* (1996). Mechanisms of acid resistance in enterohemorrhagic *Escherichia coli*. *Appl. Environ. Microbiol.* 62:3094–3100.
93. Price, S. B., Cheng, C. M., Kaspar, C. W., *et al.* (2000). Role of *rpoS* in acid resistance and fecal shedding of *Escherichia coli* O157:H7. *Appl. Environ. Microbiol.* 66:632–637.
94. Castanie-Cornet, M. P., Penfound, T. A., Smith, D., *et al.* (1999). Control of acid resistance in *Escherichia coli*. *J. Bacteriol.* 181:3525–3535.
95. Cai, Y., and Ni, Y. (1996). Purification, characterization, and pathogenicity of urease produced by *Vibrio parahaemolyticus*. *J. Clin. Lab. Anal.* 10:70–73.
96. Dunn, B. E., and Phadnis, S. H. (1998). Structure, function and localization of *Helicobacter pylori* urease. *Yale J. Biol. Med.* 71:63–73.
97. Johnson, D. E., Russell, R. G., Lockatell, C. V., *et al.* (1993). Contribution of *Proteus mirabilis* urease to persistence, urolithiasis, and acute pyelonephritis in a mouse model of ascending urinary tract infection. *Infect. Immun.* 61:2748–2754.
98. de Koning-Ward, T. F., and Robins-Browne, R. M. (1996). Analysis of the urease gene complex of members of the genus *Yersinia*. *Gene* 182:225–228.
99. Neyrolles, O., Ferris, S., Behbahani, N., *et al.* (1996). Organization of *Ureaplasma urealyticum* urease gene cluster and expression in a suppressor strain of *Escherichia coli*. *J. Bacteriol.* 178:647–655.

100. Mulrooney, S. B., and Hausinger, R. P. (1990). Sequence of the *Klebsiella aerogenes* urease genes and evidence for accessory proteins facilitating nickel incorporation. *J. Bacteriol.* 172:5837–5843.
101. Fuenmayor, S. L., Wild, M., Boyes, A. L., *et al.* (1998). A gene cluster encoding steps in conversion of naphthalene to gentisate in *Pseudomonas* sp. strain U2. *J. Bacteriol.* 180:2522–2530.
102. Williams, P. A., and Sayers, J. R. (1994). The evolution of pathways for aromatic hydrocarbon oxidation in *Pseudomonas. Biodegradation* 5:195–217.
103. Werwath, J., Arfmann, H. A., Pieper, D. H., *et al.* (1998). Biochemical and genetic characterization of a gentisate 1,2-dioxygenase from *Sphingomonas* sp. strain RW5. *J. Bacteriol.* 180:4171–4176.
104. Harwood, C. S., Nichols, N. N., Kim, M. K., *et al.* (1994). Identification of the *pcaRKF* gene cluster from *Pseudomonas putida*: involvement in chemotaxis, biodegradation, and transport of 4-hydroxybenzoate. *J. Bacteriol.* 176:6479–6488.
105. Tobe, T., Tatsuno, I., Katayama, E., *et al.* (1999). A novel chromosomal locus of enteropathogenic *Escherichia coli* (EPEC), which encodes a *bfpT*-regulated chaperone-like protein, TrcA, involved in microcolony formation by EPEC. *Mol. Microbiol.* 33:741–752.
106. Zadik, P. M., Chapman, P. A., and Siddons, C. A. (1993). Use of tellurite for the selection of verocytotoxigenic *Escherichia coli* O157. *J. Med. Microbiol.* 39:155–158.
107. Whelan, K. F., Colleran, E., and Taylor, D. E. (1995). Phage inhibition, colicin resistance, and tellurite resistance are encoded by a single cluster of genes on the IncHI2 plasmid R478. *J. Bacteriol.* 177:5016–5027.
108. Taylor, D. E. (1999). Bacterial tellurite resistance. *Trends Microbiol.* 7:111–115.
109. Miki, T., Chang, Z. T., and Horiuchi, T. (1984). Control of cell division by sex factor F in *Escherichia coli:* II. Identification of genes for inhibitor protein and trigger protein on the 42.84–43.6 F segment. *J. Mol. Biol.* 174:627–646.
110. Sukupolvi, S., and O'Connor, C. D. (1990). TraT lipoprotein, a plasmid-specified mediator of interactions between gram-negative bacteria and their environment. *Microbiol. Rev.* 54:331–341.
111. Wang, L., and Reeves, P. R. (1998). Organization of *Escherichia coli* O157 O-antigen gene cluster and identification of its specific genes. *Infect. Immun.* 66:3545–3551.
112. Yamasaki, S., Shimizu, T., Hoshino, K., *et al.* (1999). The genes responsible for O-antigen synthesis of *Vibrio cholerae* O139 are closely related to those of *Vibrio cholerae* O22. *Gene* 237:321–332.
113. Heinrichs, D. E., Yethon, J. A., and Whitfield, C. (1998). Molecular basis for structural diversity in the core regions of the lipopolysaccharides of *Escherichia coli* and *Salmonella enterica. Mol. Microbiol.* 30:221–232.
114. Reeves, P. R., Hobbs, M., Valvano, M. A., *et al.* (1996). Bacterial polysaccharide synthesis and gene nomenclature. *Trends Microbiol.* 4:495–503.
115. Jenkins, L. S., and Nunn, W. D. (1987). Genetic and molecular characterization of the genes involved in short-chain fatty acid degradation in *Escherichia coli*: the *ato* system. *J. Bacteriol.* 169:42–52.
116. Pellicer, M. T., Badia, J., Aguilar, J., *et al.* (1996). *glc* locus of *Escherichia coli:* Characterization of genes encoding the subunits of glycolate oxidase and the *glc* regulator protein. *J. Bacteriol.* 178:2051–2059.
117. Cairns, J., Stent, G. S., and Watson, J. D. (1966). *Phage and the Origins of Molecular Biology*. Cold Spring Harbor Laboratory of Quantitative Biology, Cold Spring Harbor, NY.
118. Campbell, A. M. (1996). Bacteriophages. In *Escherichia coli and Salmonella: Cellular and Molecular Biology* (F. C. Neidhardt, R. Curtis, III, J. L. Ingraham, *et al.*, eds.), Vol. 2, pp. 2325–2338. ASM Press, Washington.
119. Campbell, A. M. (1996). Cryptic prophages. In *Escherichia coli and Salmonella: Cellular and Molecular Biology* (F. C. Neidhardt, R. Curtis, III, J. L. Ingraham, *et al.*, eds.), Vol. 2, pp. 2041–2046. ASM Press, Washington.

120. Campbell, A. M., and Botstein, D. (1983). Evolution of the lambdoid phages. In *Lambda II* (R. W. Hendrix, J. W. Roberts, F. W. Stahl, *et al.*, eds.), pp. 365–380. Cold Spring Harbor Laboratory, Cold Spring Harbor, NY.
121. Hendrix, R. W., Smith, M. C., Burns, R. N., *et al.* (1999). Evolutionary relationships among diverse bacteriophages and prophages: All the world's a phage. *Proc. Natl. Acad. Sci. USA* 96:2192–2197.
122. Vander Byl, C., and Kropinski, A. M. (2000). Sequence of the genome of *Salmonella* bacteriophage P22. *J. Bacteriol.* 182:6472–6481.
123. Brussow, H., and Desiere, F. (2001). Comparative phage genomics and the evolution of Siphoviridae: Insights from dairy phages. *Mol. Microbiol.* 39:213–222.
124. Hershey, A. D., ed. (1971). *The Bacteriophage Lambda.* Cold Spring Harbor Laboratory, Cold Spring Harbor, NY.
125. Hendrix, R. W., Roberts, J. W., Stahl, F. W., *et al.*, eds. (1983). *Lambda II.* Cold Spring Harbor Laboratory, Cold Spring Harbor, NY.
126. Campbell, A. M. (1993). Thirty years ago in *Genetics*: Prophage insertion into bacterial chromosomes. *Genetics* 133:433–438.
127. Smith, H. W., Green, P., and Parsell, Z. (1983). Vero cell toxins in *Escherichia coli* and related bacteria: transfer by phage and conjugation and toxic action in laboratory animals, chickens and pigs. *J. Gen. Microbiol.* 129:3121–3137.
128. O'Brien, A. D., Newland, J. W., Miller, S. F., *et al.* (1984). Shiga-like toxin-converting phages from *Escherichia coli* strains that cause hemorrhagic colitis or infantile diarrhea. *Science* 226:694–696.
129. O'Brien, A. D., Marques, L. R., Kerry, C. F., *et al.* (1989). Shiga-like toxin converting phage of enterohemorrhagic *Escherichia coli* strain 933. *Microb. Pathog.* 6:381–390.
130. Strockbine, N. A., Marques, L. R., Newland, J. W., *et al.* (1986). Two toxin-converting phages from *Escherichia coli* O157:H7 strain 933 encode antigenically distinct toxins with similar biologic activities. *Infect. Immun.* 53:135–140.
131. Paros, M., Tarr, P. I., Kim, H., *et al.* (1993). A comparison of human and bovine *Escherichia coli* O157:H7 isolates by toxin genotype, plasmid profile, and bacteriophage lambda-restriction fragment length polymorphism profile. *J. Infect. Dis.* 168:1300–1303.
132. Chang, C. C., Gilsdorf, J. R., DiRita, V. J., *et al.* (2000). Identification and genetic characterization of *Haemophilus influenzae* genetic island 1. *Infect. Immun.* 68:2630–2637.
133. Betley, M. J., and Mekalanos, J. J. (1985). Staphylococcal enterotoxin A is encoded by phage. *Science* 229:185–187.
134. Bishai, W. R., and Murphy, J. R. (1988). Bacteriophage gene products that cause human disease. In *The Bacteriophages* (R. Calender, ed.), Vol. 2, pp. 683–724. Plenum Press, New York.
135. Cheetham, B. F., and Katz, M. E. (1995). A role for bacteriophages in the evolution and transfer of bacterial virulence determinants. *Mol. Microbiol.* 18:201–208.
136. Waldor, M. K., and Mekalanos, J. J. (1996). Lysogenic conversion by a filamentous phage encoding cholera toxin. *Science* 272:1910–1914.
137. Karaolis, D. K., Somara, S., Maneval, D. R., Jr., *et al.* (1999). A bacteriophage encoding a pathogenicity island, a type-IV pilus and a phage receptor in cholera bacteria. *Nature* 399: 375–379.
138. Allison, G. E., and Verma, N. K. (2000). Serotype-converting bacteriophages and O-antigen modification in *Shigella flexneri*. *Trends Microbiol.* 8:17–23.
139. Yamaguchi, T., Hayashi, T., Takami, H., *et al.* (2000). Phage conversion of exfoliative toxin A production in *Staphylococcus aureus. Mol. Microbiol.* 38:694–705.
140. Kropinski, A. M. (2000). Sequence of the genome of the temperate, serotype-converting, *Pseudomonas aeruginosa* bacteriophage D3. *J. Bacteriol.* 182:6066–6074.
141. Figueroa-Bossi, N., Uzzau, S., Maloriol, D., *et al.* (2001). Variable assortment of prophages provides a transferable repertoire of pathogenic determinants in *Salmonella. Mol. Microbiol.* 39: 260–271.

142. Walterspiel, J. N., Ashkenazi, S., Morrow, A. L., *et al.* (1992). Effect of subinhibitory concentrations of antibiotics on extracellular Shiga-like toxin I. *Infection* 20:25–29.
143. Neely, M. N., and Friedman, D. I. (1998). Functional and genetic analysis of regulatory regions of coliphage H-19B: location of shiga-like toxin and lysis genes suggest a role for phage functions in toxin release. *Mol. Microbiol.* 28:1255–1267.
144. Zhang, X., McDaniel, A. D., Wolf, L. E., *et al.* (2000). Quinolone antibiotics induce Shiga toxin-encoding bacteriophages, toxin production, and death in mice. *J. Infect. Dis.* 181:664–670.
145. McDonough, M. A., and Butterton, J. R. (1999). Spontaneous tandem amplification and deletion of the Shiga toxin operon in *Shigella dysenteriae* 1. *Mol. Microbiol.* 34:1058–1069.
146. Unkmeir, A., and Schmidt, H. (2000). Structural analysis of phage-borne *stx* genes and their flanking sequences in Shiga toxin-producing *Escherichia coli* and *Shigella dysenteriae* type 1 strains. *Infect. Immun.* 68:4856–4864.
147. Yokoyama, K., Makino, K., Kubota, Y., *et al.* (2000). Complete nucleotide sequence of the prophage VT1-Sakai carrying the Shiga toxin 1 genes of the enterohemorrhagic *Escherichia coli* O157:H7 strain derived from the Sakai outbreak. *Gene* 258:127–139.
148. Muniesa, M., Recktenwald, J., Bielaszewska, M., *et al.* (2000). Characterization of a Shiga toxin 2e–converting bacteriophage from an *Escherichia coli* strain of human origin. *Infect. Immun.* 68:4850–4855.
149. Lindler, L. E., Plano, G. V., Burland, V., *et al.* (1998). Complete DNA sequence and detailed analysis of the *Yersinia pestis* KIM5 plasmid encoding murine toxin and capsular antigen. *Infect. Immun.* 66:5731–5742.
150. Hu, P., Elliott, J., McCready, P., *et al.* (1998). Structural organization of virulence-associated plasmids of *Yersinia pestis. J. Bacteriol.* 180:5192–5202.
151. Souza, V., Rocha, M., Valera, A., *et al.* (1999). Genetic structure of natural populations of *Escherichia coli* in wild hosts on different continents. *Appl. Environ. Microbiol.* 65:3373–3385.
152. Boyd, E. F., Hill, C. W., Rich, S. M., *et al.* (1996). Mosaic structure of plasmids from natural populations of *Escherichia coli. Genetics* 143:1091–1100.
153. Tobe, T., Hayashi, T., Han, C. G., *et al.* (1999). Complete DNA sequence and structural analysis of the enteropathogenic *Escherichia coli* adherence factor plasmid. *Infect. Immun.* 67:5455–5462.
154. Buchrieser, C., Glaser, P., Rusniok, C., *et al.* (2000). The virulence plasmid pWR100 and the repertoire of proteins secreted by the type III secretion apparatus of *Shigella flexneri. Mol. Microbiol.* 38:760–771.
155. Venkatesan, M., Goldberg, M., Rose, D., *et al.* (2001). Complete DNA sequence and analysis of the large virulence plasmid of *Shigella flexneri. Infect. Immun.* 69:3271–3285.
156. Perry, R. D., Straley, S. C., Fetherston, J. D., *et al.* (1998). DNA sequencing and analysis of the low-Ca^{2+} -response plasmid pCD1 of *Yersinia pestis* KIM5. *Infect. Immun.* 66:4611–4623.
157. Sherburne, C. K., Lawley, T. D., Gilmour, M. W., *et al.* (2000). The complete DNA sequence and analysis of R27, a large IncHI plasmid from *Salmonella typhi* that is temperature sensitive for transfer. *Nucleic Acids Res.* 28:2177–2186.
158. Schmidt, H., Beutin, L., and Karch, H. (1995). Molecular analysis of the plasmid-encoded hemolysin of *Escherichia coli* O157:H7 strain EDL 933. *Infect. Immun.* 63:1055–1061.
159. Gorbach, S. L., Banwell, J. G., Chatterjee, B. D., *et al.* (1971). Acute undifferentiated human diarrhea in the tropics: I. Alterations in intestinal microflora. *J. Clin. Invest.* 50:881–889.
160. Czeczulin, J. R., Whittam, T. S., Henderson, I. R., *et al.* (1999). Phylogenetic analysis of enteroaggregative and diffusely adherent *Escherichia coli. Infect. Immun.* 67:2692–2699.
161. Valvano, M. A., Silver, R. P., and Crosa, J. H. (1986). Occurrence of chromosome- or plasmid-mediated aerobactin iron transport systems and hemolysin production among clonal groups of human invasive strains of *Escherichia coli* K1. *Infect. Immun.* 52:192–199.
162. Haarmann, C., Karch, H., Frosch, M., *et al.* (1998). A 3.3-kb plasmid of enterohemorrhagic *Escherichia coli* O157:H7 is closely related to the core region of the *Salmonella typhimurium* antibiotic resistance plasmid NTP16. *Plasmid* 39:134–140.

163. Topp, E. M., Moenne-Loccoz, Y., Pembroke, T., *et al.* (2000). Phenotypic traits conferred by plasmids. In *The Horizontal Gene Pool: Bacterial Plasmids and Gene Spread* (C. M. Thomas, ed.), pp. 249–286. Harwood Academic Publishers, Amsterdam.
164. Fica, A., Fernandez-Beros, M. E., Aron-Hott, L., *et al.* (1997). Antibiotic-resistant *Salmonella typhi* from two outbreaks: few ribotypes and IS200 types harbor Inc HI1 plasmids. *Microb. Drug. Resist.* 3:339–343.
165. Rowe, B., Ward, L. R., and Threlfall, E. J. (1997). Multidrug-resistant *Salmonella typhi*: a worldwide epidemic. *Clin. Infect. Dis.* 24 (Suppl. 1):S106–109.
166. Gray, C. H., and Tatum, E. L. (1944). X-ray induced growth factor requirements in bacteria. *Proc. Natl. Acad. Sci. USA* 30:404–410.
167. Berlyn, M. K. (1998). Linkage map of *Escherichia coli* K-12, edition 10: The traditional map. *Microbiol. Mol. Biol. Rev.* 62:814–984.
168. O'Farrell, P. H. (1975). High-resolution two-dimensional electrophoresis of proteins. *J. Biol. Chem.* 250:4007–4021.
169. VanBogelen, R. A., Schiller, E. E., Thomas, J. D., *et al.* (1999). Diagnosis of cellular states of microbial organisms using proteomics. *Electrophoresis* 20:2149–2159.
170. Jacob, F., Ullman, A., and Monod, J. (1965). Deletions fusionnant l'operon lactose et un operon purine chez *E. coli. J. Mol. Biol.* 13:704.
171. Bassford, P., Beckwith, J., Berman, M., *et al.* (1980). Genetic fusions of the *lac* operon: A new approach to the study of biological processes. In *The Operon* (J. H. Miller and W. S. Reznikoff, eds.), pp. 245–261. Cold Spring Harbor Laboratory, Cold Spring Harbor, NY.
172. Chuang, S. E., Daniels, D. L., and Blattner, F. R. (1993). Global regulation of gene expression in *Escherichia coli. J. Bacteriol.* 175:2026–2036.
173. Van Dyk, T. K., Wei, Y., Hanafey, M. K., *et al.* (2001). A genomic approach to gene fusion technology. *Proc. Natl. Acad. Sci. USA* 98:2555–2560.
174. Richmond, C. S., Glasner, J. D., Mau, R., *et al.* (1999). Genome-wide expression profiling in *Escherichia coli* K-12. *Nucleic Acids Res.* 27:3821–3835.
175. Selinger, D. W., Cheung, K. J., Mei, R., *et al.* (2000). RNA expression analysis using a 30 base pair resolution *Escherichia coli* genome array. *Nature Biotechnol.* 18:1262–1268.
176. Houry, W. A., Frishman, D., Eckerskorn, C., *et al.* (1999). Identification of *in vivo* substrates of the chaperonin GroEL. *Nature* 402:147–154.
177. Murphy, K. C., Campellone, K. G., and Poteete, A. R. (2000). PCR-mediated gene replacement in *Escherichia coli. Gene* 246:321–330.
178. Datsenko, K. A., and Wanner, B. L. (2000). One-step inactivation of chromosomal genes in *Escherichia coli* K-12 using PCR products. *Proc. Natl. Acad. Sci. USA* 97:6640–6645.
179. Yu, D., Ellis, H. M., Lee, E. C., *et al.* (2000). An efficient recombination system for chromosome engineering in *Escherichia coli. Proc. Natl. Acad. Sci. USA* 97:5978–5983.
180. Mori, H., Isono, K., Horiuchi, T., *et al.* (2000). Functional genomics of *Escherichia coli* in Japan. *Res. Microbiol.* 151:121–128.
181. Edwards, J. S., and Palsson, B. O. (2000). Metabolic flux balance analysis and the *in silico* analysis of *Escherichia coli* K-12 gene deletions. *BMC Bioinformatics* 1:1.
182. Edwards, J. S., and Palsson, B. O. (2000). The *Escherichia coli* MG1655 *in silico* metabolic genotype: Its definition, characteristics, and capabilities. *Proc. Natl. Acad. Sci. USA* 97:5528–5533.
183. Edwards, J. S., and Palsson, B. O. (2000). Robustness analysis of the *Escherichia coli* metabolic network. *Biotechnol. Prog.* 16:927–939.
184. Edwards, J. S., Ibarra, R. U., and Palsson, B. O. (2001). *In silico* predictions of *Escherichia coli* metabolic capabilities are consistent with experimental data. *Nature Biotechnol.* 19:125–130.
185. Roth, F. P., Hughes, J. D., Estep, P. W., *et al.* (1998). Finding DNA regulatory motifs within unaligned noncoding sequences clustered by whole-genome mRNA quantitation. *Nature Biotechnol.* 16:939–945.

186. McGuire, A. M., Hughes, J. D., and Church, G. M. (2000). Conservation of DNA regulatory motifs and discovery of new motifs in microbial genomes. *Genome Res.* 10:744–757.
187. Barbosa, T. M., and Levy, S. B. (2000). Differential expression of over 60 chromosomal genes in *Escherichia coli* by constitutive expression of MarA. *J. Bacteriol.* 182:3467–3474.
188. Wei, Y., Lee, J. M., Smulski, D. R., *et al.* (2001). Global impact of *sdiA* amplification revealed by comprehensive gene expression profiling of *Escherichia coli. J. Bacteriol.* 183:2265–2272.
189. Wendisch, V. F., Zimmer, D. P., Khodursky, A., *et al.* (2001). Isolation of *Escherichia coli* mRNA and comparison of expression using mRNA and total RNA on DNA microarrays. *Anal. Biochem.* 290:205–213.
190. Arnold, C. N., McElhanon, J., Lee, A., *et al.* (2001). Global analysis of *Escherichia coli* gene expression during the acetate-induced acid tolerance response. *J. Bacteriol.* 183:2178–2186.
191. Arfin, S. M., Long, A. D., Ito, E. T., *et al.* (2000). Global gene expression profiling in *Escherichia coli* K12: The effects of integration host factor. *J. Biol. Chem.* 275:29672–29684.
192. Khodursky, A. B., Peter, B. J., Schmid, M. B., *et al.* (2000). Analysis of topoisomerase function in bacterial replication fork movement: use of DNA microarrays. *Proc. Natl. Acad. Sci. USA* 97:9419–9424.
193. Gill, R. T., Valdes, J. J., and Bentley, W. E. (1999). Reverse transcription-PCR differential display analysis of *Escherichia coli* global gene regulation in response to heat shock. *Appl. Environ. Microbiol.* 65:5386–5393.
194. Tao, H., Bausch, C., Richmond, C., *et al.* (1999). Functional genomics: expression analysis of *Escherichia coli* growing on minimal and rich media. *J. Bacteriol.* 181:6425–6440.
195. Wei, Y., Lee, J. M., Richmond, C., *et al.* (2001). High-density microarray-mediated gene expression profiling of *Escherichia coli. J. Bacteriol.* 183:545–556.
196. Brocklehurst, K. R., and Morby, A. P. (2000). Metal-ion tolerance in *Escherichia coli*: Analysis of transcriptional profiles by gene-array technology. *Microbiology* 146:2277–2282.
197. Khodursky, A. B., Peter, B. J., Cozzarelli, N. R., *et al.* (2000). DNA microarray analysis of gene expression in response to physiological and genetic changes that affect tryptophan metabolism in *Escherichia coli. Proc. Natl. Acad. Sci. USA* 97:12170–12175.

CHAPTER 2

Evolution of Pathogenic Escherichia coli

James R. Johnson
Veterans Affairs Medical Center, and Department of Medicine, University of Minnesota, Minneapolis, Minnesota

INTRODUCTION

Escherichia coli is a genetically diverse species of considerable scientific, economic, and medical importance. It includes both nonpathogenic variants, i.e., strains with little ability to cause disease, and pathogenic variants, i.e., strains that are able to cause enteric disease or diverse types of extraintestinal infection in human and animal hosts. Efforts to elucidate the evolutionary origins of the pathogenic variants have proceeded in parallel with the detailed molecular and physiologic characterization of these pathogens, as discussed in subsequent chapters. This chapter provides an evolutionary introduction by reviewing current understandings of the emergence of pathogenic variants of *E. coli*, including *Shigella*, in relation to the evolution of the species as a whole. Specific topics addressed include general mechanisms of evolution, mechanisms of evolution of pathogenic *E. coli*, and the reason for the emergence of pathogenic variants.

Escherichia coli: Virulence Mechanisms of a Versatile Pathogen
ISBN 0-12-220751-3

EVOLUTION IN *E. COLI*

Both the overall generation of diversity within *E. coli* and the emergence specifically of pathogenic variants result from two fundamental evolutionary processes: mutation and horizontal transfer [1–9]. Mutation involves the sporadic introduction of single-nucleotide substitutions or deletions of various sizes throughout the genome. Mutations are transmitted vertically from generation to generation, resulting in a clonal population structure [4,5,8,10]. Progressive diversification by mutation occurs in a dichotomously branching fashion through the successive accumulation of mutations within lineages, with each new mutant derivative serving as the ancestor for a new branch of the phylogenetic tree.

Mutations may be silent (hence neutrally selected) or have functional consequences. Those with functional consequences can be subject to positive, negative, or neutral selection depending on the nature of the mutation and the particular selective environment [5,6,11]. Clones with positively selected characteristics will expand, whereas those with negatively selected characteristics will recede or die out altogether unless they encounter a more favorable niche that allows their persistence. Periodic selection resulting from an environmental change that strongly favors a particular clone may lead to the purging of diversity in the species through elimination of less-fit clones; i.e., it may create a clonal sweep [12,13]. Mutational diversification then resumes from the new starting point.

In contrast to the incremental nature of mutation, horizontal transfer involves the en bloc exchange of substantial amounts of DNA between different *E. coli* lineages or between another species and *E. coli* [1,2,6,9]. Large stretches of DNA (up to hundreds of kilobases) can be transferred in a single genetic event such that in one step multiple genes or significant portions of individual genes may be substituted for their counterparts in the recipient strain, or novel alien genes may be introduced. Horizontally acquired genomic DNA fragments can be introduced by bacterial conjugation, phage-mediated transduction, or passive transformation. Recombination of such fragments into the genome remodels the genomic backbone [9,10]. In contrast, horizontally acquired plasmids, which often have an evolutionary history independent of that of the genome [14–17], also can alter the recipient's total genetic content considerably but usually do not directly affect the genome. Horizontal transmission produces a web of genetic similarity relationships by creating cross-links between distinct branches of the clonally derived phylogenetic tree [7]. Horizontal exchange between *E. coli* lineages thus serves to blur clonal distinctions by creating similarities between otherwise distant lineages. This reshuffling of genetic material tends to homogenize the species as a whole, whereas it increases diversity within the recipient lineage [2]. Similarly, horizontal exchange that involves transfer of genetic material from a different species to *E. coli* [9,18–20] decreases the distance between the two species but increases overall diversity within *E. coli*.

Repeated introduction of exogenous DNA into the genome by recombination gradually breaks down the "clonal frame" of the recipient *E. coli* lineage, ultimately obscuring the clonal origins of the genomic backbone [10]. Yet, despite this obliterating effect of recombination, and although recombination may play a greater net evolutionary role within *E. coli* than does mutation [2], recombination still is not so rapid or extensive as to obscure at any point in time the underlying clonal nature of the *E. coli* population [2,12].

The combination of mutation and horizontal transfer has shaped the overall phylogenetic structure of *E. coli,* i.e., has given rise to the currently-recognized four major phylogenetic groups (see below and [21]) and their respective constituent lineages, within which the diverse pathogenic variants have emerged. Thus an appreciation of global evolutionary relationships within *E. coli* is helpful in understanding the origins of and the relationships among the various pathogenic variants. Likewise, an appreciation of mutation and horizontal transfer as significant evolutionary processes within *E. coli* in general helps in understanding their roles as well in the emergence of pathogens.

VIRULENCE FACTORS AND VIRULENCE

Pathogenic *E. coli* variants derive their ability to cause intestinal or extraintestinal disease largely from their expression of multiple specialized *virulence factors* (VFs), including diverse adhesins, toxins, siderophores, secretion systems, and so on [22–30]. These so-called virulence factors are not required for vegetative replication or simple commensalism, as evidenced by their absence from most successful commensal *E. coli* strains, e.g., those which constitute the bulk of the facultative intestinal flora of healthy human and animal hosts (but see the section "Why Pathogenic *E. coli*?"). In contrast, the virulence factors clearly do provide the organism with an enhanced ability to colonize specific host surfaces, to avoid or subvert host defense systems, to stimulate a noxious host inflammatory response, and to directly injure host cells and tissues, all of which can contribute to disease causation. (Whether or not the pathogenic functions of virulence factors are also evolutionarily relevant is discussed later under "Why Pathogenic *E. coli*?"). Multiple functional categories of virulence factors and multiple representatives within each functional category typically are present in successful pathogenic *E. coli* strains, with the strain's net virulence potential varying in relation to the number and quality of the virulence factors it possesses [31,32]. Thus the evolution of pathogenic *E. coli* variants can be understood in terms of (1) the background evolution of the species as a whole, (2) the evolution of individual virulence factors, and (3) the interactions of the virulence factors with the genomic background, leading to focal concentrations of specific virulence factors within particular lineages of *E. coli.*

PHYLOGENETIC RELATIONSHIPS IN *E. COLI*

As discussed earlier, *E. coli* is a highly clonal species. Four major phylogenetic groups (A, B1, B2, and D) are recognized on the basis of similarity relationships, as revealed by techniques such as multilocus enzyme electrophoresis (MLEE) and multilocus sequence typing (MLST) [21,33–35] (Figs. 1 and 2). These techniques, which sample multiple, presumably selection-neutral housekeeping genes that are distributed about the genome, provide a minimally biased approximation of a strain's whole genome history [12,35]. Alternative phylotyping methods, such as randomly amplified polymorphic DNA (RAPD) analysis [8,13], repetitive-element polymerase chain reaction (PCR) [36], and ribotyping [8], provide similar, although less reliable, information. Trees of varying topologies can be derived from the same phylogenetic data set by using different cluster-analysis methods or different evolutionary assumptions within a given analysis method or by applying the same cluster-analysis method to different data sets for the same strains [8,33,34,37–41]. For example, phylogenetic group B2 can appear as a late-splitting sister group of phylogenetic group B1 (see, e.g., Figs. 1 and 2: MLEE) or as the earliest-splitting branch within the *E. coli* tree (see [33]: MLST). The same is true, in reverse, for phylogenetic group A (see Fig. 1 versus [33]). Likewise, inclusion of different enzyme loci or use of different types of gels in MLEE analysis can reverse the apparent branching order of groups A and D relative to groups B1 and B2 (see Figs. 1 and 2). Consequently, although there is consensus regarding the validity of the four major *E. coli* phylogenetic groups, the true branching order of these groups and their fine internal phylogenetic structure remain unresolved issues.

EVOLUTION OF INDIVIDUAL VFs

The ultimate evolutionary origin of most *E. coli* VFs is obscure. For many VFs, an immediate proximal source can be inferred from sequence analysis or patterns of phylogenetic distribution [31,42–44]. This source can be a specific lineage or major phylogenetic group within *E. coli* or a source outside *E. coli*, whether defined [as for the *Yersinia*-derived yersiniabactin siderophore system [18–20]) or undefined (as for the seemingly non-*E. coli hly* (α-hemolysin) operon] [45,46]. However, although such inferences provide insights into the recent histories of the various virulence factors, they leave obscure the ultimate origins of the VFs, e.g., whether the VFs evolved from housekeeping genes or other building blocks, and what functions the putative ancestral proto-VFs may have subserved. Intermediate forms that could allow current VFs to be traced back to their earlier ancestors are lacking. Instead, most currently recognized VFs appear to have emerged full-fledged essentially in their present form. Thus evolutionary analysis of VFs is largely limited to assessments of their phylogenetic distribution and recent microevolution, whereas ultimate origins remain enigmatic.

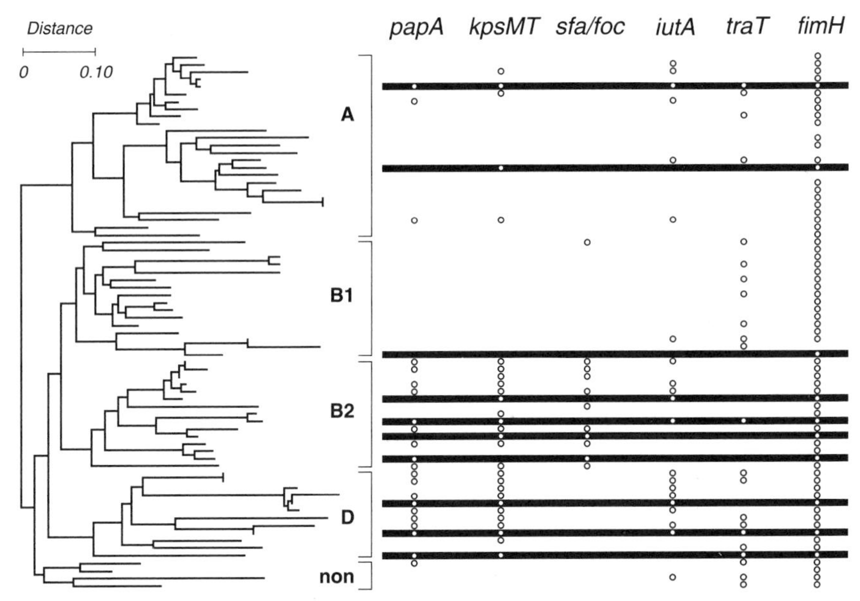

FIGURE 1 Phylogenetic distribution of *Escherichia coli* extraintestinal virulence genes in relation to clinical source. Dendrogram at left depicts phylogenetic relationships for the 72 members of the *E. coli* reference (ECOR) collection, as inferred based on multilocus enzyme electrophoresis using 38 metabolic enzymes [21]. The four major phylogenetic groups (A, B1, B2, and D) of *E. coli* and the nonaligned strains (non) are bracketed and labeled. Bullets at right indicate presence of putative virulence genes (*papA*, P fimbriae; *kpsMT*, group II capsule synthesis; *sfa/foc*, S and F1C fimbriae; *iutA*, aerobactin receptor; *traT*, serum resistance; and *fimH*, type 1 fimbriae). Horizontal bars at right indicate the 10 ECOR strains that are symptomatic infection isolates, all from humans with urinary tract infection (UTI). The remaining strains, excepting for one asymptomatic bacteriuria isolate, are fecal isolates from healthy human or animal hosts. Note the concentration of (chromosomal) virulence genes *papA*, *kpsMT*, and *sfa/foc* within phylogenetic groups B2 and D but their occasional joint appearance also in distant lineages, consistent with coordinate horizontal transfer. The more scattered phylogenetic distribution of *traT* is consistent with this gene's plasmid location [128], whereas *fimH* is nearly universally prevalent, consistent with its presence in other species of Enterobacteriaciae, presumably reflecting an origin in a shared enterobacterial ancestor. Note the concentration of UTI isolates within phylogenetic groups B2 and D and the association of virulence genes with the UTI isolates. Note that group A forms an outgroup, with groups B2 and B1 representing sister groups within a cluster that includes group D. (Neighbor-joining tree adapted from [21] with permission.)

DIVERSIFYING AND DIRECTIONAL SELECTION

Comparative sequence analysis of variants of certain VF genes from different *E. coli* strains and lineages has provided insights into their recent evolution. Genes whose products are surface-exposed and antigenic, e.g., the P fimbrial structural subunit gene *papA*, and hence presumably are subject to selective pressure from

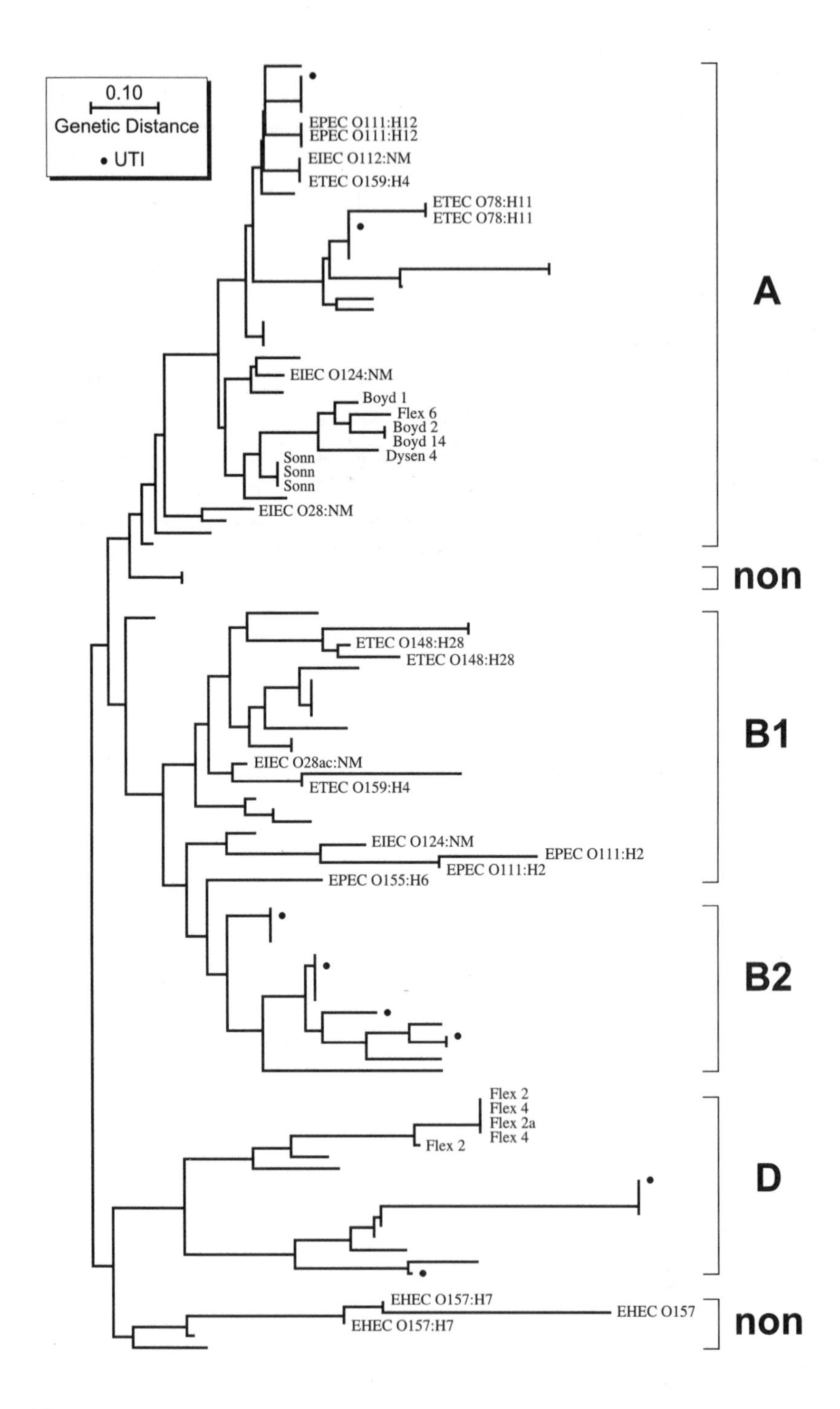
0.10
Genetic Distance
• UTI
EPEC O111:H12
EPEC O111:H12
EIEC O112:NM
ETEC O159:H4
ETEC O78:H11
ETEC O78:H11
EIEC O124:NM
Boyd 1
Flex 6
Boyd 2
Boyd 14
Dysen 4
Sonn
Sonn
Sonn
EIEC O28:NM
ETEC O148:H28
ETEC O148:H28
EIEC O28ac:NM
ETEC O159:H4
EIEC O124:NM
EPEC O111:H2
EPEC O111:H2
EPEC O155:H6
Flex 2
Flex 4
Flex 2a
Flex 4
Flex 2
EHEC O157:H7
EHEC O157
EHEC O157:H7
A
non
B1
B2
D
non

the host immune response, often exhibit evidence of diversifying selection, i.e., a higher rate of protein-sequence-modifying mutations than expected under neutral selection [11]. The regions showing the highest variability typically are those which are most surface-exposed and least functionally constrained [11,47]. Other genes, notably those encoding the fimbrial tip adhesins of P fimbriae (PapG/*papG*) and type 1 fimbriae (FimH/*fimH*), appear to be undergoing selection-driven diversification not from immune pressure but because the altered receptor repertoire exhibited by the mutants confers increased fitness in specific environments [48–50]. Interestingly, diverse point mutations distributed throughout *fimH*, most of which result in amino acid substitutions at positions that are remote from the putative mannose-binding domain in the FimH peptide, all converge to the same "pathogenic" adherence phenotype. These mutations confer the ability to bind to monomannose as well as trimannose receptors on host cells, which is adaptive for urinary tract colonization but maladaptive for intestinal colonization, whereas the (ancestral) trimannose-binding phenotype is adaptive for intestinal colonization but not urinary tract colonization [50–53]. Interestingly, one point mutation in *fimH* results in a single-amino-acid substitution (Ser62Ala) that confers both monomannose binding and a novel collagen-binding phenotype, which may be relevant to the pathogenesis of both urinary tract infection (UTI) and neonatal meningitis [54,55]. In contrast to these highly selected VFs, VF genes or gene regions whose peptide products are shielded from the host immune response or are subject to tight functional constraints exhibit little genetic diversity [11,47,48].

PHYLOGENETIC DISTRIBUTION OF VFs

Vertical transmission, the predominant mode of inheritance of VFs within *E. coli*, results in the linkage of particular VFs with specific phylogenetic lineages [31,42,43,56]. VFs are passed from parent to progeny within these lineages,

---

FIGURE 2 Phylogenetic distribution of intestinal pathogenic *Escherichia coli* variants, including *Shigella*. Dendrogram at left depicts phylogenetic relationships for selected members of the *E. coli* reference (ECOR) collection and selected diarrheagenic strains, as inferred based on multilocus enzyme electrophoresis using 10 metabolic enzymes [37]. The four major phylogenetic groups (A, B1, B2, and D) of *E. coli* and the nonaligned strains (non) are bracketed and labeled. Bullets denote the eight included ECOR strains that are symptomatic urinary tract infection (UTI) isolates from humans. The remaining ECOR strains (unlabeled) are fecal isolates from healthy human or animal hosts. The diarrheagenic strains are labeled as to pathotype (enteropathogenic *E. coli*, EPEC; enteroinvasive *E. coli*, EIEC; enterotoxigenic *E. coli*, ETEC; enterohemorrhagic *E. coli*, EHEC; and *Shigella* spp. Boyd, *boydi*; Flex, *flexneri*; Dysen, *dysenterii*; Sonn, *sonnei*) and, where applicable, O:H serotype. Not shown is a second EHEC cluster (serotypes O26:H11 and O111:H8) that is part of the EPEC O111:H2 cluster within phylogenetic group B1. Note that group D and the nonaligned strains form an outgroup, with groups B1 and B2 representing sister neighbors within a cluster that includes group A. (Neighbor-joining tree adapted from [37] with permission.)

which, because of their constituent VFs, behave as and hence are recognized as virulent clones [43,57]. However, horizontal transmission has been appreciated increasingly as an important contributor to the creation of new virulent clones [34,35,42,44,48,58,59]. VFs can be transferred horizontally between lineages by diverse genetic mechanisms and can move either as individual genes or as clusters of linked genes. Direct genetic linkage of multiple VF genes occurs on plasmids [16,17,60–62] and within pathogenicity-associated islands (PAIs) [63].

PATHOGENICITY-ASSOCIATED ISLANDS

Pathogenicity-associated islands (PAIs), a recently described phenomenon of considerable current interest to students of pathogenesis, are encountered in multiple gram-negative and gram-positive pathogens [18,64–71]. PAIs are discrete genomic (or plasmidic [72]) regions of from several to over 200 kb in length that contain multiple known or putative virulence genes, many of which may exhibit a GC content and/or codon usage pattern atypical for the host bacterial species. PAIs are flanked by inverted repeats, contain integrase genes, and often are inserted in the genome near tRNA loci, features consistent with mobility and specifically phage-mediated transfer [63,71,73]. Pathogenic strains commonly contain multiple different PAIs of varying sizes scattered about the genome and/or on their plasmids, with varying degrees of overlap among the various PAIs with respect to specific VF content [67–69,74,75].

PAIs are speculated to have evolved because the close physical proximity of their constituent VF genes provides a selective advantage in case of recombination by allowing the (presumably interactive) VFs to move together as a group rather than individually [63,64,72]. (As mentioned earlier for the evolution of VFs per se, the nature of the selective pressure that has driven the evolution of PAIs is undefined. Since enhanced virulence is not necessarily the operative driving force even among classic pathogens, and since some nonpathogenic bacteria contain analogous clusters of linked genes, the inclusive terms *genomic island* and *fitness island* have been proposed for these structures [64,71].) The hypothesized en bloc horizontal transfer of intact PAIs, although supported by the finding of highly similar PAIs in phylogenetically distant lineages, has not been demonstrated directly in *E. coli* [72]. However, spontaneous total excision of PAIs has been observed during *in vitro* passage [65], which demonstrates the lability of PAIs and is consistent with the mobility hypothesis.

Molecular epidemiologic evidence suggests that PAIs also may participate in the horizontal transfer of VFs by providing genomic "platforms" that are especially receptive to the acquisition and release of individual VFs [56,66]. The typical presence within PAIs of numerous insertion sequences and other recombination-promoting elements is consistent with the inferred high degree of internal reassortment and remodeling [63,68,69,72]. Since the *pap* (P fimbriae) operon is known to occur within PAIs, an illustration of the presumed dynamic nature of PAIs and their constituent operons and genes is provided by the observed shuffling of various *papA* (structural subunit) and *papG* (adhesin) alleles within and among different clonal groups [43,56,76] and the occurrence of chimeric recombined variants of *papG* [48].

Similarities between certain intestinal versus extraintestinal pathogenic *E. coli* with respect to their PAIs, including the insertion of quite different PAIs at the same genomic tRNA locus in different strains [63,64,71] and the inclusion within otherwise distinct PAIs of some of the same VFs [67,77,78], suggest possible evolutionary commonality between these two major pathotypic families of *E. coli* and their respective PAIs despite their numerous dissimilarities. Complete sequencing of PAIs facilitates comparisons between PAIs from different strains, pathotypes, and genera, thereby clarifying evolutionary relationships [20,66–69,74,79]. It also invariably reveals many novel putative VF genes in need of further analysis. Although certain archetypal *E. coli* PAIs [e.g., the locus of enterocyte effacement (LEE) of enteropathogenic and enterohemorrhagic *E. coli* and the yersinial high-pathogenicity island] are widely accepted as operational taxonomic units [18,20,66], no general taxonomy for the PAIs of *E. coli* has emerged, and understandings of the evolutionary relationships among PAIs are still rudimentary.

VIRULENCE PLASMIDS

Virulence plasmids, which play an important role in the virulence of extraintestinal and intestinal pathogenic *E. coli* and *Shigella*, function somewhat as episomal PAIs [72]. Each plasmid has its own evolutionary history. Although this is usually independent of the history of the host strain genome [14–17], in some instances virulence plasmids and genome appear to have coevolved. Coevolution of plasmids and genome is suggested when each encodes different components of a coordinated system [72,80], when plasmid genes regulate expression of chromosomal VF genes [81,82], or when appearance of a particular virulence plasmid, even in distant lineages, is accompanied consistently by inactivation or deletion of specific chromosomal genes [34,83]. Likewise, the consistent association of certain virulence plasmids with particular lineages (e.g., ColV—aerobactin plasmids and one subclone within the *E. coli* O18:K1:H7 clonal group [61]) suggests a possible affinity between the plasmid and the particular genome.

Like PAIs, virulence plasmids appear to undergo internal remodeling as well as horizontal transfer [16,17]. In some instances, virulence plasmids may exchange VFs with the genome [80] or actually may integrate into the genome, thereby in essence transforming into a genomic PAI [72,84]. In extraintestinal pathogenic *E. coli*, certain virulence plasmids, e.g., those containing the aerobactin system, also may contain antibiotic resistance genes, which can provide a selective advantage in compromised hosts who are more likely to be exposed to antibiotics [85,86].

REPEATED EMERGENCE OF VIRULENT CLONES

Acquisition by distantly related lineages of the combinations of VFs required, respectively, for intestinal or extraintestinal pathogenicity results in the emergence of new virulent clones [35,42,43]. This phenomenon appears to have occurred independently on multiple occasions in *E. coli*, usually as a multistep process [33,34,44]. An example of this phenomenon is provided by the parallel evolution in two distant lineages of both enterohemorrhagic *E. coli* (EHEC) and enteropathogenic *E. coli* (EPEC), in each instance from a common ancestor [44,87,88]. Within each of these lineages, an EPEC-like ancestor that already possessed the LEE island next either acquired the EPEC-associated EAF virulence plasmid, to become a full-fledged EPEC strain, or was lysogenized by an *stx*-encoding bacteriophage and acquired the (hemolysin-encoding) EHEC virulence plasmid, to become a full-fledged EHEC strain [44,87,88]. Subsequent alterations within these two distant EHEC lineages as they continued to evolve included reshuffling of O- and H-antigen-encoding genes, metabolic gene inactivation (e.g., to give sorbitol and β-glucuronidase negativity in *E. coli* O157:H7), loss of motility (to give *E. coli* O157:H-), and reshuffling of *stx* variants [44,87,88].

Presumably, the selective advantage conferred by the co-occurrence of particular VFs has repeatedly driven the sequential accumulation of additional virulence genes within clones that have, by whatever mechanism, acquired a partial complement of the VFs required for pathogenicity. This may occur because the initial acquisition event allows the clone to occupy a new niche, e.g., the pathogenic niche. Within the new niche, acquisition of additional niche-specific fitness-promoting factors will be favored, leading ultimately to the accumulation of a host of such factors in combinations that may be niche-specific rather than lineage-specific. Whatever its origins, this convergence-to-pathogenicity phenomenon, in which pathotypically similar strains occur in phylogenetically distant lineages, illustrates the supremacy of VF repertoire over genomic background in determining virulence potential [32]. However, genomic background nonetheless contributes significantly to the emergence of pathogens, as discussed next.

GENOMIC BACKGROUND AND VIRULENCE

The genome has several known or suspected roles in the evolution of pathogenic strains. Genome size is greater among extraintestinal pathogenic *E. coli* than among nonpathogenic wild-type *E. coli* or laboratory strains, a consequence largely of the extra DNA contained in the multiple PAIs and other VFs present in the pathogenic strains [75,89,90]. However, the net increase in genome size is less than would be predicted based on the total amount of extra DNA present, evidence that these strains have deleted some of their generic background DNA in addition to gaining new pathogen-specific DNA [75]. Consistent with the hypothesis that such deletions are not random, extraintestinal pathogenic strains J96 (O4:H5; pyelonephritis) and RS218 (O18:K1:H7; neonatal meningitis) exhibit similar large deletions of the same region of the *E. coli* K-12 chromosome [75]. An analogous phenomenon occurs in *Shigella* and enteroinvasive *E. coli*, in which, irrespective of phylogenetic background, certain housekeeping genes are consistently inactivated or deleted, creating so-called genomic black holes [83,91].

The basis for these loss-of-function mutations in pathogens is unclear. One hypothesis is that they improve pathogenic fitness, i.e., that the inactivated or deleted regions contain (commensal-specific) antivirulence genes that not only are dispensable but also are frankly disadvantageous once the organism has acquired certain VFs and can adopt a pathogenic "lifestyle" [75,83,91]. Another hypothesis is that such deletions help regulate genome size, maintaining it below a postulated upper limit imposed by constraints such as replication time, chromosome packaging, and so on [90]. This may include the maintenance of chromosomal symmetry on either side of the origin of replication [90].

It has been proposed that the observed concentration of many extraintestinal VFs in phylogenetic groups B2 and D reflects a special receptivity of the B2/D genomic background to acquisition and retention of these VFs [32,33]. This hypothesis is particularly appealing in view of the seemingly sequential or stepwise addition of VFs to certain lineages within phylogenetic group B2, which cannot be explained by a single acquisition event involving an early-group B2 ancestor and a PAI containing the full complement of group B2-associated VFs [43,56,92]. One mechanism that could explain such selective receptivity to VFs by different phylogenetic groups is a difference between the phylogenetic groups with respect to their intrinsic ability to undergo the balancing mutations (i.e., genomic fine-tuning) necessary to accommodate the added genetic material represented by the VFs, as described earlier [32]. Alternatively (or in addition), principles such as "like attracts like" or "the rich become richer" may be operative. That is, an initial (chance) acquisition event involving VF genes and/or the associated PAI framework would directly favor subsequent VF acquisitions through a combination of increased fitness (through synergy among different VFs) and direct genetic receptivity. These hypotheses are in need of experimental evaluation.

PHYLOGENETIC DISTRIBUTION OF PATHOGENIC *E. COLI*

The preceding sections describe the evolutionary phenomena that have led to the emergence of the various pathogenic *E. coli* variants, including *Shigella*. This section summarizes the phylogenetic distribution of these pathotypes within the species.

Most extraintestinal pathogenic *E. coli*, including those with the most robust VF repertoires and those which are best able to infect noncompromised hosts, derive from phylogenetic group B2 [43,56,93–97] (see Fig. 1). Most currently recognized extraintestinal VFs are concentrated in or confined to group B2, including specific lineages thereof [42,43,56,92] (see Fig. 1). Likewise, group B2 accounts for most of the traditionally recognized virulent clones of extraintestinal pathogenic *E. coli*, including serotypes O18:K1:H7, O4:K12:H5, O6:K2:H1, O75:K100:H-, and several others [43,56].

Group D is the second largest contributor of extraintestinal pathogenic *E. coli*. It accounts for several additional familiar virulent clones, including O7:K1:H-, O15:K52:H1, and the recently described multi-drug-resistant clonal group A (unpublished data and [43,56,98]). Extraintestinal isolates from group D typically have both somewhat fewer VFs and a different mix of VFs than group B2 isolates [43,56,76,92] (see Fig. 1).

In contrast to group B2 and D strains, strains from groups A and B1 uncommonly cause extraintestinal infection. Those which do usually either represent the exceptional members of these groups that by horizontal transfer have acquired sufficient extraintestinal VFs to cause disease in noncompromised hosts [42,43,56] or are intrinsically low-virulence strains whose ability to cause invasive disease is largely limited to compromised hosts (unpublished data and [99]) (see Fig. 1).

Among extraintestinal pathogens, with few exceptions, isolates from specific clinical syndromes, anatomic sites of infection, and host populations overlap considerably with respect to phylogenetic background [43,55,56,95,100–106]. Conversely, no one *E. coli* phylogenetic group or lineage thereof is specific to any one clinical context. The same is true largely for the various extraintestinal VFs [43,55,56,95,100–106]. This suggests that, for the most part, the selection pressures that have driven the evolution of extraintestinal pathogens and their VFs are independent of the specific clinical context. Such independence could mean that the relevant selective pressures derive mainly from phenomena that are broadly prevalent in diverse extraintestinal contexts and hosts, e.g., certain conserved host defense systems [107]. Alternatively, it is possible that they derive from a nonpathogenic selective environment, as discussed later (in the section "Why Pathogenic *E. coli*?").

INTESTINAL PATHOGENIC *E. COLI*

Analogous conclusions regarding the evolution of intestinal pathogenic strains are supported by available phylogenetic data regarding enteroinvasive *E. coli* (EIEC), enterotoxigenic *E. coli* (ETEC), enteropathogenic *E. coli* (EPEC), enterohemorrhagic *E. coli* (EHEC), and *Shigella*. Regardless of the molecular typing method used (e.g., MLEE, single- or multilocus sequence typing, ribotyping, or DNA-DNA hybridization), each of these pathotypes clearly is present in multiple phylogenetically distant lineages [31,34,35,37,108–110] (see Fig. 2). Many such lineages actually are phylogenetically closer to other lineages containing nonpathogens, extraintestinal pathogenic *E. coli*, or intestinal pathogenic *E. coli* of a different pathotype than they are to lineages containing members of the same intestinal pathotype [31,34,35,37,108–110] (see Fig. 2). Certain representatives of different diarrheagenic pathotypes even exhibit identical housekeeping gene sequences at a particular locus (e.g., *mdh*, for selected ETEC, EPEC, and EIEC isolates) and hence occupy the same terminal branch of the corresponding gene tree [37]. Taken together, these observation indicate that each diarrheagenic pathotype has emerged independently multiple times within *E. coli* by horizontal acquisition of virulence genes (and the possible accompanying genomic adjustments), with convergent evolution yielding similar phenotypes despite the underlying phylogenetic diversity [34,35] (see Fig. 2).

Of particular interest is *Shigella*, which based on phenotypic criteria traditionally has been regarded as a separate genus with four putative species (*dysenteriaie, flexneri, boydii*, and *sonnei*). However, multiple lines of genetic evidence unambiguously establish that with the exception of *S. dysenteriae* serotype 13, which appears to represent a (distant) unique species [34], all *Shigella* actually fall well within *E. coli* [3,12,34,37,110,111] (see Fig. 2). Moreover, the four putative *Shigella* "species" do not have a valid genetic basis, as demonstrated by their phylogenetic intermingling with one another and with non-*Shigella E. coli* [34] (see Fig. 2).

Shigella and EIEC exhibit numerous pathotypic and phenotypic similarities and in some instances contain identical virulence plasmids, evidence of evolutionary commonality [16,34,83]. Although no EIEC strains have been identified within the major *Shigella* clusters [34] or have been shown to be phylogenetically identical to nonclustered *Shigella* strains, certain EIEC and *Shigella* do occur in nearby lineages within phylogenetic groups A and/or B1 [37,110] (see Fig. 2). Consequently, it has been proposed that *Shigella* are actually a variant of EIEC, with the major *Shigella* clusters representing particularly successful EIEC clones [16]. The evolution of EIEC and *Shigella* forms of *E. coli* involves gain of a pINV (invasiveness) plasmid and loss of motility and specific catabolic functions [16]. Interestingly, *S. dysenteriae* type 1, which is the type form of *Shigella* and a prominent cause of disease outbreaks, is distinct from other *Shigella* and EIEC in both chromosomal genes and pINV genes, as well as in uniquely producing Shiga toxin [16], a property it shares instead with *E. coli* O157:H7 and other EHEC and shigatoxigenic *E. coli* [31].

To summarize, the various diarrheagenic pathotypes occur almost exclusively in *E. coli* phylogenetic groups A, B1, and D and among the ungrouped strains, with the inferred distribution varying somewhat depending on the phylogenetic typing method used [34,37,110] (see Fig. 2). They rarely derive from phylogenetic group B2 [37,110] (see Fig. 2). This is almost precisely the reverse of the phylogenetic distribution of extraintestinal pathogenic *E. coli*, which are concentrated in group B2 and to a lesser extent group D and are sparsely represented within groups A and B1 [22,43,56,112] (see Fig. 1). The basis for this reversed phylogenetic distribution of diarrheagenic versus extraintestinal pathogenic strains is unknown.

WHY PATHOGENIC *E. COLI*?

There is little question that both intestinal and extraintestinal pathogenic *E. coli* possess an enhanced ability to cause disease as compared with typical commensal strains and that specific properties of these strains (their so-called VFs) are responsible for this enhanced virulence capability. Moreover, the evidence strongly suggests that the VFs have been acquired by and are conserved within certain lineages because they confer an important fitness advantage, particularly when present in combination. However, although these strains clearly have a fitness advantage within the pathogenic niche, and although their VFs play an essential role in pathogenesis, it is not at all clear that the pathogenic niche is the actual selective environment responsible for the evolution of the pathogens and their VFs. Stated differently, it is not clear that the pathogenic strains and their respective VFs actually have evolved because of their ability to cause disease, i.e., that these strains are "intentional" or "professional" pathogens, notwithstanding that this is a common perspective shared by many investigators of pathogenesis, clinical microbiologists, and clinicians.

For intestinal pathogenic *E. coli*, the ability to cause diarrhea might provide a theoretical selective advantage to the organism with respect to replication and opportunities for dissemination to new hosts, whether directly or via food, water, fomites, or insect or animal vectors. Thus it is conceivable that the various diarrheagenic pathotypes of *E. coli*, including *Shigella*, may have evolved at least in part because of their special ability to cause the clinical syndromes for which they are best known. However, this assumption is unproven, and it remains possible (and perhaps even probable) that intestinal virulence is an "unintended" consequence of the combined actions of bacterial properties that actually evolved because they serve some unrelated advantageous function for the organism, possibly in a different environmental niche than the human intestinal tract. An example may be EHEC such as *E. coli* O157:H7, for which the primary environmental niche seems to be the bovine intestinal tract, within which these organisms function as innocuous commensals.

This consideration applies all the more to extraintestinal pathogenic *E. coli*. Nearly all such strains, including most members of even the most virulent clones, never have an opportunity to cause disease during their life span but function instead as harmless intestinal commensals in healthy hosts. The more severe types of infections these strains cause, e.g., sepsis and meningitis, often literally represent a dead end for both the host and the pathogen. Less severe extraintestinal syndromes such as cystitis, although permitting host survival, provide no obvious means of transmission to new hosts and offer no clear reproductive advantage over intestinal commensalism. Thus why extraintestinal pathogenic variants of *E. coli* have evolved, and not just once but repeatedly in different branches of the species, is puzzling.

This conundrum has prompted the suggestion that the various so-called extraintestinal VFs of *E. coli* actually have evolved not because they enable the organism to cause extraintestinal disease but because they confer fitness in some other niche, with extraintestinal virulence occurring as an unintended (but medically significant) side effect. Indeed, several lines of epidemiologic and experimental evidence support the hypothesis that P fimbriae, and possibly some group II capsules, may promote intestinal colonization, thereby contributing to commensalism, which has clear selective advantages to the organism [113–119]. Likewise, evidence against certain VFs as having evolved specifically because they promote virulence includes the surprisingly high prevalence among fecal *E. coli* from healthy Australian mammals of the so-called invasion of brain endothelium gene *ibeA* (unpublished data and personal communication, David Gordon), which was discovered and named because of its contribution to the pathogenesis of hematogenous neonatal meningitis [120]. Likewise, fecal *E. coli* from dasyurids (small carnivorous marsupials) are significantly less likely to possess *malX* (a marker of uncertain function for a PAI from human pyelonephritis isolate CFT073 [43,121,122]) than are fecal *E. coli* from other Australian mammals such as kangaroos or rodents, evidence suggesting that host-specific, commensalism-related variables are important selection factors for this and possibly other VFs.

However, the intestinal colonization hypothesis in particular fails to account for several observations regarding extraintestinal VFs. First, most putative extraintestinal VFs have known or postulated mechanisms through with they could or do contribute to virulence but in contrast have no documented or suspected mechanism for promoting intestinal colonization [123]. This is particularly true for the pathoadaptive monomannose-binding FimH variants, which while exhibiting enhanced fitness in the pathogenic niche actually exhibit reduced fitness in the commensal niche [50]. Second, most putative extraintestinal VFs occur in PAIs that closely resemble the PAIs of acknowledged pathogens such as *E. coli* O157:H7, enteropathogenic *E. coli*, and *Shigella* and that even include some of the same genes as do these pathogens' PAIs [67,77]. By analogy, this supports the inference that the PAIs of extraintestinal pathogenic *E. coli* also have evolved because they promote virulence. Third, in cross-sectional point-prevalence studies of fecal colonization, strains resembling extraintestinal pathogenic *E. coli* occur as the predominant fecal *E. coli* strain in only a minority of healthy humans [124–126]. In contrast, if such strains actually have a fitness advantage as intestinal colonizers over typical commensal strains, they would be expected to outcompete the avirulent commensals and establish fecal predominance in a majority of the population. Finally, in a competition model of intestinal colonization in monkeys, a wild-type P fimbriated strain exhibited no competitive advantage over an isogenic *papG* (adhesin) mutant [127]. Nonetheless, each of these objections to the intestinal colonization hypothesis can be countered with further arguments. Thus the explanation for the evolution of both intestinal and extraintestinal pathogenic *E. coli* remains uncertain.

SUMMARY

Pathogenic *E. coli* variants have emerged on multiple occasions within diverse segments of the highly clonal *E. coli* population. Different phylogenetic groups are favored as the background for the development of intestinal versus extraintestinal pathogens, respectively, which are largely mutually exclusive pathotypic categories. Virulent clones emerge and diversify by the acquisition of relevant VFs via horizontal transfer (including in association with PAIs and virulence plasmids), by mutation of vertically inherited genes, and by inactivation or deletion of unnecessary and/or disadvantageous genes. Although advances in comparative sequence analysis permit increasingly refined inferences regarding phylogenetic relationships among pathogenic strains and their VFs in comparison with nonpathogenic *E. coli*, the biologic basis for the observed relationships often remains unclear, and many hypotheses await experimental validation.

ACKNOWLEDGMENTS

This work was supported by the Office of Research and Development, Medical Research Service, Department of Veterans Affairs VA Merit Review, National Institutes of Health Grant DK-47504 and National Research Initiative (NRI) Competitive Grants Program/U.S. Department of Agriculture Grant 00-35212-9408. Daniel Dykhuizen and David Gordon provided helpful suggestions.

REFERENCES

1. Ochman, H., Lawrence, J. G., and Groisman, E. A. (2000). Lateral gene transfer and the nature of bacterial innovation. *Nature* 405:299–304.
2. Guttman, D. S., and Dykhuizen, D. E. (1994). Clonal divergence in *Escherichia coli* as a result of recombination, not mutation. *Science* 266:1380–1383.
3. Ochman, H., Whittam, T. S., Caugant, D. A., and Selander, R. K. (1983). Enzyme polymorphisms and genetic population structure in *Escherichia coli* and *Shigella*. *J. Gen. Microbiol.* 129:2715–2726.
4. Ochman, H., and Selander, R. K. (1984). Evidence for clonal population structure in *Escherichia coli*. *Proc. Natl. Acad. Sci. USA* 81:198–201.

5. Achtman, M. (1986). Clonal analysis of descent and virulence among selected *Escherichia coli*. *Annu. Rev. Microbiol.* 40:185–210.
6. Hartl, D. L., and Dykhuizen, D. E. (1984). The population genetics of *Escherichia coli*. *Annu. Rev. Genet.* 18:31–68.
7. Maynard Smith, J., Smith, N. H., O'Rourke, N. H., and Spratt, B. G. (1993). How clonal are bacteria? *Proc. Natl. Acad. Sci. USA*. 90:4384–4388.
8. Desjardins, P., Picard, B., Kaltenbock, B., Elion, J., and Denamur, E. (1995). Sex in *Escherichia coli* does not disrupt the clonal structure of the population: Evidence from random amplified polymorphic DNA and restriction-fragment-length polymorphism. *J. Mol. Evol.* 41:440–448.
9. Lawrence, J. G., and Ochman, H. (1998). Molecular archaeology of the *Escherichia coli* genome. *Evolution* 95:9413–9417.
10. Milkman, R., and Bridges, M. M. (1990). Molecular evolution of the *Escherichia coli* chromosome: III. Clonal frames. *Genetics* 126:505–517.
11. Boyd, E. F., and Hartl, D. L. (1998). Diversifying selection governs sequence polymorphisms in the major adhesin proteins FimA, PapA, and SfaA of *Escherichia coli*. *J. Mol. Evol.* 47:258–267.
12. Selander, R. K., and Caugant, D. A., and Whittam, T. S. (1987). Genetic structure and variation in natural populations of *Escherichia coli*. In F. C. Neidhardt, K. L. Ingraham, B. Magasanik, *et al.* (eds.), *Escherichia coli* and *Salmonella typhimurium*: Cellular and Molecular Biology. Washington, American Society for Microbiology, 1625–1648.
13. Gordon, D. M. (1997). The genetic structure of *Escherichia coli* populations in feral house mice. *Microbiology* 143:2039–2046.
14. Souza, V., and Eguiarte, L. E. (1997). Bacteria gone native vs bacteria gone awry? Plasmidic transfer and bacterial evolution. *Proc. Natl. Acad. Sci. USA* 94:5501–5503.
15. Boyd, E. F., Hill, C. W., Rich, S. M., and Hartl, D. L. (1996). Mosaic structure of plasmids from natural populations of *Escherichia coli*. *Genetics* 143:1091–1100.
16. Lan, R., Lumb, B., Ryan, D., and Reeves, P. (2001). Molecular evolution of large virulence plasmid in *Shigella* clones and enteroinvasive *Escherichia coli*. *Infect. Immun.* 69:6303–6309.
17. Boerlin, P., Chen, S., Colbourne, J. K., *et al.* (1998). Evolution of enterohemorrhagic *Escherichia coli* hemolysin plasmids and the locus for enterocyte effacement in Shiga toxin–producing *E. coli*. *Infect. Immun.* 66:2553–2561.
18. Schubert, S., Rakin, A., Karch, H., *et al.* (1998). Prevalence of the "high-pathogenicity island" of *Yersinia* species among *Escherichia coli* strains that are pathogenic to humans. *Infect. Immun.* 66:480–485.
19. Schubert, S., Cuenca, S., Fischer, D., and Heesemann, J. (2000). High-pathogenicity island of *Yersinia pestis* in Enterobacteriaceae isolated from blood cultures and urine samples: Prevalence and functional expression. *J. Infect. Dis.* 182:1268–1271.
20. Karch, H., Schubert, S., Zhang, D., *et al.* (1999). A genomic island, termed high-pathogenicity island, is present in certain non-O157 Shiga toxin–producing *Escherichia coli* clonal lineages. *Infect. Immun.* 67:5994–6001.
21. Herzer, P. J., Inouye, S., Inouye, M., and Whittam, T. S. (1990). Phylogenetic distribution of branched RNS-linked multicopy single-stranded DNA among natural isolates of *Escherichia coli*. *J. Bacteriol.* 172:6175–6181.
22. Picard, B., Sevali Garcia, J., Gouriou, S., *et al.* (1999). The link between phylogeny and virulence in *Escherichia coli* extraintestinal infection. *Infect. Immun.* 67:546–553.
23. Donnenberg, M. S., and Welch, R. A. (1996). Virulence determinants of uropathogenic *Escherichia coli*. In H. L. T. Mobley, J. W. Warren (eds.), Urinary Tract Infections: Molecular Pathogenesis and Clinical Management. Washington, ASM Press, 135–174.
24. Eisenstein, B. I., and Jones, G. W. (1988). The spectrum of infections and pathogenic mechanisms of *Escherichia coli*. *Adv. Intern Med.* 33:231–252.
25. Siitonen, A. (1994). What makes *Escherichia coli* pathogenic? *Ann. Med.* 26:229–231.

26. Orskov, I., and Orskov, F. (1985). *Escherichia coli* in extra-intestinal infections. *J. Hyg. (Lond)* 95:551–575.
27. Mobley, H. L. T., Island, M. D., and Massad. G. (1994). Virulence determinants of uropathogenic *Escherichia coli* and *Proteus mirabilis*. Kidney *Int.* 46(suppl. 47):S129–136.
28. Arduino, R. C., and DuPont, H. L. (1999). Enteritis, enterocolitis and infectious diarrhea syndromes. In D. Armstrong, J. Cohen (eds.), Infectious Diseases, Vol. 1. London, Mosby, 1–10.
29. Guerrant, R. L., and Steiner, T. S. (2000). Principles and syndromes of enteric infection. In G. L. Mandell, J. E. Bennett, and R. Dolin, (eds.), Principles and Practice of Infectious Diseases, Vol. 1. New York, Churchill-Livingstone, 1076–1093.
30. Nataro, J. P., and Kaper, J. B. (1998). Diarrheagenic *Escherichia coli* [published erratum appears in Clin Microbiol Rev. 1998; 11:403]. *Clin. Microbiol. Rev.* 11:142–201.
31. Schmidt, H., Geitz, C., Tarr, P. I., *et al.* (1999). Non-O157:H7 pathogenic Shiga toxin–producing *Escherichia coli*: phenotypic and genetic profiling of virulence traits and evidence for clonality. *J. Infect. Dis.* 179:115–123.
32. Johnson, J. R., Kuskowski, M., Denamur, E., *et al.* (2000). Clonal origin, virulence factors, and virulence (Letter and Reply). *Infect. Immun.* 68:424–425.
33. Lecointre, G., Rachdi, L., Darlu, P., and Denamur, E. (1998). *Escherichia coli* molecular phylogeny using the incongruence length difference test. *Mol. Biol. Evol.* 15:1685–1695.
34. Pupo, G., Lan, R., and Reeves, P. R. (2000). Multiple independent origins of Shigella clones of *Escherichia coli* and convergent evolution of many of their characteristics. *Proc. Natl. Acad. Sci. USA* 97:10567–10572.
35. Reid, S. D., Herbelin, C. J., Bumbaugh, A. C., *et al.* (2000). Parallel evolution of virulence in pathogenic *Escherichia coli. Nature* 406:64–67.
36. Johnson, J. R., and O'Bryan, T. T. (2000). Improved repetitive-element (rep-) polymerase chain reaction (rep-PCR) fingerprinting for resolving pathogenic and nonpathogenic phylogenetic groups within *Escherichia coli. Clin. Diagn. Lab. Immunol.* 7:265–273.
37. Pupo, G. M., Karaolis, D. K. R., Lan, R., and Reeves, P. R. (1997). Evolutionary relationships among pathogenic and nonpathogenic *Escherichia coli* strains inferred from multilocus enzyme electrophoresis and *mdh* sequence studies. *Infect. Immun.* 65:2685–2692.
38. Huelsenbeck, J. P., and Rannala, B. (1997). Phylogenetic methods come of age: testing hypothesis in an evolutionary context. *Science* 276:227–232.
39. Felsenstein, J. (1988). Phylogenies from molecular sequences: inference and reliability. *Annu. Rev. Genet* 22:521–565.
40. Feng, D. F., and Doolittle, R. F. (1987). Progressive sequence alignment as a prerequisite to correct phylogenetic trees. *J. Mol. Evol.* 25:351–360.
41. Saitou, N., and Nei, M. (1987). The neighbor-joining method: A new method for reconstructing phylogenetic trees. *Mol. Biol. Evol.* 4:406–425.
42. Boyd, E. F., Hartl, D. L. (1998). Chromosomal regions specific to pathogenic isolates of *Escherichia coli* have a phylogenetically clustered distribution. *J. Bacteriol.* 180:1159–1165.
43. Johnson, J. R., Delavari, P., Kuskowski, M., and Stell, A. L. (2001). Phylogenetic distribution of extraintestinal virulence-associated traits in *Escherichia coli. J. Infect. Dis.* 183:78–88.
44. Whittam, T. S. (1998). Evolution of *Escherichia coli* O157:H7 and other Shiga toxin–producing *E. coli* strains In J. B. Kaper, A. D. O'Brien (eds.), *Escherichia coli* O157:H7 and Other Shiga Toxin–Producing *E. coli* Strains. Washington, American Society for Microbiology, 195–209.
45. Felmlee, T., Pellett, S., Welch, R. A. (1985). Nucleotide sequence of an *Escherichia coli* chromosomal hemolysin. *J. Bacteriol.* 163:94–105.
46. Welch, R. A. (1991). Pore-forming cytolysins of gram-negative bacteria. *Mol. Microbiol.* 5:521–528.
47. Schmidt, M. A., O'Hanley, P., and Schoolnik, G. (1984). Gal-Gal pyelonephritis *Escherichai coli* pili linear immunogenic and antigenic epitopes. *J. Exp. Med.* 161:705–717.

48. Marklund, B. I., Tennent, J. M., Garcia, E., *et al.* (1992). Horizontal gene transfer of the *Escherichia coli pap* and *prs* pili operons as a mechanism for the development of tissue-specific adhesive properties. *Mol. Microbiol.* 6:2225–2242.
49. Johnson, J. R., Stell, A. L., Kaster, N., *et al.* (2001). Novel molecular variants of allele I of the *Escherichia coli* P fimbrial adhesin gene *papG*. *Infect. Immun.* 69:2318–2327.
50. Sokurenko, E. V., Chesnokova, V., Dykhuizen, D. E., *et al.* (1998). Pathogenic adaptation of *Escherichia coli* by natural variation of the FimH adhesin. *Proc. Natl. Acad. Sci. USA* 95:8922–8926.
51. Sokurenko, E. V., Courtney, H. S., Ohman, D. E., *et al.* (1994). FimH family of type 1 fimbrial adhesins: Functional heterogeneity due to minor sequence variations among *fimH* genes. *J. Bacteriol.* 176.
52. Sokurenko, E. V., Courtney, H. S., Maslow, J., *et al.* (1995). Quantitative differences in adhesiveness of type 1 fimbriated *Escherichia coli* due to structural differences in *fimH* genes. *J. Bacteriol.* 177:3680–3686.
53. Sokurenko, E. V., Chesnokova, V., Doyle, R. J., Hasty, D. L. (1997). Diversity of the *Escherichia coli* type 1 fimbrial lectin: Differential binding to mannosides and uroepithelial cells. *J. Biol. Chem.* 272:17880–17886.
54. Pouttu, R., Puustinen, T., Virkola, R., *et al.* (1999). Amino acid residue Ala-62 in the FimH fimbrial adhesins is critical for the adhesiveness of meningitis-associated *Escherichia coli* to collagens. *Mol. Microbiol.* 31:1747–1757.
55. Johnson, J. R., Weissman, S. J., Stell, A. L., *et al.* (2001). Clonal and pathotypic analysis of archetypal *Escherichia coli* cystitis isolate NU14. *J. Infect. Dis.* 184:1556–1565.
56. Johnson, J. R., O'Bryan, T. T., Kuskowski, M. A., Maslow, J. N. (2001). Ongoing horizontal and vertical transmission of virulence genes and *papA* alleles among *Escherichia coli* blood isolates from patients with diverse-source bacteremia. *Infect. Immun.* 69:5363–5374.
57. Vaisanen-Rhen, V., Elo, J., Vaisanen, E., *et al.* (1984). P-fimbriated clones among uropathogenic *Escherichia coli* strains. *Infect. Immun.* 43:149–155.
58. Plos, K., Hull, S. I, Hull, R. A., *et al.* (1989). Distribution of the p-associated-pilus (*pap*) region among *Escherichia coli* from natural sources: Evidence for horizontal gene transfer. *Infect. Immun.* 57:1604–1611.
59. Arthur, M., Arbeit, R. D., Kim, C., *et al.* (1990). Restriction fragment length polymorphisms among uropathogenic *Escherichia coli* isolates: *pap*-related sequences compared with *rrn* operons. *Infect. Immun.* 58:471–479.
60. Tobe, T., Hayashi, T., Han, C.-G., *et al.* (1999). Complete DNA sequence and structural analysis of the enteropathogenic *Escherichia coli* adherence factor plasmid. *Infect. Immun.* 67:5455–5462.
61. Valvano, M. A., Silver, R. P., and Crosa, J. H. (1986). Occurrence of chromosome- or plasmid-mediated aerobactin iron transport systems and hemolysin production among clonal group of human invasive strains of *Escherichia coli* K1. *Infect. Immun.* 52:192–199.
62. Binns, M. M., Mayden, J., and Levine, R. P. (1982). Further characterization of complement resistance conferred on *Escherichia coli* by the plasmid genes *traT* of R100 and *iss* of ColV, I-K94. *Infect. Immun.* 35:654–659.
63. Groisman, E. A., and Ochman, H. (1996). Pathogenicity islands: Bacterial evolution in quantum leaps. *Cell* 87:791–794.
64. Hacker, J., and Carniel, E. (2001). Ecological fitness, genomic islands and bacterial pathogenicity: A Darwinian view of the evolution of microbes. *EMBO Rep.* 2:376–381.
65. Blum, G., Ott, M., Lischewski, A., *et al.* (1994). Excision of large DNA regions termed pathogenicity islands from tRNA-specific loci in the chromosome of an *Escherichia coli* wild-type pathogen. *Infect. Immun.* 62:606–614.
66. Perna, N. T., Mayhew, G. F., Posfai, G., *et al.* (1998). Molecular evolution of a pathogenicity island from enterohemorrhagic *Escherichia coli* O157:H7. *Infect. Immun.* 66:3810–3817.

67. Tarr, P. I., Bilge, S. S., Vary, J. C., Jr., *et al.* (2000). Iha: A novel *Escherichia coli* O157:H7 adherence-conferring molecule encoded on a chromosomal region of conserved structure. *Infect. Immun.* 68:1400–1407.
68. Rasko, D. A., Phillips, J. A. Li, X., and Mobley, H. L. T. (2001). Identification of DNA sequences from a second pathogenicity island of uropathogenic *Escherichia coli* CFT073: probes specific for uropathogenic populations. *J. Infect. Dis.* 184:1041–1049.
69. Dobrindt, U., Blum-Oehler, G., Hartsch, T., *et al.* (2001). S-fimbria-encoding determinant sfa_I is located on pathogenicity island III_{536} of uropathogenic *Escherichia coli* strain 536. *Infect. Immun.* 69:4248–4256.
70. Bonacorsi, S. P. P., Clermont, O., Tinsley, C., *et al.* (2000). Identification of regions of the *Escherichia coli* chromosome specific for neonatal meningitis-associated strains. *Infect. Immun.* 68:2096–2101.
71. Hacker, J., Kaper, J. B. (2000). Pathogenicity islands and the evolution of microbes. *Annu. Rev. Microbiol.* 641–679.
72. Mecsas, J., and Strauss, E. J. (1996). Molecular mechanisms of bacterial virulence: type III secretion and pathogenicity islands. *Emerg. Infect. Dis.* 2:271–280.
73. Ritter, A., Blum, G., Emody, L., *et al.* (1995). tRNA genes and pathogenicity islands: Influence on virulence and metabolic properties of uropathogenic *Escherichia coli. Mol. Microbiol.* 17:109–121.
74. Swenson, D. L., Bukanov, N. O., Berg, D. E., and Welch, R. A. (1996). Two pathogenicity islands in uropathogenic *Escherichia coli* J96: Cosmid cloning and sample sequencing. *Infect. Immun.* 64:3736–3743.
75. Rode, C. K., Melkerson-Watson, L. J., Johnson, A. T., Bloch, C. A. (1999). Type-specific contributions to chromosome size differences in *Escherichia coli. Infect. Immun.* 19:230–236.
76. Johnson, J. R., Stell, A. L., Scheutz, F., *et al.* (2000). Analysis of F antigen–specific *papA* alleles of extraintestinal pathogenic *Escherichia coli* using a novel multiplex polymerase chain reactions–based assay. *Infect. Immun.* 68:1587–1599.
77. Johnson, J. R., Russo, T. A., Tarr, P. I., *et al.* (2000). Molecular epidemiological and phylogenetic associations of two novel putative virulence genes, *iha* and $iroN_{E. coli}$, among *Escherichia coli* isolates from patients with urosepsis. *Infect. Immun.* 68:3040–3047.
78. Ye, C., Xu, J. (2001). Prevalence of iron transport gene on pathogenicity-associated island of uropathogenic *Escherichia coli* in *E. coli* O157:H7 containing Shiga toxin gene. *J. Clin. Microbiol.* 39:2300–2305.
79. Zhu, C., Agin, T. S., Elliott, S. J., *et al.* (2001). Complete nucleotide sequence and analysis of the locus of enterocyte effacement from rabbit diarrheagenic *Escherichia coli* RDEC-1. *Infect. Immun.* 69:2107–2115.
80. Wang, L., Qu, W., and Reeves, P. R. (2001). Sequence analysis of four *Shigella boydii* O-antigen loci: implication for *Escherichia coli* and *Shigella* relationship. *Infect. Immun.* 69:6923–6930.
81. Gomez-Duarte, O. G., and Kaper, J. B. (1995). A plasmid-encoded regulatory region activates chromosomal *eaeA* in enteropathogenic *Escherichia coli. Infect. Immun.* 63:1767–1776.
82. Tatsuno, I., Horie, M., Abe, H., *et al.* (2001). *toxB* gene on pO157 of enterohemorrhagic *Escherichia coli* O157:H7 is required for full epithelial cell adherence phenotype. *Infect. Immun.* 69:6660–6669.
83. Maurelli, A. T., Fernandez, R. I., Bloch, C. A., *et al.* (1998). "Black holes" and bacterial pathogenicity: A large genomic deletion that enhances the virulence of *Shigella* spp. and enteroinvasive *Escherichia coli. Proc. Natl. Acad. Sci. USA* 95:3943–3948.
84. Valvano, M. A., and Crosa, J. H. (1984). Aerobactin iron transport genes commonly encoded by certain ColV plasmids occur in the chromosome of a human invasive strain of *Escherichia coli* K1. *Infect. Immun.* 46:159–167.
85. Johnson, J. R., Moseley, S., Roberts, P., Stamm, W. E. (1988). Aerobactin and other virulence factor genes among strains of *Escherichia coli* causing urosepsis, association with patient characteristics. *Infect. Immun.* 56:405–412.

86. Phillips, I., Eykyn, S., King, A., *et al.* (1988). Epidemic multiresistant *Escherichia coli* infection in West Lambeth health district. *Lancet* 1:1038–1041.
87. Feng, P., Lampel, K. A., Karch, H., and Whittam, T. S. (1998). Genotypic an phenotypic changes in the emergence of *Escherichiaq coli* O157:H7. *J. Infect. Dis.* 177.
88. Donnenberg, M. S., and Whittam, T. S. (2001). Pathogenesis and evolution of virulence in enteropathogenic and enterohemorrhagic *Escherichia coli. J. Clin. Invest.* 107:539–548.
89. Bergthorsson, U., Ochman, H. (1995). Heterogeneity of genome sizes among natural isolates of *Escherichia coli. J. Bacteriol.* 177:5784–5789.
90. Bergthorsson, U., and Ochman, H. (1998). Distribution of chromosome length variation in natural isolates of *Escherichia coli. Mol. Biol. Evol.* 15:6–16.
91. Day, W. A., Fernandez, R. E., and Maurelli, A. T. (2001). Pathoadaptive mutations that enhance virulence: Genetic organization of the *cadA* regions of *Shigella* spp. *Infect. Immun.* 69:7471–7480.
92. Johnson, J. R., and Stell, A. L. (2000). Extended virulence genotypes of *Escherichia coli* strains from patients with urosepsis in relation to phylogeny and host compromise. *J. Infect. Dis.* 181:261–272.
93. Goullet, P., and Picard, B. (1990). Electrophoretic type B_2 of carboxylesterase B for characterization of highly pathogenic *Escherichia coli* strains from extra-intestinal infections. *J. Gen. Microbiol.* 33:1–6.
94. Johnson, J. R., Goullet, P. H., Picard, B., *et al.* (1991). Association of carboxylesterase B electrophoretic pattern with presence and expression of urovirulence factor determinants and antimicrobial resistance among strains of *Escherichia coli* causing urosepsis. *Infect. Immun.* 59:2311–2315.
95. Johnson, J. R., O'Bryan, T. T., Delavari, P., *et al.* (2001). Clonal relationships and extended virulence genotypes among *Escherichia coli* isolates from women with first episode or recurrent cystitis. *J. Infect. Dis.* 183:1508–1517.
96. Bingen, E., Picard, B., Brahimi, N., *et al.* (1998). Phylogenetic analysis of *Escherichia coli* strains causing neonatal meningitis suggests horizontal gene transfer from a predominant pool of highly virulent B2 group strains. *J. Infect. Dis.* 177:642–650.
97. Johnson, J. R., Oswald, E., and O'Bryan, T. T., *et al.* (2002). Phylogenetic distribution of virulence-associated genes among neonatal meningitis isolates of *Escherichia coli* from The Netherlands. *J. Infect. Dis.* 185:774–784.
98. Manges, A. R., Johnson, J. R., Foxman, B., *et al.* (2001). Widespread distribution of urinary tract infections caused by a multidrug-resistant *Escherichia coli* clonal group. *New Engl. J. Med.* 345:1007–1013.
99. Johnson, J. R., Orskov, I., Orskov. F., *et al.* (1994). O, K, and H antigens predict virulence factors, carboxylesterase B pattern, antimicrobial resistance, and host compromise among *Escherichia coli* strains causing urosepsis. *J. Infect. Dis.* 169:119–126.
100. Whittam, T. S., Wolfe, M. L., Wilson, R. A. (1989). Genetic relationships among *Escherichia coli* isolates causing urinary tract infections in humans and animals. *Epidemiol. Infect.* 102:37–46.
101. Johnson, J. R., Delavari, P., Stell, A. L., *et al.* (2001). Molecular comparison of extraintestinal *Escherichia coli* isolates from the same electrophoretic lineages from humans and domestic animals. *J. Infect. Dis.* 183:154–159.
102. Johnson, J. R., Delavari, P., O'Bryan, T. (2001). *Escherichia coli* O18:K1:H7 isolates from acute cystitis and neonatal meningitis exhibit common phylogenetic origins and virulence factor profiles. *J. Infect. Dis.* 183:425–434.
103. Johnson, J. R., Stell, A. L., Delavari, P., *et al.* (2001). Phylogenetic and pathotypic similarities between *Escherichia coli* isolates from urinary tract infections in dogs and extraintestinal infections in humans. *J. Infect. Dis.* 183:897–906.
104. Johnson, J. R., Stell, A., and Delavari, P. (2001). Canine feces as a reservoir of extraintestinal pathogenic *Escherichia coli. Infect. Immun.* 69:1306–1314.

105. Johnson, J. R., O'Bryan, T. T., Low, D. A., *et al.* (2000). Evidence of commonality between canine and human extraintestinal pathogenic *Escherichia coli* that express *papG* allele III. *Infect. Immun.* 68:3327–3336.
106. Dozois, C. M., Harel, J., and Fairbrother, J. M. (1996). P-fimbriae-producing septicaemic *Escherichia coli* from poultry possess *fel*-related gene clusters whereas *pap*-hybridizing P-fimbriae-negative strains have partial or divergent P fimbrial gene clusters. *Microbiology* 142:2759–2766.
107. Russo, T. A., Johnson, J. R. (2000). A proposal for an inclusive designation for extraintestinal pathogenic *Escherichia coli*: ExPEC. *J. Infect. Dis.* 181:1753–1754.
108. Brenner, D. J., Fanning, G. R., Skerman, F. J., Falkow, S. (1972). Polynucleotide sequence divergence among strains of *Escherichia coli* and closely related organisms. *J. Bacteriol.* 109:953–965.
109. Wang, G., Whittam, T. S., Berg, C. M., and Berg, D. E. (1993). RAPD (arbitrary primer) PCR is more sensitive than multilocus enzyme electrophoresis for distinguishing related bacterial strains. *Nucleic Acids Res.* 21:5930–5933.
110. Rolland, K., Lambert-Zechovsky, N., Picard, B., Denamur, E. (1998). *Shigella* and enteroinvasive *Escherichia coli* strains are derived from distinct ancestral strains of *E. coli*. *Microbiology* 144:2667–2672.
111. Whittam, T. S., Ochman, H., and Selander, R. K. (1983). Multilocus genetic structure in natural populations of *Escherichia coli*. *Population Biol.* 80:1751–1755.
112. Picard, B., Journet-Mancy, C., Picard-Pasquier, N., Goullet, P. (1993). Genetic structures of the B_2 and B_1 *Escherichia coli* strains responsible for extra-intestinal infections. *J. Gen. Microbiol.* 139:3079–3088.
113. Wold, A. E., Thorssen, M., Hull, S., and Svanborg-Eden, C. (1988). Attachment of *Escherichia coli* via mannose- or Galα1-4Galβ–containing receptors to human colonic epithelial cells. *Infect. Immun.* 56:2531–2537.
114. Wold, A. E., Caugant, D. A., Lidin-Janson, G., *et al.* (1992). Resident colonic *Escherichia coli* strains frequently display uropathogenic characteristics. *J. Infect. Dis.* 165:46–52.
115. Mahmood, A., Engle, M. J., Hultgren, S. J., *et al.* (2000). Role of intestinal surfactant-like particles as a potential reservoir of uropathogenic *Escherichia coli*. *Biochim. Biophys. Acta.* 1519:49–55.
116. Herías, M. V., Midtvedt, T., Hanson, L. A., and Wol, A. (1997). *Escherichia coli* K5 capsule expression enhances colonization of the large intestine in the gnotobiotic rat. *Infect. Immun.* 65:531–536.
117. Herías, M. V., Midtvedt, T., Hanson, L. Å., and Wold, A. E. (1995). Role of *Escherichia coli* P fimbriae in intestinal colonization in gnotobiotic rats. *Infect. Immun.* 63:4781–4789.
118. Adlerberth, I., Hanson, L. A., Svanborg, C., *et al.* (1995). Adhesins of *Escherichia coli* associated with extra-intestinal pathogenicity confer binding to colonic epithelial cells. *Microb. Pathogenet.* 18:373–385.
119. Plos, K., Connell, H., Jodal, U., *et al.* (1995). Intestinal carriage of P fimbriated *Escherichia coli* and the susceptibility to urinary tract infection in young children. *J. Infect. Dis.* 171:625–631.
120. Huang, S.-H., Wass, C., Fu, Q., *et al.* (1995). *Escherichia coli* invasion of brain microvascular endothelial cells *in vitro* and *in vivo*: Molecular cloning and characterization of invasion gene *ibe10*. *Infect. Immun.* 63:4470–4475.
121. Guyer, D. M., Kao, J.-S., and Mobley, H. L. T. (1998). Genomic analysis of a pathogenicity island in uropathogenic *Escherichia coli* CFT073: Distribution of homologous sequences among isolates from patients with pyelonephritis, cystitis, and catheter-associated bacteriuria and from fecal samples. *Infect. Immun.* 66:4411–4417.
122. Kao, J.-S., Stucker, D. M., Warren, J. W., and Mobley, H. L. T. (1997). Pathogenicity island sequences of pyelonephritogenic *Escherichia coli* CFT073 are associated with virulent uropathogenic strains. *Infect. Immun.* 65:2812–2820.
123. Dozois, C., Johnson, J. (2000). The place of P fimbriae in the emergence of *Escherichia coli* extraintestinal pathotypes. *ASM News* 66:262–263.

124. Johnson, J. R. (1991). Virulence factors in *Escherichia coli* urinary tract infection. *Clin. Microbiol. Rev.* 4:80–128.

125. Siitonen, A. (1992). *Escherichia coli* in fecal flora of healthy adults: Serotypes, P and type 1C fimbriae, non-P mannose-resistant adhesins, and hemolytic activity. *J. Infect. Dis.* 166:1058–1065.

126. Muhldorfer, I., Blum, G., Donohue-Rolfe, A., *et al.* (1996). Characterization of *Escherichia coli* strains isolated from environmental water habitats and from stool samples of healthy volunteers. *Res. Microbiol.* 147:625–635.

127. Winberg, J., Möllby, R., Bergström, J., *et al.* (1995). The PapG-adhesin at the tip of P-fimbriae provides *Escherichia coli* with a competitive edge in experimental bladder infections of cynomolgus monkeys. *J. Exp. Med.* 182:1695–1702.

128. Timmis, K. N., Boulnois, G. J., Bitter-Suermann, D., and Cabello, F. C. (1985). Surface components of *Escherichia coli* that mediate resistance to the bactericidal activities of serum and phagocytes. *Curr. Topics Microbiol. Immunol.* 118:197–218.

PART II

Escherichia coli Pathotypes

Enteropathogenic Escherichia coli

T. Eric Blank

Division of Infectious Diseases, Department of Medicine, University of Maryland School of Medicine, Baltimore, Maryland

Jean-Philippe Nougayrède

Division of Infectious Diseases, Department of Microbiology and Immunology, University of Maryland School of Medicine, Baltimore, Maryland

Michael S. Donnenberg

Division of Infectious Diseases, Department of Medicine, University of Maryland School of Medicine, Baltimore, Maryland

CHARACTERISTICS OF EPEC AND EPEC INFECTIONS

Definition of EPEC

Enteropathogenic strains of *Escherichia coli* (EPEC) are defined as those having each of the following characteristics: (1) the ability to cause diarrhea, (2) the ability to produce a histopathology on the intestinal epithelium known as the *attaching and effacing* (A/E) *lesion*, and (3) the inability to produce Shiga toxins (verocytotoxins) [1]. The second characteristic distinguishes EPEC from most other pathotypes of diarrheagenic *E. coli*, including enteroaggregative *E. coli* (EAEC), enteroinvasive *E. coli* (EIEC), and enterotoxigenic *E. coli* (ETEC) [2] (see Chaps. 5 through 7). The third characteristic distinguishes EPEC from Shiga toxin—producing *E. coli* (STEC or VTEC) and enterohemorrhagic *E. coli* (EHEC), a subset of STEC that, like EPEC, produces A/E lesions (see Chap. 4). Typical EPEC strains causing human infections carry a large virulence plasmid [the EPEC adhesion factor (EAF) plasmid] that allows them to produce bundle-forming pili and attach to epithelial cells in a characteristic pattern termed

Escherichia coli: Virulence Mechanisms of a Versatile Pathogen
ISBN 0-12-220751-3

localized adherence. Other EPEC strains, known as atypical EPEC, lack these properties [1].

Clinical Manifestations of EPEC Infections

EPEC infection leads to a watery diarrhea that may contain mucus but not blood. Associated symptoms may include fever, malaise, vomiting, food intolerance, dehydration, and weight loss [2–4]. EPEC diarrhea often lasts 5–15 days but can become chronic [5,6] and may result in a mortality rate as high as 50%. EPEC are thought to inhabit primarily the small intestine rather than the colon, which is the habitat of commensal *E. coli* and EHEC strains. However, children with severe EPEC disease may have A/E bacteria simultaneously in the duodenum and rectum [7].

Epidemiology of EPEC Infections

Human EPEC infections are common in the developing world, with occasional outbreaks in nurseries and day-care centers in developed countries [2,4,8,9]. Symptoms most often are associated with infants 6 months of age or younger. EPEC diarrhea can be elicited in adult volunteers by a large inoculum of 10^8 to 10^{10} organisms [10]. EPEC occasionally have been isolated from acquired immunodeficiency syndrome (AIDS) patients with diarrhea [11,12].

E. coli fulfilling the definition of EPEC have been isolated from mammals and birds with diarrhea [13–18]. Such strains often belong to different serotypes than EPEC causing human infections and in some cases possess different adhesins. Animals that have served as important models for understanding features of EPEC infections include rabbits and piglets. Mice are naturally infected by *Citrobacter rodentium*, an EPEC-like pathogen that forms A/E lesions and causes colonic hyperplasia [19]. In this chapter we will focus on EPEC strains causing human infections, referring to the zoonotic EPEC literature only in cases that shed light on general EPEC virulence mechanisms.

EPEC Clonal Diversity

EPEC were classified originally as having specific combinations of O (somatic, LPS) and H (flagellar) antigens that differ from those of other pathogenic and nonpathogenic *E. coli* strains. The specific serotypes included in the EPEC fold have been revised continually; for current lists, see refs. 2 and 4. While strains having these serotypes often exhibit EPEC virulence properties [20,21], it is now appreciated that serotype, while somewhat indicative of strain clonality and virulence properties, is an imperfect predictor of EPEC. A strain having a traditional EPEC serotype may or may not display virulence properties that fit the current definition of EPEC. Conversely, a strain that has EPEC characteristics may not have a common EPEC serotype. Multilocus sequence typing (MLST) and multilocus enzyme electrophoresis (MLEE) are more precise methods than serotyping for determining *E. coli* strain clonality because they rely on the comparison of multiple genes or proteins. Using such methods, it has been determined that the majority of EPEC strains fall into two major clonal groups known as EPEC 1 and EPEC 2 [22–25] (see Chap. 2). Both groups have similar virulence characteristics yet are thought to derive from separate acquisitions of EPEC virulence genes [26]. Our current understanding of EPEC pathogenesis comes primarily from studies on a few prototype strains. Recently, however, it has been recognized that different strains within a single EPEC clone or serotype can carry different arsenals of virulence factors [27–31]. Therefore, an underappreciated diversity of pathogenic mechanisms may exist amongst EPEC strains.

EPEC MOLECULAR PATHOGENESIS

Phenotypes Characteristic of EPEC

Localized Adherence (LA) to Epithelial Cells

Typical EPEC strains adhere to HeLa, HEp-2, and other cell lines and to organ cultures *in vitro* in a distinctive pattern of three-dimensional clusters or microcolonies [32–38] (Fig. 1*A*). The LA pattern is characteristic only of EPEC strains of *E. coli* and therefore has been used widely as a diagnostic tool (see below). A similar adherence pattern has been seen in tissue biopsies of EPEC-infected humans [7,39]. EPEC microcolonies in tissue culture form and then disperse over a period of 6 hours [40], but the role of such colony dynamics in the pathogenicity of EPEC is currently unclear. *E. coli* strains fitting the definition of EPEC but having non-LA adherence patterns have been noted [29,31,41–44].

LA to cultured epithelial cells depends on an EPEC surface structure known as a *bundle-forming pilus* (BFP) (see below). In contrast, it has been reported that EPEC strains lacking BFP remain able to adhere to pediatric small intestinal tissue explants, although they form smaller colonies that lack three-dimensional structure [38]. Other surface components contributing to EPEC adherence include intimin (see below) and the EspA filament (see below and Chap. 13). The identities of EPEC receptors on host cells, with the exception of the intimin receptor Tir (see below), remain unknown. EPEC adherence is inhibited by

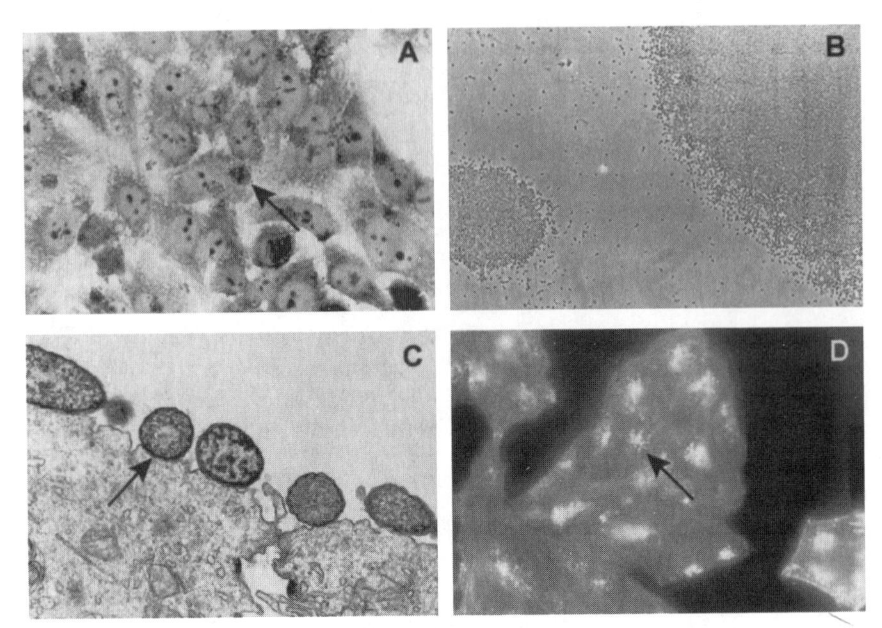

FIGURE 1 EPEC phenotypes. (*A*) Localized adherence (LA) of EPEC on HEp-2 epithelial cells in tissue culture. Arrow indicates a representative cluster of EPEC. (*B*) Autoaggregation of EPEC in DMEM tissue culture medium. Portions of two aggregates are shown. (*C*) Attaching and effacing (A/E) of T84 cells by EPEC. The arrow indicates a site of intimate adherence between a bacterium and host cell. (Photo courtesy of Barry McNamara.) (*D*) Fluorescent-actin-staining (FAS) assay of EPEC adhering to HEp-2 cells. FITC-labeled phalloidin was used to identify highly concentrated actin filaments localizing beneath EPEC. (Photo courtesy of Colin O'Connell.)

various sugar moieties, including galactose [45], *N*-acetylgalactosamine [46], *N*-acetyllactosamine [47], and fucosylated oligosaccharides and gangliosides from milk [48–50]. These saccharides could be moieties of host cell glycoproteins that serve as EPEC receptors [51,52]. EPEC also binds to phosphatidylethanolamine, a component of cell membranes [53].

Autoaggregation/Disaggregation

When grown in tissue culture medium at 37°C, typical EPEC strains aggregate into large clusters that may contain hundreds or thousands of bacteria [54] (see Fig. 1*B*). These autoaggregates are readily visible under a low-power microscope or even to the naked eye. They are unstable, dispersing quickly when moved to nonoptimal conditions [55]. Autoaggregates are reminiscent of, and presumably dependent on, similar mechanisms as the LA clusters that form on the host epithelium. Like LA, autoaggregation requires BFP (see below).

Attaching and Effacing (A/E), Intimate Attachment, and Pedestal Formation

EPEC have the ability to strikingly alter the surface of the cells to which they attach. The characteristic phenotype noted in EPEC infection is the A/E lesion [56], in which brush-border microvilli are sloughed off the apical surface of enterocytes (effacement) and transiently replaced by elongated microvillus-like processes [57] and ultimately by prominent cuplike pedestals and elongated (up to 10 μm) pseudopod structures [58] to which the bacteria attach in very close (intimate) apposition to the host cell membrane (see Fig. 1*C*). The pedestal structures are dynamic, being able to change length, shape, and position over time, and move the attached EPEC along the cell surface [59]. A/E lesions are observed in model EPEC infections with cultured cells and mucosal explants [36,38,60], as well as in intestinal biopsies from EPEC-infected infants [6,7,39,61,62] or animals [13,56,63–67]. A/E lesions are not formed on formalin-fixed epithelial cells [68]. The ability to form A/E lesions is shared between EPEC and other pathogens, including EHEC and *C. rodentium* [19,69].

Invasion of Epithelial Cells

The ability of EPEC to be internalized by epithelial cells has been noted both in tissue culture [70–74] and in small intestinal biopsies from an EPEC-infected infant [75]. Several EPEC virulence factors were identified initially as mutants that lacked the ability to invade HEp-2 cells [76]. The ability of EPEC to invade epithelial cells is in marked contrast to their capacity to evade phagocytic cells (see below). Invasion requires the same virulence factors as A/E (see below); therefore, it has been suggested that invasion is a by-product of the cytoskeletal rearrangements that occur during the A/E process [58]. Interestingly, however, an intimin mutant of EPEC that performs A/E but does not invade HEp-2 cells has been described recently [77]. Despite its invasive potential, EPEC is considered to be a pathogen of the cell surface, not an invasive pathogen.

EPEC Virulence Factors and Genomics

Multiple proven and potential virulence factors have been identified in EPEC, some only recently. The genes encoding these factors are listed in Table I along with the phenotypes of EPEC strains carrying mutations in these genes.

The EAF Plasmid

Typical EPEC strains carry a large (50–70 MDa) extrachromosomal element known as the *EPEC adherence factor* (EAF) *plasmid* [34,78–81]. Curing of the EAF plasmid from EPEC abolishes the LA and autoaggregation phenotypes

TABLE I Comparisons of the Phenotypes of Various EPEC Mutants: EPEC Virulence Factors

		Mutant phenotypes								
Gene	Protein	LA	Intimate Adherence and A/E	Protein secretion/ translocation	Host cell cytoskeletal remodeling	Invasion	Decrease in TEER	Other	Volunteer diarrhea attack rate mutant vs. wild-type	Virulence in Rabbit
bfpA	bundlin	−	+	NT	+	−	NT		2/16 vs. 11/13	NA
bfpF	BfpF	+	+	+	+	+	NT	[a]	4/13 vs. 11/13	NA
eae	intimin	+	−	+	±	−	+	[b]	4/11 vs. 11/11	−
EAF plasmid	BFP and Per	−	+	±	+	−	−		2/9 vs. 9/10	NA
esc/sep	TTSS	+	−	−	−	−	−		NT	−
espA	EspA	+	−	±	−	−	−		NT	−
espB	EspB	+	−	±	−	−	−		1/10 vs. 9/10	−
espC	EspC	+	+	+	+	+	NT	[c]	NT	NT
EspD	EspD	+	−	↓/−	−	−	−		NT	−/+
EspF	EspF	+	+	+	+	+	−	[d]	NT	NT
EspG	EspG	+	+	+	+	±	+	[e]	NT	+
Ler	Ler	Altered	−	−	NT	NT	NT		NT	−
LifA	lymphostatin	+	+	NT	NT	NT	NT		NT	NT
orf19	Map	NT	NT	NT	NT	NT	NT	[f]	NT	NT
perA (bfpT)	PerA (BfpT)	Altered	NT	NT	NT	NT	NT		3/14 vs. 11/13	NT
Tir	Tir	NT	−	+	−	NT	−		NT	−

Note: + indicates phenotype similar to wild-type EPEC strain; − indicates defective phenotype; NT = not tested; NA = not applicable.
[a]Defective in disaggregation; increased BFP piliation and LA.
[b]Fails to focus Tir and cytoskeletal components beneath EPEC.
[c]Increases the potential difference (PD) and total tissue conductance (I_{sc}) *in vitro*.
[d]Defective in causing host cell death.
[e]Slightly attenuated in intracellular invasion and persistence; slightly impaired colonization in rabbit.
[f]Defective in disruption of mitochondrial membrane potential.

[48,54,60,78,82], and leads to attenuated virulence in colostrum-deprived piglets [78] and in volunteers [83]. However, plasmid-cured EPEC retain the ability to form A/E lesions [36,60,84]. Transfer of the EAF plasmid to non-EPEC strains endows them with the ability to carry out LA [60,78]. Two loci important for pathogenicity have been located on the EAF plasmid: the *bfp* gene cluster specifying BFP (see below) and the *per* locus specifying a transcriptional activator (see Chap. 13). The entire 68.8-kb EAF plasmid of the prototype EPEC-2 strain B171 has been sequenced recently [85], revealing additional genes with likely involvement in plasmid maintenance and replication, enzymatic functions, and perhaps virulence. This EAF plasmid also contains many complete or partial insertion sequences, suggesting a complicated history of DNA rearrangements. Some EAF plasmids can be mobilized to other *E. coli* strains by conjugation [60,78–80,86,87], whereas others cannot be because they appear to lack transfer genes [80,85,88].

Bundle-Forming Pili (BFPs)

BFPs form ropelike bundles extending from the cell surface of typical EPEC strains [89]. BFP is a member of the type IV-B (or 4B) class of fimbriae that are also produced by ETECs, *Salmonella typhi*, and *Vibrio cholerae* (see Chap. 12). BFPs appear to be composed of a single structural subunit, known as a *bundlin*, that is encoded by the *bfpA* gene located on the EAF plasmid [90,91]. The *bfpA* sequence varies between different EPEC strains in a manner that does not correlate with the overall strain lineage, suggesting that EAF plasmid shuffling has taken place between EPEC strains in the relatively recent past [92]. Mutation of the *bfpA* gene leads to loss of both LA and autoaggregation (see above) [55,90,93], clustering phenotypes that are thought to result from the interbundling of BFP fibers from multiple bacterial cells [89]. While BFPs are involved in EPEC-EPEC interactions, it is not clear whether they are also capable of directly mediating attachment of EPEC to host cells and, if so, whether a specific epithelial cell receptor is involved. Anti-BFP antiserum partially blocks EPEC adherence to HEp-2 cells, supporting the notion of such a receptor [89]. Contacts between BFPs and the membranes of host cells have been noted in electron micrographs [40,89]. Although the precise role of BFPs in EPEC pathogenicity remains to be determined, its importance in virulence is indicated by a volunteer study in which a *bfpA* mutant was reduced significantly in its ability to cause diarrhea [55].

Thirteen additional *bfp* genes are found directly downstream of *bfpA* [94,95]. Most of these genes are required for BFP synthesis [55,93,96,97]. Mutations in the *bfpF* gene produce especially intriguing phenotypes. These EPEC mutants synthesize BFP, carry out LA, and form autoaggregates; however, the LA clusters and autoaggregates fail to disperse [40,55,98]. A *bfpF* mutant is also unable to produce conspicuous changes in BFP morphology that occur over time in a tissue culture infection model [40]. The fact that a *bfpF* mutant exhibits a decreased diarrheal attack rate in volunteers suggests that BFP dynamics leading to dispersal of bacterial aggregates play a role in EPEC pathogenesis [55,99,100,101]. For more detailed information, see Chap. 11.

The LEE Pathogenicity Island, the Type III Secretion/Translocation System (TTSS), and Translocated Proteins

The *l*ocus of *e*nterocyte *e*ffacement (LEE) is a 35.6-kb pathogenicity island of EPEC containing genes sufficient for the formation of the A/E lesion when introduced into a nonpathogenic *E. coli* strain [102]. The EPEC LEE contains 41 genes, most of them being organized into five major operons [103–106]. Many of the LEE genes encode proteins that are components of a type III protein secretion system (TTSS), which is capable of transporting proteins across three membranes: the cytoplasmic and outer membranes of EPEC and the cytoplasmic membrane of the host cell to which the EPEC is attached. The TTSS initially was found to secrete multiple EPEC proteins into the culture supernatant [42,99,107,108], although the relevant destination for many of these proteins is believed to be within the host cell. Mutations in genes of the EPEC TTSS cause deficiencies in A/E, invasion, and host cell cytoskeletal rearrangements, in addition to the expected deficiencies in protein secretion [76,99,107,109–112]. The former phenotypes are apparently due to the lack of secretion of effector proteins that act within the host cell. One exceptional TTSS mutant (*sepZ1*::Tn*phoA*) exhibits a delayed A/E and protein translocation phenotype while lacking detectable protein secretion or invasion ability [99,100,101]. For more detailed information on the LEE and TTSS, see Chap. 12.

To date, seven proteins have been identified that are encoded by genes of the LEE and are secreted or translocated via the TTSS. These proteins can be separated into two functional groups: (1) components of an extracellular translocation apparatus (EspA, EspB, and EspD) and (2) translocated effector proteins (Tir, EspF, EspG, Map, and possibly EspB). Additional unknown proteins are also secreted in a type III—dependent manner [99,107,108]. The *espA*, *espB* (previously *eaeB*), and *espD* mutants are each deficient in the translocation of one or more of the effector proteins into host cells [113–117], indicating that EspA, EspB, and EspD proteins are each part of a translocation apparatus. Each of these mutants is also deficient in A/E, invasion, and host cell cytoskeletal reorganization [76,118–125]. The latter phenotypes presumably are a result of being unable to translocate specific effector proteins into the host cell. EspA, EspB, and EspD each has been shown to be required for virulence in a rabbit model [124] (J. P. Nougayréde, Ph.D. thesis). EspA is a component of a prominent filamentous structure that appears to connect EPEC and host cells [114]. EspB and EspD are both localized to the host cell membrane after infection [114,116,126]. Detailed information on EspA, EspB, and EspD can be found in Chap. 14.

The remaining EPEC-secreted proteins are thought to be effectors, performing a specific function or attacking specific targets after introduction into and sublocalization in the host cell. The best understood effector, Tir, will be discussed separately (see below). In addition to its role in the translocation apparatus, EspB also may serve as an effector protein. EspB accumulates beneath attached EPEC in the cytoplasmic compartment of host cells, as well as in the host cell membrane [115,116]. Translocation of EspB is impaired in the absence of intimate attachment [116]. EspB cannot be supplied in *trans* form by an *espA* mutant bacterium to functionally complement another (*espB* mutant) bacterium for translocation or invasion activities [116,120]. This indicates that EspB translocation is required for activity; secretion is not sufficient. The assignment of EspB as an effector is supported by the finding that ectopic expression of *espB* within HeLa cells results in profound changes in cell morphology concurrent with a reduction in actin stress fibers [127]. Whether EspB plays a direct role in the cytoskeletal rearrangements that occur during normal EPEC infection requires further investigation. A pyridoxal-phosphate-binding motif has been noted in EspB [118], but mutation of this motif did not inactivate A/E activity of EPEC [128]. Regardless of its role, the overall importance of EspB in the scheme of human EPEC pathogenesis is indicated by studies in which an *espB* mutant was markedly attenuated for virulence in volunteers [129].

The identification and functional characterization of additional EPEC translocated effector proteins, including EspF, EspG, and Map, currently are the subjects of intensive research. These proteins are the products of the *espF* (*orf30/ORFD4*), *espG* (*rorf2*), and *map* (*orf19*) genes of the LEE, respectively. Where tested, *espF* and *espG* mutants show no defect in A/E, LA, FAS, Tir phosphorylation, invasion, or hemolysis assays [130–133]. Therefore, they have phenotypes distinct from the *espA*, *espB*, and *espD* mutants. Novel cell biology methods may be necessary to fully understand the intracellular actions of these proteins.

EPEC EspF is a 206-amino-acid protein with three identical repeats of a proline-rich sequence that resembles those recognized by eukaryotic signaling proteins containing SH3 domains [112]. Interactions between proline-rich proteins and SH3 domains are involved in many signaling pathways, suggesting that EspF might elicit host cell signal transduction. EspF is translocated by the TTSS into host cells, as demonstrated by confocal microscopy and an adenylate cyclase reporter system, making it a candidate effector protein [133]. Although initial studies were unable to identify a phenotype associated with *espF* mutant EPEC [130], further studies coupling bacterial genetics with host physiologic investigations revealed that EspF expression is required for the full impact of EPEC on disruption of host barrier function [133] (see below). In addition,

EspF was recently implicated as an effector of host cell apoptosis [134] (see below).

The role of the EspG protein in EPEC pathogenesis is currently unclear. EspG is found in the Triton-soluble fraction of infected host cells, suggesting an intracellular location [132]. Unlike EspF, EspG does not affect transepithelial electrical resistance (TEER). EspG is similar in sequence to both the product of the *orf3* gene of the EPEC *espC* pathogenicity island [132] and the VirA protein of *Shigella*. The cloned *espG* or *orf3* genes can complement a *virA* mutant of *Shigella*, fully restoring its capacity to invade and persist in HeLa cells [132]. However, the *espG* single and *espG/orf3* double mutants of EPEC are only slightly deficient in their invasion ability. An *espG* mutant is not significantly attenuated for virulence in a rabbit EPEC model but appears to be slightly impaired in colonization [135].

Map (product of the *orf19* gene of the LEE), or mitochondrial-associated protein, is targeted to the host cell mitochondrial membrane, where it disrupts the membrane potential [117]. While the implications of Map activity for pathogenesis are unclear, it may serve to inhibit ATP production as well as release of proapoptotic factor. Curiously, two genes similar to *orf19* have been found in EPEC 2 strain B171: *trcA* on a potential pathogenicity islet inserted in the chromosomal *potB* gene [136] and *trcP* on the EAF plasmid [85]. TrcA and Map may have divergent functions despite their sequence similarity. A *trcA* (*bfpT*-regulated chaperone) mutant of B171 exhibits reduced microcolony formation in the LA assay and reduced levels of bundlin [136]. By far western blotting, the TrcA protein was demonstrated to bind to many EPEC and host cell proteins, including bundlin and intimin [117,136]. This capacity has led to the suggestion that TrcA may act as a chaperone, although the lack of binding specificity lends doubt to its significance.

Intimin and Tir: Adhesin and Receptor

Intimin is the adhesin protein required for intimate adherence of EPEC to host cells at the sites of A/E lesions. It is required for full virulence in human and rabbit infection [137–139]. Intimin is encoded by the *eae* (*eaeA*, *E. coli* attaching and effacing) gene of the LEE [87] and localizes to the EPEC outer membrane [140]. In addition to being defective in A/E [74,76,87,125,141], *eae* mutants of EPEC are unable to invade the host cell [74,76], do not promote pseudopod formation [58], and are unable to bind to human mucosal explants [38]. Unlike TTSS mutants, *eae* mutants retain the ability to redistribute host cell actin but are unable to focus host cell cytoskeletal elements normally underneath EPEC [58,76,109,138]. Intimin is not strictly required for EPEC protein secretion or translocation of EspB or Tir into host cells [99,108–110,116,123]. However, in the absence of intimin, the amount of EspB translocation is decreased greatly [116]. Mutant *eae* strains of EPEC can complement strains carrying mutations in *espA*, *espB*, or genes of the TTSS, allowing them to establish an infection

[109,119,120]. Preinfection with an *eae* mutant also allows a laboratory *E. coli* strain carrying the cloned intimin gene [58] or beads coated with intimin [142] to bind to epithelial cells. Such *trans* complementation is possible because the *eae* mutant remains capable of introducing Tir, the translocated intimin receptor, into the host cell membrane.

Originally thought to be a native host cell protein called Hp90 [58,109], Tir has been convincingly demonstrated to be an EPEC product [113]. After injection into the host cell via the TTSS/translocon, Tir becomes integrated into the host cell membrane in a protease-sensitive conformation [58,113]. Tir exhibits a hairpin topology [143–145], with N-terminal and C-terminal domains that project into the host cell cytoplasm serving as the focus for the actin reorganization that occurs underneath the EPEC pedestal (see below). The extracellular loop of Tir binds to intimin [144–146], establishing intimate attachment of the EPEC to the host cell. On introduction into the host cell by an EPEC, Tir is modified by mechanisms that include tyrosine phosphorylation [58,113,145,147]. As expected, a mutation in the *tir* gene results in the loss of the A/E phenotype [113,138] and a loss of virulence in a rabbit model [138]. Tir remains the primary example of a receptor in the cell membrane generated by a pathogenic microbe.

The N-terminus of intimin anchors the protein in the EPEC outer membrane and exhibits little sequence variation, whereas the C-terminal end of intimin extends from the EPEC surface, binds to Tir [144], and exhibits extreme sequence variation that may be involved in tissue tropism [148–150]. The structure of the C-terminal end of intimin, alone and in complex with Tir, has been determined recently by nuclear magnetic resonance (NMR) [151,152] and crystallography [153]. This region of intimin is composed of three tandem immunoglobulin-like domains followed by a C-type lectin-like domain that contains the Tir binding site (Fig. 2). The lectin-like domain contains a disulfide bridge that is required for A/E lesion formation and binding to intestinal mucosae [38,77] but not for binding to Tir [144].

Intimin or C-terminal intimin fragments are reported to bind HEp-2 epithelial cells independently of EPEC [144,154–156], indicating the presence of intimin receptors other than Tir. In contrast to the binding of intimin to Tir, which has a constant (K_a) of $3.2 \times 10^6 M^{-1}$ [153], the strength of this Tir-independent binding has not been measured. Such binding requires the C-terminal disulfide bridge (cysteine 937 residue) of intimin [144,155,157]. Intimin has sequence similarity to the invasins of *Yersinia* spp. [87,158], which use β_1 integrins as receptors [159]. Likewise, in one study, purified intimin derivatives were shown to bind purified β_1 integrins or integrins expressed on T cells [157]. In contrast, other laboratories have been unable to reproduce these findings, reporting that binding of intimin to host cells and EPEC-mediated actin condensation require Tir translocation but not β_1 integrins and that purified β_1 integrins fail to coprecipitate with intimin or localize underneath EPEC [58,113,160].

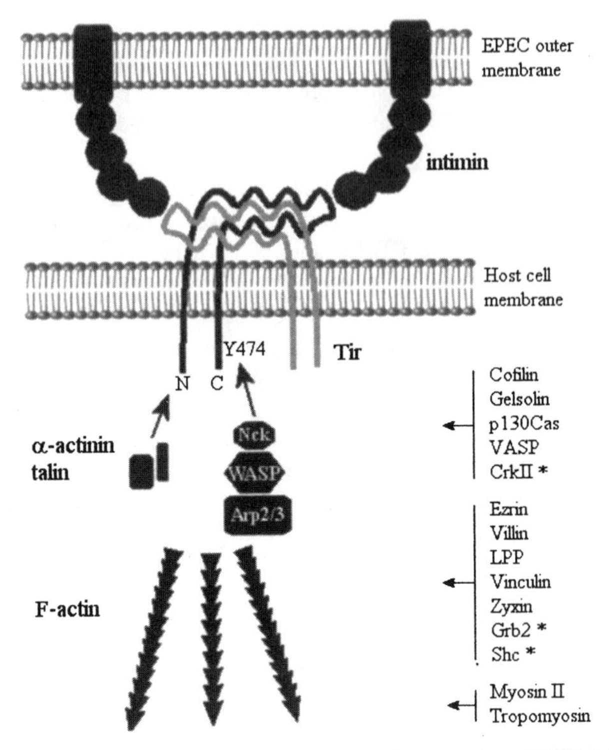

FIGURE 2 The intimin-Tir interaction and cytoskeletal proteins in an EPEC-induced pedestal. On delivery in the host cell by the EPEC TTSS, Tir is inserted into the host membrane and forms a dimer that serves as the receptor for the EPEC outer membrane adhesin intimin. Tir is phosphorylated on Y174 and binds the adaptor Nck. N-WASP and the Arp2/3 complex are recruited to the tip of the pedestal and stimulate the formation of actin filaments. Alpha-actinin and talin are also recruited to the pedestal, bind Tir at its N-terminus, and may link Tir directly to the actin cytoskeleton. Cytoskeletal proteins and adaptors (*asterisk*) found at the tip and along the length (cofilin, gelsolin, p130Cas, VASP, CrkII), along the length only (ezrin, villin, LPP, vinculin, zyxin, Grb2, Shc), or at the base of the pedestal (myosin II and tropomyosin) are indicated.

Whether non-Tir intimin receptors are relevant to *in vivo* infection is currently a matter of debate [113,151,161]. One possibility is that intimin might bind non-Tir receptors at the initial stage of EPEC infection, before Tir has been translocated to the host cell membrane, subsequently shifting to a potentially more stable interaction with Tir [151,161].

Lymphostatin

Lymphostatin is an EPEC protein that inhibits the expression of multiple lymphokines, including interleukin 2, interleukin 4, and interferon-γ at the level of transcription, and inhibits lymphocyte proliferation [162–165]. A similar activity is produced by other A/E pathogens, including EHEC O157:H7 strains and *C. rodentium*. The significance of this activity and its potential role in inhibition of the host immune response by EPEC infection are not yet tested *in vivo*. The *lifA* gene, encoding lymphostatin, is the largest reported gene in an *E. coli* strain, and its product contains sequence similarity to the glycosyltransferase region of the large clostridial cytotoxins [165]. The same gene, called *efa1* in this case, has been implicated in adhesion of an EHEC strain to epithelial cells [166]. The specific chromosomal location of the *lifA* gene is not yet defined, excluding in the rabbit EPEC strain RDEC-1, where it is located adjacent to the LEE [167]. A similar locus is found on the large plasmid of EHECs [168,169], and a partial locus has been found on the large plasmid of EPEC [85].

EspC Autotransporter Protease and EspC Island

EspC is a large (110 kDa) secreted protein of EPEC that does not require the TTSS for export [99,107]. As a member of the SPATE (serine protease autotransporters of the Enterobacteriaceae) family of autotransporter proteins (reviewed in ref. 170), EspC encodes its own transport mechanism. Such proteins possess (1) an N-terminal signal sequence that promotes secretion through the inner membrane via the *sec* apparatus and (2) a C-terminal domain that forms a β-barrel pore in the outer membrane and exports (3) a central "passenger" domain of the protein to the bacterial cell surface. The protease activity of EspC does not appear to be required for release of the passenger domain from the cell [171]. Proteases similar to EspC in the SPATE family are enterotoxins capable of being internalized by and damaging epithelial cells [170] (see Chaps. 6 and 8). Likewise, EspC has been shown to increase the potential difference (PD) and total tissue conductance (I_{sc}) across rat jejunal tissue *in vitro* [135], which may be relevant to known EPEC phenotypes (see below). The relevant substrate(s) of EspC and the role of this protease in EPEC virulence remain unknown. No distinct phenotype has yet been noted for an *espC* mutant [171]. The *espC* gene is located on a 15.1-kb chromosomal island specific to EPEC 1 strains [135]. This island contains 13 additional genes, including those encoding putative transposases and potential virulence factors.

Other Toxins

Some EPEC strains produce a heat-stable enterotoxin named EAST1, and a survey of diarrheogenic *E. coli* strains reported that 14 of 65 EPEC strains tested (22%) hybridized with an EAST1 probe [172]. E2348/69, the prototype EPEC strain often used for volunteer studies, is reported to contain two copies of the *east1* gene, one in the chromosome and one in the EAF plasmid.

Scott and Kaper have cloned and characterized a gene from an EPEC strain that encodes cytolethal distending toxin (CDT) [173]. The mechanism of action of this toxin involves chromatin disruption, which leads to G_2/M-phase growth arrest of the target cell and ultimately cell death [174]. However, its mechanism of action in diarrhea is not known yet. There are other sporadic reports of production of CDT by EPEC. A study of CDT-producing *E. coli* in Bangladeshi children found that although CDT-positive EPEC strains were isolated from more children with diarrhea than from healthy controls, this difference did not reach statistical significance [175]. The prototype EPEC strain E2348/69 does not encode CDT. The significance of CDT and EAST1 toxins in EPEC pathogenesis remains unknown.

Other Fimbriae and Pili

Some EPEC strains elaborate pili or fimbriae other than or in addition to BFP [176,177]. Type I pili of EPEC have been found to be antigenic in volunteer studies [178]. However, type I pili do not have a role in adherence to epithelial cells *in vitro* [179]. The expression of long, fine fimbriae; rigid, bent fimbriae; and short, fine fimbriae is derepressed in the *ler* mutant of EPEC strain E2348/69 [180]. The role of these various fimbriae in EPEC pathogenesis, if any, has yet to be understood.

Effects of EPEC on Host Cells

Actin Cytoskeleton Reorganization

On EPEC infection and insertion of Tir into a host cell membrane, Tir becomes clustered at the tip of pedestals (see above), where it serves as the receptor for intimin. F-actin and other host cytoskeletal proteins accumulate in the pedestals [36,181,182]. The localized F-actin accumulation can be demonstrated using fluorescent probes, and this forms the basis of a diagnostic test (see below). Recent advances have begun to decipher the detailed molecular architecture of the pedestals and how EPEC remodels the host cytoskeleton at the cell surface. EPEC Tir is phosphorylated on Y174 by an unidentified kinase; this event is essential for actin remodeling and pedestal formation by EPEC [58,145] but is not required for the membrane insertion of Tir [183] or intimin binding [113,144]. Given their central role in actin dynamics, the role of Rho small GTP-binding proteins has been investigated, but RhoA-, Rac1-, and Cdc42-dependent pathways are not involved in pedestal formation [184,185]. Recent studies have implicated the neural-Wiskott-Aldrich syndrome protein (N-WASP) and the actin-related proteins 2 and 3 (Arp2/3) complex. A phosphorylated tyrosine in Tir (Y194) binds the adaptor protein Nck, which recruits N-WASP and Arp2/3 to the tip of the pedestals, where they nucleate actin and stimulate the formation of actin filaments [186,187]. α-Actinin and talin are also recruited to pedestals and were shown to bind the N-terminus of Tir independently of its tyrosine phosphorylation [188,189]. In addition, numerous signaling and actin-associated/binding proteins are recruited to the site of bacterial intimate adhesion (see Fig. 2) [182,190,191]. Interestingly, many of these proteins are involved in focal adhesions, the structures found at cell attachment sites that link the extracellular matrix to the cytoskeleton via integrins. This suggests that EPEC subvert fundamental host cell functions to build focal adhesion-like structures that anchor the bacteria to the host cell cytoskeleton via Tir. It is possible that binding of Tir to intimin focuses Tir beneath bacteria in a manner analogous to integrin clustering by the extracellular matrix. When Tir is delivered to host cell membranes but not clustered (using an *eae* mutant), it still recruits cytoskeletal proteins but remains unfocused, appearing as a diffuse structure beneath bacteria. Not all cytoskeletal proteins found in the pedestals are focal adhesion proteins. Nonmuscle isoforms of myosin II and tropomyosin also are found at the base of the pedestals [59], suggesting a similarity to the brush-border microvilli. It is tempting to hypothesize that the effacement of microvilli (which also may result from the EPEC-induced cytoskeleton rearrangements) provides the materials used to form pedestals.

Several studies indicated that EPEC infection induces the release of Ca^{2+} from inositol triphosphate (IP_3)—sensitive intracellular stores [192–194]. IP_3 fluxes were reported on EPEC infection, a TTSS- and EspB-dependent effect [110,194]. Furthermore, the phospholipase Cγ1 (PLC-γ1) is stimulated (tyrosine phosphorylated) in an intimin-dependent manner on EPEC infection [123]. PLC-γ1 activation could be responsible for the increase in IP_3 and subsequent increase of Ca^{2+} concentrations. Given that Ca^{2+} can induce microvilli vesiculation and effacement (Gelsolin severs F-actin in response to Ca^{2+}), such events were considered to participate in A/E lesion formation. However, the most recent data do not show any increase in Ca^{2+} level [195]. This discrepancy may be explained by different infection procedures, some leading to general intoxication of the cells and nonspecific release of Ca^{2+}. The involvement of Ca^{2+} in EPEC-induced signaling and its role in A/E lesion formation remain unresolved.

In conclusion, EPEC usurp components of focal contacts and the actin cytoskeleton to manipulate actin dynamics. Besides providing a strong attachment of EPEC to the host, the role of the pedestals in infection is unknown. The loss of absorptive microvilli in the A/E lesion probably contributes to EPEC-induced diarrhea (malabsorption). However, in experimental human EPEC infections, the incubation period before the onset of diarrhea is only 4 hours, suggesting that an active secretion response also is involved [137].

Tight Junction Disruption and Alteration of Barrier Function

EPEC infection of polarized cultured intestinal cells (Caco-2 or T84) increases the permeability of the monolayer, as reflected by a decrease in TEER and the flux of paracellular markers (such as mannitol). These permeability changes result from a modification of the tight junction (TJ) structure [196–198]. EPEC seem to perturb TJs by various mechanisms. During EPEC infection, a protein involved in regulating TJ permeability, the myosin light chain (MLC), is phosphorylated and associates with the cytoskeleton [199]. MLC phosphorylation may occur via protein kinase C (PKC), which has been shown to be activated in EPEC-infected cells [200,201]. Another kinase, the MLC kinase, is also involved in MLC phosphorylation [199,202]. EPEC also induce the phosphorylation of ezrin and its association with the cytoskeleton, a phenomenon that may be involved in TJ disruption, since expression of a dominant-negative ezrin diminishes the disruption of TJs [203].

Besides these biochemical alterations, recent studies have demonstrated that EPEC infection of intestinal epithelial monolayers perturbs the phosphorylation state and distribution of occludin, a component of the TJ [204]. On EPEC infection, occludin is dephosphorylated, dissociates from the TJ, and assumes a cytoplasmic location. EspF was recently implicated as a bacterial effector of the reduction in TJ integrity because it is required in a dose-dependent fashion for redistribution of occludin, for the loss of TEER, and for increased paracellular permeability [133]. The effect of EPEC infection on another TJ protein, zonula occludens (ZO-1), was also reported [198], but this observation remains controversial because another study did not show a change in ZO-1 distribution [196]. Altogether these effects ultimately lead to the disruption of TJ integrity and an increase in paracellular permeability and represent a putative mechanism underlying EPEC-induced diarrhea.

Alteration of Ion Secretion

Alteration of electrolyte transport during EPEC infection *in vitro* has also been reported. Stein and colleagues [205] demonstrated using a patch-clamp technique that EPEC infection of HeLa and Caco-2 epithelial cells reduces the cell resting membrane potential. EPEC infection of Caco-2 cell monolayers also stimulates ion secretion activity, as demonstrated by a rapid and transient increase in short-circuit current (I_{sc}) that was dependent on a functional TTSS and intimate adhesion [206,207]. Part of this response was attributable to chloride (Cl^-) secretion. PKC, which is activated on EPEC infection [201], could contribute to the activation of chloride secretion through the cystic fibrosis transmembrane conductance regulator. On the other hand, Hecht and Koutsouris [208] showed that EPEC altered bicarbonate (HCO_3^-) transport rather than chloride secretion in T84 intestinal epithelial cells monolayers. In addition, EPEC increase the expression of the galanin receptor in T84 cells (via the activation of NF-κB [209]), which on activation causes Cl^- secretion, thus allowing for increased fluid secretion [210]. Altogether these effects indicate that EPEC-induced diarrhea involves transcellular pathways. In conclusion, the mechanism that leads to diarrhea on EPEC infection is not yet unraveled but most probably is multifactorial, related to changes in ion secretion and intestinal barrier function (reduction of TJ integrity) together with the loss of microvilli and absorptive surfaces.

Host Cell Death (Apoptosis)

EPEC induce death in cultured cells (T84, HeLa), with mixed features of necrosis and apoptosis [211]. EPEC TTSS, *espA*, *espB*, *espD*, and *espF* mutants are attenuated in host cell killing, and transfection experiments implicated EspF as a mediator of apoptosis [134]. Crane and colleagues [134] described morphologic modifications (cell shrinkage, membrane blebbing, nuclear condensation), early expression of phosphatidyl serine on the cell surface, internucleosomal cleavage, and cleavage of cytokeratin-18 by caspases, all of which are indicators of apoptosis. On the other hand, experiments in a naturally occurring rabbit infection model suggest that rabbit pathogenic strains of EPEC do not promote apoptosis *in vivo* [212]. In addition, EPEC activate known antiapoptotic pathways, including PKC [201], MAP kinases [213], and transcription of NF-κB [209], suggesting that EPEC have developed strategies to slow rather than enhance host cell killing. How EPEC-induced cell death seen *in vitro* relates to epithelial damage *in vivo* remains to be determined.

EPEC Interactions with the Host Immune System

In addition to interaction with the intestinal epithelium, EPEC also encounter host defenses. Histologic examination of EPEC-infected tissue demonstrates that EPEC, like many other pathogens, trigger an inflammatory response in the host, with neutrophils and macrophages recruited to the sites of infection [61,138,214]. In cultured epithelial cells, EPEC induce the transcription of NF-κB, which then triggers the expression of interleukin 8 (IL-8) [209,215]. In addition, EPEC induce the activation of the ERK1/2, JNK, and p38 MAP kinase cascade in T84 cultured cells, an activity that depends on bacterial adhesion and expression of the TTSS. These pathways contribute to the stimulation of IL-8 expression but do not participate in A/E lesion formation or TEER alteration [213,216]. Both host responses (activation of NF-κB and MAP kinase pathways) require bacterial adhesion and the TTSS and are not merely due to LPS. Activation of NF-κB and IL-8 expression is associated with the transmigration of polymorphonuclear leukocytes (PMNs) through the epithelial cell monolayer [215]. Recruitment of PMNs to the site of infection is also observed *in vivo*. The epithelial inflammatory response could cause tissue damage (through the release of toxic inflammatory factors by inflammatory cells), an increase in paracellular permeability, and a stimulation of ion secretion (as a consequence of the infiltration of PMNs, which release the secretagogue adenosine-5′-monophosphate), contributing to the duration and severity of the disease. These data, however, should be interpreted with caution. Bacterial flagellin alone can stimulate IL-8 secretion [217,218]. This response could reflect a nonspecific host response to bacterial adhesion.

In the intestinal Peyer's patches, M cells transcytose bacteria and other particles from the lumen to the underlying lymphoid cells to initiate the mucosal immune response. Rabbit experimental infections indicate that EPEC resist transcytosis by M cells [219]. This finding indicates that EPEC have evolved strategies of phagocytosis avoidance. Indeed, EPEC actively inhibits its own phagocytosis by macrophages *in vitro*. This inhibition requires a functional TTSS and the expression of EspA, EspB, and EspD but not Tir and is correlated with the tyrosine dephosphorylation of several unidentified host proteins [220]. Antiphagocytosis occurs by inhibition of the phosphatidyl inositol 3 (PI_3) kinase—dependent actin rearrangements required for bacterial uptake [221].

Regulation of Virulence Factors

EPEC virulence factors are produced only in small amounts when EPEC are grown in standard laboratory medium. However, these factors, as well as the phenotypes associated with them, are highly expressed within a few hours after the introduction of EPEC into medium intended for tissue culture [54]. Optimal conditions for the production of BFP [89,90,96,222], EspA filaments [114], and intimin [68,223] and for the secretion of EPEC proteins (including EspC) [42,99,108,121,224] include an EPEC culture in exponential-phase growth, a temperature of 37°C, neutral pH, the presence of calcium and sodium bicarbonate, appropriate salt and iron levels, and the absence of ammonium. Overall, the conditions are strikingly similar to those likely to be encountered in the small intestine. How these environmental conditions act at the molecular level to influence the expression or activity of EPEC virulence factors is a matter of ongoing investigation. Many of these factors influence the transcription of *per* regulatory genes. Per, Ler, and other factors that regulate the expression of the *bfp* gene cluster and the operons of the LEE are described in detail in Chaps. 11 and 12.

Intriguing evidence suggests that EPEC can recognize the presence of host cells and respond accordingly to adjust their export and display of virulence factors in the sequence necessary to establish and maintain a successful infection.

The function of type III secretion systems generally is stimulated by contact with the host cells (see Chap. 12). In accord, it has been demonstrated that the presence of HeLa cells increases EspB expression and EspB and EspD secretion [116,126]. The expression and presentation of EPEC virulence factors appear to be reduced or abrogated after they are no longer needed. EspA filaments disappear [114] and intimin is downregulated [68] after the formation of A/E lesions. The factor(s) responsible for this effect are unknown.

INTEGRATION OF EPEC PATHOGENIC FEATURES

EPEC pathogenesis has been formalized as a multiple-step model that has evolved with increased understanding of this bacterium [36,38,58,114,225,226] (Fig. 3). The precise order of the steps has been a matter of debate. In any case, the expression of the virulence factors and phenotypes of EPEC is coregulated, and they must act cooperatively for EPEC to cause diarrhea. In the earliest step,

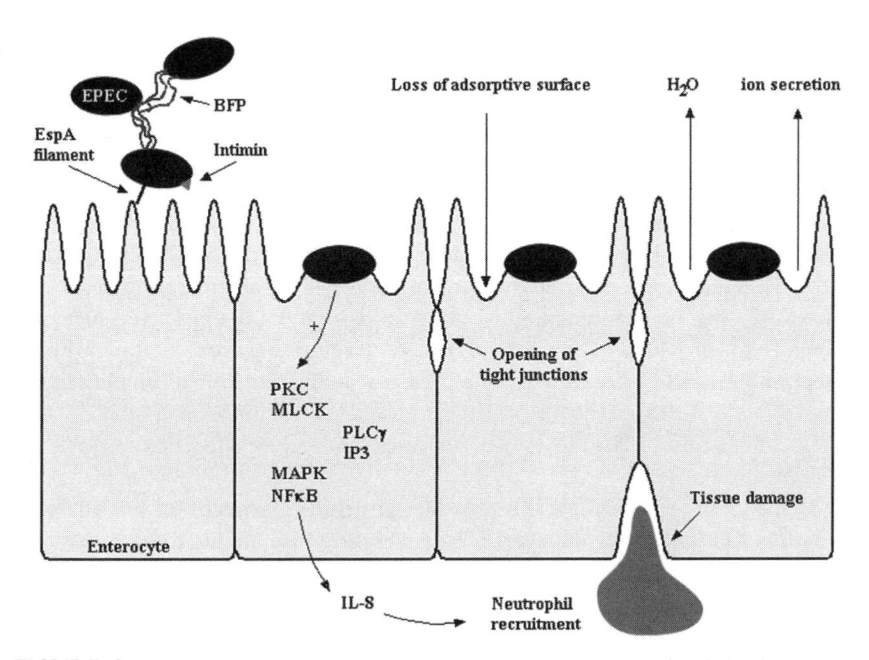

FIGURE 3 Pathogenic scheme of EPEC and mechanisms of EPEC-mediated diarrhea. EPEC interact with the host intestinal epithelium through EspA filaments (and perhaps through BFP and/or intimin), and BFP serves to aggregate individual bacteria into clusters. EPEC intimately attach to the host cell through intimin-Tir binding. EPEC infection activates host cell signaling pathways, leading to the alteration of physiologic functions: increase in epithelial permeability, alterations in ion secretion (Cl^- and HCO_3^-), and reduction in tight junction integrity. Other structural changes include loss of absorptive surfaces and inflammation-induced tissue damage.

EPEC encounter conditions in the small intestine that promote the expression of virulence genes. As a result, intimin is synthesized and exported to the EPEC outer membrane, and BFP and the TTSS/translocation apparatus are assembled within and/or extending from the cell envelope. BFP serves to aggregate individual bacteria into clusters, either before or after interaction of individual bacteria with the host intestinal epithelium through EspA filaments (and perhaps through BFP). The clustering of bacteria on the host cell surface may provide for the injection of a requisite dose of effector proteins into a localized area of the host cell and/or promote regulation by quorum sensing. Next, EPEC effector proteins are introduced into the host cell, probably through the EspA filament, and localize to the membrane (EspB, EspD, and Tir), cytoplasm (EspF, EspB, and EspG), or mitochondria (Map). Intimin in the EPEC membrane binds to Tir in the host cell membrane, forming a tight interaction that holds the EPEC close to the host cell surface and further stimulates the secretion of effector proteins. Tir (and perhaps EspB) promotes cytoskeletal reorganization through interactions with WASP and other proteins, ultimately leading to the effacement of microvilli and the production of pedestals. The ensuing stages of EPEC pathogenesis are less clear. Recent findings indicate that both BFP and EPEC pedestals are dynamic structures and that EPEC clusters disperse over time. EPEC dispersal may encourage the spread of EPEC within a host or to other hosts or may influence the disease process in some way that we do not yet understand. EspF, Map, and any remaining secreted effector proteins alter the metabolism and signaling pathways of the host cell in ways that may promote the availability of nutrients for EPEC and/or may allow EPEC to evade the host immune response. Additional secreted factors such as EspC and lymphostatin presumably also play a role in these efforts. EPEC interfere with three major intestinal epithelium functions: (1) the transport of ions and solutes, (2) the barrier function provided by TJs, and (3) the surveillance and response to the lumen content, ultimately leading to diarrhea, the primary symptom of EPEC infection. From a bacterial point of view, this whole process may provide EPEC with a selective advantage: Intimate adhesion to the enterocyte permits EPEC to remain attached while competing flora are flushed away by diarrhea.

Tremendous progress has been made in EPEC research in the previous two decades, with new discoveries being reported at an increasing pace. We fully expect that many additional intriguing discoveries are in store for EPEC. These may include (1) increased understanding of the pathways through which EPEC virulence factors are regulated by environmental cues, (2) elucidation of the structure and mechanisms of operation of the cellular machineries that synthesize BFP, the TTSS, the translocation apparatus, and the individual components of these systems, (3) the identification of novel effector proteins, and (4) increased understanding of the mechanisms by which effector proteins alter host cell physiology.

DIAGNOSIS, PREVENTION, AND CONTROL OF EPEC INFECTIONS

Phenotypic Diagnostics

EPEC strains traditionally were identified by serotyping (see above) by agglutination with antisera raised against various standard *E. coli* strains. However, serotyping is inaccurate and should be abandoned. A related method is biotyping, in which strains are tested for their ability to metabolize particular sugars or other biochemicals [20]. Of note, strains of the EPEC 2 clone appear to be unique in their ability to metabolize phenylpropionic acid [227].

EPEC strains often have been distinguished from other *E. coli* by their ability to adhere in a localized fashion to human epithelial cells in tissue culture [20,228,229] (see above). Case-control studies have shown that there is a correlation between LA-positive strains and diarrhea [230–234]. The propensity of EPEC to autoaggregate when grown in tissue culture medium also should provide a facile method of distinguishing typical EPEC strains from other *E. coli*, although this notion has not been tested in diagnostic studies (see above). The limitation of both the adherence and autoaggregation assays is that the phenotypes being studied depend on BFP, and therefore, the assays are unable to identify atypical EPEC (see above).

In the fluorescent-actin staining (FAS) assay, fluorescein isothiocyanate (FITC)—phalloidin or other actin-binding compounds are used to detect the actin that accumulates under EPEC adhesion pedestals [235] (see Fig. 1*D*). The FAS assay has been used rarely for diagnostic studies [41], being somewhat more laborious than other methods. Its usefulness in identifying EPEC is limited by the inability to distinguish EPEC from other A/E bacteria, including EHECs. A variation of this assay uses anti-phosphotyrosine antibodies to identify phosphorylated proteins (i.e., Tir) that accumulate at sites of EPEC but not EHEC attachment [109,236].

Molecular Diagnostics

DNA probe hybridization and polymerase chain reaction (PCR) have become standard methods to identify EPEC-specific genetic material [2]. These methods have been used in numerous studies but are limited by their inability to identify all EPEC strains. The first such diagnostic to be developed was an approximately 1-kb DNA probe specific for the EAF plasmid [34,237,238], also developed in oligonucleotide and PCR versions [239,240]. The EAF probe does not seem to identify any virulence factor or even coding DNA [240], and in the prototype EPEC 2 strain B171, it contains an insertion sequence [85]. However, it been used extensively to identify EPEC (e.g., see ref. 82). Several case-control studies have shown that there is a positive correlation between EAF-positive strains and diarrhea [21,230,241,242]. However, it should be noted that EAF-negative EPEC strains have been isolated frequently from patients with diarrhea [230,232,243,244].

A DNA probe to diagnose the presence of the *eae* gene encoding intimin has been developed [87,245]. Multiple *eae* alleles have been identified among the various A/E pathogens [19,87,154,156,158,246–252] that differ in antigenicity as well as in sequence. The N-terminal end of intimin displays high sequence conservation, whereas the C-terminal end is sequence-variable. It has been possible to develop PCR primers and antisera that identify all known intimin types [253], as well those which identify specific types [248,249,254,255]. To date most EPEC 1 strains express an α_1 type of intimin, whereas most EPEC 2 strains express a β_1 type. However, some EHEC and *C. rodentium* strains also express these intimin types, making it difficult to identify EPEC conclusively using intimin diagnostics. To rule out EHECs, a probe for Shiga toxins should be employed.

Several probes and primers also have been developed to identify the *bfpA* gene encoding bundlin [2,91,254,256–260]. As for EAF, *bfpA* diagnostics fail to identify atypical EPEC strains. Furthermore, multiple alleles of *bfpA* have been identified recently [92], suggesting that some current PCR methods may fail to identify all *bfpA*-positive EPEC strains. Note that the presence of a gene does not necessarily indicate that its product is functional (e.g., see refs. 92 and 261). Ideally, molecular diagnostics used to detect virulence factor genes should be coupled with phenotypic assays for the activities of the virulence factors themselves to properly analyze the pathogenic potential of new EPEC strains.

Control and Prevention

Immune Response to EPEC

Many of the known EPEC virulence factors, including bundlin, intimin, EspA, EspB, EspC, and Tir, induce an immune response. Serum antibodies recognizing these proteins have been demonstrated in volunteers experimentally infected with EPEC [83,107,140,262] and in children naturally infected or living in EPEC-endemic areas [263–265].

Vaccines

Theorically, some combination of EPEC virulence factors that elicit a protective immune response without causing diarrheal disease could be expressed in an appropriate bacterial vector as a vaccine [266]. A preliminary study has used bundlin in such a manner [267]. An alternative strategy would be to develop a recombinant, attenuated EPEC strain that fails to cause diarrhea but retains adherence factors that allow it to bind to the intestinal lining and antigens that elicit protection against EPEC infection. However, it remains a formidable challenge to deliver any potentially effective vaccines to the target population most likely to benefit from them.

Protective Effects of Breast-Feeding

Numerous studies have suggested that breast-feeding provides excellent protection against EPEC infection. Interestingly, secretory IgA antibodies against bundlin, intimin, EspA, EspB, EspC, Tir, and other EPEC proteins have been identified in human colostrum obtained from mothers living in an EPEC-endemic areas [48,249,264,265,268,269]. In addition to serving in the immune response against EPEC, these antibodies appear to be partially responsible for the ability of colostrum and milk to inhibit EPEC adherence to cultured epithelial cells [48,270,271]. Other components of milk, notably oligosaccharides, are also capable of inhibiting or modifying EPEC adherence [48–50]. These findings support the importance of breast-feeding as an essential preventive measure in EPEC-endemic areas.

Therapeutics

Routine treatment of EPEC diarrhea includes oral rehydration therapy to correct fluid and electrolyte and nutritional imbalances, although in severe cases treatment may include parenteral rehydration or total parenteral nutrition [2,4]. Bismuth subsalicylate has proven to be an effective additional therapy in one study [272]. Racecadotril is also an effective and safe additional treatment for acute watery diarrhea in young children and may prove useful in EPEC diarrhea [273]. Treatment with epidermal growth factor prevented the occurrence of EPEC diarrhea in a rabbit model [274]. Mixed results have been obtained in studies in which EPEC-infected infants were treated by passive immunization with anti-EPEC bovine immunoglobulin concentrate [275,276]. EPEC can be attacked successfully with antibiotics [6,277]; however, it should be noted that many EPEC strains exhibit resistance to multiple antibiotics [86,229,278–280]. Probiotic therapy with nonpathogenic organisms is another potentially effective therapy [216]. Theoretically, EPEC infections could be treated by novel therapeutics that would inhibit EPEC autoaggregation (see above), adhesion to the intestinal mucosa (see above), the functions of EPEC virulence factors (see above), or the regulatory pathways that control virulence factor production (see above).

REFERENCES

1. Kaper, J. B. (1996). Defining EPEC. *Rev. Microbiol. Sao Paulo* 27(suppl. 1):130–133.
2. Nataro, J. P., and Kaper, J. B. (1998). Diarrheagenic *Escherichia coli*. *Clin. Microbiol. Rev.* 11:142–201.
3. Levine, M. M. (1987). *Escherichia coli* that cause diarrhea: Enterotoxigenic, enteropathogenic, enteroinvasive, enterohemorrhagic, and enteroadherent. *J. Infect. Dis.* 155:377–389.
4. Donnenberg, M. S. (1995). Enteropathogenic *Escherichia coli*. In *Infections of the Gastrointestinal Tract* (M. J. Blaser, P. D. Smith, J. I. Ravdin, H. B. Greenberg, and R. L. Guerrant, eds.), pp. 709–726. New York: Raven Press.
5. Lacroix, J., Delage, G., Gosselin, F., and Chicoine, L. (1984). Severe protracted diarrhea due to multiresistant adherent *Escherichia coli*. *Am. J. Dis. Child.* 138:693–696.
6. Hill, S. M., Phillips, A. D., and Walker-Smith, J. A. (1991). Enteropathogenic *Escherichia coli* and life-threatening chronic diarrhoea. *Gut* 32:154–158.
7. Rothbaum, R. J., Partin, J. C., Saalfield, K., and McAdams, A. J. (1983). An ultrastructural study of enteropathogenic *Escherichia coli* infection in human infants. *Ultrastruct. Pathol.* 4:291–304.
8. Levine, M. M., and Edelman, R. (1984). Enteropathogenic *Escherichia coli* of classic serotypes associated with infant diarrhea: Epidemiology and pathogenesis. *Epidemiol. Rev.* 6:31–51.
9. Robins-Browne, R. M. (1987). Traditional enteropathogenic *Escherichia coli* of infantile diarrhea. *Rev. Infect. Dis.* 9:28–53.
10. Levine, M. M., Bergquist, E. J., Nalin, D. R., *et al.* (1978). *Escherichia coli* strains that cause diarrhoea but do not produce heat-labile or heat-stable enterotoxins and are non-invasive. *Lancet* 1:1119–1122.
11. Kotler, D. P., and Orenstein, J. M. (1993). Chronic diarrhea and malabsorption associated with enteropathogenic bacterial infection in a patient with AIDS. *Ann. Intern. Med.* 119: 127–128.
12. Polotsky, Y., Nataro, J. P., Kotler, D., *et al.* (1997). HEp-2 cell adherence patterns, serotyping, and DNA analysis of *Escherichia coli* isolates from eight patients with AIDS and chronic diarrhea. *J. Clin. Microbiol.* 35:1952–1958.
13. Peeters, J. E., Charlier, G. J., and Halen, P. H. (1984). Pathogenicity of attaching effacing enteropathogenic *Escherichia coli* isolated from diarrheic suckling and weanling rabbits for newborn rabbits. *Infect. Immun.* 46:690–696.
14. Drolet, R., Fairbrother, J. M., Harel, J., and Hélie, P. (1994). Attaching and effacing and enterotoxigenic *Escherichia coli* associated with enteric colibacillosis in the dog. *Can. J. Vet. Res.* 58:87–92.
15. Zhu, C., Harel, J., Jacques, M., *et al.* (1994). Virulence properties and attaching-effacing activity of *Escherichia coli* O45 from swine postweaning diarrhea. *Infect. Immun.* 62:4153–4159.
16. Beaudry, M., Zhu, C., Fairbrother, J. M., and Harel, J. (1996). Genotypic and phenotypic characterization of *Escherichia coli* isolates from dogs manifesting attaching and effacing lesions. *J. Clin. Microbiol.* 34:144–148.
17. Schremmer, C., Lohr, J. E., Wastlhuber, U., *et al.* (1999). Enteropathogenic *Escherichia coli* in *Psittaciformes*. *Avian Pathol.* 28:349–354.
18. Mansfield, K. G., Lin, K. C., Newman, J., *et al.* (2001). Identification of enteropathogenic *Escherichia coli* in simian immunodeficiency virus—infected infant and adult rhesus macaques. *J. Clin. Microbiol.* 39:971–976.
19. Schauer, D. B., and Falkow, S. (1993). Attaching and effacing locus of a *Citrobacter freundii* biotype that causes transmissible murine colonic hyperplasia. *Infect. Immun.* 61:2486–2492.
20. Scaletsky, I. C. A., Silva, M. L. M., Toledo, M. R. F., *et al.* (1985). Correlation between adherence to HeLa cells and serogroups, serotypes, and bioserotypes of *Escherichia coli*. *Infect. Immun.* 49:528–532.

21. Gomes, T. A. T., Vieira, M. A. M., Wachsmuth, I. K., *et al.* (1989). Serotype-specific prevalence of *Escherichia coli* strains with EPEC adherence factor genes in infants with and without diarrhea in São Paulo, Brazil. *J. Infect. Dis.* 160:131–135.
22. Ørskov, F., Whittam, T. S., Cravioto, A., and Ørskov, I. (1990). Clonal relationships among classic enteropathogenic *Escherichia coli* (EPEC) belonging to different O groups. *J. Infect. Dis.* 162:76–81.
23. Whittam, T. S., Wolfe, M. L., Wachsmuth, I. K., *et al.* (1993). Clonal relationships among *Escherichia coli* strains that cause hemorrhagic colitis and infantile diarrhea. *Infect. Immun.* 61:1619–1629.
24. Whittam, T. S., and McGraw, E. A. (1996). Clonal analysis of EPEC serogroups. *Rev. Microbiol. Sao Paulo* 27:7–16.
25. Reid, S. D., Herbelin, C. J., Bumbaugh, A. C., *et al.* (2000). Parallel evolution of virulence in pathogenic *Escherichia coli*. *Nature* 406:64–67.
26. Donnenberg, M. S., and Whittam, T. S. (2001). Pathogenesis and evolution of virulence in enteropathogenic and enterohemorrhagic *Escherichia coli*. *J. Clin. Invest* 107:539–548.
27. Beutin, L., Ørskov, I., Ørskov, F., *et al.* (1990). Clonal diversity and virulence factors in strains of *Escherichia coli* of the classic enteropathogenic serogroup O114. *J. Infect. Dis.* 162:1329–1334.
28. Campos, L. C., Whittam, T. S., Gomes, T. A. T., *et al.* (1994). *Escherichia coli* serogroup O111 includes several clones of diarrheagenic strains with different virulence properties. *Infect. Immun.* 62:3282–3288.
29. Rodrigues, J., Scaletsky, I. C. A., Campos, L. C., *et al.* (1996). Clonal structure and virulence factors in strains of *Escherichia coli* of the classic serogroup O55. *Infect. Immun.* 64:2680–2686.
30. Gonçalves, A. G., Campos, L. C., Gomes, T. A. T., *et al.* (1997). Virulence properties and clonal structures of strains of *Escherichia coli* O119 serotypes. *Infect. Immun.* 65:2034–2040.
31. Pelayo, J. S., Scaletsky, I. C. A., Pedroso, M. Z., *et al.* (1999). Virulence properties of atypical EPEC strains. *J. Med. Microbiol.* 48:41–49.
32. Clausen, C. R., and Christie, D. L. (1982). Chronic diarrhea in infants caused by adherent enteropathogenic *Escherichia coli*. *J. Pediatr.* 100:358–361.
33. Scaletsky, I. C. A., Silva, M. L. M., and Trabulsi, L. R. (1984). Distinctive patterns of adherence of enteropathogenic *Escherichia coli* to HeLa cells. *Infect. Immun.* 45:534–536.
34. Nataro, J. P., Scaletsky, I. C. A., Kaper, J. B., *et al.* (1985). Plasmid-mediated factors conferring diffuse and localized adherence of enteropathogenic *Escherichia coli*. *Infect. Immun.* 48:378–383.
35. Nataro, J. P., Kaper, J. B., Robins-Browne, R., *et al.* (1987). Patterns of adherence of diarrheagenic *Escherichia coli* to HEp-2 cells. *Pediatr. Infect. Dis. J.* 6:829–831.
36. Knutton, S., Lloyd, D. R., and McNeish, A. S. (1987). Adhesion of enteropathogenic *Escherichia coli* to human intestinal enterocytes and cultured human intestinal mucosa. *Infect. Immun.* 55:69–77.
37. Yamamoto, T., Koyama, Y., Matsumoto, M., *et al.* (1992). Localized, aggregative, and diffuse adherence to HeLa cells, plastic, and human small intestines by *Escherichia coli* isolated from patients with diarrhea. *J. Infect. Dis.* 166:1295–1310.
38. Hicks, S., Frankel, G., Kaper, J. B., *et al.* (1998). Role of intimin and bundle-forming pili in enteropathogenic *Escherichia coli* adhesion to pediatric intestinal tissue *in vitro*. *Infect. Immun.* 66:1570–1578.
39. Rothbaum, R., McAdams, A. J., Giannella, R., and Partin, J. C. (1982). A clinicopathological study of enterocyte-adherent *Escherichia coli*: A cause of protracted diarrhea in infants. *Gastroenterology* 83:441–454.
40. Knutton, S., Shaw, R. K., Anantha, R. P., *et al.* (1999). The type IV bundle-forming pilus of enteropathogenic *Escherichia coli* undergoes dramatic alterations in structure associated with bacterial adherence, aggregation and dispersal. *Mol. Microbiol.* 33:499–509.
41. Knutton, S., Phillips, A. D., Smith, H. R., *et al.* (1991). Screening for enteropathogenic *Escherichia coli* in infants with diarrhea by the fluorescent-actin staining test. *Infect. Immun.* 59:365–371.

42. Beinke, C., Laarmann, S., Wachter, C., *et al.* (1998). Diffusely adhering *Escherichia coli* strains induce attaching and effacing phenotypes and secrete homologs of Esp proteins. *Infect. Immun.* 66:528–539.
43. Scaletsky, I. C. A., Pedroso, M. Z., Oliva, C. A. G., *et al.* (1999). A localized adherence-like pattern as a second pattern of adherence of classic enteropathogenic *Escherichia coli* to HEp-2 cells that is associated with infantile diarrhea. *Infect. Immun.* 67:3410–3415.
44. Scaletsky, I. C. A., Pedroso, M. Z., and Silva, R. M. (1999). Phenotypic and genetic features of *Escherichia coli* strains showing simultaneous expression of localized and diffuse adherence. *FEMS Immunol. Med. Microbiol.* 23:181–188.
45. Vanmaele, R. P., and Armstrong, G. D. (1997). Effect of carbon source on localized adherence of enteropathogenic *Escherichia coli. Infect. Immun.* 65:1408–1413.
46. Scaletsky, I. C. A., Milani, S. R., Trabulsi, L. R., and Travassos, L. R. (1988). Isolation and characterization of the localized adherence factor of enteropathogenic *Escherichia coli. Infect. Immun.* 56:2979–2983.
47. Vanmaele, R. P., Heerze, L. D., and Armstrong, G. D. (1999). Role of lactosyl glycan sequences in inhibiting enteropathogenic *Escherichia coli* attachment. *Infect. Immun.* 67:3302–3307.
48. Cravioto, A., Tello, A., Villafán, H., *et al.* (1991). Inhibition of localized adhesion of enteropathogenic *Escherichia coli* to HEp-2 cells by immunoglobulin and oligosaccharide fractions of human colostrum and breast milk. *J. Infect. Dis.* 163:1247–1255.
49. Jagannatha, H. M., Sharma, U. K., Ramaseshan, T., *et al.* (1991). Identification of carbohydrate structures as receptors for localised adherent enteropathogenic *Escherichia coli. Microb. Pathog.* 11:259–268.
50. Idota, T., and Kawakami, H. (1995). Inhibitory effects of milk gangliosides on the adhesion of *Escherichia coli* to human intestinal carcinoma cells. *Biosci. Biotechnol. Biochem.* 59:69–72.
51. Vanmaele, R. P., Finlayson, M. C., and Armstrong, G. D. (1995). Effect of enteropathogenic *Escherichia coli* on adherent properties of Chinese hamster ovary cells. *Infect. Immun.* 63:191–198.
52. Manjarrez-Hernandez, A., Gavilanes-Parra, S., Chavez-Berrocal, M. E., *et al.* (1997). Binding of diarrheagenic *Escherichia coli* to 32- to 33-kilodalton human intestinal brush border proteins. *Infect. Immun.* 65:4494–4501.
53. Foster, D. B., Philpott, D., Abul-Milh, M., *et al.* (1999). Phosphatidylethanolamine recognition promotes enteropathogenic *E. coli* and enterohemorrhagic *E. coli* host cell attachment. *Microb. Pathog.* 27:289–301.
54. Vuopio-Varkila, J., and Schoolnik, G. K. (1991). Localized adherence by enteropathogenic *Escherichia coli* is an inducible phenotype associated with the expression of new outer membrane proteins. *J. Exp. Med.* 174:1167–1177.
55. Bieber, D., Ramer, S. W., Wu, C. Y., *et al.* (1998). Type IV pili, transient bacterial aggregates, and virulence of enteropathogenic *Escherichia coli. Science* 280:2114–2118.
56. Moon, H. W., Whipp, S. C., Argenzio, R. A., *et al.* (1983). Attaching and effacing activities of rabbit and human enteropathogenic *Escherichia coli* in pig and rabbit intestines. *Infect. Immun.* 41:1340–1351.
57. Phillips, A. D., Giron, J., Hicks, S., *et al.* (2000). Intimin from enteropathogenic *Escherichia coli* mediates remodelling of the eukaryotic cell surface. *Microbiology* 146(6):1333–1344.
58. Rosenshine, I., Ruschkowski, S., Stein, M., *et al.* (1996). A pathogenic bacterium triggers epithelial signals to form a functional bacterial receptor that mediates actin pseudopod formation. *EMBO J.* 15:2613–2624.
59. Sanger, J. M., Chang, R., Ashton, F., *et al.* (1996). Novel form of actin-based motility transports bacteria on the surface of infected cells. *Cell Motil. Cytoskel.* 34:279–287.
60. Knutton, S., Baldini, M. M., Kaper, J. B., and McNeish, A. S. (1987). Role of plasmid-encoded adherence factors in adhesion of enteropathogenic *Escherichia coli* to HEp-2 cells. *Infect. Immun.* 55:78–85.

61. Ulshen, M. H., and Rollo, J. L. (1980). Pathogenesis of *Escherichia coli* gastroenteritis in man: Another mechanism. *New Engl. J. Med.* 302:99–101.
62. Taylor, C. J., Hart, A., Batt, R. M., *et al.* (1986). Ultrastructural and biochemical changes in human jejunal mucosa associated with enteropathogenic *Escherichia coli* (0111) infection. *J. Pediatr. Gastroenterol. Nutr.* 5:70–73.
63. Staley, T. E., Jones, E. W., and Corley, L. D. (1969). Attachment and penetration of *Escherichia coli* into intestinal epithelium of the ileum in newborn pigs. *Am. J. Pathol.* 56:371–392.
64. Cantey, J. R., and Blake, R. K. (1977). Diarrhea due to *Escherichia coli* in the rabbit: a novel mechanism. *J. Infect. Dis.* 135:454–462.
65. Polotsky, Y. E., Dragunskaya, E. M., Seliverstova, V. G., *et al.* (1977). Pathogenic effect of enterotoxigenic *Escherichia coli* and *Escherichia coli* causing infantile diarrhoea. *Acta Microbiol. Acad. Sci. Hung.* 24:221–236.
66. Tzipori, S., Robins-Browne, R. M., Gonis, G., *et al.* (1985). Enteropathogenic *Escherichia coli* enteritis: Evaluation of the gnotobiotic piglet as a model of human infection. *Gut* 26:570–578.
67. Embaye, H., Batt, R. M., Saunders, J. R., *et al.* (1989). Interaction of enteropathogenic *Escherichia coli* 0111 with rabbit intestinal mucosa *in vitro*. *Gastroenterology* 96:1079–1086.
68. Knutton, S., Adu-Bobie, J., Bain, C., *et al.* (1997). Downregulation of intimin expression during attaching and effacing enteropathogenic *Escherichia coli* adhesion. *Infect. Immun.* 65:1644–1652.
69. Tzipori, S., Wachsmuth, I. K., Chapman, C., *et al.* (1986). The pathogenesis of hemorrhagic colitis caused by *Escherichia coli* O157:H7 in gnotobiotic pigs. *J. Infect. Dis.* 154:712–716.
70. Andrade, J. R., Da Veiga, V. F., De Santa Rosa, M. R., and Suassuna, I. (1989). An endocytic process in HEp-2 cells induced by enteropathogenic *Escherichia coli*. *J. Med. Microbiol.* 28:49–57.
71. Donnenberg, M. S., Donohue-Rolfe, A., and Keusch, G. T. (1989). Epithelial cell invasion: an overlooked property of enteropathogenic *Escherichia coli* (EPEC) associated with the EPEC adherence factor. *J. Infect. Dis.* 160:452–459.
72. Miliotis, M. D., Koornhof, H. J., and Phillips, J. I. (1989). Invasive potential of noncytotoxic enteropathogenic *Escherichia coli* in an *in vitro* Henle 407 cell model. *Infect. Immun.* 57:1928–1935.
73. Donnenberg, M. S., Donohue-Rolfe, A., and Keusch, G. T. (1990). A comparison of HEp-2 cell invasion by enteropathogenic and enteroinvasive *Escherichia coli*. *FEMS Microbiol. Lett.* 57:83–86.
74. Francis, C. L., Jerse, A. E., Kaper, J. B., and Falkow, S. (1991). Characterization of interactions of enteropathogenic *Escherichia coli* O127:H6 with mammalian cells *in vitro*. *J. Infect. Dis.* 164:693–703.
75. Fagundes-Neto, U., Freymuller, E., Gatti, M. S. V., *et al.* (1995). Enteropathogenic *Escherichia coli* O111ab:H2 penetrates the small bowel epithelium in an infant with acute diarrhoea. *Acta Paediatr.* 84:453–455.
76. Donnenberg, M. S., Calderwood, S. B., Donohue-Rolfe, A., *et al.* (1990). Construction and analysis of Tn*phoA* mutants of enteropathogenic *Escherichia coli* unable to invade HEp-2 cells. *Infect. Immun.* 58:1565–1571.
77. Frankel, G., Philips, A. D., Novakova, M., *et al.* (1998). Generation of *Escherichia coli* intimin derivatives with differing biological activities using site-directed mutagenesis of the intimin C-terminus domain. *Mol. Microbiol.* 29:559–570.
78. Baldini, M. M., Kaper, J. B., Levine, M. M., *et al.* (1983). Plasmid-mediated adhesion in enteropathogenic *Escherichia coli*. *J. Pediatr. Gastroenterol. Nutr.* 2:534–538.
79. Nataro, J. P., Maher, K. O., Mackie, P., and Kaper, J. B. (1987). Characterization of plasmids encoding the adherence factor of enteropathogenic *Escherichia coli*. *Infect. Immun.* 55:2370–2377.
80. McConnell, M. M., Chart, H., Scotland, S. M., *et al.* (1989). Properties of adherence factor plasmids of enteropathogenic *Escherichia coli* and the effect of host strain on expression of adherence to HEp-2 cells. *J. Gen. Microbiol.* 135:1123–1134.

81. Senerwa, D., Olsvik, O., Mutanda, L. N., *et al.* (1989). Enteropathogenic *Escherichia coli* serotype O111:HNT isolated from preterm neonates in Nairobi, Kenya. *J. Clin. Microbiol.* 27:1307–1311.
82. Chart, H., Scotland, S. M., Willshaw, G. A., and Rowe, B. (1988). HEp-2 adhesion and the expression of a 94-kDa outer-membrane protein by strains of *Escherichia coli* belonging to enteropathogenic serogroups. *J. Gen. Microbiol.* 134:1315–1321.
83. Levine, M. M., Nataro, J. P., Karch, H., *et al.* (1985). The diarrheal response of humans to some classic serotypes of enteropathogenic *Escherichia coli* is dependent on a plasmid encoding an enteroadhesiveness factor. *J. Infect. Dis.* 152:550–559.
84. Tzipori, S., Gibson, R., and Montanaro, J. (1989). Nature and distribution of mucosal lesions associated with enteropathogenic and enterohemorrhagic *Escherichia coli* in piglets and the role of plasmid-mediated factors. *Infect. Immun.* 57:1142–1150.
85. Tobe, T., Hayashi, T., Han, C. G., *et al.* (1999). Complete DNA sequence and structural analysis of the enteropathogenic *Escherichia coli* adherence factor plasmid. *Infect. Immun.* 67:5455–5462.
86. Laporta, M. Z., Silva, M. L. M., Scaletsky, I. C. A., and Trabulsi, L. R. (1986). Plasmids coding for drug resistance and localized adherence to HeLa cells in enteropathogenic *Escherichia coli* O55:H- and O55:H6. *Infect. Immun.* 51:715–717.
87. Jerse, A. E., Yu, J., Tall, B. D., and Kaper, J. B. (1990). A genetic locus of enteropathogenic *Escherichia coli* necessary for the production of attaching and effacing lesions on tissue culture cells. *Proc. Natl. Acad. Sci. USA* 87:7839–7843.
88. Hales, B. A., Hart, C. A., Batt, R. M., and Saunders, J. R. (1992). The large plasmids found in enterohemorrhagic and enteropathogenic *Escherichia coli* constitute a related series of transfer-defective Inc F-IIA replicons. *Plasmid* 28:183–193.
89. Girón, J. A., Ho, A. S. Y., and Schoolnik, G. K. (1991). An inducible bundle-forming pilus of enteropathogenic *Escherichia coli. Science* 254:710–713.
90. Donnenberg, M. S., Girón, J. A., Nataro, J. P., and Kaper, J. B. (1992). A plasmid-encoded type IV fimbrial gene of enteropathogenic *Escherichia coli* associated with localized adherence. *Mol. Microbiol.* 6:3427–3437.
91. Sohel, I., Puente, J. L., Murray, W. J., *et al.* (1993). Cloning and characterization of the bundle-forming pilin gene of enteropathogenic *Escherichia coli* and its distribution in *Salmonella* serotypes. *Mol. Microbiol.* 7:563–575.
92. Blank, T. E., Zhong, H., Bell, A. L., *et al.* (2000). Molecular variation among type IV pilin (*bfpA*) genes from diverse enteropathogenic *Escherichia coli* strains. *Infect. Immun.* 68:7028–7038.
93. Anantha, R. P., Stone, K. D., and Donnenberg, M. S. (2000). Effects of *bfp* mutations on biogenesis of functional enteropathogenic *Escherichia coli* type IV pili. *J. Bacteriol.* 182:2498–2506.
94. Stone, K. D., Zhang, H.-Z., Carlson, L. K., and Donnenberg, M. S. (1996). A cluster of fourteen genes from enteropathogenic *Escherichia coli* is sufficient for biogenesis of a type IV pilus. *Mol. Microbiol.* 20:325–337.
95. Sohel, I., Puente, J. L., Ramer, S. W., *et al.* (1996). Enteropathogenic *Escherichia coli*: Identification of a gene cluster coding for bundle-forming pilus morphogenesis. *J. Bacteriol.* 178:2613–2628.
96. Ramer, S. W., Bieber, D., and Schoolnik, G. K. (1996). BfpB, an outer membrane lipoprotein required for the biogenesis of bundle-forming pili in enteropathogenic *Escherichia coli. J. Bacteriol.* 178:6555–6563.
97. Blank, T. E., and Donnenberg, M. S. (2001). Novel topology of BfpE, a cytoplasmic membrane protein required for type IV fimbrial biogenesis in enteropathogenic *Escherichia coli. J. Bacteriol.* 183:4435–4450.
98. Anantha, R. P., Stone, K. D., and Donnenberg, M. S. (1998). The role of BfpF, a member of the PilT family of putative nucleotide-binding proteins, in type IV pilus biogenesis and in interactions between enteropathogenic *Escherichia coli* and host cells. *Infect. Immun.* 66:122–131.

99. Kenny, B., and Finlay, B. B. (1995). Protein secretion by enteropathogenic *Escherichia coli* is essential for transducing signals to epithelial cells. *Proc. Natl. Acad. Sci. USA* 92:7991–7995.
100. Rabinowitz, R. P., Lai, L.-C., Jarvis, K., *et al.* (1996). Attaching and effacing of host cells by enteropathogenic *Escherichia coli* in the absence of detectable tyrosine kinase mediated signal transduction. *Microb. Pathog.* 21:157–171.
101. DeVinney, R., Nisan, I., Ruschkowski, S., *et al.* (2001). Tir tyrosine phosphorylation and pedestal formation are delayed in enteropathogenic *Escherichia coli sepZ*::Tn*phoA* mutant 30–5-1(3). *Infect. Immun.* 69:559–563.
102. McDaniel, T. K., and Kaper, J. B. (1997). A cloned pathogenicity island from enteropathogenic *Escherichia coli* confers the attaching and effacing phenotype on K-12 *E. coli*. *Mol. Microbiol.* 23:399–407.
103. Elliott, S. J., Wainwright, L. A., McDaniel, T. K., *et al.* (1998). The complete sequence of the locus of enterocyte effacement (LEE) of enteropathogenic *E. coli* E2348/69. *Mol. Microbiol.* 28:1–4.
104. Mellies, J. L., Elliott, S. J., Sperandio, V., *et al.* (1999). The Per regulon of enteropathogenic *Escherichia coli*: Identification of a regulatory cascade and a novel transcriptional activator, the locus of enterocyte effacement (LEE)—encoded regulator (Ler). *Mol. Microbiol.* 33:296–306.
105. Elliott, S. J., Hutcheson, S. W., Dubois, M. S., *et al.* (1999). Identification of CesT, a chaperone for the type III secretion of Tir in enteropathogenic *Escherichia coli*. *Mol. Microbiol.* 33:1176–1189.
106. Sanchez-SanMartin, C., Bustamante, V. H., Calva, E., and Puente, J. L. (2001). Transcriptional regulation of the *orf19* gene and the tir-cesT-eae operon of enteropathogenic *Escherichia coli*. *J Bacteriol.* 183:2823–2833.
107. Jarvis, K. G., Girón, J. A., Jerse, A. E., *et al.* (1995). Enteropathogenic *Escherichia coli* contains a putative type III secretion system necessary for the export of proteins involved in attaching and effacing lesion formation. *Proc. Natl. Acad. Sci. USA* 92:7996–8000.
108. Haigh, R., Baldwin, T., Knutton, S., and Williams, P. H. (1995). Carbon dioxide regulated secetion of the EaeB protein of enteropathogenic *Escherichia coli*. *FEMS Microbiol. Lett.* 129:63–67.
109. Rosenshine, I., Donnenberg, M. S., Kaper, J. B., and Finlay, B. B. (1992). Signal exchange between enteropathogenic *Escherichia coli* (EPEC) and epithelial cells: EPEC induce tyrosine phosphorylation of host cell protein to initiate cytoskeletal rearrangement and bacterial uptake. *EMBO J.* 11:3551–3560.
110. Foubister, V., Rosenshine, I., and Finlay, B. B. (1994). A diarrheal pathogen, enteropathogenic *Escherichia coli* (EPEC), triggers a flux of inositol phosphates in infected epithelial cells. *J. Exp. Med.* 179:993–998.
111. De Rycke, J., Comtet, E., Chalareng, C., *et al.* (1997). Enteropathogenic *Escherichia coli* O103 from rabbit elicits actin stress fibers and focal adhesions in HeLa epithelial cells, cytopathic effects that are linked to an analog of the locus of enterocyte effacement. *Infect. Immun.* 65:2555–2563.
112. Donnenberg, M. S., Lai, L. C., and Taylor, K. A. (1997). The locus of enterocyte effacement pathogenicity island of enteropathogenic *Escherichia coli* encodes secretion functions and remnants of transposons at its extreme right end. *Gene* 184:107–114.
113. Kenny, B., DeVinney, R., Stein, M., *et al.* (1997). Enteropathogenic *E. coli* (EPEC) transfers its receptor for intimate adherence into mammalian cells. *Cell* 91:511–520.
114. Knutton, S., Rosenshine, I., Pallen, M. J., *et al.* (1998). A novel EspA-associated surface organelle of enteropathogenic *Escherichia coli* involved in protein translocation into epithelial cells. *EMBO J.* 17:2166–2176.
115. Taylor, K. A., O'Connell, C. B., Luther, P. W., and Donnenberg, M. S. (1998). The EspB protein of enteropathogenic *Escherichia coli* is targeted to the cytoplasm of infected HeLa cells. *Infect. Immun.* 66:5501–5507.

116. Wolff, C., Nisan, I., Hanski, E., *et al.* (1998). Protein translocation into host epithelial cells by infecting enteropathogenic *Escherichia coli. Mol. Microbiol.* 28:143–155.
117. Kenny, B., and Jepson, M. (2000). Targeting of an enteropathogenic *Escherichia coli* (EPEC) effector protein to host mitochondria. *Cell. Microbiol.* 2:579–590.
118. Donnenberg, M. S., Yu, J., and Kaper, J. B. (1993). A second chromosomal gene necessary for intimate attachment of enteropathogenic *Escherichia coli* to epithelial cells. *J. Bacteriol.* 175:4670–4680.
119. Foubister, V., Rosenshine, I., Donnenberg, M. S., and Finlay, B. B. (1994). The *eaeB* gene of enteropathogenic *Escherichia coli* is necessary for signal transduction in epithelial cells. *Infect. Immun.* 62:3038–3040.
120. Kenny, B., Lai, L.-C., Finlay, B. B., and Donnenberg, M. S. (1996). EspA, a protein secreted by enteropathogenic *Escherichia coli* (EPEC), is required to induce signals in epithelial cells. *Mol. Microbiol.* 20:313–323.
121. Abe, P., Kenny, B., Stein, M., and Finlay, B. B. (1997). Characterization of two virulence proteins secreted by rabbit enteropathogenic *Escherichia coli*, EspA and EspB, whose maximal expression is sensitive to host body temperature. *Infect. Immun.* 65:3547–3555.
122. Lai, L. C., Wainwright, L. A., Stone, K. D., and Donnenberg, M. S. (1997). A third secreted protein that is encoded by the enteropathogenic *Escherichia coli* pathogenicity island is required for transduction of signals and for attaching and effacing activities in host cells. *Infect. Immun.* 65:2211–2217.
123. Kenny, B., and Finlay, B. B. (1997). Intimin-dependent binding of enteropathogenic *Escherichia coli* to host cells triggers novel signaling events, including tyrosine phosphorylation of phospholipase C-gamma1. *Infect. Immun.* 65:2528–2536.
124. Abe, A., Heczko, U., Hegele, R. G., and Finlay, B. B. (1998). Two enteropathogenic *Escherichia coli* type III secreted proteins, EspA and EspB, are virulence factors. *J. Exp. Med.* 188:1907–1916.
125. Nougayréde, J. P., Marchés, O., Boury, M., *et al.* (1999). The long-term cytoskeletal rearrangement induced by rabbit enteropathogenic *Escherichia coli* is Esp dependent but intimin independent. *Mol. Microbiol.* 31:19–30.
126. Wachter, C., Beinke, C., Mattes, M., and Schmidt, M. A. (1999). Insertion of EspD into epithelial target cell membranes by infecting enteropathogenic *Escherichia coli. Mol. Microbiol.* 31:1695–1707.
127. Taylor, K. A., Luther, P. W., and Donnenberg, M. S. (1999). Expression of the EspB protein of enteropathogenic *Escherichia coli* within HeLa cells affects stress fibers and cellular morphology. *Infect. Immun.* 67:120–125.
128. Taylor, K. A., O'Connell, C. B., Thompson, R., and Donnenberg, M. S. (2001). The role of pyridoxal phosphate in the function of EspB, a protein secreted by enteropathogenic *Escherichia coli. FEBS Lett.* 488:55–58.
129. Tacket, C. O., Sztein, M. B., Losonsky, G., *et al.* (2000). Role of EspB in experimental human enteropathogenic *Escherichia coli* infection. *Infect. Immun.* 68:3689–3695.
130. McNamara, B. P., and Donnenberg, M. S. (1998). A novel proline-rich protein, EspF, is secreted from enteropathogenic *Escherichia coli* via the type III export pathway. *FEMS Microbiol. Lett.* 166:71–78.
131. Warawa, J., Finlay, B. B., and Kenny, B. (1999). Type III secretion-dependent hemolytic activity of enteropathogenic *Escherichia coli. Infect. Immun.* 67:5538–5540.
132. Elliott, S. J., Krejany, E. O., Mellies, J. L., *et al.* (2001). EspG, a novel type III secreted protein from enteropathogenic *E. coli* with similarities to VirA of *Shigella. Infect. Immun.* 69:4027–4033.
133. McNamara, B. P., Koutsouris, A., O'Connell, C. B., *et al.* (2001). Translocated EspF protein from enteropathogenic *Escherichia coli* disrupts host intestinal barrier function. *J. Clin. Invest.* 107:621–629.
134. Crane, J. K., McNamara, B. P., and Donnenberg, M. S. (2001). Role of EspF in host cell death induced by enteropathogenic *Escherichia coli. Cell. Microbiol.* 3:197–211.

135. Mellies, J. L., Navarro-Garcia, F., Okeke, I., *et al.* (2001). *espC* pathogenicity island of enteropathogenic *Escherichia coli* encodes an enterotoxin. *Infect. Immun.* 69:315–324.
136. Tobe, T., Tatsuno, I., Katayama, E., *et al.* (1999). A novel chromosomal locus of enteropathogenic *Escherichia coli* (EPEC), which encodes a *bfpT*-regulated chaperone-like protein, TrcA, involved in microcolony formation by EPEC. *Mol. Microbiol.* 33:741–752.
137. Donnenberg, M. S., Tacket, C. O., James, S. P., *et al.* (1993). The role of the *eaeA* gene in experimental enteropathogenic *Escherichia coli* infection. *J. Clin. Invest.* 92:1412–1417.
138. Marchés, O., Nougayréde, J. P., Boullier, S., *et al.* (2000). Role of Tir and intimin in the virulence of rabbit enteropathogenic *Escherichia coli* serotype O103:H2. *Infect. Immun.* 68:2171–2182.
139. Krejany, E. O., Grant, T. H., Bennett-Wood, V., *et al.* (2000). Contribution of plasmid-encoded fimbriae and intimin to capacity of rabbit-specific enteropathogenic *Escherichia coli* to attach to and colonize rabbit intestine. *Infect. Immun.* 68:6472–6477.
140. Jerse, A. E., and Kaper, J. B. (1991). The *eae* gene of enteropathogenic *Escherichia coli* encodes a 94-kilodalton membrane protein, the expression of which is influenced by the EAF plasmid. *Infect. Immun.* 59:4302–4309.
141. Donnenberg, M. S., and Kaper, J. B. (1991). Construction of an *eae* deletion mutant of enteropathogenic *Escherichia coli* by using a positive-selection suicide vector. *Infect. Immun.* 59:4310–4317.
142. Liu, H., Magoun, L., Luperchio, S., *et al.* (1999). The Tir-binding region of enterohaemorrhagic *Escherichia coli* intimin is sufficient to trigger actin condensation after bacterial-induced host cell signalling. *Mol. Microbiol.* 34:67–81.
143. De Grado, M., Abe, A., Gauthier, A., *et al.* (1999). Identification of the intimin-binding domain of Tir of enteropathogenic *Escherichia coli*. *Cell. Microbiol.* 1:7–17.
144. Hartland, E. L., Batchelor, M., Delahay, R. M., *et al.* (1999). Binding of intimin from enteropathogenic *Escherichia coli* to Tir and to host cells. *Mol. Microbiol.* 32:151–158.
145. Kenny, B. (1999). Phosphorylation of tyrosine 474 of the enteropathogenic *Escherichia coli* (EPEC) Tir receptor molecule is essential for actin nucleating activity and is preceded by additional host modifications. *Mol. Microbiol.* 31:1229–1241.
146. Abe, A., De Grado, M., Pfuetzner, R. A., *et al.* (1999). Enteropathogenic *Escherichia coli* translocated intimin receptor, Tir, requires a specific chaperone for stable secretion. *Mol. Microbiol.* 33:1162–1175.
147. Kenny, B., and Warawa, J. (2001). Enteropathogenic *Escherichia coli* (EPEC) Tir receptor molecule does not undergo full modification when introduced into host cells by EPEC-independent mechanisms. *Infect. Immun.* 69:1444–1453.
148. Tzipori, S., Gunzer, F., Donnenberg, M. S., *et al.* (1995). The role of the *eaeA* gene in diarrhea and neurological complications in a gnotobiotic piglet model of enterohemorrhagic *Escherichia coli* infection. *Infect. Immun.* 63:3621–3627.
149. Phillips, A. D., and Frankel, G. (2000). Intimin-mediated tissue specificity in enteropathogenic *Escherichia coli* interaction with human intestinal organ cultures. *J. Infect. Dis.* 181:1496–1500.
150. Reece, S., Simmons, C. P., Fitzhenry, R. J., *et al.* (2001). Site-directed mutagenesis of intimin alpha modulates intimin-mediated tissue tropism and host specificity. *Mol. Microbiol.* 40:86–98.
151. Kelly, G., Prasannan, S., Daniell, S., *et al.* (1999). Structure of the cell-adhesion fragment of intimin from enteropathogenic *Escherichia coli*. *Nature Struct. Biol.* 6:313–318.
152. Batchelor, M., Prasannan, S., Daniell, S., *et al.* (2000). Structural basis for recognition of the translocated intimin receptor (Tir) by intimin from enteropathogenic *Escherichia coli*. *EMBO J.* 19:2452–2464.
153. Luo, Y., Frey, E. A., Pfuetzner, R. A., *et al.* (2000). Crystal structure of enteropathogenic *Escherichia coli* intimin-receptor complex. *Nature* 405:1073–1077.
154. Frankel, G., Candy, D. C. A., Everest, P., and Dougan, G. (1994). Characterization of the C-terminal domains of intimin-like proteins of enteropathogenic and enterohemorrrhagic *Escherichia coli*, *Citrobacter freundii*, and *Hafnia alvei*. *Infect. Immun.* 62:1835–1842.

155. Frankel, G., Candy, D. C. A., Fabiani, E., *et al.* (1995). Molecular characterization of a carboxy-terminal eukaryotic-cell-binding domain of intimin from enteropathogenic *Escherichia coli*. *Infect. Immun.* 63:4323–4328.
156. An, H. Y., Fairbrother, J. M., Dubreuil, J. D., and Harel, J. (1997). Cloning and characterization of the *eae* gene from a dog attaching and effacing *Escherichia coli* strain 4221. *FEMS Microbiol. Lett.* 148:239–244.
157. Frankel, G., Lider, O., Hershkoviz, R., *et al.* (1996). The cell binding domain of intimin from enteropathogenic *Escherichia coli* binds to β_1 integrins. *J. Biol. Chem.* 271:20359–20364.
158. Yu, J., and Kaper, J. B. (1992). Cloning and characterization of the *eae* gene of enterohemorrhagic *Escherichia coli* O157:H7. *Mol. Microbiol.* 6:411–417.
159. Isberg, R. R., and Leong, J. M. (1990). Multiple β_1 chain integrins are receptors for Invasin, a protein that promotes bacterial penetration into mammalian cells. *Cell* 60:861–871.
160. Liu, H., Magoun, L., and Leong, J. M. (1999). β_1-Chain integrins are not essential for intimin-mediated host cell attachment and enteropathogenic *Escherichia coli*-induced actin condensation. *Infect. Immun.* 67:2045–2049.
161. Frankel, G., Phillips, A. D., Trabulsi, L. R., *et al.* (2001). Intimin and the host cell: Is it bound to end in Tir(s)? *Trends Microbiol.* 9:214–218.
162. Klapproth, J.-M., Donnenberg, M. S., Abraham, J. M., *et al.* (1995). Products of enteropathogenic *Escherichia coli* inhibit lymphocyte activation and lymphokine production. *Infect. Immun.* 63:2248–2254.
163. Klapproth, J.-M., Donnenberg, M. S., Abraham, J. M., and James, S. P. (1996). Products of enteropathogenic *E. coli* inhibit lymphokine production by gastrointestinal lymphocytes. *Am. J. Physiol. Gastrointest. Liver Physiol.* 271:G841–G848.
164. Malstrom, C., and James, S. (1998). Inhibition of murine splenic and mucosal lymphocyte function by enteric bacterial products. *Infect. Immun.* 66:3120–3127.
165. Klapproth, J.-M., Scaletsky, I. C. A., McNamara, B. P., *et al.* (2000). A large toxin from pathogenic *Escherichia coli* strains that inhibits lymphocyte activation. *Infect. Immun.* 68:2148–2155.
166. Nicholls, L., Grant, T. H., and Robins-Browne, R. M. (2000). Identification of a novel genetic locus that is required for in vitro adhesion of a clinical isolate of enterohaemorrhagic *Escherichia coli* to epithelial cells. *Mol. Microbiol.* 35:275–288.
167. Zhu, C., Agin, T. S., Elliott, S. J., *et al.* (2001). Complete nucleotide sequence and analysis of the locus of enterocyte effacement from rabbit diarrheagenic *Escherichia coli* RDEC-1. *Infect. Immun.* 69:2107–2115.
168. Burland, V., Shao, Y., Perna, N. T., *et al.* (1998). The complete DNA sequence and analysis of the large virulence plasmid of *Escherichia coli* O157:H7. *Nucleic Acids Res.* 26:4196–4204.
169. Makino, K., Ishii, K., Yasunaga, T., *et al.* (1998). Complete nucleotide sequences of 93-kb and 3.3-kb plasmids of an enterohemorrhagic *Escherichia coli* O157:H7 derived from Sakai outbreak. *DNA Res.* 5:1–9.
170. Henderson, I. R., and Nataro, J. P. (2001). Virulence functions of autotransporter proteins. *Infect. Immun.* 69:1231–1243.
171. Stein, M., Kenny, B., Stein, M. A., and Finlay, B. B. (1996). Characterization of EspC, a 110-kilodalton protein secreted by enteropathogenic *Escherichia coli* which is homologous to members of the IgA protease-like family of secreted proteins. *J. Bacteriol.* 178:6546–6554.
172. Savarino, S. J., McVeigh, A., Watson, J., *et al.* (1996). Enteroaggregative *Escherichia coli* heat-stable enterotoxin is not restricted to enteroaggregative *E. coli*. *J. Infect. Dis.* 173:1019–1022.
173. Scott, D. A., and Kaper, J. B. (1994). Cloning and sequencing of the genes encoding *Escherichia coli* cytolethal distending toxin. *Infect. Immun.* 62:244–251.
174. Lara-Tejero, M., and Galan, J. E. (2000). A bacterial toxin that controls cell cycle progression as a deoxyribonuclease I-like protein. *Science* 290:354–357.
175. Albert, M. J., Faruque, S. M., Faruque, A. S., *et al.* (1996). Controltoxin-producing *Escherichia coli* infections in Bangladeshi children. *J. Clin. Microbiol.* 34:717–719.

176. Bradley, D. E., and Thompson, C. R. (1992). Synthesis of unusual thick pili by *Escherichia coli* of EPEC serogroup O119. *FEMS Microbiol. Lett.* 94:31–36.
177. Girón, J. A., Ho, A. S. Y., and Schoolnik, G. K. (1993). Characterization of fimbriae produced by enteropathogenic *Escherichia coli. J. Bacteriol.* 175:7391–7403.
178. Karch, H., Heesemann, J., Laufs, R., *et al.* (1987). Serological response to type 1—like somatic fimbriae in diarrheal infection due to classical enteropathogenic *Escherichia coli. Microb. Pathog.* 2:425–434.
179. Elliott, S. J., and Kaper, J. B. (1997). Role of type 1 fimbriae in EPEC infections. *Microb. Pathog.* 23:113–118.
180. Elliott, S. J., Sperandio, V., Giron, J. A., *et al.* (2000). The locus of enterocyte effacement (LEE)—encoded regulator controls expression of both LEE- and non-LEE-encoded virulence factors in enteropathogenic and enterohemorrhagic *Escherichia coli. Infect. Immun.* 68:6115–6126.
181. da Silva, M. L., Mortara, R. A., Barros, H. C., *et al.* (1989). Aggregation of membrane-associated actin filaments following localized adherence of enteropathogenic *Escherichia coli* to HeLa cells. *J. Cell Sci.* 93:439–446.
182. Finlay, B. B., Rosenshine, I., Donnenberg, M. S., and Kaper, J. B. (1992). Cytoskeletal composition of attaching and effacing lesions associated with enteropathogenic *Escherichia coli* adherence to HeLa cells. *Infect. Immun.* 60:2541–2543.
183. Gauthier, A., De Grado, M., and Finlay, B. B. (2000). Mechanical fractionation reveals structural requirements for enteropathogenic *Escherichia coli.* Tir insertion into host membranes. *Infect. Immun.* 68:4344–4348.
184. Ben-Ami, G., Ozeri, V., Hanski, E., *et al.* (1998). Agents that inhibit Rho, Rac, and Cdc42 do not block formation of actin pedestals in HeLa cells infected with enteropathogenic *Escherichia coli. Infect. Immun.* 66:1755–1758.
185. Ebel, F., Von Eichel-Streiber, C., Rohde, M., and Chakraborty, T. (1998). Small GTP-binding proteins of the Rho- and Ras-subfamilies are not involved in the actin rearrangements induced by attaching and effacing *Escherichia coli. FEMS Microbiol. Lett.* 163:107–112.
186. Gruenheid, S., DeVinney, R., Bladt, F., *et al.* (2001). Enteropathogenic *E. coli.* Tir binds Nck to initiate actin pedestal formation in host cells. *Nature Cell Biol* 3:856–859.
187. Kalman, D., Weiner, O. D., Goosney, D. L., *et al.* (1999). Enteropathogenic *E. coli.* acts through WASP and Arp2/3 complex to form actin pedestals. *Nature Cell Biol.* 1:389–391.
188. Goosney, D. L., DeVinney, R., Pfuetzner, R. A., *et al.* (2000). Enteropathogenic *E. coli.* translocated intimin receptor, Tir, interacts directly with α-actinin. *Curr. Biol.* 10:735–738.
189. Cantarelli, V. V., Takahashi, A., Yanagihara, I., *et al.* (2001). Talin, a host cell protein, interacts directly with the translocated intimin receptor, Tir, of enteropathogenic *Escherichia coli*, and is essential for pedestal formation. *Cell Microbiol.* 3:745–751.
190. Cantarelli, V. V., Takahashi, A., Akeda, Y., *et al.* (2000). Interaction of enteropathogenic or enterohemorrhagic *Escherichia coli* with HeLa cells results in translocation of cortactin to the bacterial adherence site. *Infect. Immun.* 68:382–386.
191. Goosney, D. L., DeVinney, R., and Finlay, B. B. (2001). Recruitment of cytoskeletal and signaling proteins to enteropathogenic and enterohemorrhagic *Escherichia coli* pedestals. *Infect. Immun.* 69:3315–3322.
192. Baldwin, T. J., Ward, W., Aitken, A., *et al.* (1991). Elevation of intracellular free calcium levels in HEp-2 cells infected with enteropathogenic *Escherichia coli. Infect. Immun.* 59: 1599–1604.
193. Baldwin, T. J., Lee-Delaunay, M. B., Knutton, S., and Williams, P. H. (1993). Calcium-calmodulin dependence of actin accretion and lethality in cultured HEp-2 cells infected with enteropathogenic *Escherichia coli. Infect. Immun.* 61:760–763.
194. Dytoc, M., Fedorko, L., and Sherman, P. M. (1994). Signal transduction in human epithelial cells infected with attaching and effacing *Escherichia coli in vitro. Gastroenterology* 106:1150–1161.

195. Bain, C., Keller, R., Collington, G. K., *et al.* (1998). Increased levels of intracellular calcium are not required for the formation of attaching and effacing lesions by enteropathogenic and enterohemorrhagic *Escherichia coli. Infect. Immun.* 66:3900–3908.
196. Canil, C., Rosenshine, I., Ruschkowski, S., *et al.* (1993). Enteropathogenic *Escherichia coli* decreases the transepithelial electrical resistance of polarized epithelial monolayers. *Infect. Immun.* 61:2755–2762.
197. Spitz, J., Yuhan, R., Koutsouris, A., *et al.* (1995). Enteropathogenic *Escherichia coli* adherence to intestinal epithelial monolayers diminishes barrier function. *Am. J. Physiol. Gastrointest. Liver Physiol.* 268:G374–G379.
198. Philpott, D. J., McKay, D. M., Sherman, P. M., and Perdue, M. H. (1996). Infection of T84 cells with enteropathogenic *Escherichia coli* alters barrier and transport functions. *Am. J. Physiol. Gastrointest. Liver Physiol.* 270:G634–G645.
199. Manjarrez-Hernandez, H. A., Baldwin, T. J., Williams, P. H., *et al.* (1996). Phosphorylation of myosin light chain at distinct sites and its association with the cytoskeleton during enteropathogenic *Escherichia coli* infection. *Infect. Immun.* 64:2368–2370.
200. Baldwin, T. J., Brooks, S. F., Knutton, S., *et al.* (1990). Protein phosphorylation by protein kinase C in HEp-2 cells infected with enteropathogenic *Escherichia coli* [published erratum appears in Infect. Immun. 1990;58(6):2024]. *Infect. Immun.* 58:761–765.
201. Crane, J. K., and Oh, J. S. (1997). Activation of host cell protein kinase C by enteropathogenic *Escherichia coli. Infect. Immun.* 65:3277–3285.
202. Yuhan, R., Koutsouris, A., Savkovic, S. D., and Hecht, G. (1997). Enteropathogenic *Escherichia coli*—induced myosin light chain phosphorylation alters intestinal epithelial permeability. *Gastroenterology* 113:1873–1882.
203. Simonovic, I., Arpin, M., Koutsouris, A., *et al.* (2001). Enteropathogenic *Escherichia coli* activates ezrin, which participates in disruption of tight junction barrier function. *Infect. Immun.* 69:5679–5688.
204. Simonovic, I., Rosenberg, J., Koutsouris, A., and Hecht, G. (2000). Enteropathogenic *Escherichia coli* dephosphorylates and dissociates occludin from intestinal epithelial tight juctions. *Cell. Microbiol.* 2:305–315.
205. Stein, M. A., Mathers, D. A., Yan, H., *et al.* (1996). Enteropathogenic *Escherichia coli* (EPEC) markedly decreases the resting membrane potential of Caco-2 and HeLa human epithelial cells. *Infect. Immun.* 64:4820–4825.
206. Collington, G. K., Booth, I. W., and Knutton, S. (1998). Rapid modulation of electrolyte transport in Caco-2 cell monolayers by enteropathogenic *Escherichia coli* (EPEC) infection. *Gut* 42:200–207.
207. Collington, G. K., Booth, I. W., Donnenberg, M. S., *et al.* (1998). Enteropathogenic *Escherichia coli* virulence genes encoding secreted signalling proteins are essential for modulation of Caco-2 cell electrolyte transport. *Infect. Immun.* 66:6049–6053.
208. Hecht, G., and Koutsouris, A. (1999). Enteropathogenic *E. coli* attenuates secretagogue-induced net intestinal ion transport but not Cl^- secretion. *Am. J. Physiol. Gastrointest. Liver Physiol.* 276:G781–G788.
209. Savkovic, S. D., Koutsouris, A., and Hecht, G. (1997). Activation of NF-kappaB in intestinal epithelial cells by enteropathogenic *Escherichia coli. Am. J. Physiol.* 273:C1160–C1167.
210. Hecht, G., Marrero, J. A., Danilkovich, A., *et al.* (1999). Pathogenic *Escherichia coli* increase Cl^- secretion from intestinal epithelia by upregulating galanin-1 receptor expression. *J. Clin. Invest.* 104:253–262.
211. Crane, J. K., Majumdar, S., and Pickhardt, D. F., III (1999). Host cell death due to enteropathogenic *Escherichia coli* has features of apoptosis. *Infect. Immun.* 67:2575–2584.
212. Heczko, U., Carthy, C. M., O'Brien, B. A., and Finlay, B. B. (2001). Decreased apoptosis in the ileum and ileal Peyer's patches: A feature after infection with rabbit enteropathogenic *Escherichia coli* O103. *Infect. Immun.* 69:4580–4589.

213. Czerucka, D., Dahan, S., Mograbi, B., *et al.* (2001). Implication of mitogen-activated protein kinases in T84 cell responses to enteropathogenic *Escherichia coli* infection. *Infect. Immun.* 69:1298–1305.
214. Higgins, L. M., Frankel, G., Connerton, I., *et al.* (1999). Role of bacterial intimin in colonic hyperplasia and inflammation. *Science* 285:588–591.
215. Savkovic, S. D., Koutsouris, A., and Hecht, G. (1996). Attachment of a noninvasive enteric pathogen, enteropathogenic *Escherichia coli*, to cultured human intestinal epithelial monolayers induces transmigration of neutrophils. *Infect. Immun.* 64:4480–4487.
216. Czerucka, D., Dahan, S., Mograbi, B., *et al.* (2000). *Saccharomyces boulardii* preserves the barrier function and modulates the signal transduction pathway induced in enteropathogenic *Escherichia coli*—infected T84 cells. *Infect. Immun.* 68:5998–6004.
217. Steiner, T. S., Nataro, J. P., Poteet-Smith, C. E., *et al.* (2000). Enteroaggregative *Escherichia coli* expresses a novel flagellin that causes IL-8 release from intestinal epithelial cells. *J. Clin. Invest* 105:1769–1777.
218. Gewirtz, A. T., Simon, P. O., Jr., Schmitt, C. K., *et al.* (2001). *Salmonella typhimurium* translocates flagellin across intestinal epithelia, inducing a proinflammatory response. *J. Clin. Invest* 107:99–109.
219. Inman, L. R., and Cantey, J. R. (1983). Specific adherence of *Escherichia coli* (strain RDEC-1) to membranous (M) cells of the Peyer's patch in Escherichia coli diarrhea in the rabbit. *J. Clin. Invest* 71:1–8.
220. Goosney, D. L., Celli, J., Kenny, B., and Finlay, B. B. (1999). Enteropathogenic *Escherichia coli* inhibits phagocytosis. *Infect. Immun.* 67:490–495.
221. Celli, J., Olivier, M., and Finlay, B. B. (2001). Enteropathogenic *Escherichia coli* mediates antiphagocytosis through the inhibition of PI_3-kinase-dependent pathways. *EMBO J.* 20:1245–1258.
222. Puente, J. L., Bieber, D., Ramer, S. W., *et al.* (1996). The bundle-forming pili of enteropathogenic *Escherichia coli*: Transcriptional regulation by environmental signals. *Mol. Microbiol.* 20:87–100.
223. Rosenshine, I., Ruschkowski, S., and Finlay, B. B. (1996). Expression of attaching effacing activity by enteropathogenic *Escherichia coli* depends on growth phase, temperature, and protein synthesis upon contact with epithelial cells. *Infect. Immun.* 64:966–973.
224. Kenny, B., Abe, A., Stein, M., and Finlay, B. B. (1997). Enteropathogenic *Escherichia coli* protein secretion is induced in response to conditions similar to those in the gastrointestinal tract. *Infect. Immun.* 65:2606–2612.
225. Donnenberg, M. S., and Kaper, J. B. (1992). Minireview: Enteropathogenic *Escherichia coli. Infect. Immun.* 60:3953–3961.
226. Donnenberg, M. S., Kaper, J. B., and Finlay, B. B. (1997). Interactions between enteropathogenic *Escherichia coli* and host epithelial cells. *Trends Microbiol.* 5:109–114.
227. Monteiro-Neto, V., and Trabulsi, L. R. (1999). Phenylpropionic acid metabolism: a marker for enteropathogenic *Escherichia coli* clonal group 2 strains. *J. Clin. Microbiol.* 37:2121–2121.
228. Cravioto, A., Gross, R. J., Scotland, S. M., and Rowe, B. (1979). An adhesive factor found in strains of *Escherichia coli* belonging to the traditional infantile enteropathogenic serotypes. *Current Microbiol.* 3:95–99.
229. Moyenuddin, M., Wachsmuth, I. K., Moseley, S. L., *et al.* (1989). Serotype, antimicrobial resistance, and adherence properties of *Escherichia coli* strains associated with outbreaks of diarrheal illness in children in the United States. *J. Clin. Microbiol.* 27:2234–2239.
230. Levine, M. M., Prado, V., Robins-Browne, R., *et al.* (1988). Use of DNA probes and HEp-2 cell adherence assay to detect diarrheagenic *Escherichia coli. J. Infect. Dis.* 158:224–228.
231. Gomes, T. A. T., Blake, P. A., and Trabulsi, L. R. (1989). Prevalence of *Escherichia coli* strains with localized, diffuse, and aggregative adherence to HeLa cells in infants with diarrhea and matched controls. *J. Clin. Microbiol.* 27:266–269.

232. Cravioto, A., Tello, A., Navarro, A., *et al.* (1991). Association of *Escherichia coli* HEp-2 adherence patterns with type and duration of diarrhoea. *Lancet* 337:262–264.
233. Echeverria, P., Serichantalerg, O., Changchawalit, S., *et al.* (1992). Tissue culture-adherent *Escherichia coli* in infantile diarrhea. *J. Infect. Dis.* 165:141–143.
234. Begaud, E., Jourand, P., Morillon, M., *et al.* (1993). Detection of diarrheogenic *Escherichia coli* in children less than ten years old with and without diarrhea in New Caledonia using seven acetylaminofluorene-labeled DNA probes. *Am. J. Trop. Med. Hyg.* 48:26–34.
235. Knutton, S., Baldwin, T., Williams, P. H., and McNeish, A. S. (1989). Actin accumulation at sites of bacterial adhesion to tissue culture cells: basis of a new diagnostic test for enteropathogenic and enterohemorrhagic *Escherichia coli*. *Infect. Immun.* 57:1290–1298.
236. Ismaili, A., Philpott, D. J., Dytoc, M. T., and Sherman, P. M. (1995). Signal transduction responses following adhesion of verocytotoxin-producing *Escherichia coli*. *Infect. Immun.* 63:3316–3326.
237. Nataro, J. P., Baldini, M. M., Kaper, J. B., *et al.* (1985). Detection of an adherence factor of enteropathogenic *Escherichia coli* with a DNA probe. *J. Infect. Dis.* 152:560–565.
238. Baldini, M. M., Nataro, J. P., and Kaper, J. B. (1986). Localization of a determinant for HEp-2 adherence by enteropathogenic *Escherichia coli*. *Infect. Immun.* 52:334–336.
239. Jerse, A. E., Martin, W. C., Galen, J. E., and Kaper, J. B. (1990). Oligonucleotide probe for detection of the enteropathogenic *Escherichia coli* (EPEC) adherence factor of localized adherent EPEC. *J. Clin. Microbiol.* 28:2842–2844.
240. Franke, J., Franke, S., Schmidt, H., *et al.* (1994). Nucleotide sequence analysis of enteropathogenic *Escherichia coli* (EPEC) adherence factor probe and development of PCR for rapid detection of EPEC harboring virulence plasmids. *J. Clin. Microbiol.* 32:2460–2463.
241. Cravioto, A., Reyes, R. E., Trujillo, F., *et al.* (1990). Risk of diarrhea during the first year of life associated with initial and subsequent colonization by specific enteropathogens. *Am. J. Epidemiol.* 131:886–904.
242. Echeverria, P., Orskov, F., Orskov, I., *et al.* (1991). Attaching and effacing enteropathogenic *Escherichia coli.* as a cause of infantile diarrhea in Bangkok. *J. Infect. Dis.* 164:550–554.
243. Scotland, S. M., Smith, H. R., and Rowe, B. (1991). *Escherichia coli* O128 strains from infants with diarrhea commonly show localized adhesion and positivity in the fluorescent-actin staining test but do not hybridize with an enteropathogenic *E. coli* adherence factor probe. *Infect. Immun.* 59:1569–1571.
244. Scotland, S. M., Smith, H. R., Said, B., *et al.* (1991). Identification of enteropathogenic *Escherichia coli* isolated in Britain as enteroaggregative or as members of a subclass of attaching-and-effacing *E. coli.* not hybridising with the EPEC adherence-factor probe. *J. Med. Microbiol.* 35:278–283.
245. Jerse, A. E., Gicquelais, K. G., and Kaper, J. B. (1991). Plasmid and chromosomal elements involved in the pathogenesis of attaching and effacing *Escherichia coli*. *Infect. Immun.* 59:3869–3875.
246. Beebakhee, G., Louie, M., De Azavedo, J., and Brunton, J. (1992). Cloning and nucleotide sequence of the *eae* gene homologue from enterohemorrhagic *Escherichia coli* serotype O157: H7. *FEMS Microbiol. Lett.* 91:63–68.
247. Agin, T. S., Cantey, J. R., Boedeker, E. C., and Wolf, M. K. (1996). Characterization of the *eaeA* gene from rabbit enteropathogenic *Escherichia coli* strain RDEC-1 and comparison to other *eaeA* genes from bacteria that cause attaching-effacing lesions. *FEMS Microbiol. Lett.* 144:249–258.
248. Agin, T. S., and Wolf, M. K. (1997). Identification of a family of intimins common to *Escherichia coli* causing attaching-effacing lesions in rabbits, humans and swine. *Infect. Immun.* 65:320–326.
249. Adu-Bobie, J., Trabulsi, L. R., Carneiro-Sampaio, M. M. S., *et al.* (1998). Identification of immunodominant regions within the C-terminal cell binding domain of intimin α and intimin β from enteropathogenic *Escherichia coli*. *Infect. Immun.* 66:5643–5649.

250. Voss, E., Paton, A. W., Manning, P. A., and Paton, J. C. (1998). Molecular analysis of shiga toxigenic *Escherichia coli* O111:H- proteins which react with sera from patients with hemolytic-uremic syndrome. *Infect. Immun.* 66:1467–1472.
251. McGraw, E. A., Li, J., Selander, R. K., and Whittam, T. S. (1999). Molecular evolution and mosaic structure of α, β, and γ intimins of pathogenic *Escherichia coli*. *Mol. Biol. Evol.* 16:12–22.
252. Oswald, E., Schmidt, H., Morabito, S., *et al.* (2000). Typing of intimin genes in human and animal enterohemorrhagic and enteropathogenic *Escherichia coli*: characterization of a new intimin variant. *Infect. Immun.* 68:64–71.
253. Batchelor, M., Knutton, S., Caprioli, A., *et al.* (1999). Development of a universal intimin antiserum and PCR primers. *J. Clin. Microbiol.* 37:3822–3827.
254. Wieler, L. H.,Vieler, E., Erpenstein, C., *et al.* (1996). Shiga toxin-producing *Escherichia coli* strains from bovines: Association of adhesion with carriage of *eae* and other genes. *J. Clin. Microbiol.* 34:2980–2984.
255. Adu-Bobie, J., Frankel, G., Bain, C., *et al.* (1998). Detection of intimins alpha, beta, gamma, and delta, four intimin derivatives expressed by attaching and effacing microbial pathogens. *J. Clin. Microbiol.* 36:662–668.
256. Girón, J. A., Donnenberg, M. S., Martin, W. C., *et al.* (1993). Distribution of the bundle-forming pilus structural gene (*bfpA*) among enteropathogenic *Escherichia coli*. *J. Infect. Dis.* 168:1037–1041.
257. Gunzburg, S. T., Tornieporth, N. G., and Riley, L. W. (1995). Identification of enteropathogenic *Escherichia coli* by PCR-based detection of the bundle-forming pilus gene. *J. Clin. Microbiol.* 33:1375–1377.
258. Tornieporth, N. G., John, J., Salgado, K., *et al.* (1995). Differentiation of pathogenic *Escherichia coli* strains in Brazilian children by PCR. *J. Clin. Microbiol.* 33:1371–1374.
259. Nagayama, K., Bi, Z., Oguchi, T., *et al.* (1996). Use of an alkaline phosphatase-conjugated oligonucleotide probe for the gene encoding the bundle-forming pilus of enteropathogenic *Escherichia coli*. *J. Clin. Microbiol.* 34:2819–2821.
260. Tsukamoto, T. (1996). [PCR methods for detection of enteropathogenic *Escherichia coli* (localized adherence) and enteroaggregative *Escherichia coli*]. *Kansenshogaku Zasshi.* 70:569–573.
261. Bortolini, M. R., Trabulsi, L. R., Keller, R., *et al.* (1999). Lack of expression of bundle-forming pili in some clinical isolates of enteropathogenic *Escherichia coli* (EPEC) is due to a conserved large deletion in the *bfp* operon. *FEMS Microbiol. Lett.* 179:169–174.
262. Donnenberg, M. S., Tacket, C. O., Losonsky, G., *et al.* (1998). Effect of prior experimental human enteropathogenic *Escherichia coli* infection on illness following homologous and heterologous rechallenge. *Infect. Immun.* 66:52–58.
263. Martinez, M. B., Taddei, C. R., Ruiz-Tagle, A., *et al.* (1999). Antibody response of children with enteropathogenic *Escherichia coli* infection to the bundle-forming pilus and locus of enterocyte effacement-encoded virulence determinants. *J. Infect. Dis.* 179:269–274.
264. Parissi-Crivelli, A., Parissi-Crivelli, J. M., and Girón, J. A. (2000). Recognition of enteropathogenic *Escherichia coli* virulence determinants by human colostrum and serum antibodies. *J. Clin. Microbiol.* 38:2696–2700.
265. Sanches, M. I., Keller, R., Hartland, E. L., *et al.* (2000). Human colostrum and serum contain antibodies reactive to the intimin- binding region of the enteropathogenic *Escherichia coli* translocated intimin receptor. *J. Pediatr. Gastroenterol. Nutr.* 30:73–77.
266. Levine, M. M. (1996). Vaccines against enteropathogenic *Escherichia coli*. *Rev. Microbiol. Sao. Paulo.* 27:126–129.
267. Schriefer, A., Maltez, J. R., Silva, N., *et al.* (1999). Expression of a pilin subunit BfpA of the bundle-forming pilus of enteropathogenic *Escherichia coli* in an *aroA* live salmonella vaccine strain. *Vaccine* 17:770–778.
268. Loureiro, I., Frankel, G., Adu-Bobie, J., *et al.* (1998). Human colostrum contains IgA antibodies reactive to enteropathogenic *Escherichia coli* virulence-associated proteins: intimin, BfpA, EspA, and EspB. *J. Pediatr. Gastroenterol. Nutr.* 27:166–171.

269. Manjarrez-Hernandez, H. A., Gavilanes-Parra, S., Chavez-Berrocal, E., *et al.* (2000). Antigen detection in enteropathogenic *Escherichia coli* using secretory immunoglobulin A antibodies isolated from human breast milk. *Infect. Immun.* 68:5030–5036.
270. Camara, L. M., Carbonare, S. B., Silva, M. L. M., and Carneiro-Sampaio, M. M. S. (1994). Inhibition of enteropathogenic *Escherichia coli* (EPEC) adhesion to HeLa cells by human colostrum: Detection of specific sIgA related to EPEC outer-membrane proteins. *Int. Arch. Allergy Immunol.* 103:307–310.
271. Delneri, M. T., Carbonare, S. B., Silva, M. L., *et al.* (1997). Inhibition of enteropathogenic *Escherichia coli* adhesion to HEp-2 cells by colostrum and milk from mothers delivering low-birth-weight neonates. *Eur. J. Pediatr.* 156:493–498.
272. Figueroa-Quintanilla, D., Salazar-Lindo, E., Sack, R. B., *et al.* (1993). A controlled trial of bismuth subsalicylate in infants with acute watery diarrheal disease. *New Engl. J. Med.* 328:1653–1658.
273. Salazar-Lindo, E., Santisteban-Ponce, J., Chea-Woo, E., and Gutierrez, M. (2000). Racecadotril in the treatment of acute watery diarrhea in children. *New Engl. J. Med.* 343:463–467.
274. Buret, A., Olson, M. E., Gall, D. G., and Hardin, J. A. (1998). Effects of orally administered epidermal growth factor on enteropathogenic *Escherichia coli.* infection in rabbits. *Infect. Immun.* 66:4917–4923.
275. Mietens, C., Keinhorst, H., Hilpert, H., *et al.* (1979). Treatment of infantile *E. coli.* gastroenteritis with specific bovine anti-*E. coli.* milk immunoglobulins. *Eur. J. Pediatr.* 132:239–252.
276. Casswall, T. H., Sarker, S. A., Faruque, S. M., *et al.* (2000). Treatment of enterotoxigenic and enteropathogenic *Escherichia coli*—induced diarrhoea in children with bovine immunoglobulin milk concentrate from hyperimmunized cows: A double-blind, placebo- controlled, clinical trial. *Scand. J. Gastroenterol.* 35:711–718.
277. Thorén, A., Wolde-Mariam, T., Stintzing, G., *et al.* (1980). Antibiotics in the treatment of gastroenteritis caused by enteropathogenic *Escherichia coli. J. Infect. Dis.* 141:27–31.
278. Gomes, T. A. T., Rassi, V., Macdonald, K. L., *et al.* (1991). Enteropathogens associated with acute diarrheal disease in urban infants in São Paulo, Brazil. *J. Infect. Dis.* 164:331–337.
279. Senerwa, D., Mutanda, L. N., Gathuma, J. M., and Olsvik, O. (1991). Antimicrobial resistance of enteropathogenic *Escherichia coli* strains from a nosocomial outbreak in Kenya. *APMIS* 99:728–734.
280. Torres, M. E., Pirez, M. C., Schelotto, F., *et al.* (2001). Etiology of children's diarrhea in Montevideo, Uruguay: Associated pathogens and unusual isolates. *J. Clin. Microbiol.* 39:2134–2139.

CHAPTER 4

Enterohemorrhagic and Other Shiga Toxin–Producing Escherichia coli

Cheleste M. Thorpe
Division of Geographic Medicine and Infectious Diseases, Tufts-New England Medical Center, Boston, Massachusetts

Jennifer M. Ritchie
Division of Geographic Medicine and Infectious Diseases, Tufts-New England Medical Center, Boston, Massachusetts

David W. K. Acheson
Department of Epidemiology and Preventive Medicine, University of Maryland School of Medicine, Baltimore, Maryland

BACKGROUND

History

Following an epidemic of severe dysentery in Japan in the late 1800s, the now-famous microbiologist Kiyoshi Shiga first described the causative agent of this outbreak [1]. Later, Shiga's bacillus would be given the name *Shigella dysenteriae*, and from this bacillus, Shiga toxin would be first described [2,3]. However, it was not until the late 1970s that *S. dysenteriae* was associated with the clinical condition known as *hemolytic-uremic syndrome* (HUS), a triad of renal failure, thrombocytopenia, and hemolytic anemia [4]. Around this time, it was discovered that certain strains of *Escherichia coli* could produce Shiga toxins.

The discovery of Shiga toxin–producing *E. coli* (STEC) began when Konowalchuk and colleagues reported that culture filtrates of several different strains of *E. coli* were cytotoxic for Vero cells [5]. This cytolethal activity was heat labile but not neutralized by antiserum to the classic cholera-like *E. coli* heat-labile enterotoxin. Subsequently, analysis of 136 *E. coli* strains revealed that 10 of

Escherichia coli: Virulence Mechanisms of a Versatile Pathogen
ISBN 0-12-220751-3

these strains were capable of producing this Vero cell cytotoxin [6]. Following these observations, Wade and colleagues [7] in England noted the presence of cytotoxin-producing *E. coli* O26 strains in association with bloody diarrhea. Similar clinical observations followed from other parts of the United Kingdom and New Zealand [8,9].

In 1983, an association between a then-rare serotype of *E. coli* (O157:H7) and the development of hemorrhagic colitis (HC) was reported [10]. In the same year, Karmali and colleagues made the association between the presence of Shiga toxins produced by *E. coli* and the development of HUS [11]. These observations initiated a widespread interest in STEC. Since then, over 200 different types of *E. coli* have been reported to make Shiga toxins, many of which have been associated with human disease [12].

STEC Diversity

It is now appreciated that STEC are a heterogeneous group of organisms linked by a single feature: the ability to produce Shiga toxins. Shiga toxins are considered to be the principal virulence factors involved in the thrombotic microangiopathic disorders associated with STEC infection, but the precise mechanisms by which Shiga toxins cause these disorders remain unclear. The effects of Shiga toxin on various cell types thought to be involved in HUS pathogenesis will be discussed later in detail.

Aside from the production of Shiga toxins, STEC are known to be diverse with respect to other known virulence determinants and are thought to be diverse in their capacity to cause disease. There is some overlap between certain types of Shiga toxin–producing *E. coli* and enteropathogenic *E. coli*. STEC, by definition, are Shiga toxin–positive, but within this group there is a subgroup known as enterohemorrhagic *E. coli* (EHEC) that, like EPEC, have a locus for enterocyte effacement (LEE) and contain a large plasmid. Whittam has attempted to classify STEC into four groups based on clonal analysis [13]. The major features of these groups are summarized in Table I. EHEC 1 contains organisms of the O157:H7 serotype and closely related O55 serotypes. EHEC 1 organisms commonly are associated with both sporadic cases and human outbreaks. This

TABLE I The Diversity of Shiga Toxin–Producing *E. coli*

Group name	Common serotypes	LEE status	Clinical features
EHEC 1	O157:H7, O55	Positive	Human outbreaks and sporadic cases
EHEC 2	O111, O26, etc.	Positive	Human outbreaks and sporadic cases
STEC 1	Many different	Negative	Occasional sporadic cases
STEC 2	Many different	Negative	Not (yet) associated with disease

group bears the LEE pathogenicity island, or LEE region. We discuss the role of this pathogenicity island later in detail. EHEC 2 is comprised of the non-O157 serotypes, such as O111 and O26, that are LEE-positive, and these organisms are also commonly associated with disease. The STEC 1 group contains many different serotypes; STEC 1 members are LEE-negative, and these strains occasionally are associated with sporadic cases of human disease, including HUS. We have relegated all other STEC to the STEC 2 group, consisting of a divergent group of organisms that includes many of the STEC that have been found in meat and animals and have not yet been associated with human disease. Although imperfect, this classification provides a descriptive framework that allows one to appreciate the diversity of the STEC family.

Nomenclature

Following the original description of the cytotoxic effects of Shiga toxins on Vero cells, these toxins were given the name *verotoxins*, and Shiga toxin–producing *E. coli* are also referred to as *verotoxin-producing E. coli* (VTEC). Others used the terminology *Shiga-like toxins* (SLTs) to refer to the newly described *E. coli* cytotoxins with similarity to Shiga toxin from *S. dysenteriae*. However, over the decade following the discovery of STEC, it was appreciated that Shiga toxins shared a common mechanism of action. In 1996, an international group of investigators decided simply to call this group of biologically homogeneous toxins, irrespective of their bacterial origin, the *Shiga toxins* (Stx) after the original description nearly 100 years ago [14]. The gene designation (*stx*) was already well established, and the new nomenclature has maintained the *stx* gene designation. Currently, there are seven members of this family, as shown in Table II. The toxins are divided into two main groups: Shiga toxin from *S. dysenteriae* type 1 and Shiga toxin 1 form one group, and the Shiga toxin 2 family forms the other group. The characteristics of these proteins are discussed later.

TABLE II The Shiga Toxin Family

Name	Gene	Protein	Receptor	Comments
Shiga toxin	*stx*	Stx	Gb3	From *S. dysenteriae*
Shiga toxin 1[a]	stx_1	Stx1	Gb3	Virtually identical to Stx
Shiga toxin 2[a]	stx_2	Stx2	Gb3	Immunologically distinct from Stx1
Shiga toxin 2c[a]	stx_{2c}	Stx2c	Gb3	Nearly identical to Stx2
Shiga toxin 2d[a]	stx_{2d}	Stx2d	Gb3	Mucus activatable
Shiga toxin 2e[a]	stx_{2e}	Stx2e	Gb4	Associated with porcine edema disease
Shiga toxin 2f[a]	stx_{2f}	Stx2f	?	Associated with feral pigeons

[a]From *Escherichia coli*.

EPIDEMIOLOGY OF STEC

Serotype Distribution

Since the first associations between *E. coli* O157:H7 and thrombotic microangiopathy [10,11], the incidence of disease attributed to O157:H7 in particular and STEC in general has increased [15–17]. In many parts of the world, *E. coli* O157:H7 is the predominant STEC reported as causing HUS in humans [18]. One prospective study reports that throughout the United States, STEC could be implicated in 72% of cases of HUS and that *E. coli* O157:H7 was the likely cause in more than 80% of patients [19]. While this study was conducted over a 3-year period from 1987 to 1991, more recent data suggest that *E. coli* O157:H7 is still the predominant cause of HUS [20]. Additionally, serotype O157:H7 appears to be responsible for the largest outbreaks, including the 1996 outbreak in Sakai, Japan, with over 8000 cases traced to the consumption of contaminated radish sprouts [21].

The inability of the O157:H7 serotype to ferment sorbitol has facilitated the differentiation and hence detection of this serotype from other *E. coli* in routine testing, perhaps leading to an overestimation of the relative importance of this serotype in the world of STEC. In recent years, there has been growing awareness for the role of non-O157 serogroups as a cause of human disease. Standardized tests to screen for all STEC by methods that do not rely on the detection of a single serotype but rather assess the ability of an *E. coli* isolate to produce Shiga toxins are necessary. These methods include polymerase chain reaction (PCR) detection for *stx* genes and direct detection of toxins in feces and will provide a truer picture of the distribution and importance of the different serotypes in human disease [22]. Several reports have documented the isolation of multiple serotypes from a single patient [23–25]. In such cases, the contribution of an individual serotype to the pathogenesis of disease will be difficult to determine [16]. There is also the possibility that *in vivo* transduction of *stx* genes in the host is the explanation for isolation of STEC of different serotypes from the same host at the same time, rather than someone being infected with two strains simultaneously.

A number of studies have implicated STEC of many different O serogroups as human pathogens [16,26]. Among these non-O157 serogroups, O6, O26, O91, O103, O111, O113, and OX3 have been associated with HC and HUS in humans [24,27–30]. The frequency of disease associated with a particular serotype varies from country to country [31]. In contrast to North America, Japan, and the British Isles, serogroup O157 is relatively rare in Australia. There, O111 strains predominate as the most common serogroup associated with sporadic HUS [32,33]. Similarly, in Argentina, non-O157 STEC have been reported to be isolated more frequently from patients with HUS than O157:H7 [34]. Interestingly, Argentina has the highest incidence of HUS in children (<4 years of age) in the world, amounting to 22 cases per 100,000 children per year (compared with typically 2 to 4 cases per 100,000 children per year in the United States, United Kingdom, or Canada). The widespread consumption of undercooked meat by children in infancy (beginning around 3 months of age) may play a role in explaining these findings [34].

Over the last 10 years, sorbitol-fermenting O157:H negative isolates have emerged in Germany as an important cause of HUS and diarrhea [35,36]. These isolates have now been found elsewhere in central Europe [37]. Cattle have been implicated as the reservoir for this organism [38]. While it was generally thought that non-O157 isolates were less capable of causing severe symptoms than O157 serogroups, the increasing incidence of severe disease attributed to non-O157 isolates challenges this idea.

The global prevalence of STEC infection is difficult to quantify due to a lack of uniform surveillance and reporting systems. Within the United States, some studies have suggested that *E. coli* O157:H7 infection is more common in northern compared with southern states [31]. Similarly, the distribution of O157:H7 infection in the United Kingdom also indicates a greater incidence in the northern regions [39]. In both northern and southern hemispheres, human infection with *E. coli* O157:H7 has been found to be more common in the warm summer months [40,41].

Sources of STEC

In the environment, the main reservoir of STEC appears to be domestic animals, especially ruminants such as cattle, sheep, deer, and goats [42–44]. The proportion of STEC carriers in cattle herds varies greatly from between 1 and 80%. Most epidemiologic studies have assessed cattle, where more than 100 different O:H types of STEC were detected in animals spread over geographically diverse locations. Cattle surveys in North America and the United Kingdom consistently have shown that around 4% of herds are positive for O157:H7 at any given time [15]. However, following a review of several epidemiologic studies in the United States and Europe, Meyer-Broseta and colleagues concluded that they could not draw any firm conclusions as to the effects of geographic location and season on the prevalence of STEC in cattle [45]. The shedding of STEC from individual animals in feces is transient and episodic [46] and affected by diet, age, feeding, and levels of stress [47]. Calves at weaning have been identified as the cattle group most likely to be shedding STEC [48].

STEC are shed into the environment in animal feces and have been reported to survive for a number of months in the soil environment and as long as 21 months in manure [49,50]. Numerous outbreaks separated temporally and geographically in Scotland can be attributed to one recurrent clonal type of *E. coli* O157:H7 named the *West Lothian clonal type* [51]. Repeated isolation of this clone from animals, food, the environment, and humans over the last 5 years highlights the persistence of this organism over time. *E. coli* O157:H7 has been found to survive for considerable periods of time in water [52], and some reports have documented isolation of STEC in cattle water troughs and natural water supplies [53,54]. Contamination of a private water supply was directly responsible for subsequent human infection in one outbreak in a rural area of Scotland [55]. Similarly, a large O157:H7 outbreak occurred in Walkerton, Canada, in 2000, when the municipal water supply became contaminated [56].

It is not yet clear if all STEC occurring in animals are pathogenic for humans. The presence of other potential virulence markers, such as colonization factors or hemolysin production, has been found in a proportion of isolates obtained from animal sources [57–59]. The precise combination of virulence genes required for STEC infection in humans is not completely understood, and recently, at least one HUS cluster has been attributed to an *eae*-negative STEC O113 strain [60].

Transmission Routes of STEC

Human infections with STEC usually are linked to the consumption of contaminated and undercooked beef, unpasteurized milk, or feces-contaminated vegetables, water, or apple cider [16,52,61,62]. Jones reviewed the potential risks associated with the persistence of *E. coli* O157:H7 in agricultural environments [47]. At the farm level, control measures may include restricting the application of animal fecal wastes to crops used for direct human consumption, as well as employing measures that improve animal welfare. Secondary person-to-person spread has been well documented during outbreaks, particularly in institutional settings [63,64]. Given the low infectious dose of some O157:H7 strains (<100 organisms), this type of spread is not unexpected.

Role of Shiga Toxin Type in Disease

There is mounting evidence supporting the hypothesis that Stx2-producing strains are potentially more virulent than strains that produce Stx1 only or that produce both Stx1 and Stx2. In humans, epidemiologic data suggest that *E. coli* O157:H7 strains that express Stx2 are more important than Stx1 in the development of HUS and may result in increased disease severity [26,31,65–68]. Interestingly, the most common toxin profiles of STEC strains in the United States are those which make both Stx1 and Stx2, contrasting with the predominance of Stx2-only producers in the United Kingdom [69]. In one study in France, all non-O157 serotypes associated with HUS contained the Stx2-encoding gene only [27].

CLINICAL MANIFESTATIONS OF STEC INFECTION

It is through epidemiologic data that we know that Shiga toxins from *E. coli* are involved in the pathogenesis of thrombotic microangiopathy (TMA). Human challenge studies are unethical to perform due of the severity of the disease. Unfortunately, there is no well-characterized animal model of STEC infection that perfectly reproduces the clinicopathologic features of STEC-associated HUS. Animal models that reproduce some of the manifestations of STEC infection are reviewed in ref. 70. In one primate model of HUS, studies have been performed with intravenous injection of purified Stx1 in baboons [71,72]. While the HUS manifestations of STEC infection have not yet been reproduced following oral challenge in the baboon model, recent data indicate that monkeys challenged orally with O157:H7 develop gastrointestinal lesions reminiscent of human disease [73].

Typically, STEC infection begins with a watery diarrhea that is associated frequently with abdominal pain and occasionally with nausea and vomiting. Fever is not usually a prominent feature. The watery diarrhea may or may not progress to bloody diarrhea within 1–2 days. In its most severe form, STEC-associated diarrhea can present as HC, which in some cases may be difficult to distinguish from inflammatory bowel disease.

It has been appreciated recently that humans with STEC infection frequently are found to have fecal leukocytes as well as a peripheral leukocytosis [74]. Somewhat surprisingly, in this study, fecal leukocytes were seen more frequently in the STEC O157:H7–infected patients than in those infected with the invasive pathogens *Salmonella*, *Campylobactor*, or *Shigella*. The limited number of histologic studies undertaken on gastrointestinal tissue from patients infected with STEC support the idea that intestinal mucosal neutrophil infiltration is a significant component of the disease process [75–78].

Most people infected with STEC recover without incident. However, in some patients, STEC infection can progress to HUS, characterized histopathologically by endothelial cell swelling and detachment and deposition of fibrin-platelet thrombi in the microvasculature of affected tissues. HUS is a serious and sometimes fatal disease. From outbreaks, it has been estimated that 5–10% of infected individuals will develop HUS. Affected organs include the kidney, gastrointestinal tract, brain, and occasionally, the lung and pancreas.

It is not clear why some individuals develop HUS and others do not. However, some risk factors for developing HUS have been identified. Although individuals of any age can develop HUS, it is more likely to occur in children and the elderly [17]. Severity of diarrhea appears to be correlated with an increased likelihood of developing HUS [65]. As discussed earlier, infection with an Stx2-producing organism has been linked with a higher likelihood of developing HUS. Experiments in mice and gnotobiotic pigs support this observation

[79–81]. In the Sakai, Japan, outbreak, three clinical observations were associated with increased risk of developing HUS: elevated C-reactive protein, elevated peripheral blood neutrophil count, and fever higher than 38°C [82]. High blood peripheral neutrophil counts have been associated with the risk of developing HUS in multiple other studies [83–85]. Finally, use of certain antimicrobials for treatment of STEC diarrhea has been associated with HUS (reviewed in ref. 86), prompting recent recommendations that treatment with antimicrobials should be avoided in STEC infection [85,86].

For patients who develop HUS, the prognosis can be extremely poor. Approximately 5–10% of HUS patients will either die or develop major complications such as stroke, and approximately 50% of the HUS survivors will have some evidence of permanent renal damage [87]. HUS is now the most common cause of acute renal failure among U.S. children. In 20–30% of cases, HUS is associated with encephalopathy. Encephalopathy in the setting of HUS is a poor prognostic sign and is associated with increased morbidity and mortality [87,88]. At present, care of HUS patients is entirely supportive; there are no clinically proven effective therapies to prevent or treat STEC-associated HUS.

SHIGA TOXINS

In this section we discuss the principal virulence factor of STEC, Shiga toxins, paying particular attention to molecular structure, genetics, cell receptors, and cell trafficking.

Molecular Structure

The family of Shiga toxins consists of several structurally and functionally similar protein toxins. The prototype of the family is the Shiga toxin (Stx) elaborated by *S. dysenteriae* type 1; the Shiga toxins expressed by *E. coli* have a numerical designation following the name of Shiga toxin, e.g., Shiga toxin 1. Shiga toxin 1 (Stx1) differs from Shiga toxin from *S. dysenteriae* by a single amino acid. Shiga toxin 2 (Stx2) is approximately 50% homologous with Stx1 at the protein level and is immunologically distinct. Other forms of Shiga toxins are listed in Table II.

Shiga toxin family proteins inhibit protein synthesis by blocking the elongation factor 1–dependent binding of aminoacyl-tRNA to ribosomes. All Shiga toxins known to date are composed of an enzymatically active A subunit and a pentamer of identical B subunits that mediates specific binding activity. The A subunit functions as a glycohydrolase, cleaving a specific adenine from the 28S rRNA and irreversibly inhibiting ribosomal function. This activity is identical to the enzymatic activity observed for the plant toxin ricin [89].

X-ray crystallographic analyses of Shiga toxin and Stx1 B subunits have confirmed the AB5 structure, revealing a pentameric B subunit ring surrounding a C-terminal α-helix of the A subunit [90,91], which is similar to the AB5 structure of cholera toxin and *E. coli* heat-labile toxin. When the A subunit is nicked with trypsin and reduced, an A1 portion of approximately 28 kDa and an A2 peptide of approximately 4 kDa are separated. The A1 fragment contains the enzymatically active portion of the toxin molecule, and the A2 component is required to noncovalently bind the whole A subunit to the B pentamer [92]. The A1 and A2 fragments are linked by a disulfide bond. The A2 fragment may be important in holotoxin assembly [92], and the disulfide bridge between A1 and A2 appears necessary for pentamer formation [93].

The disulfide loop contains the sequence Arg-X-X-Arg, a consensus motif for cleavage by the membrane-anchored protease furin. Cleavage of Shiga toxin at this site, resulting in the formation of A1 and A2 fragments, appears to be important for cellular intoxication [94]. Interestingly, Stx2d is activated by intestinal mucus; this characteristic distinguishes Stx2d from other Stx2 family members. The substance from mucus that activates Stx2d has been identified as an elastase that cleaves the C-terminal two amino acids of the A2 subunit [95], but elegant studies with hybrid Shiga toxins have demonstrated that this activation also depends on the structure of the B pentamer [96].

Structural features important to the enzymatic activity of Shiga toxins have been defined. Stx1, Stx2, and Stx2e and the plant toxin ricin, all of which inhibit protein synthesis by the same mechanism of action, share two areas of homology in their A subunits [97]. In area 1, glutamic acid 167 is critical for biologic activity of Stx1 [98]. Similar experiments in Stx2 have shown similar results, affirming the importance of this glutamic acid residue in the active site [97]. In area 2, deletion of amino acids 202 through 213 of the Stx2 A subunit still allows holotoxin assembly but not cytotoxicity [97].

Genetics

Shiga toxins are either phage-encoded (Stx1 and Stx2) or chromosomally encoded (Stx and Stx2e) [99–101]. In the early 1970s it was first reported that lysates of an *E. coli* O26:H11 strain isolated from an outbreak of infantile diarrhea could transfer enterotoxigenicity to a nonpathogenic *E. coli in vitro* [102]. This was the first indication that in STEC, Shiga toxins were encoded by bacteriophages. From lysates of this strain, the phage known as H-19B subsequently was isolated. This phage was shown to encode Stx1 and to have DNA sequence homology with the phage lambda (λ) [103,104]. A second lambdoid phage encoding Stx2, designated 933W, was isolated from a clinical O157:H7 isolate responsible for an outbreak of hemorrhagic colitis in 1982 [10,99,105]. Since

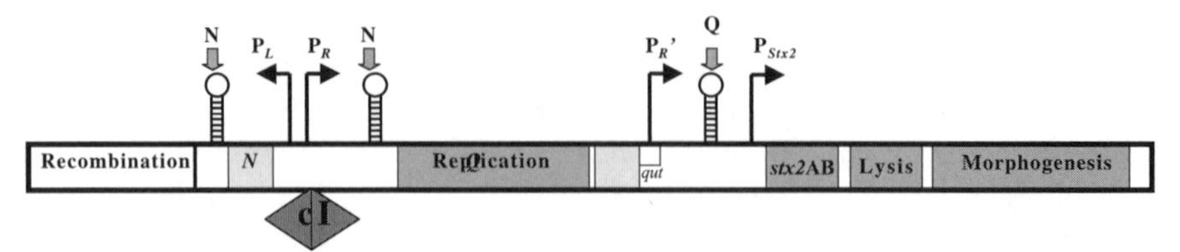

FIGURE 1 Diagram of Stx2-encoding phages. Shown are some of the relevant genes, promoters, terminators, and operators, not drawn to scale. Induction inactivates the repressor cI, resulting in transcription initiating at the early promoters P_L and P_R and expression of the N antiterminator. N modification of RNA polymerase allows read-through of N antiterminator sites, resulting in production of proteins catalyzing excision and replication of the phage genome and synthesis of Q. Q is an antiterminator that binds to *qut* at the site of the late $P_{R'}$ promoter, allowing read-through of the *stx* genes as well as the genes of the phage lysis cassette. Stx1 phages are similarly arranged, except that the Stx1 promoter replaces the Stx2 promoter. The Stx1 promoter is under the control of the *fur* gene and therefore is regulated by iron, as described in the text. (Courtesy of Drs. P. L. Wagner and M. K. Waldor.) See Color Plate 5.

these initial observations, it is now appreciated that Stx-encoding phages are lambda-like and that the regulatory components relating to induction and phage gene control appear to be similar to those in lambda. Clinical STEC isolates may contain a single phage or multiple phages. Many clinical O157:H7 isolates harbor distinct Stx1 and Stx2 bacteriophages.

Various groups have published the nucleotide sequences of the *stx* genes [106–111]. The different *stx* operons have a similar structure and are composed of a single transcriptional unit consisting of one copy of the A subunit gene followed by the B subunit gene. It was believed initially that the B subunit had its own promoter [112]; later it was appreciated that a single promoter transcribes A and B subunits; B subunit translation is augmented due to a stronger ribosomal binding site compared with A subunit translation. This results in more B subunit translation, providing more B subunits for the A1:B5 structure of the Shiga toxin holotoxin [113].

Comparison of the nucleotide sequence of the A and B subunits of Stx1 and Stx2 reveals 57% and 60% homology, respectively, with 55% and 57% amino acid homology [109]. Despite this degree of homology, Stx1 and Stx2 are immunologically distinct, and neither is cross-neutralized by polyclonal antibody raised to the other toxin. Stx2c is very similar to Stx2; the A subunits are identical, and the B subunits share 97% amino acid homology. Stx2c and Stx2d have identical B subunits; the A subunits share 99% homology [114]. Stx2e has the least similarity to other Stx2 family members. Although the Stx2 and Stx2e A subunits have 93% deduced amino acid homology, the B subunits have only 84% deduced amino acid sequence homology [115]. It is not surprising therefore that the receptor binding specificity of Stx2e is different from that of other Stx2 family members. A fifth family member, Stx2f, has been described recently in

STEC isolated from populations of feral pigeons, although little is known about this variant [116].

Toxin Receptors

The search for Shiga toxin receptors on mammalian cells began in 1977, when Keusch and Jacewicz reported that toxin-sensitive cells in tissue culture removed toxin bioactivity from the medium, whereas toxin-resistant cells did not [117]. Furthermore, these studies suggested that the receptor was carbohydrate in nature and that the toxin was a sugar-binding protein or lectin. Later, a toxin-binding membrane component was extracted from toxin-sensitive HeLa cells and from rabbit jejunal microvillus membranes (MVMs) [118]. The MVM binding site was shown to be globotriaosylceramide (Gb3) [118,119]. Lindberg and colleagues reported that Stx1 bound to the P blood group active glycolipid Gb3, which consists of a trisaccharide of galactose α1-4-galactose-β-1-4-glucose linked to ceramide, and that Gb3 could inhibit biologic activity of Stx1 in cell culture systems [120]. Most Stx2 family members share a preference for Gb3 binding, except for Stx2e, which binds preferentially to globotetrosylceramide (Gb4), another neutral glycolipid, which has a subterminal Gal-α1-4-Gal disaccharide [121–123].

In many cases, the sensitivity of cells to Shiga toxins appears to be related to the number of toxin receptors present on the cell surface. Measures that increase or decrease toxin receptor expression directly alter responses to these potent toxins [124]. In addition, toxin activity can be modulated by the fatty acid composition of the lipid ceramide moiety [125], in particular fatty acid carbon chain length, which alters the intracellular uptake pathway of toxin and its biologic activity. This may explain why some cells can express Gb3 but fail to respond to toxin [126]. Arab and Lingwood [127] demonstrated the importance of the surrounding lipid environment on the availability of glycolipid carbohydrate for Shiga toxin binding. The lipid heterogeneity of Gb3 appears to be important in Shiga toxin binding and may define a growth-related signal-transduction pathway used by Shiga toxin [128].

Sensitivity of cells to Shiga toxin also can be affected by regulation of receptor expression. For example, rabbit intestinal brush-border membrane Gb3 is both developmentally and maturationally regulated via the biosynthetic Gb3-galactosyltransferase and degradative α-galactosidase enzymes [129]. Gb3 is also maturationally regulated in cultured human intestinal epithelial cells. Gb3 is induced by exposure of villus-like Caco2A cells, but not crypt-like T84 cells, to known regulators of gene transcription, e.g., sodium butyrate [130]. In Caco2A cells, butyrate-induced Gb3 expression coincides with expression of villus cell differentiation markers such as alkaline phosphatase, lactase, and sucrase. Produced by normal resident enteric flora in high concentration, butyrate effects on intestinal epithelium may be pertinent in the human colon, the site of STEC infection.

Shiga Toxin Trafficking

The intracellular sorting of Shiga toxins has been studied in certain cells in detail (reviewed in ref. 131). After binding to neutral glycosphingolipids with terminal Gal-α1-4-Gal via the B subunit, Shiga toxin is internalized by receptor-mediated endocytosis at clathrin-coated pits. The cytosolic signal required for endocytosis is unknown. From there, Shiga toxins appear to enter the endosome. Generally, proteins that are endocytosed into the endosome can undergo one of four possible fates: They may be (1) returned to the cell surface of entry, (2) targeted to lysosomes for subsequent degradation, (3) transcytosed through the cell to the opposite surface in polarized epithelia, or (4) delivered to the trans-Golgi network (TGN). Shiga toxins have been observed to traffic to the TGN, and routing to the TGN appears to be important for intoxication because treatment with Brefeldin A, a drug that disrupts the Golgi apparatus, protects against intoxication by Shiga toxins [94]. Interestingly, Shiga toxins can be observed to traffic retrograde from the Golgi network to the endoplasmic reticulum (ER) and, in some sensitive cells, to the nuclear envelope. Shiga toxins do not contain a KDEL sequence, which is the usual retention signal for ER proteins. Recent work suggests that Shiga toxins may usurp a preexisting cellular pathway that mediates retrograde transport independent of KDEL-mediated retrieval, regulated in part by a small GTPase called Rab6 [132]. It is thought that the B subunit determines this trafficking ability and that toxin translocation occurs from the ER to the cytosol, where ribosomal intoxication occurs.

MOLECULAR PATHOGENESIS

In this section we discuss the host-pathogen interactions believed to occur during STEC infection leading to disease, including the effects of Shiga toxins on various cells thought to be involved in HUS pathogenesis.

Gut Colonization: Acid Tolerance and the LEE Pathogenicity Island

To cause disease, STEC must be ingested, survive the harsh acidic milieu of the upper gastrointestinal tract, and colonize the lower gastrointestinal tract. A unique acid-tolerance response (ATR) system has been described for both *E. coli* and *Shigella*. These organisms contain an ATR system consisting of a glutamate decarboxylase and glutamate/γ-aminobutyrate antiporter encoded by *gadB* and *gadC*, respectively. In this system, glutamate is transported into the cell, where decarboxylation occurs. The decarboxylation reaction consumes a proton. The γ-aminobutyrate antiporter then exports the product, thereby helping to maintain a neutral cytoplasmic pH when extracellular pH drops [133]. *GadCB* are regulated by the σ_factor *rpoS*, expressed during stationary phase. One study of both clinical and environmental STEC of multiple different serotypes showed that approximately 75% of isolates (45 of 58) were acid-tolerant. Further analysis of the non-acid-tolerant strains suggested that mutations in *rpoS* play an important role in loss of acid tolerance [134].

The production of colanic acid is another factor that has been linked with acid tolerance in STEC. Colanic acid is an exopolysaccharide that, when produced, results in the formation of mucoid colonies. The hypothesized mechanism by which colanic acid mediates acid tolerance is enhancement of the bacterial cell's ability to buffer extracellular protons by virtue of the negatively charged mucoid matrix [135].

Once STEC have reached the lower gastrointestinal tract, colonization may occur. The mechanism by which adherence occurs is best described for STEC containing the LEE pathogenicity island. The LEE pathogenicity island encodes genes responsible for the bacterial interactions with intestinal epithelial cells called *attaching and effacing lesions* (A/E lesions). Originally described for EPEC, this histopathology is also seen with STEC infection in intestinal epithelial cell lines and some animals, but it has not yet been documented in STEC-infected humans. The details of the molecular events involved in formation of the A/E lesion are discussed elsewhere in this book (see Chap. 13). Briefly, there are five polycistronic regions encoded by the LEE pathogenicity island. Three regions of genes in this pathogenicity island, *LEE1*, *LEE2*, and *LEE3*, encode a type III secretion system. *LEE4* encodes several secreted proteins that cause alterations in host signal transduction, such as EspA, EspB, and EspD. The *tir* operon encodes for the translocated intimin receptor, its chaperone protein CesT, and intimin. The translocated intimin receptor (Tir) is injected into the host cell via the type III secretion system, where it acts as a receptor for intimin, an outer membrane protein present on the bacterium. It is the close association between Tir and intimin that mediates, in part, the formation of the A/E lesion.

It has been recognized recently that various intimin subtypes exist and that intimin subtype may play a role in the tissue tropism of LEE-positive STEC. The intimin subtype associated with *E coli* O157:H7 is subtype γ. In one model of STEC colonization using *in vitro* organ culture, intimin subtype γ conferred a selective tropism for Peyer's patches on the host organism, whereas intimin receptor subtype α, associated with certain EPEC, conferred a less selective tropism that allowed adherence to both Peyer's patches and non-follicle-associated epithelium [136].

Quorum Sensing

Recent studies from the laboratory of Kaper have shown that in EHEC, *LEE1* and *LEE2* are regulated by quorum sensing [137]. One of the proteins encoded by *LEE1*, the LEE-encoded regulator (Ler), then acts to regulate *LEE3* and *tir*. In this way, four of the five operons in the LEE operon are directly or indirectly under the control of a quorum-sensing mechanism. The autoinducer molecule has yet to be identified biochemically but is encoded by a member of the *luxS* gene family described by Surette and colleagues [138]. Interestingly, nonpathogenic *E. coli* can express *luxS*. The role of quorum sensing in STEC pathogenesis is at present unclear. However, it has been hypothesized that quorum sensing via an auto-inducer molecule expressed by nonpathogenic *E. coli* may stimulate expression of LEE pathogenicity island genes once STEC are present in the colonic milieu and may in part explain the low infectious dose observed with these organisms [137].

Other Virulence Factors

STEC adherence mechanisms not associated with the LEE pathogenicity island have only been described at the phenotypic level. The significance of these mechanisms for STEC pathogenesis is presently unclear; these adherence mechanisms have been described recently in reference [16].

There are several other potential virulence factors that have been found in STEC strains, although at present there is no clear role for these factors in modulating STEC adherence or HUS pathogenesis. Two of these putative virulence factors are encoded by a 60-MDa plasmid (pO157) that is present in some STEC strains. One of these factors is an α-hemolysin (EHEC-Hly); the other is a secreted serine protease called EspP, with significant sequence homology with EspC from EPEC. The third factor, chromosomally encoded, is the enterotoxin EAST1, found in many different *E. coli* associated with diarrhea, especially enteroaggregative *E. coli*. Data regarding the potential virulence roles of these factors have been reviewed recently [16].

Toxin Production in the Gut

Because STEC are noninvasive, it is presumed that Shiga toxin produced by bacteria in the gut is then absorbed into the systemic circulation, resulting in the systemic sequelae associated with STEC infection. Therefore, both the amount of Shiga toxin produced in the gut and the amount of Shiga toxin that is absorbed from the gut may determine who is at risk for developing the systemic sequelae of STEC infection.

As described previously (in the genetics section), the *stx* genes of all studied STEC are located on the genomes of lambda-like prophages [104,105,139]. This genetic arrangement may be applicable to all STEC. However, the physical proximity of *stx* genes to phage genes does not necessarily mean that the toxin is encoded by intact, functional phages [140]. While gene regulation in the lambda prophage is now largely understood [141], our understanding of the mechanisms and controls of Shiga toxin production both *in vitro* and, more important, *in vivo* is still in its infancy.

The DNA sequences of a limited number of Stx1 (e.g., part of H-19B, VT1-Sakai) and Stx2 phages (e.g., 933W, VT2-Sa and VT2-Sakai) have been completed to date [140,142–145]. Comparisons between the lambda phage and STEC phages have revealed a common arrangement of genes and strategies to govern gene expression [140,146]. The general position of the *stx* genes appears to be conserved, located between the genes encoding the transcriptional antiterminator Q protein with its site of action, *qut*, at the associated p_R' promoter, and the genes of the lysis cassette. STEC phage induction is thought to regulate toxin expression and release in several ways: (1) via replication bringing a concomitant increase in toxin gene copy number, (2) via a phage-encoded regulatory molecule that increases transcription from phage promoters typically repressed during lysogeny, and (3) via linkage of toxin production to cell lysis [146].

Initial investigations with Stx1 toxin from *E. coli* and Stx from *S. dysenteriae* led to the concept that there was a family of Shiga toxins [148]. Subsequent gene analysis found that Stx1 phages share similar gene sequences [106] and that high levels of iron repress toxin expression in both [101,149]. In Stx1-producing organisms, a promoter is located just upstream of the toxin structural genes [109]. Regulation of the Stx1 promoter (p_{Stx1}) occurs through the *fur* gene product, which was found to bind to a potential *fur* operator binding site located upstream of the *stx* operon [111,149]. The genes encoding *fur* are located on the chromosome of the host bacterium. Hence linkage of *fur*-controlled toxin expression and iron regulation in the bacterial cell illustrates the coordinated nature of bacterial pathogenesis. The bacterial cell uses the low iron availability typically present in mammalian cells as a trigger to express virulence determinants, although it is unclear what role iron regulation plays *in vivo*.

More recently, Neely and Friedman have provided further insight into Stx1 and Stx2 regulation [150,151]. Following DNA analysis of part of the Stx1 phage and of the published sequence of the Stx2 phage, these investigators found considerable similarity between regulation of Stx phages and lambda phages. In both, Q acts as a transcriptional antiterminator at *qut* to modify RNA polymerase to a highly processive form that reads through downstream terminators and switches on toxin expression. Thus, in Stx1-producing phages, toxin expression can be regulated via two mechanisms: (1) via the *fur* protein acting on the p_{Stx1} and (2) via the action of the Q protein on p_R'.

In contrast, Stx2 phages are not iron-regulated via the *fur* gene protein (Fig. 1). Instead, these phages have a different promoter named p_{Stx2} located immediately upstream of the toxin genes [152]. The activity of this promoter is low, however, and similar to that of the iron-repressed Stx1 promoter. Muhldorfer and colleagues originally postulated that a phage-encoded factor had a positive regulatory impact on Stx2 expression, and this now has been shown [146,153]. Wagner and colleagues [146] observed a significant reduction in toxin production in an Stx2-producing isolate carrying a mutation in the p_R' and Q regions. Additionally, these authors suggest that Stx2 phages direct the production of toxin as part of their lytic cycle and hence contradict the prevailing assumption that phages serve merely as agents for virulence gene transfer.

Phage induction occurs as a consequence of the activation of the bacterial SOS response to DNA damage [154]. Some factors that trigger phage induction *in vitro* are well documented and include ultraviolet (UV) radiation and other DNA-damaging agents (e.g., mitomycin C, certain classes of antibiotics). However, less is known about the factors that may trigger phage induction within the human patient. *In vitro*, certain antibiotics (e.g., quinolones, trimethoprim) have been found to be potent inducers of Shiga toxin production [155,156]. These findings were confirmed *in vivo* in experiments carried out in a murine model [157]. The consequence of these findings with respect to the treatment of patients with STEC infection is discussed earlier. Wagner and colleagues [158] found that both H_2O_2 (antibacterial molecules released by human cells) and neutrophils augmented Stx2 production *in vitro* when an Stx2-producing organism was co-cultured with human neutrophils. These authors suggested that the production of reactive-oxygen intermediates by neutrophils induced the bacterial SOS response, thereby stimulating Stx2 production.

In addition to their role in virulence gene production, bacteriophages also can play an important role in dissemination of virulence genes within the environment [159]. For example, the use of phage-inducing antimicrobials in farm animals may enhance the dissemination of the *stx* genes and lead to the development of new STEC pathotypes [160]. The evolutionary pathways of different *E. coli* pathotypes reflect the huge genetic variation present within the population and the potential for the reintroduction of virulence factors into established pathogens [161].

Toxin Uptake from Gut

Until recently, relatively little was known about how Shiga toxins cross intestinal epithelial cells (IECs), although the trafficking of Shiga toxins in other cell types has been well studied, as described earlier. Acheson *et al.* have demonstrated that Shiga toxins cross-polarized IECs by an energy-dependent transcellular process [162]. At least a portion of Shiga toxins are transcytosed in their active form to the basolateral surface [162,163]. It is unknown whether this transcytosed toxin is taken up by the Gb3 receptor or via nonspecific pinocytotic mechanisms. However, the data suggest that translocation of Shiga toxin across intestinal epithelial cells is not Golgi-dependent [164]. Furthermore, it appears that Stx1 and Stx2 use different routes across IECs.

As discussed earlier, neutrophil transmigration across the polarized intestinal epithelium with resulting accumulation in the gastrointestinal lumen occurs in many inflammatory diseases of the gastrointestinal (GI) tract, including STEC infection [74]. This migration process results in transient epithelial barrier disruption with subsequent leakage of lumenal contents into the systemic circulation [165–168]. In rabbits, STEC infection has been shown to cause an intestinal epithelial cell disruption that is prevented by pretreatment with a monoclonal antibody to the leukocyte adhesion molecule CD18, suggesting that the host response to infection *per se* may be important to damaging the intestine [169]. Recently, STEC-induced neutrophil migration has been shown to enhance Shiga toxin movement across intestinal epithelial cell monolayers *in vitro* [170]. These data suggest that inflammation of the intestinal tract during STEC infection may promote systemic absorption of Shiga toxins.

HUS PATHOGENESIS AND EFFECTS OF SHIGA TOXIN ON VARIOUS CELL TYPES

Shiga toxins have been appreciated classically as ribosome-intoxicating proteins, mediating their effects on cells by blocking new protein synthesis and thereby preventing synthesis of critical host proteins needed by the cell to survive and/or function properly. Initial studies on cytotoxicity in the 1960s and 1970s revealed that Shiga toxins were cytotoxic to some cells and not others; cytotoxicity was related to the expression of appropriate cell-surface receptors. However, these studies were undertaken in epithelial cell types that had little relevance to the likely *in vivo* target cells.

Recently, investigators have focused on evaluating the effects of Shiga toxins on more relevant cell types based on the assumption that Shiga toxins produced in the intestinal lumen gain access to the systemic circulation by traversing the intestinal epithelium. During passage from lumen to sensitive organ beds such as the kidney and brain, Shiga toxins are thought to encounter various cell types that may be involved in pathogenesis, including the intestinal epithelium, circulating blood cells, and endothelium of target organs. The data regarding interactions between these cell types and Shiga toxins are discussed in detail later. As a result of investigations of the Shiga toxin–host cell interaction, data supporting two alternative roles for Shiga toxins in HUS pathogenesis have emerged. One role is for Shiga toxin as a stimulus of cytokine release by various cell types, and the mechanism(s) by which this occurs is an area of very active research interest. A second role is for Shiga toxin as a proapoptotic stimulus, an idea initiated in 1993 by investigations on the effects of Shiga toxin on Burkitt lymphoma cells [171]. This is also an area of very active research interest. However, our understanding of the mechanism(s) by which these events occur and their relevance to pathogenesis is not yet complete.

Intestinal Epithelial Cells

Infection with STEC clearly stimulates an intestinal inflammatory response, but the role of Shiga toxins in inducing this host response has not been defined clearly. Early work in adult rabbit loop models failed to demonstrate that Shiga toxins stimulated an inflammatory response [172,173]. However, in an infant rabbit model, oral challenge with purified Stx2 resulted in histologic changes in the middle to distal colon similar to those observed with *E. coli* O157:H7 infection, including lamina propria neutrophil infiltration [174]. More recently, in a rabbit model of STEC-induced hemorrhagic colitis, animals infected with a rabbit enteropathogenic *E. coli* strain engineered to produce Stx1 (RDEC H19A) developed more severe intestinal inflammation than rabbits infected with the non-Stx1-producing parent RDEC-1 strain [175]. Differences observed included increased fluid secretion, increased histopathologic inflammatory changes, and elevated mucosal interleukin 1β (IL-1β) levels in the rabbits infected with the Stx1-producing strain compared with the non-Stx1-producing parent strain, suggesting that Stx1 itself, either directly or indirectly, may have a significant role in promoting inflammation. In the primate model of HUS, parenteral Stx1 resulted in intestinal epithelial cell damage localized to the microvillus tip, where Gb3 is concentrated. Accompanying this lesion, there also were varying degrees of mucosal and submucosal congestion, hemorrhage, and necrosis [71].

In vitro, Shiga toxins have been demonstrated to induce the neutrophil chemokine IL-8 from multiple different intestinal epithelial cell lines [176–179]. Thorpe et al. [178] further characterized this IL-8 response, linking IL-8 expression to a response called the *ribotoxic stress response*. The ribotoxic stress response, first described by Iordanov and colleagues [180], is a host cellular response to 28S rRNA damage by certain protein synthesis inhibitors, resulting in activation of host MAP kinase pathways and induction of *c-fos* and *c-jun* mRNA. Thorpe *et al.* have demonstrated in an intestinal epithelial cell line that Shiga toxins stimulated IL-8 expression via the ribotoxic stress response, resulting in induction of *c-jun* mRNA and activation of one of the three best-characterized MAP kinase cascades. Interestingly, IL-8 expression occurred despite profound inhibition of protein synthesis. These observations are supported by the work of Yamasaki and colleagues [179], who demonstrated in a different intestinal epithelial cell line that Shiga toxin–induced IL-8 expression did not occur when cells were challenged with an Shiga toxin A subunit active site mutant, suggesting that ribosomal intoxication is required for IL-8 induction. The ribotoxic stress response appears to be one mechanism by which Shiga toxins can stimulate cytokine expression.

Further work in intestinal epithelial cells *in vitro* has demonstrated that multiple C-X-C chemokines are induced by Shiga toxin and that at least one mechanism by which this induction occurs is by enhanced mRNA stabilization [181]. In addition, the phenomenon of *superinduction* has been observed with Shiga toxins. Superinduction occurs when cells are exposed to a known stimulus in the presence of a protein synthesis inhibitor, resulting in accumulation of massive amounts of mRNA compared with that observed with the stimulus alone [182]. Thorpe and colleagues have observed superinduction of multiple C-X-C chemokines when intestinal epithelial cells are exposed to tumor necrosis factor α (TNF-α) and Stx1 *in vitro*; superinduction can occur at very low concentrations of Shiga toxin [178,181]. However, at present, it is completely unknown whether Shiga toxins can cause superinduction *in vivo* under any circumstances.

Intestinal epithelial cell apoptosis may occur in response to Shiga toxins. Very early studies in rabbit intestinal loops using Shiga toxin from *S. dysenteriae* suggested that Shiga toxin could mediate intestinal epithelial damage directly [172]. Later, Shiga toxins from STEC were demonstrated to mediate intestinal epithelial cell damage by cellular apoptosis in the same model system [173]. In the infant rabbit model, feeding of purified Shiga toxin resulted in apoptosis, as well as inflammation, in the surface epithelium of the middle to distal colon [174]. *In vitro* studies have shown that Shiga toxins can mediate apoptosis of Gb3-expressing intestinal epithelial cells [183]. Apoptosis was associated with enhanced expression of Bax, a member of the Bcl-2 family of proteins. These investigators demonstrated that apoptosis could be mediated by the B subunit alone, which is consistent with the findings of other investigators in other cell types [171,184]. Finally, recent studies from Barnett-Foster and colleagues [185] demonstrate that in some intestinal epithelial cell lines, both Shiga toxin and STEC may play a role in inducing apoptosis.

Blood Components and Blood Transport

There is limited direct evidence that Shiga toxins can gain access to the systemic circulation following STEC infection. In a murine oral challenge model of STEC infection, it has been shown recently that Shiga toxin can be detected in mouse brain following oral challenge with STEC [186]. There are some data in humans. Shiga toxin has been detected by immunohistochemistry in both lung and renal tissue obtained at autopsy from a patient who died from HUS [187,188]. In more recent studies, te Loo and colleagues have demonstrated Shiga toxins bound to neutrophils from peripheral blood in patients with HUS [189]. Together these data support the idea that Shiga toxins may gain access to the systemic circulation from the gut and then travel to distal sites such as the kidney and brain.

There is evidence that neutrophils may be important in HUS pathogenesis. High peripheral blood neutrophil counts at the time of presentation with HUS are associated with a worse prognosis [190–192]. Elevated levels of the neutrophil chemoattractant IL-8 and evidence of increased neutrophil activation have been demonstrated in some patients with HUS and are associated with worse outcomes [191,193,194]. How neutrophils become activated in STEC-associated HUS is unknown, as is the tissue source(s) of the neutrophil chemoattractant IL-8.

There are few studies that examine direct effects of Shiga toxins on neutrophils. Stx1 treatment of polymorphonuclear cells results in release of reactive intermediates from PMNs and causes a reduction in their phagocytosis without inducing apoptosis or necrosis [195]. The mechanism by which this occurs has not been elucidated. Stx2 treatment of neutrophils has been shown to delay spontaneous neutrophil apoptosis through a mechanism that involves protein kinase C [196].

Direct interactions between Shiga toxins and macrophage-like cells have been well studied *in vitro*. These data support a role for Shiga toxins in promoting cytokine expression. Barrett and colleagues [197] found that exposure of murine peritoneal macrophages to Stx2 resulted in an increase in TNF-α bioactivity. Tesh and colleagues showed that murine peritoneal macrophages, human peripheral blood monocytes, and human monocytic cell lines were relatively refractory to the cytotoxic action of Stxs but responded by secreting the proinflammatory cytokines TNF-α and IL-1β in a dose-dependent manner [198,199]. Another group of investigators reported that human peripheral blood monocytes make TNF-α, IL-1β, IL-6, and IL-8 in response to Stx1 [200]. These findings are particularly interesting in light of the role proinflammatory cytokines have in sensitizing endothelial cells to the cytotoxic effects of Shiga toxins (discussed below). Furthermore, *in vitro* and *in vivo* studies have implicated a role for TNF-α in promoting the renal cell injury associated with Shiga toxins (reviewed in ref. 201).

The mechanism by which Shiga toxins cause TNF-α release from macrophages is being unraveled. Tesh and colleagues have demonstrated that Stx1 can induce both the transcriptional activators NF-κB and AP-1 to translocate to the nucleus; this is associated with TNF-α mRNA expression [202]. Furthermore, toxin enzymatic activity is required for TNF-α expression, as is phosphokinase C activity [203]. These data have prompted some speculation as to whether Shiga toxin–induced TNF-α expression may be a form of the ribotoxic stress response [203].

Erythrocytes have the capacity to bind Shiga toxins of various types through association with various P blood group antigens, which are glycolipid in nature. Two contradictory hypotheses have emerged for the role of erythrocytes in HUS pathogenesis. One hypothesis supports a role for erythrocytes binding and sequestering Shiga toxins, thereby preventing Shiga toxin from reaching target organs. The other hypothesis supports a role for erythrocytes in the delivery of Shiga toxin to sensitive organs. Data supporting these opposing hypotheses are reviewed in ref. 16; the significance of the erythrocyte–Shiga toxin interaction is unknown and has not been demonstrated *in vivo*.

There has been some controversy as to whether Shiga toxins directly interact with platelets and, if this interaction occurs, whether it plays a role in the TMA seen in STEC-associated HUS. The Shiga toxin receptor Gb3 is known to be present on the platelet membrane [204]. Recently, it was shown that Shiga toxin could bind Gb3 isolated from platelets, as well as a minor platelet glycolipid termed band *0.03*. The significance of these data is not yet known, but band 0.03 appears to be expressed in approximately 20% of donors and was associated with increased Gb3 expression [205]. Several groups have reported that Shiga toxins do not affect aggregation of human platelets in platelet-rich plasma, either alone or in the presence of other agonists, as measured by aggregometry [206–208]. Thorpe and colleagues showed that Stx1 or Stx2 did not induce alpha-granule secretion of purified human platelets or aggregation, as determined by an aggregometer [207]. Recently, however, Karpman and colleagues have demonstrated that treatment of human platelets with either Stx1 or the purified B subunit of Stx1 (Stx1B) resulted in increased platelet aggregation, as measured by confocal microscopy [209]. Furthermore, Stx1 and Stx1B induced the binding of platelets to TNF-α-treated human umbilical vein endothelial cells (HUVECs) and increased fibrinogen binding to platelets. These results suggest a direct role for Shiga toxins in the altered platelet function seen in TMA. In summary, while the activation of platelets is clearly an important step in HUS pathogenesis, the precise way in which Shiga toxin exposure results in formation of platelet thrombi remains to be elucidated.

Endothelium

Obrig and colleagues were among the first to recognize that HUS might be due to direct toxin effects on endothelial cells [210,211]. Since these initial studies, many investigators have reported on the sensitivity of various types of endothelial cells to the effects of Shiga toxins. Four themes emerge from these data: (1) the tissue of origin, (2) the size of the vessel from which the cells were derived, (3) interindividual variability, and (4) preincubation with proinflammatory cytokines, all appear to be important in determining how sensitive endothelial cells will be to the cytotoxic effects of Shiga toxins. In general, endothelial cells derived from microvascular sources appear to have increased sensitivity to the cytotoxic effects of Shiga toxins compared with endothelial cells from large-vessel sources. Furthermore, proinflammatory cytokines such as TNF-α and IL-1β have been shown in some instances to increase the sensitivity of certain endothelial cells to the cytotoxic effects of Shiga toxins, making the cytokine-releasing effects of Shiga toxins on various cell types of particular interest.

Obrig and coworkers reported that preincubation of HUVECs with lipopolysaccharide (LPS) or the LPS-induced cytokines IL-1β or TNF-α, converted the relatively Shiga toxin–resistant HUVECs into relatively responsive cells and was associated with upregulation of Gb3 [210]. These investigators later reported that human glomerular endothelial cells (GECs) constitutively produced Gb3 and were not further induced by cytokines, suggesting a reason why these cells may be a preferred target for Shiga toxin [212]. Interestingly, these cells were more sensitive to Stx2 than Stx1. The significance of constitutive Gb3 induction by GECs is now uncertain because conflicting data have been reported. Monnens and colleagues [213] demonstrated that GECs required preexposure to TNF-α to become sensitive to Shiga toxin. Since these experiments were performed on different primary cell lines using different isolation techniques, interindividual variation or other factors are possible explanations for these differing observations.

Hutchinson and colleagues examined the effects of Stx1 on cerebral endothelial cells (CECs) derived from patients of unknown ages with severe refractory epilepsy [214]. They found that CECs were very sensitive to the cytotoxic effects of Stx1. However, Ramegowda and colleagues [215], using primary brain microvascular endothelial cells from a young male trauma patient, found recently that cells from this origin were much less sensitive to the effects of Stx1. Similar results were observed by Eisenhauer and colleagues [216] in a separate study. Again, interindividual variations may play a role. Human intestinal microvascular endothelial cells have been shown to be very sensitive to both Stx1 and Stx2 and were not made more sensitive by exposure to proinflammatory cytokines or LPS [217]. Like the primary GECs used in ref. 212, these cells also were more sensitive to the cytotoxic effects of Stx2 than Stx1. These *in vitro* data may explain in part observations of increased virulence associated with infection with Stx2-producing strains.

There have been some observations in endothelial cells of large-vessel origin suggesting that the ribotoxic stress response may occur in these cells in response to Shiga toxins. In bovine aortic endothelial cells, which are relatively insensitive to Shiga toxin cytotoxic effects, Shiga toxin can stimulate preproendothelin-1 mRNA, the precursor mRNA of endothelin-1 [218]. Shiga toxin treatment also resulted in enhanced expression of endothelin-1 protein. At least one mechanism involved is enhanced stabilization of preproendothelin-1 mRNA by Shiga toxin.

Apoptosis may play a role in the endothelial cell injury seen in HUS. Shiga toxin—induced apoptosis has now been demonstrated in endothelial cells of different origins [186,219–221]. In some studies, this response is augmented by preexposure to proinflammatory cytokines such as TNF-α. Also, injury to endothelial cells by circulating activated neutrophils may play a role in HUS. Shiga toxin treatment of HUVECs under flow conditions has been shown to increase adhesion of neutrophils, which was inhibited by blocking neutrophil receptors such as E-selectin, ICAM-1, and VCAM-1 [222]. One recent *in vitro* study has demonstrated that Shiga toxin can bind to neutrophils, following which the Shiga toxin activity is then transferred to glomerular microvascular endothelial cells, resulting in endothelial cell death [223].

Kidney

Based on renal histopathology from patients with HUS, it was appreciated early on that the renal endothelial cell is a possible target of Shiga toxins. The *in vitro* data, as described earlier, regarding the relative sensitivity of endothelial cells of renal origin support this idea. However, other renal cell types may be involved in HUS pathogenesis. Clinical research has shown that early in disease, proteins in the urine that are consistent with renal tubular damage are observed, as well as histopathologic damage to proximal tubules [224]. Shiga toxins have been shown to bind to renal tubule epithelium in renal histologic sections [225]. *In vitro*, proximal tubular cells are very sensitive to Shiga toxins, can express proinflammatory cytokines in response to Shiga toxins, and can undergo apoptosis [226–228]. Clinical studies have demonstrated elevated proinflammatory cytokines in the urine of children with HUS; whether the renal tubular epithelium is the source of these cytokines is unknown [229]. The newly described baboon model of HUS shows TNF-α and IL-6 in the urine, but not the serum, associated with predominantly proximal tubular and glomerular endothelial histopathology [71,72]. A recent renal biopsy from one patient with predominantly renal tubular dysfunction following STEC infection demonstrated the presence of Stx2 in the renal tubular epithelium and showed evidence of apoptotic cell death [230]. Other renal cell types shown to be sensitive to the cytotoxic effects of Shiga toxins are glomerular epithelial cells, cortical tubular epithelial cells, and mesangial cells [231,232]. Glomerular epithelial cells also have been shown to produce proinflammatory cytokines in response to Shiga toxin [233].

PROSPECTS FOR PREVENTION AND CONTROL

Once the human host has been infected with STEC, there are no established procedures to prevent HUS. However, recent prevention efforts have focused on three main mechanisms: (1) immunization with Shiga toxin components, (2) use of toxin sequestration mechanisms in the GI tract, and (3) use of probiotic strains to prevent toxin uptake [234–241]. Interestingly, one probiotic strain under investigation is not intended simply as a competitor of STEC colonization but instead is a recombinant bacterium expressing a Shiga toxin receptor mimic on its surface, thereby acting to sequester Shiga toxins from absorption across the GI epithelium [241]. At present, however, none of these methods has been demonstrated to be effective in humans. Therefore, lack of therapy to prevent or treat HUS highlights the need for effective measures to control STEC infection of humans.

Proper food handling is one of the key components in the prevention of STEC infection. This involves maintaining the correct storage temperatures for both cooked and uncooked foods and cooking food to the appropriate temperatures (160°F for ground beef). Preventing cross-contamination between raw products, such as ground beef, and prepared foods, such as salads, is also important in both home and commercial environments. Finally, ensuring that safeguards are in place to prevent person-to-person spread when an individual is sick with STEC infection is critical in preventing secondary cases, especially within families. Given the lack of a specific therapy for HUS and HC, coupled with concerns regarding the role of antibiotic therapy in promoting HUS in STEC-infected individuals, primary prevention of STEC infection is a key public health measure to minimize the numbers of new cases of HUS.

REFERENCES

1. Shiga, K. (1898). Ueber den Dysenteriebacillus (*Bacillus dysenteriae*). *Zentrabl. Bakt. Parasit. Abt. 1 Orig.* 24:817–824.
2. Flexner, S. (1900). On the etiology of tropical dysentery. *Bull. Johns Hopkins Hosp.* 11:231–242.
3. Conradi, H. (1903). Über lösliche, durch aseptische Autolyste erhatlene Giftstoffe von Ruhr- und typhus-bazillen. *Dtsch. Med. Wochenschr.* 20:26–28.
4. Koster, F., Levin, J., Walker, L., *et al.* (1978). Hemolytic-uremic syndrome after shigellosis. *New Engl. J. Med.* 298:927–933.
5. Konowalchuk, J., Speirs, J. I., and Stavric, S. (1977). Vero response to a cytotoxin of *Escherichia coli*. *Infect. Immun.* 18:775–779.
6. Konowalchuk, J., Dickie, N., Stavric, S., and Speirs, J. I. (1978). Properties of an *Escherichia coli* cytotoxin. *Infect. Immun.* 20:575–577.
7. Wade, W. G., Thom, B. T., and Evens, N. (1979). Cytotoxic enteropathogenic *Escherichia coli*. *Lancet* 2:1235–1236.
8. Scotland, S. M., Day, N. P., and Rowe, B. (1979). Production by strains of *Escherichia coli* of a cytotoxin (VT) affecting Vero cells. *Soc. Gen. Microbiol. Q.* 6:156–157.
9. Wilson, M. W., and Bettelheim, K. A. (1980). Cytotoxic *Escherichia coli* serotypes. *Lancet* 2:201.
10. Riley, L. W., Temis, R. S., Helgerson, S. D., *et al.* (1983). Hemorrhagic colitis associated with a rare *Escherichia coli* serotype. *New Engl. J. Med.* 308:681–685.
11. Karmali, M. A., Steele, B. T., Petric, M., and Lim, C. (1983). Sporadic cases of hemolytic-uremic syndrome associated with faecal cytotoxin and cytotoxin-producing *Escherichia coli* in stools. *Lancet* 2:619–620.
12. Acheson, D. W. K., and Keusch, G. T. (1996). Which Shiga toxin—producing types of *E. coli* are important? *ASM News.* 62:302–306.
13. Whittam, T. S. (1998). Evolution of *Escherichia coli* O157:H7 and other Shiga toxin—producing *E. coli* strains. In J. B. Kaper and A. D. O'Brien (eds.), *Escherichia coli O157:H7 and*

Other Shiga Toxin—Producing E. coli Strains, pp. 195–212. Washington: American Society for Microbiology.

14. Calderwood, S. B., Acheson, D. W. K., Keusch, G. T., *et al.* (1996). Proposed new nomenclature for SLT (VT) family. *ASM News* 62:118–119.
15. Coia, J. E. (1998). Clinical, microbiological and epidemiological aspects of *Escherichia coli* O157:H7 infection. *FEMS Immun. Med. Microbiol.* 20:1–9.
16. Paton, J. C., and Paton, A. W. (1998). Pathogenesis and diagnosis of Shiga toxin—producing *Escherichia coli* infections. *Clin. Microbiol. Rev.* 11:450–479.
17. Su, C., and Brandt, L. J. (1995). *Escherichia coli* O157:H7 infection in humans. *Ann. Intern. Med.* 123:698–714.
18. O'Brien, A. D., and Kaper, J. B. (1998). Shiga toxin—producing *Escherichia coli*: Yesterday, today and tomorrow. In J. B. Kaper and A. D. O'Brien (eds.), *Escherichia coli O157:H7 and Other Shiga Toxin—Producing E. coli strains.* Washington: American Society for Microbiology.
19. Banatvala, N., Griffin, P. M., Greene, K. D., *et al.* and the Hemolytic Uremic Syndrome Study Collaborators. (2001). The United States National Prospective Hemolytic Uremic Syndrome Study: Microbiological, serological, clinical and epidemiological findings. *J. Infect. Dis.* 183:1063–1070.
20. Mead, P., Bender, I., Dembek, Z., *et al.* (1998). Active surveillance for hemolytic uremic syndrome at selected sites, United States (1997). In *Abstracts of the International Conference on Emerging Infectious Diseases (Atlanta).* Abstract P-3.12.
21. Michino, H., Araki, K., Minami, S., *et al.* (1999). Massive outbreak of *Escherichia coli* O157:H7 infection in schoolchildren in Sakai City, Japan, associated with consumption of white radish sprouts. *Am. J. Epidemiol.* 150:797–803.
22. Goldwater, P. N., and Bettelheim, K. A. (1998). New perspectives on the role of *Escherichia coli* O157:H7 and other enterohaemorrhagic *E. coli* serotypes in human disease. *J. Med. Microbiol.* 47:1039–1045.
23. Goldwater, P. N., and Bettelheim, K. A. (2000). *Escherichia coli* O group serology of a haemolytic uraemic syndrome (HUS) epidemic. *Scand. J. Infect. Dis.* 32:385–394.
24. Bielaszewska, M., Janda, J., Blahova, K., *et al.* (1996). Verocytotoxin producing *Escherichia coli* in children with hemolytic uremic syndrome in the Czech Republic. *Clin. Nephrol.* 46:42–44.
25. Thomas, A., Cheasty, T., Chart, H., and Rowe, B. (1994). Isolation of Vero cytotoxin—producing *Escherichia coli* serotypes O9ab:H- and O101:H-carrying VT2 variant gene sequences from a patient with hemolytic uraemic syndrome. *Eur. J. Clin. Microbiol. Infect. Dis.* 13:1074–1076.
26. Karmali, M. A. (1989). Infection by Verocytotoxin-producing *Escherichia coli. Microbiol. Rev.* 2:15–38.
27. Bonnet, R., Souweine, B., Gauthier, G., *et al.* (1998). Non-O157:H7 Stx2-producing *Escherichia coli* strains associated with sporadic cases of hemolytic-uremic syndrome in adults. *J. Clin. Microbiol.* 36:1777–1780.
28. Keskimaki, M., Ikaheimo, R., Karkkainen, P., *et al.* (1997). Shiga toxin—producing *Escherichia coli* serotype OX3:H21 causing hemolytic uremic syndrome. *Clin. Infect. Dis.* 24:1278–1279.
29. Schmidt, H., and Karch, H. (1996). Enterohemolytic phenotypes and genotypes of Shiga toxin—producing *Escherichia coli* O111 strains from patients with diarrhea and hemolytic-uremic syndrome. *J. Clin. Microbiol.* 34:2364–2367.
30. Goldwater, P. N., and Bettelheim, K. A. (1994). The role of enterohaemorrhagic *E. coli* serotypes other than O157:H7 as causes of disease. In M. A. Karmali and A. G. Goglio (eds.), *Recent Advances in Verocytotoxin-Producing Escherichia coli Infections*, pp. 57–60. New York: Elsevier Science.
31. Griffin, P. M., and Tauxe, R. V. (1991). The epidemiology of infections caused by *Escherichia coli* O157:H7, other enterohaemorrhagic *E. coli* and the associated haemolytic uraemic syndrome. *Epidemiol. Rev.* 13:60–98.
32. Robins-Browne, R. M., Elliot, E., and Desmarchelier, P. (1998). Shiga toxin—producing *Escherichia coli* in Australia. In J. B. Kaper and A. D. O'Brien (eds.), *Escherichia coli*

O157:H7 and Other Shiga Toxin—Producing E. coli Strains. Washington: American Society for Microbiology.

33. Elliott, E. J., Robins-Browne, R. M., O'Loughlin, E. V., *et al.* and contributors to the Australian Paediatric Surveillance Unit. (2001). Nationwide study of haemolytic uremic syndrome: Clinical, microbiological, and epidemiological features. *Arch. Dis. Child.* 85:125–131.
34. Lopez, E. L., Contrini, M. M., and de Rosa, M. F. (1998). Epidemiology of Shiga toxin—producing *Escherichia coli* in South America. In J. B. Kaper and A. D. O'Brien (eds.), *Escherichia coli O157:H7 and Other Shiga Toxin—Producing E. coli Strains*. Washington: American Society for Microbiology.
35. Gunzer, F., Bohm, H., Russmann, M., *et al.* (1992). Molecular detection of sorbitol-fermenting *Escherichia coli* O157:H7 in patients with hemolytic uremic syndrome. *J. Clin. Microbiol.* 30:1807–1810.
36. Ammon, A., Peterson, L. R., and Karch, H. (1999). A large outbreak of hemolytic uremic syndrome caused by an unusual sorbitol-fermenting strain O157:H-. *J. Infect. Dis.* 179:1274–1277.
37. Bielaszewska, M., Schmidt, H., Karmali, M. A., *et al.* (1998). Isolation and characterisation of sorbitol-fermenting Shiga toxin—producing *Escherichia coli* O157:H- strains in the Czech republic. *J. Clin. Microbiol.* 36:21235–2137.
38. Bielaszewska, M., Schmidt, H., Liesegang, A., *et al.* (2000). Cattle can be a reservoir of sorbitol-fermenting Shiga toxin—producing *Escherichia coli* O157:H- strains and a source of human disease. *J. Clin. Microbiol.* 38:3470–3473.
39. Smith, H. R., Rowe, B., Adak, G. K., and Reilly, W. J. (1998). Shiga toxin (verocytotoxin)—producing *Escherichia coli* in the United Kingdom. In J. B. Kaper and A. D. O'Brien (eds.), *Escherichia coli O157:H7 and Other Shiga Toxin—Producing E. coli Strains*. Washington: American Society for Microbiology.
40. Coia, J. E., Sharp, J. C. M., Curnow, J., and Reilly, W. J. (1994). Ten years of *Escherichia coli* O157:H7 in Scotland (1984–1993). In M. A. Karmali and A. G. Goglio (eds.), *Recent Advances in Verocytotoxin-Producing Escherichia coli Infections*, pp. 41–44. New York: Elsevier Science.
41. Waters, J. R., Sharp, J. C. M., and Dev, V. J. (1994). Infection caused by *Escherichia coli* O157:H7 in Alberta, Canada, and in Scotland: A five-year review. *Clin. Infect. Dis.* 19:834–843.
42. Beutin, L., Geier, D., Steinruck, H., *et al.* (1993). Prevalence and some properties of verotoxin (Shiga-like toxin)—producing *Escherichia coli* in seven different species of healthy domestic animals. *J. Clin. Microbiol.* 31:2483–2488.
43. Beutin, L., Geier, D., Steinruck, H., *et al.* (1997). Epidemiological relatedness and clonal types of natural populations of *Escherichia coli* strains producing Shiga toxins in separate populations of cattle and sheep. *Appl. Environ. Microbiol.* 63:2175–2180.
44. Kudva, I. T., Hatfield, P. G., and Hovde, C. J. (1997). Characterisation of *Escherichia coli* O157:H7 and other Shiga toxin—producing *E. coli* serotypes isolated from sheep. *J. Clin. Microbiol.* 35:892–899.
45. Meyer-Broseta, S., Bastian, S. N., Arne, P. D., *et al.* (2001). Review of epidemiological surveys on the prevalence of contamination of healthy cattle with *Escherichia coli* O157:H7. *Int. J. Hyg. Environ. Health* 203:347–361.
46. Sargeant, J. M., Gillespie, J. R., Oberst, R. D., *et al.* (2000). Results of a longitudinal study of the prevalence of *Escherichia coli* O157:H7 on cow-calf farms. *Am. J. Vet. Res.* 61:1375–1379.
47. Jones, D. L. (1999). Potential health risks associated with the persistence of *Escherichia coli* O157:H7 in agricultural environments. *Soil Use Manag.* 15:76–83.
48. Cobbold, R., and Desmarchelier, P. (2000). A longitudinal study of Shiga toxigenic *Escherichia coli* (STEC) prevalence in three Australian diary herds. *Vet. Microbiol.* 71:125–137.
49. Fukushima, H., Hoshina, K., and Gomyoda, M., (1999). Long-term survival of Shiga toxin—producing *Escherichia coli* O26, O111 and O157:H7 in bovine feces. *Appl. Environ. Microbiol.* 65:5177–5181.

50. Maule, A. (1997). Survival of the verotoxigenic strain *E. coli* O157:H7 in laboratory-scale microcosms. In D. Kay and C. Fricker (eds.), *Coliforms and E. coli: Problem or Solution?* pp. 61–65. Cambridge, England: The Royal Society of Chemistry.
51. Allison, L. J., Carter, P. E., and Thomson-Carter, F. M. (2000). Characterisation of a recurrent clonal type of *Escherichia coli* O157:H7 causing major outbreaks of infection in Scotland. *J. Clin. Microbiol.* 38:1632–1635.
52. Wang, G. D., and Doyle, M. P. (1998). Survival of enterohemorrhagic *Escherichia coli* O157:H7 in water. *J. Food Protect.* 61:662–667.
53. Hancock, D. D., Besser, T. E., Rice, D. H., *et al.* (1998). Multiple sources of *Escherichia coli* O157:H7 in feedlots and diary farms in the northwestern United States. *Prevent. Vet. Med.* 35:11–19.
54. Dargatz, D. A., Wells, S. J., Thomas, L. A., *et al.* (1997). Factors associated with the presence of *Escherichia coli* O157 in feces of feedlot cattle. *J. Food Protect.* 60:466–470.
55. Licence, K., Oates, K. R., Synge, B. A., and Reid, T. M. (2001). An outbreak of *E. coli* O157:H7 infection with evidence of spread from animals to man through contamination of a private water supply. *Epidemiol. Infect.* 126:135–138.
56. Anonymous (2000). Waterborne outbreak of gastroenteritis associated with a contaminated municipal water supply, Walkerton, Ontario, May–June 2000. *Can. Commun. Dis. Rep.* 26: 170–173.
57. Wieler, L. H., Vieler, E., Erpenstein, C., *et al.* (1996). Shiga toxin—producing *Escherichia coli* strains from bovines: association of adhesion with carriage of *eae* and other genes. *J. Clin. Microbiol.* 34:2980–2984.
58. Galland, J. C., Hyatt, D. R., Crupper, S. S., and Acheson, D. W. (2001). Prevalence, antibiotic susceptibility, and diversity of *Escherichia coli* O157:H7 isolates from a longitudinal study of beef cattle feedlots. *Appl. Environ. Microbiol.* 67:1619–1627.
59. Kobayashi, H., Shimada, J., Nakazawa, M., *et al.* (2001). Prevalence and characteristics of Shiga toxin—producing *Escherichia coli* from healthy cattle in Japan. *Appl. Environ. Microbiol.* 67: 484–489.
60. Paton, A. W., Woodrow, M. C., Doyle, R. M., *et al.* (1999). Molecular characterisation of a Shiga toxigenic *Escherichia coli* O113:H21 strain lacking *eae* responsible for a cluster of cases of hemolytic-uremic syndrome. *J. Clin. Microbiol.* 37:3357–3361.
61. Abdul-Raouf, U. M., Beuchat, L. R., and Ammar, M. S. (1993). Survival and growth of *Escherichia coli* O157:H7:H7 on salad vegatables. *Appl. Environ. Microbiol.* 59:1999–2006.
62. Besser, R. E., Lett, S. M., Weber, J. T., *et al.* (1993). An outbreak of diarrhea and hemolytic uremic syndrome from *Escherichia coli* O157:H7 in fresh-pressed apple cider. *JAMA* 269:2217–2220.
63. Belongia, E. A., Osterholm, M. T., Soler, J. T., *et al.* (1993). Transmission of *Escherichia coli* O157:H7 infection in Minnesota child day-care facilities. *JAMA* 269:883–888.
64. Ryan, C. A., Tauxe, R. V., Hosek, G. W., *et al.* (1986). *Escherichia coli* O157:H7 diarrhea in a nursing home: Clinical, epidemiological, and pathological findings. *J. Infect. Dis.* 154:631–638.
65. Griffin, P. M. (1995). *Escherichia coli* O157:H7 and other enterohaemorrhagic *Escherichia coli.* In M. J. Blaser, P. D. Smith, J. I. Rovin JI, *et al.* (eds.), *Infections of the Gastrointestinal Tract*, pp. 739–761. New York: Raven Press.
66. Boerlin, P., McEwan, S. A., Boerlin-Petzold, F., *et al.* (1999). Associations between virulence factors of Shiga toxin—producing *Escherichia coli* and disease in humans. *J. Clin. Microbiol.* 37:497–503.
67. Ostroff, S. M., Tarr, P. L., Neill, M. A., *et al.* (1989). Toxin genotypes and plasmid profiles as determinants of systemic sequelae in *Escherichia coli* O157:H7 infections. *J. Infect. Dis.* 160:994–998.

68. Scotland, S. M., Willshaw, G. A., Smith, H. R., and Rowe, B. (1987). Properties of strains of *Escherichia coli* belonging to serogroup O157:H7 with special reference to production of vero cytotoxins VT1 and VT2. *Epidemiol. Infect.* 99:613–624.
69. Neill, M. A. (1994). Pathogenesis of *Escherichia coli* O157:H7 infection. *Curr. Opin. Infect. Dis.* 7:295–303.
70. Moxley, R. A., and Francis, D. H. (1998). Overview of animal models. In J. B. Kaper and A. D. O'Brien (eds.), *Escherichia coli O157:H7 and Other Shiga Toxin—Producing E. coli Strains*, pp. 249–260. Washington: American Society for Microbiology.
71. Taylor, F. B., Tesh, V. L., DeBault, L., *et al.* (1999). Characterization of the baboon responses to Shiga-like toxin: Descriptive study of a new primate model of toxin responses to Stx-1. *Am. J. Pathol.* 154:1285–1298.
72. Siegler, R. L., Pysher, T. J., Tesh, V. L., and Taylor, F. B., Jr. (2001). Response to single and divided doses of Shiga toxin-1 in a primate model of hemolytic uremic syndrome. *J. Am. Soc. Nephrol.* 12:1458–1467.
73. Kang, G., Pulimood, A. B., Koshi, R., *et al.* (2001). A monkey model for enterohemorrhagic *Escherichia coli* infection. *J. Infect. Dis.* 184:206–210.
74. Slutsker, L., Ries, A. A., Greene, K. D., *et al.* (1997). *Escherichia coli* O157:H7 diarrhea in the United States: Clinical and epidemiologic features. *Ann. Intern. Med.* 126:505–513.
75. Kelly, J. K., Pai, C. H., Jadusingh, I. H., *et al.* (1987). The histopathology of rectosigmoid biopsies from adults with bloody diarrhea due to verotoxin-producing *Escherichia coli. Am. J. Clin. Pathol.* 88:78–82.
76. Griffin, P. M., Olmstead, L. D., and Petras, R. E. (1990). *Escherichia coli* O157:H7- associated colitis. *Gastroenterology* 99:142–149.
77. Richardson, S. E., Karmali, M. A., Becker, L. E., and Smith, C. R. (1988). The histopathology of the hemolytic uremic syndrome associated with verotoxin-producing *Escherichia coli* infections. *Hum. Pathol.* 19:1102–1108.
78. Kelly, J., Oryshak, A., Wenetsek, M., *et al.* (1990). The colonic pathology of *Escherichia coli* O157:H7 infection. *Am. J. Surg. Pathol.* 14:87–92.
79. Francis, D. H., Moxley, R. A., and Andraos, C. Y. (1989). Edema disease—like brain lesions in gnotobiotic piglets infected with *Escherichia coli* serotype O157:H7. *Infect. Immun.* 57:1339–1342.
80. Tesh, V. L., Burris, J. A., Owens, J. W., *et al.* (1993). Comparison of the relative toxicities of Shiga-like toxin type 1 and type 2 for mice. *Infect. Immun.* 61:3392–3402.
81. Wadolkowski, E. A., Sung, L. M., Burris, J. A., *et al.* (1990). Acute renal tubular necrosis and death of mice orally infected with *Escherichia coli* strains that product Shiga-like toxin type 2. *Infect. Immun.* 58:3959–3965.
82. Ikeda, K., Ida, O., Kimoto, K., *et al.* (2000). Predictors for the development of haemolytic uraemic syndrome with *Escherichia coli* O157:H7 infections: with focus on the day of illness. *Epidemiol. Infect.* 124:343–349.
83. Buteau, C., Proulx, F., Chaibou, M., *et al.* (2000). Leukocytosis in children with *Escherichia coli* O157:H7 enteritis developing the hemolytic uremic syndrome. *Pediatr. Infect. Dis. J.* 16:642–647.
84. Bell, P., Griffin, P. M., Lozano, P., *et al.* (1997). Predictors of hemolytic uremic syndrome in children during a large outbreak of *Escherichia coli* O157:H7 infections. *Pediatrics* 100:E12.
85. Wong, C. S., Jelacic, S., Habeeb, R. L., *et al.* (2000). The risk of the hemolytic uremic syndrome after antibiotic treatment of *Escherichia coli* O157:H7 infections. *New Engl. J. Med.* 323:1930–1936.
86. Guerrant, R. L., van Gilder, T., Steiner, T. S., *et al.* (2001). Practice guidelines for the management of infectious diarrhea. *Clin. Infect. Dis.* 32:331–351.
87. Siegler, R. L., Pavia, A. T., Christoffeson, R. D., *et al.* (1994). A 20-year population-based study of post-diarrheal hemolytic uremic syndrome in Utah. *J. Pediatr.* 94:35–40.
88. Robson, W. L., Leung, A. K., and Montgomery, M. D. (1991). Causes of death in hemolytic uremic syndrome. *Child. Nephrol. Urol.* 11:228–233.

89. Endo, Y., Tsurugi, K., Yutsudo, T., *et al.* (1988). Site of action of vero toxin (VT2) from *Escherichia coli* O157:H7 and of Shiga toxin on eukaryotic ribosomes: RNA *N*-glycosidase activity of the toxins. *Eur. J. Biochem.* 171:45–50.
90. Stein, P. E., Boodhoo, A., Tyrrell, J., *et al.* (1992). Crystal structure of the cell-binding B oligomer of verotoxin-1 from *E. coli*. *Nature* 355:748–750.
91. Fraser, M. E., Chernaia, M. M., Kozlov, Y. V., and James, M. N. (1994). Crystal structure of the holotoxin from *Shigella dysenteriae* at 2.5 A resolution. *Nature Struct. Biol.* 1:59–64.
92. Austin, P. R., Jablonski, P. E., Bohach, G. A., *et al.* (1994). Evidence that the A2 fragment of Shiga-like toxin type I is required for holotoxin integrity. *Infect. Immun.* 62:1768–1775.
93. Jackson, M. P., Wadolkowski, E. A. Weinstein, D. L., *et al.* (1990b). Functional analysis of the Shiga toxin and Shiga-like toxin type II variant binding subunit by using site-directed mutagenesis. *J. Bacteriol.* 172:653–658.
94. Garred, Dubinina, E., Holm, P. K., *et al.* (1995). Role of processing and intracellular transport for optimal toxicity of Shiga toxin and toxin mutants. *Exp. Cell Res.* 218:39–49.
95. Kokai-Kun, J. F., Melton-Celsa, A. R., and O'Brien, A. D. (2000). Elastase in intestinal mucus enhances the cytotoxicity of Shiga toxin type 2d. *J. Biol. Chem.* 275:3713–3721.
96. Melton-Celsa, A., Kokai-Kun, J., and O'Brien, A. D. (2000). Shiga toxins: Activatable or not is just a clip of the tail. In *Proceedings of the 4th International Symposium and Workshop on Shiga Toxin (Verocytotoxin)—Producing Escherichia coli Infections*, October 29–November 2, 2000, Kyoto, Japan.
97. Jackson, M. P., Deresiewicz, R. L., and Calderwood, S. B., (1990). Mutational analysis of the Shiga toxin and Shiga-like toxin II enzymatic subunits. *J. Bacteriol.* 172:3346–3350.
98. Hovde, C. J., Calderwood, S. B., Mekalanos, J. J., and Collier, R. J. (1988). Evidence that glutamic acid 167 is an active-site residue of Shiga-like toxin I. *Proc. Natl. Acad. Sci. USA* 85:2568–2572.
99. Newland, J. W., Strockbine, N. A., and Neill, R. J. (1987). Cloning of genes for the production of *Escherichia coli* Shiga-like toxin type II. *Infect. Immun.* 5:2675–2680.
100. O'Brien, A. D., Marques, L. R. M., Kerry, C. F., *et al.* (1989). Shiga-like toxin converting phage of enterohemorrhagic *Escherichia coli* strain 933. *Microb. Pathog.* 6:381–390.
101. Weinstein, D. L., Holmes, R. K., and O'Brien, A. D. (1988). Effects of iron and temperature on Shiga-like toxin I production by *Escherichia coli*. *Infect. Immun.* 56:106–111.
102. Smith, H. W., and Linggood, M. A. (1971). The transmissible nature of enterotoxin production in a human enteropathogenic strain of *Escherichia coli*. *J. Med. Microbiol.* 4:301–305.
103. Huang, A., De Grandis, S., Friesen, J., *et al.* (1986). Cloning and expression of the gene specifying Shiga-like toxin production in *Escherichia coli*. H19. *J. Bacteriol.* 166:375–379.
104. Huang, A., Friesen, J., and Brunton, J. L. (1987). Characterization of a bacteriophage that carries the genes for production of Shiga-like toxin I in *Escherichia coli*. *J. Bacteriol.* 169:4308–4312.
105. O'Brien, A. D., Newland, J. W., Miller, S. F., *et al.* (1984). Shiga-like toxin—converting phages from *Escherichia coli* strains that cause hemorrhagic colitis or infantile diarrhea. *Science* 226:694–696.
106. Calderwood, S. B., Auclair, F., Donohue-Rolfe, A., *et al.* (1987). Nucleotide sequence of the Shiga-like toxin genes of *Escherichia coli*. *Proc. Natl. Acad. Sci. USA* 84:4364–4368.
107. DeGrandis, S., Ginsberg, J., Toone, M., *et al.* (1987). Nucleotide sequence and promoter mapping of the *Escherichia coli* Shiga-like toxin operon of bacteriophage H19B. *J. Bacteriol.* 169: 4313–4319.
108. Jackson, M. P., Newland, J. W., Holmes, R. K., and O'Brien, A. D. (1987). Nucleotide sequence analysis of the structural genes for Shiga-like toxin I encoded by bacteriophage 933J from *Escherichia coli*. *Microb. Pathog.* 2:147–153.
109. Jackson, M. P., Neill, R. J., O'Brien, A. D., *et al.* (1987). Nucleotide sequence analysis and comparison of the structural genes for Shiga-like toxin I and Shiga-like toxin II encoded by bacteriophages from *Escherichia coli* 933. *FEMS Microbiol. Lett.* 44:109–114.

110. Kozlov, Y. V., Kabishev, A. A., Lukyanov, E. V., and Bayev, A. A. (1988). The primary structure of the operons coding for *Shigella dysenteriae* toxin and temperate phage H30 Shiga-like toxin. *Gene* 67:213–221.
111. Strockbine, N. A., Jackson, M. P., Sung, L. M., *et al.* (1988). Cloning and sequencing of the genes for Shiga toxin from *Shigella dysenteriae* type I. *J. Bacteriol.* 179:1116–1122.
112. Habib, N. F., and Jackson, M. P. (1992). Identification of a B subunit gene promoter in the Shiga toxin operon of *Shigella dysenteriae* 1. *J. Bacteriol.* 174:6498–6507.
113. Habib, N. F., and Jackson, M. P. (1993). Roles of a ribosome-binding site and mRNA secondary structure in differential expression of Shiga toxin genes. *J. Bacteriol.* 175:597–603.
114. Melton-Celsa, A. R., and O'Brien, A. D. (1998). Structure, biology, and relative toxicity of Shiga toxin family members for cells and animals. In J. B. Kaper and A. D. O'Brien (eds.), *Escherichia coli O157:H7 and Other Shiga Toxin—Producing E. coli Strains*, pp. 1211128. Washington: American Society for Microbiology.
115. Weinstein, D. L., Jackson, M. P., Samuel, J. E., *et al.* (1988). Cloning and sequencing of a Shiga-like toxin type II variant from an *Escherichia coli* strain responsible for edema disease of swine. *J. Bacteriol.* 170:4223–4230.
116. Morabito, S., Dell'Omo, G., Agrimi, U., *et al.* (2001). Detection and characterization of Shiga toxin—producing *Escherichia coli* in feral pigeons. *Vet. Microbiol.* 82:275–283.
117. Keusch, G. T., and Jacewicz, M. (1977). Pathogenesis of *Shigella* diarrhea: VII. Evidence for a cell membrane toxin receptor involving α1-4 linked *N*-acetyl-d-glucosamine oligomers. *J. Exp. Med.* 146:535–546.
118. Jacewicz, M., Clausen, H., Nudelman, E., *et al.* (1986). Pathogenesis of *Shigella* diarrhea: XI. Isolation of a *Shigella* toxin—binding glycolipid from rabbit jejunum and HeLa cells and its identification as globotriosylceramide. *J. Exp. Med.* 163:1391–1404.
119. Mobassaleh, M., Donohue-Rolfe, A., Jacewicz, M., *et al.* (1988). Pathogenesis of *Shigella* diarrhea: Evidence for a developmentally regulated glycolipid receptor for *Shigella* toxin involved in the fluid secretory response of rabbit small intestine. *J. Infect. Dis.* 157:1023–1031.
120. Lindberg, A. A., Brown, J. E., Stromberg, N., *et al.* (1987). Identification of the carbohydrate receptors for Shiga toxin produced by *Shigella dysenteriae* type 1. *J. Biol. Chem.* 262:1779–1785.
121. Weinstein, D. L., Jackson, M. P., Perera, L. P., *et al.* (1989). In vivo formation of hybrid toxins comprising Shiga toxin and the Shiga-like toxins and a role of the B subunit in localization and cytotoxic activity. *Infect. Immun.* 57:3743–3750.
122. DeGrandis, S., Law, H., Brunton, J., *et al.* (1989). Globotetrosylceramide is recognized by the pig edema disease toxin. *J. Biol. Chem.* 264:12520–12525.
123. Samuel, J. E., Perera, L. P., Ward, S., *et al.* (1990). Comparison of the glycolipid receptor specificities of Shiga-like toxin type II and Shiga-like toxin type II variants. *Infect. Immun.* 58:611–618.
124. Jacewicz, M. S., Mobassaleh, M., Gross, S. K., *et al.* (1994). Pathogenesis of *Shigella* diarrhea: XVII. A mammalian cell membrane glycolipid, Gb3, is required but not sufficient to confer sensitivity to Shiga toxin. *J. Infect. Dis.* 169:538–546.
125. Kiarash, A., Boyd, G., and Lingwood, C. A. (1994). Glycosphinogolipd receptor function is modified by fatty acid content. Verotoxin 1 and Verotoxin 2c perferentially recognize different globotriaosyl ceramide fatty acid homologues. *J. Biol. Chem.* 269:1139–1146.
126. Sandvig, K., Garred, O., van Helvoort, A., *et al.* (1996). Importance of glycolipid synthesis for butyric acid-induced sensitisation to Shiga toxin and intracellular sorting of toxin in A431 cells. *Mol. Biol. Cell.* 7:1391–1404.
127. Arab, S., and Lingwood, C. A. (1996). Influence of phospholipid chain length on Verotoxin/globotriaosyl ceramide binding in model membranes: Comparison of a supported bilayer film and liposomes. *Glycoconj. J.* 13:159–166.
128. Lingwood, C. A. (1996). Role of verotoxin receptors in pathogenesis. *Trends Microbiol.* 4:147–153.

129. Mobassaleh, M., Koul, O., Mishra, K., *et al.* (1994). Developmentally regulated Gb3 galactosyltransferase and alpha-galactosidase determine Shiga toxin receptors in the intestine. *Am. J. Physiol.* 267:G618–624.
130. Jacewicz, M. S., Mobassaleh, M., Acheson, D. W. K., *et al.* (1995). Maturational regulation of globotriaosylceramide, the Shiga-like toxin I receptor, by butyrate in intestinal epithelial lines. *J. Clin. Invest.* 96:1328–1335.
131. Sandvig, K., and van Deurs, B. (1996). Endocytosis, intracellular transport, and cytotoxic action of Shiga toxin and ricin. *Physiol. Rev.* 76:949–966.
132. White, J., Johannes, L., Mallard, F., *et al.* (1999). Rab6 coordinates a novel Golgi to ER retrograde transport pathway in live cells. *J. Cell Biol.* 147:743–759.
133. Small, P. L., and Waterman, S. R. (1998). Acid stress, anaerobiosis, and gadCB: Lessons from *Lactococcus lactis* and *Escherichia coli. Trends Microbiol.* 6:214–216.
134. Waterman, S. R., and Small, P. L. (1996). Characterization of the acid resistance phenotype and rpoS alleles of Shiga-like toxin—producing *Escherichia coli. Infect. Immun.* 64:2808–2811.
135. Mao, Y., Doyle, M. P., and Chen, J. (2001). Insertion mutagenesis of *wca* reduces acid and heat tolerance of enterohemorrhagic *Escherichia coli* O157:H7. *J. Bacteriol.* 183:3811–3815.
136. Phillips, A. D., and Frankel, G. (2000). Intimin-mediated tissue specificity in enteropathogenic *Escherichia coli* interaction with human intestinal organ cultures. *J. Infect. Dis.* 181:1496–1500.
137. Sperandio, V., Mellies, J. L., Nguyen, W., *et al.* (1999). Quorum sensing controls expression of the type III secretion gene transcription and protein secretion in enterohemorrhagic and enteropathogenic *Escherichia coli. Proc. Natl. Acad. Sci. USA* 96:15196–15201.
138. Surette, M. G., Miller, M. B., and Bassler, B. L. (1999). Quorum sensing in *Escherichia coli, Salmonella typhimurium,* and *Vibrio harveyi*: A new family of genes responsible for autoinducer production. *Proc. Natl. Acad. Sci. USA* 96:1639–1644.
139. Mizutani, S., Nakazono, N., and Sugino, Y. (1999). The so-called chromosomal verotoxin genes are actually carried by defective prophages. *DNA Res.* 6:141–143.
140. Unkmeir, A., and Schmidt, H. (2000). Structural analysis of phage-borne *stx* genes and their flanking sequences in Shiga toxin—producing *Escherichia coli* and *Shigella dysenteriae* type 1 strains. *Infect. Immun.* 68:4856–4864.
141. Ptashne, M. (1992). *A Genetic Switch: Phage Lambda and Higher Organisms,* 2d ed. New York: Blackwell Scientific Publications.
142. Plunkett, G., Rose, D. J., Durfee, T. J., and Blattner, F. R. (1999). Sequence of shiga toxin 2 phage 933W from *Escherichia coli* O157:H7: Shiga toxin as a late phage product. *J. Bacteriol.* 181:1767–1778.
143. Makino, K., Yokoyama, K., Kubota, Y., *et al.* (1999). Complete nucleotide sequence of the prophage VT2-Sakai carrying the Verotoxin 2 genes of the enterohemorrhagic *Escherichia coli* O157:H7 derived from the Sakai outbreak. *Genes Genet. Syst.* 74:227–239.
144. Miyamoto, H., Nakai, W., Yajima, N., *et al.* (1999). Sequence analysis of Stx2-converting phage VT2-Sa shows a great divergence in early regulation and replication regions. *DNA Res.* 6:235–240.
145. Yokoyama, K., Makino, K., Kubota, Y., *et al.* (2000). Complete nucleotide sequence of the prophage VT1-Sakai carrying the Shiga toxin 1 genes of the enterohemorrhagic *Escherichia coli* O157:H7 strain derived from the Sakai outbreak. *Gene* 258:127–139.
146. Wagner, P. L., Neely, M. N., Zhang, X., *et al.* (2001). Role for a phage promoter in Shiga toxin 2 expression from a pathogenic *Escherichia coli* strain. *J. Bacteriol.* 183:2081–2085.
147. Johansen, B. K., Wasteson Y., Granum, P. E., and Brynestad, S. (2001). Mosaic structure of Shiga-toxin-2—encoding phages isolated from *Escherichia coli* O157:H7 indicates frequent gene exchange between lambdoid phage genomes. *Microbiology* 147:1929–1936.
148. O'Brien, A. D., and Laveck, G. D. (1983). Purification and characterisation of a *Shigella dysenteriae* 1—like toxin produced by *Escherichia coli. Infect. Immun.* 40:675–683.

149. Calderwood, S. B., and Mekalanos, J. J. (1987). Iron regulation of Shiga-like toxin expression in *Escherichia coli* is mediated by the *fur* locus. *J. Bacteriol.* 169:4759–4764.
150. Neely, M. N., and Friedman, D. I. (1998). Arrangement and functional identification of genes in the regulatory region of lambdoid phage H-19B, a carrier of a Shiga-like toxin. *Gene* 223:105–113.
151. Neely, M. N., and Friedman, D. I. (1998). Functional and genetic analysis of regulatory regions of coliphage H-19B: Location of Shiga-like toxin and lysis genes suggest a role for phage functions in toxin release. *Mol. Microbiol.* 28:1255–1267.
152. Sung, L. M., Jackson, M. P., O'Brien, A. D., and Holmes, R. K. (1990). Transcription of the Shiga-like toxin type II and Shiga-like toxin II variant operons of *Escherichia coli. J. Bacteriol.* 172:6386–6395.
153. Muhldorfer, I., Hacker, J., Keusch, G. T., *et al.* (1996). Regulation of the Shiga-like 2 operon in *Escherichia coli. Infect. Immun.* 64:495–502.
154. Walker, G. C. (1996). The SOS response of *Escherichia coli.* In F. C. Neidhardt (ed.), *Escherichia coli and Salmonella Cellular and Molecular Biology*, pp. 1400–1416.
155. Matsushiro, A., Sato, K., Miyamoto, H., *et al.* (1999). Induction of prophages of enterohemorrhagic *Escherichia coli* O157:H7 with norfloxacin. *J. Bacteriol.* 181:2257–2260.
156. Kimmitt, P. T., Harwood, C. R., and Barer, M. R. (2000). Toxin gene expression by Shiga toxin—producing *Escherichia coli*: The role of antibiotics and the bacterial SOS response. *Emerg. Infect. Dis.* 6:458–465.
157. Zhang, X., McDaniel, A. D., Wolf, L. E., *et al.* (2000). Quinolone antibiotic induces Shiga toxin—encoding bacteriophages, toxin production and death in mice. *J. Infect. Dis.* 181:664–670.
158. Wagner, P. L., Acheson, D. W. K., and Waldor, M. K. (2001). Human neutrophils and their products induce Shiga toxin production by enterohemorrhagic *Escherichia coli. Infect. Immun.* 69:1934–1937.
159. Iyoda, S., Tamura, K., Itoh, K., *et al.* (2000). Inducible *stx*2 phages are lysogenised in the enteroaggregative and other phenotypic *Escherichia coli* O86:HNM isolated from patients. *FEMS Microbiol. Lett.* 191:7–10.
160. Kohler, B., Karch, H., and Schmidt, H. (2000). Antibacterials that are used as growth promoters in animal husbandry can affect the release of Shiga-toxin-2—converting bacteriophages and Shiga toxin 2 from *Escherichia coli* strains. *Microbiology* 146:1085–1090.
161. Donnenberg, M. S., and Whittam, T. S. (2001). Pathogenesis and evolution of virulence in enteropathogenic and enterohemorrhagic *Escherichia coli. J. Clin. Invest.* 107:539–548.
162. Acheson, D. W. K., Moore, R., De Breucker, S., *et al.* (1996). Translocation of Shiga toxin across polarized intestinal cells in tissue culture. *Infect. Immun.* 64:3294–3300.
163. Philpott, D. J., Ackerley, C. S., Kiliaan, A. J., *et al.* (1997). Translocation of verotoxin-1 across T84 monolayers: mechanism of bacterial toxin penetration of epithelium. *Am. J. Physiol.* 273:G1349–1358.
164. Hurley, B. P., Jacewicz, M., Thorpe, C. M., *et al.* (1999). Shiga toxins 1 and 2 translocate differently across polarized intestinal epithelial cells. *Infect. Immun.* 67:6670–6677.
165. Parkos, C. A., Delp, C., Arnaout, M. A., and Madara, J. L. (1991). Neutrophil migration across a cultured intestinal epithelium. *J. Clin. Invest.* 88:1605–1612.
166. Nusrat, A., Parkos, C. A., Liang, T. W., *et al.* (1997). Neutrophil migration across model intestinal epithelia: Monolayer disruption and subsequent events in epithelial repair. *Gastroenterology* 113:1489–1500.
167. Nash, S., Stafford, J., and Madara, J. L. (1987). Effects of polymorphonuclear leukocyte transmigration on barrier function of cultured intestinal epithelial monolayers. *J. Clin. Invest.* 80:1104.
168. Parsons, P. E., Sugahara, K., Cott, G. R., *et al.* (1987). The effect of neutrophil migration and prolonged neutrophil contact on epithelial permeability. *Am. J. Pathol.* 129:302.
169. Elliott, E., Li, Z., Bell, C., *et al.* (1994). Modulation of host response to *Escherichia coli* O157:H7 infection by anti-CD18 antibody in rabbits. *Gastroenterology* 106:1554–1561.

170. Hurley, B. P., Thorpe, C. M., and Acheson, D. W. K. (2001). Shiga toxin translocation across intestinal epithelial cells is enhanced by neutrophil transmigration. *Infect. Immun.* 69:6140–6147.
171. Mangeney, M., Lingwood, C. A., Taga, S., *et al.* (1993). Apoptosis induced in Burkitt's lymphoma cells via Gb3/Cd77, a glycolipid antigen. *Cancer Res.* 53:5314–5319.
172. Keusch, G. T., Grady, G. F., Takeuchi, A., and Sprinz, H. (1972). The pathogenesis of *Shigella* diarrhea: II. Enterotoxin-induced acute enteritis in the rabbit ileum. *J. Infect. Dis.* 126:92–95.
173. Keenan, K. P., Sharpnack, D. D., Collins, H., *et al.* (1986). Morphologic evaluation of the effects of Shiga toxin and *E. coli* Shiga-like toxin on the rabbit intestine. *Am. J. Pathol.* 125:69–80.
174. Pai, C. H., Kelly, J. K., and Meyers, G. L. (1986). Experimental infection of infant rabbits with Verotoxin-producing *Escherichia coli. Infect. Immun.* 52:16–23.
175. Blake, D. C. I., Russell, R. G., Santini, E., *et al.* (1996). Pro-inflammatory mucosal cytokine responses to Shiga-like toxin-1 (SLT-1). In G. T. Keusch and M. Kawakami (eds.), *Cytokines, Cholera, and the Gut*, pp. 75–82. Amsterdam: IOS Press.
176. Acheson, D. W. K. (1996). Effect of Shiga toxins on cytokine production from intestinal epithelial cells. In G. T. Keusch and M. Kawakami (eds.), *Cytokines, Cholera, and the Gut*, pp. 67–73. Amsterdam: IOS Press.
177. Ismaili, A., Philpott, D. J., McKay, D. M., *et al.* (1998). Epithelial cell responses to Shiga toxin—producing *Escherichia coli* infection. In J. B. Kaper and A. D. O'Brien (eds.), *Escherichia coli O157:H7 and Other Shiga Toxin—Producing E. coli Strains*, pp. 213–225. American Society for Microbiology, Washington, DC.
178. Thorpe, C. M., Hurley, B. P., Lincicome, L. L., *et al.* (1999). Shiga toxins stimulate secretion of IL-8 from intestinal epithelial cells. *Infect. Immun.* 67:5985–5993.
179. Yamasaki, C., Natori, Y., Zheng, X.-T., *et al.* (1999). Induction of cytokines in a human colonic epithelial cell line by Shiga toxin 1 (Stx1) and Stx2 but not by a nontoxic mutant Stx1 which lacks *N*-glycosidase activity. *FEBS Lett.* 442:231–234.
180. Iordanov, M., Pribnow, M. D., Magun, J. L., *et al.* (1997). Ribotoxic stress response: Activation of the stress-activated protein kinase JNK1 by inhibitors of the peptidyl transferase reaction and by sequence-specific damage to the -sarcin/ricin loop in the 28S rRNA. *Mol. Cell. Biol.* 17:3373–3381.
181. Thorpe, C. M., Smith, W. E., Hurley, B. P., and Acheson, D. W. K. (2001). Shiga toxins induce, superinduce, and stabilize a variety of C-X-C chemokine mRNA in intestinal epithelial cells resulting in increased chemokine expression. *Infect. Immun.* 69:6148–6155.
182. Cochran, B. H., Refuel, A. C., and Stiles, C. D. (1983). Molecular cloning of gene sequences regulated by platelet derived growth factor. *Cell* 33:939–947.
183. Jones, N. L., Islur, A., Haq, R., *et al.* (2000). *Escherichia coli* Shiga toxins induce apoptosis in epithelial cells that is regulated by the Bcl-2 family. *Am. J. Physiol. Gastrointes. Liver Physiol.* 278:G811–G819.
184. Nakagawa, I., Nakata, M., Kawahata, S., and Hamada, S. (1999). Regulated expression of the Shiga toxin B gene induces apoptosis in mammalian fibroblastic cells. *Mol. Microbiol.* 33:1190–1199.
185. Barnett-Foster, D., Abul-Milh, M., Huesca, M., and Lingwood, C. A. (2000). Enterohemorrhagic *Escherichia coli* induces apoptosis which augments bacterial binding and phosphatidylethanolamine exposure on the plasma membrane outer leaflet. *Infect. Immun.* 68:3108–3115.
186. Kita, E., Yunou, Y., Kurioka, T., *et al.* (2000). Pathogenic mechanism of mouse brain damage caused by oral infection with Shiga toxin—producing *Escherichia coli* O157:H7. *Infect. Immun.* 68:1207–1214.
187. Uchida, H., Kiyokawa, N., Horie, H., *et al.* (1999). The detection of Shiga toxins in the kidney of a patient with hemolytic uremic syndrome. *Pediatr. Res.* 45:133–137.
188. Uchida, H., Kiyokawa, N., Taguchi, T., *et al.* (1999). Shiga toxins induce apoptosis in pulmonary epithelium-derived cells. *J. Infect. Dis.* 180:1902–1911.

189. te Loo, D. M., van Hinsbergh, V. W., van den Heuvel, L. P., and Monnens, L. A. (2001). Detection of Verocytotoxin bound to circulating polymorphonuclear leukocytes of patients with hemolytic uremic syndrome. *J. Am. Soc. Nephrol.* 12:800–806.
190. Forsyth, K. D., Simpson, A. C., Fitzpatrick, N. M., *et al.* (1989). Neutrophil-mediated endothelial injury in haemolytic uraemic syndrome. *Lancet* 19:411–414.
191. Milford, D. V., Staten, J., MacGreggor, I., *et al.* (1991). Prognostic markers in diarrhoea-associated haemolytic-uremic syndrome: Initial neutrophil count, human neutrophil elastase and von Willebrand factor antigen. *Nephrol. Dial. Transplant.* 6:232–237.
192. Walters, M. D., Matthei, I. U., Kay, R., *et al.* (1989). The polymorphonuclear leukocyte count in childhood haemolytic uraemic syndrome. *Pediatr. Nephrol.* 3:130–134.
193. Fitzpatrick, N. M., Shah, V., Filler, G., *et al.* (1992). Neutrophil activation in the haemolytic uraemic syndrome: Free and complexed elastase in plasma. *Pediatr. Nephrol.* 6:50–53.
194. Fitzpatrick, M. M., Shah, V., Trompeter, R. S., *et al.* (1992). Interleukin-8 and polymorphoneutrophil leukocyte activation in hemolytic uremic syndrome of childhood. *Kidney Int.* 42:951–956.
195. King, A. J., Sundaram, S., Cendorogio, M., *et al.* (1999). Shiga toxin induces superoxide production in polymorphonuclear cells with subsequent impairment of phagocytosis and responsiveness to phorbol esters. *J. Infect. Dis.* 179:503–507.
196. Liu, J., Akahoshi, T., Sasahana, T., *et al.* (1999). Inhibition of neutrophil apoptosis by verotoxin 2 derived from *Escherichia coli* O157:H7. *Infect. Immun.* 67:6203–6205.
197. Barrett, T. J., Potter, M. E., and Strockbine, N. A. (1990). Evidence for participation of the macrophage in Shiga-like toxin II—induced lethality in mice. *Microb. Pathog.* 9:95–103.
198. Tesh, V. L., Ramegowda, B., and Samuel, J. E. (1994). Purified Shiga-like toxins induce expression of proinflammatory cytokines from murine peritoneal macrophages. *Infect. Immun.* 62:5085–5094.
199. Ramegowda, B., and Tesh, V. L. (1996). Differentiation-associated toxin receptor modulation, cytokine production, and sensitivity to Shiga-like toxins in human monocytes and moncytic cell lines. *Infect. Immun.* 64:1173–1180.
200. Van Setten, P. A., Monnens, L. A. H., Verstraten, R. G. G., *et al.* (1996). Effects of Verocytotoxin-1 on nonadherent human monocytes: Binding characteristics, protein synthesis, and induction of cytokine release. *Blood* 88:174–183.
201. Proulx, F., Seidman, E. G., and Karpman, D. (2001). Pathogenesis of Shiga toxin—associated hemolytic uremic syndrome. *Pediatr. Res.* 50:163–171.
202. Sakiri, R., Ramegowda, B., and Tesh, V. L. (1998). Shiga toxin type 1 activates tumor necrosis factor-α gene transcription and nuclear translocation of the transcriptional activators nuclear factor-κB and activator protein-1. *Blood* 92:558–566.
203. Foster, G. H., Armstrong, C. S., Sakiri, R., and Tesh, V. L. (2000). Shiga toxin—induced tumor necrosis factor alpha expression: Requirement for toxin enzymatic activity and monocyte protein kinase C and protein tyrosine kinases. *Infect. Immun.* 68:5183–5189.
204. Tao, R. V., Sweeley, C. C., and Jamieson, G. A. (1973). Sphingolipid composition of human platelets. *J. Lipid Res.* 14:16–25.
205. Cooling, L. L., Walker, K. E., Gille, T., and Koerner, T. A. (1998). Shiga toxin binds human platelets via globotriaosylceramide (Pk antigen) and a novel platelet glycosphingophospholipid. *Infect. Immun.* 66:4355–4366.
206. Yoshimura, K., Fujii, J., Yutsudo, T., *et al.* (1998). No direct effects of Shiga toixn 1 and 2 on the aggregation of human platelets in vitro. *Thromb. Haemost.* 80:529–530.
207. Thorpe, C. M., Flaumenhaft, R., Hurley, B., *et al.* (1999). Shiga toxins do not directly stimulate alpha-granule secretion or enhance aggregation of human platelets. *Acta Hematol.* 102:51–55.
208. Viisoreanu, D., Polanowska-Grabowska, R., Suttitanamongkol, S., *et al.* (2000). Human platelet aggregation is not altered by Shiga toxins 1 or 2. *Thromb. Res.* 98:403–410.

209. Karpman, D., Papadopoulou, D., Nilsson, K., *et al.* (2001). Platelet activation by Shiga toxin and circulatory factors as a pathogenetic mechanism in the hemolytic uremic syndrome. *Blood* 97:3100–3108.
210. Obrig, T. G., Del-Vecchio, P. J., Brown, J. E., *et al.* (1988). Direct cytotoxic action of Shiga toxin on human vascular endothelial cells. *Infect. Immun.* 56:2372–2378.
211. Louise, C. B., and Obrig, T. G. (1991). Shiga toxin associated hemolytic uremic syndrome: Combined cytotoxic effects of Shiga toxin IL-1, and tumor necrosis factor alpha on human vascular endothelial cells in vitro. *Infect. Immun.* 59:4173–4179.
212. Obrig, T. G., Louise, C. B., Lingwood, C. A., *et al.* (1993). Endothelial heterogeneity in Shiga toxin receptors and responses. *J. Biol. Chem.* 268:15484–15488.
213. Monnens, L., Savage, C. O., and Taylor, C. M. (1998). Pathophysiology of hemolytic-uremic syndrome. In J. B. Kaper and A. D. O'Brien (eds.), *Escherichia coli O157:H7 and Other Shiga Toxin—Producing E. coli Strains*, pp. 287–292. Washington: American Society for Microbiology.
214. Hutchinson, J. S., Stanimirovic, D., Shapiro, A., and Armstrong, G. D. (1998). Shiga toxin toxicity in human cerebral endothelial cells. In J. B. Kaper and A. D. O'Brien (eds.), *Escherichia coli O157:H7 and Other Shiga Toxin—Producing E. coli Strains*, pp. 323–328. Washington: American Society for Microbiology.
215. Ramegowda, B., Samuel, J. E., and Tesh, V. L. (1999). Interaction of Shiga toxins with human brain microvascular endothelial cells: Cytokines as sensitizing agents. *J. Infect. Dis.* 180: 1205–1213.
216. Eisenhauer, P. B., Chaturvedi, P., Fine, R. E., *et al.* (2001). Tumor necrosis factor alpha increases human cerebral endothelial cell Gb3 and sensitivity to Shiga toxin. *Infect. Immun.* 69:1889–1894.
217. Jacewicz, M. S., Acheson, D. W., Binion, D. G., *et al.* (1999). Responses of human intestinal microvascular endothelial cells to Shiga toxins 1 and 2 and pathogenesis of hemorrhagic colitis. *Infect. Immun.* 67:1439–1444.
218. Bitzan, M. M., Wang, Y., Lin, J., and Marsden, P. A. (1998). Verotoxin and ricin have novel effects on preproendothelin-1 expression but fail to modify nitric oxide synthase (ecNOS) expression and NO production in vascular endothelium. *J. Clin. Invest.* 101:372–382.
219. Yoshida, T., Fukada, M., Koide, N., *et al.* (1999). Primary cultures of human endothelial cells are susceptible to low doses of Shiga toxins and undergo apoptosis. *J. Infect. Dis.* 180:2048–2052.
220. Pijpers, A. H., van Setten, P. A., van den Heuvel, L. P., *et al.* (2001). Verocytotoxin-induced apoptosis of human microvascular endothelial cells. *J. Am. Soc. Nephrol.* 12:767–778.
221. Molostvov, G., Morris, A., Rose, P., and Basu, S. (2001). Interaction of cytokines and growth factor in the regulation of Verotoxin-induced apoptosis in cultured human endothelial cells. *Br. J. Haematol.* 113:891–897.
222. Morigi, M., Micheletti, G., Figliuzzi, M., *et al.* (1995). Verotoxin-1 promotes leukocyte adhesion to cultured endothelial cells under physiologic flow conditions. *Blood* 86:4553–4558.
223. te Loo, D. M. W. M., Monnens, L. A. H., van der Velden, T. J. A. M., *et al.* (2000). Binding and transfer of Verocytotoxin by polymorphonuclear leukocytes in hemolytic uremic syndrome. *Blood* 95:3396–3402.
224. Takeda, T., Dohi, S., Igarashi, T., *et al.* (1993). Impairment by Verotoxin of tubular function contributes to the renal damage seen in haemolytic uraemic syndrome. *J. Infect.* 27:339–341.
225. Lingwood, C. A. (1994). Verocytotoxin-binding in human renal sections. *Nephron* 66:21–28.
226. Hughes, A. K., Stricklett, P. K., and Kohan, D. E. (1998). Cytotoxic effect of Shiga toxin-1 on human proximal tubule cells. *Kidney Int.* 54:426–437.
227. Hughes, A. K., Stricklett, P. K., and Kohan, D. E. (1998). Shiga toxin-1 regulation of cytokine production by human proximal tubule cells. *Kidney Int.* 54:1093–1106.
228. Taguchi, T., Uchida, H., Kiyokawa, N., *et al.* (1998). Verotoxins induce apoptosis in human renal tubular epithelium derived cells. *Kidney Int.* 53:1681–1688.
229. Karpman, D., Andreasson, A., Thysell, G., *et al.* (1995). Cytokines in childhood hemolytic uremic syndrome and thrombotic thrombocytopenic purpura. *Pediatr. Nephrol.* 9:694–699.

230. Kaneko, K., Kiyokawa, N., Ohtomo, Y., *et al.* (2001). Apoptosis of renal tubular cells in Shiga toxin—mediated hemolytic uremic syndrome. *Nephron* 87:182–185.
231. Hughes, A. K., Stricklett, P. K., Schmid, D., and Kohan, D. E. (2000). Cytotoxic effect of Shiga toxin-1 on human glomerular epithelial cells. *Kidney Int.* 57:2650–2651.
232. Williams, J. M., Boyd, B., Nutikka, A., *et al.* (1999). A comparison of the effects of Verocytotoxin-1 on primary human renal cell cultures. *Toxicol. Lett.* 105:47–57.
233. Hughes, A. K., Stricklett, P. K., and Kohan, D. E. (2001). Shiga toxin-1 regulation of cytokine production by human glomerular epithelial cells. *Nephron* 88:14–23.
234. Matise, I., Cornick, N. A., Booher, S. L., *et al.* (2001). Intervention with Shiga toxin (Stx) antibody after infection by Stx-producing *Escherichia coli. J. Infect. Dis.* 183:347–350.
235. Marcato, P., Mulvey, G., Read, R. J., *et al.* (2001). Immunoprophylactic potential of cloned Shiga toxin 2 B subunit. *J. Infect. Dis.* 183:435–443.
236. Donohue-Rolfe, A., Kondova, I., Mukherjee, J., *et al.* (1999). Antibody-based protection of gnotobiotic piglets infected with *Escherichia coli* O157:H7 against systemic complications associated with Shiga toxin 2. *Infect. Immun.* 67:3645–3648.
237. Yamagami, S., Motoki, M., Kimura, T., *et al.* (2001). Efficacy of postinfection treatment with anti-Shiga toxin (Stx) 2 humanized monoclonal antibody TMA-15 in mice lethally challenged with Stx-producing *Escherichia coli. J. Infect. Dis.* 15:738–742.
238. Trachtman, H., and Christen, E. (1999). Pathogenesis, treatment, and therapeutic trials in hemolytic uremic syndrome. *Curr. Opin. Pediatr.* 11:162–168.
239. Takeda, T., Yoshino, K., Adachi, E., *et al.* (1999). In vitro assessment of a chemically synthesized Shiga toxin receptor analog attached to chromosorb P (Synsorb Pk) as a specific absorbing agent of Shiga toxin 1 and 2. *Microbiol. Immunol.* 43:331–337.
240. Ogawa, M., Shimizu, K., Nomoto, K., *et al.* (2001). Protective effect of Lactobacillus casei strain Shirota on Shiga toxin—producing *Escherichia coli* O157:H7 infection in infant rabbits. *Infect. Immun.* 69:1101–1108.
241. Paton, A. W., Morona, R., and Paton, J. C. (2000). A new biologic agent for treatment of Shiga toxigenic *Escherichia coli* infections and dysentery in humans. *Nature Med.* 6:265–270.

CHAPTER 5

Enterotoxigenic Escherichia coli

Eric A. Elsinghorst
Department of Clinical Laboratory Sciences, University of Kansas Medical Center, Kansas City, Kansas

Enterotoxigenic *Escherichia coli* (ETEC) are a major cause of diarrheal disease in humans and animals. ETEC-mediated diarrhea in humans is endemic in developing countries, where this pathogen is responsible for one-fifth of all severe diarrheal illnesses [1]. ETEC infections result in about 600 million cases of diarrhea worldwide annually, with an estimated 800,000 deaths in children younger than 5 years of age [2]. In animals, ETEC infections are associated with neonatal diarrhea in calves, lambs, and pigs and postweaning diarrhea in pigs. These infections cause significant morbidity and mortality in these animals with substantial economic impact on the livestock industry. ETEC are distinguished from other *E. coli* pathotypes by their production of enterotoxins LT (heat-labile enterotoxin) and ST (heat-stable enterotoxin). For an *E. coli* pathogen to be considered an ETEC strain, it must produce at least one of these enterotoxins. Another characteristic common to ETEC strains is their ability to adhere to the intestinal epithelium, a phenotype mediated by structures referred to as *colonization factors*. Colonization factor–mediated adherence allows for the delivery of enterotoxins and the subsequent host secretory response that is experienced as diarrhea.

Escherichia coli: Virulence Mechanisms of a Versatile Pathogen
ISBN 0-12-220751-3

ETEC EPIDEMIOLOGY

In developing countries, ETEC-mediated diarrhea primarily is a disease of weanling children. Fecal contamination of food and drink is very prevalent in endemic areas. Due to the high level of environmental ETEC contamination, infants are likely to become infected on weaning [3–5]. In the absence of prior exposure to the pathogen, such an infection may result in diarrhea. Epidemiologic studies in Bangladesh have shown that an average of six to seven diarrheal episodes are experienced by a child per year during the first 2 years of life [6]. Of these episodes, about two per year are the result of ETEC infection. This attack rate decreases with age until the incidence of ETEC-mediated diarrhea is low in older children and adults [7–9]. This decreased incidence of disease appears to reflect an accumulation of acquired mucosal immunity to the most commonly encountered ETEC colonization factors, as evidenced by the prevalence of anti-colonization factor antibodies in older children and adults. This correlation is supported by experimental infection studies in which challenged volunteers who had developed diarrhea were significantly protected from disease when subsequently rechallenged with the homologous ETEC strain [7]. Asymptomatic infections of immune individuals can occur, and such individuals may shed virulent organisms in their stools. In endemic areas, most ETEC-mediated diarrheal episodes are caused by ST-producing strains [10,11] and occur most frequently during warm seasons when multiplication can increase the bacterial load in contaminated foods [4,12]. Direct contact (person to person) does not seem to be a major route of transmission for ETEC infections [13].

While most ETEC-mediated diarrhea in endemic areas occurs in infants and young children, any immunologically naive person can experience disease. Consequently, ETEC remains the most common cause of diarrhea in individuals traveling from industrialized to developing countries [14] and in soldiers deployed to developing countries [15,16]. Such individuals lack prior exposure to ETEC and therefore are susceptible to infection. It has been estimated that 20–60% of travelers to ETEC-endemic areas experience diarrhea and that 20–40% of these diarrheal episodes are the result of ETEC infection [17–20]. ETEC diarrhea is most likely to be experienced by first-time travelers to an endemic area [20,21]; repeat or prolonged travel increases the chance of acquiring immunity. As with endogenous populations, ETEC infections are acquired by travelers to endemic areas through the consumption of contaminated food or drink.

Although ETEC diarrhea is less common in developed countries, outbreaks are being encountered increasingly in the United States [22,23]. Routine surveillance by the Minnesota Department of Health of stools from individuals with diarrheal illness found ETEC in 24% of specimens in which an enteric pathogen was identified [24]. Such findings suggest that ETEC may be responsible for a greater proportion of diarrheal illness in the United States than previously appreciated. In the United States, ETEC diarrhea may result from travel to endemic areas (i.e., imported traveler's diarrhea), from the consumption of foods imported from endemic areas [22], and from fecal contamination of food or water by travelers or individuals who have immigrated from endemic areas.

ETEC PATHOGENESIS

ETEC disease typically is initiated by consumption of contaminated food or water. In humans, a dose of 10^8 to 10^{10} organisms is required for disease, as determined in volunteer studies [25]. A lower infectious dose may be required for transmission in animals [26]. Bacteria transit to and colonize the proximal small intestine. In healthy individuals, the stomach, duodenum, and jejunum generally do not contain coliform bacteria [27]. However, during an infection, ETEC have been found in the stomach and throughout the small and large bowels [28]. ETEC attachment to the intestinal epithelium is mediated by adhesive fimbriae (also referred to as *colonization factor antigens*, or CFAs). Colonization of the intestinal mucosa allows for the localized delivery of enterotoxins that are thought to be responsible for the watery diarrhea typically associated with ETEC infections [29,30]. Two classes of ETEC enterotoxins have been described [heat-labile (LT) and heat-stable (ST)], and a strain may produce one or both of these types of toxins. The production of enterotoxins leads to net fluid secretion in the jejunum and ileum, with greater fluid loss occurring in the jejunum [31].

ETEC infections are characterized by the rapid onset of watery diarrhea after an incubation period of 14–50 hours [23,32,33]. The severity of ETEC-mediated diarrhea ranges from mild to severe, with severe cases being indistinguishable clinically from cholera [34,35]. In addition to diarrhea, infected individuals are likely to experience abdominal cramping [22]. Low-grade fever, nausea, and vomiting also may be experienced, but these symptoms are less common [8,22,33]. Individuals with untreated infections may have occult or gross blood in their stools [36]. Fecal leukocytes can be detected in the stools of humans and animals infected with ETEC strains [37,38], indicating intestinal inflammation during disease. In untreated infections, symptoms resolve spontaneously within a few days (typically 3 days; ranging from 1 to 11 days) [22]. Lethal ETEC infections occur as a result of severe dehydration and electrolyte imbalance.

ETEC VIRULENCE FACTORS

Two classes of virulence factors have been established as having a role in the pathogenesis of ETEC infections: adhesive fimbriae and enterotoxins. Recently, it has been shown that ETEC strains possess the capacity to penetrate human intestine epithelial cell lines. However, it has not been determined if the proteins that direct this activity have a role in human infections.

ETEC Colonization Factors

Colonization of the host intestinal epithelium is an essential step in ETEC pathogenesis. This adherence is mediated by proteinaceous surface structures that are referred to as *colonization factors* (CFs), *colonization factor antigens* (CFAs), *coli surface antigens* (CSAs), or *putative colonization factors* (PCFs). In human-specific ETEC strains, 21 different CFs have been identified. However, since some human ETEC isolates do not produce any of the known CFs, it is likely that more will be discovered. Animal-specific ETEC strains produce a variety of CFs that are distinct from those produced by human-specific isolates. Most often, CFs are rigid or flexible filamentous structures 2–7 nm in diameter that are composed of a single repeating protein subunit arranged in a helical array, although some CFs may possess more than one type of subunit protein [39,40]. Depending on the particular CF, it is thought that the subunit protein acts as the adhesin in one of two ways: (1) only the subunit at the CF terminus has a receptor binding site exposed or (2) receptor binding sites are exposed on many or all of the subunits of the CF [40]. ETEC CFs can be classified as fimbriae or fibrillae depending on their structure. The fimbrial CFs are rigid, filamentous, rodlike structures, whereas the fibrillar CFs are thinner, more flexible, and have fewer subunits in each helical turn than do fimbrial CFs. The CFs are distributed peritrichously on the organism, and a particular ETEC strain may produce one or multiple CFs. Some CFs lack a fimbrial or fibrilliar structure and are referred to as *nonfimbrial*.

The nomenclature for the human ETEC CFs is complex. To simplify the designations for these adhesins, Gaastra and Svennerholm [40] suggested a uniform system for the classification of human CFs using the designation *CS* followed by a number indicating the order in which they were discovered. Indicating both their original and CS designations, the 21 identified human CFs are listed in Table I, where they are clustered into five groups based on their genetic and structural homologies.

The CFA/I Group

Similarity in the amino acid sequences of the major subunit proteins and in the proteins required for fimbrial assembly show that the members of this group of fimbriae are closely related [61]. Although antigenically distinct, some of the members of this family share common epitopes in the fimbrial subunits [62–65]. Additionally, conserved residues involved in the biologic activity of major fimbrial subunits have been identified [66].

The CS5 Group

The amino acid sequences of the CS5, CS13, CS18, and CS20 subunits [48,50,52,67] indicate that this group is related to the fimbriae (K88, F41, 987P)

TABLE I ETEC Colonization Factors Associated with Human Infections

Original designation	CS designation	CF morphology/ diameter (nm)	Subunit mass (kDa)	Reference
CFA/I group				
CFA/I	CFA/I	Fimbrial (7 nm)	15.0	[41]
CS1	CS1	Fimbrial (7 nm)	16.5	[42]
CS2	CS2	Fimbrial (7 nm)	15.3	[43]
CS4	CS4	Fimbrial (6 nm)	17.0	[44]
PCFO166	CS14	Fimbrial (7 nm)	15.5/17.0	[45]
CS17	CS17	Fimbrial (7 nm)	17.5	[46]
CS19	CS19	Fimbrial (7 nm)	16.0	[47]
CS5 group				
CS5	CS5	Fimbrial (5 nm)	21.0	[48]
CS7	CS7	Fimbrial (3–6 nm)	21.5	[49]
PCFO9	CS13	Fibrillar	27.0	[50]
PCFO20	CS18	Fimbrial (7 nm)	25.0	[51]
CS20	CS20	Fimbrial (7 nm)	20.8	[52]
Bundle-forming group				
CFA/III	CS8	Fimbrial (7 nm)	18.0	[53]
Longus	CS21	Fimbrial (7 nm)	22.0	[54]
CS15 group				
8786	CS15	Nonfimbrial	16.3	[55]
CS22	CS22	Fibrillar	15.7	[56]
Nonhomologous				
CS3	CS3	Fibrillar (2–3 nm)	15.1	[57]
CS6	CS6	Nonfimbrial	14.5/16.0	[44]
2230	CS10	Nonfimbrial	16.0	[58]
PCFO148	CS11	Fibrillar (3 nm)	—	[59]
PCFO159	CS12	Fimbrial (7 nm)	19.0	[60]

produced by ETEC strains that infect animals [68]. CS7 also has been placed in this group based on multiple similarities with CS5: similarity in fimbrial morphology, identity in the N-terminal 20 amino acid residues, similarity in hemagglutination patterns, and antigenic cross-reactivity [40]. CS5 and CS7 fimbriae consist of two fimbrils coiled in a double-helix arrangement [69].

The Bundle-Forming Group

CS3 (CFA/III) and CS21 (Longus) are most closely related to the type IV family of fimbriae (see Chap. 11) [53,54,70,71].

The CS15 Group

CS15 (antigen 8786) and the recently described CS22 share a high degree of homology, and are related to the SEF14 fimbriae of *Salmonella enterica* serovar *enteritidis* [55,56,72,73].

Nonhomologous

The CS3, CS6, and CS10–12 fimbriae do not share homology with any known fimbriae.

While 21 CFs currently have been described, some CF types are encountered more frequently than are others [39,40]. In an analysis of 988 strains isolated from 18 geographically diverse regions, Wolf [74] found that about 75% of ETEC strains produce CFA/I, CS3, CS6, or CS17. Some CFs are found in association with the production of another CF. For example, CFA/II, a CF described in the literature, was found to consist of CS1, CS2, and CS3. Strains that express CFA/II produce CS3 alone or CS3 along with CS1 or CS2 [75]. A similar situation was encountered for the CF designated CFA/IV. CFA/IV consists of CS4, CS5, and CS6. Strains producing CFA/IV may express CS6 alone or CS6 along with CS4 or CS5 [76]. Certain CFs may be found associated with a limited number of serotypes, sometimes with only a single serotype. Some CFs are associated with ETEC strains producing only LT or ST, whereas others are found predominantly on strains that synthesize both enterotoxins [40,74].

During epidemiologic investigations, ETEC strains that do not express any of the known CFs are commonly isolated. Such strains actually may produce a known CF, but tests were not performed for all the known CFs during the study, or the detection method used during the study was insufficiently sensitive to identify a known CF. Alternatively, such CF-negative strains may produce novel CFs that remain to be identified. Interestingly, the majority of CF-negative ETEC isolates express LT alone [74].

Colonization Factor Genetics

ETEC strains typically possess multiple plasmids with a wide range of molecular masses [77,78]. The genes encoding CFs generally are found on a plasmid that also encodes ST and/or LT [68]. The best-characterized CF genes are those encoding the K88, K99, and 987P fimbriae of animal-specific ETEC strains [68]. These genes are linked in clusters of 8 to 11 genes coding for proteins that are involved in the regulation of CF gene expression, CF assembly (an outer membrane usher and a periplasmic chaperone), and the CF structure itself (major and minor subunits). In contrast, human CFs typically are encoded in gene clusters consisting of 4 genes coding for the fimbrial subunit and the proteins involved in CF assembly/transport [41,43,57,61,68,79,80]. Genes encoding proteins involved in CF gene expression may be physically separated from CF structure/assembly genes [41]. Human CF gene clusters frequently are flanked by fragments of insertion sequences and transposases, suggesting that these operons may have been part of transposable elements.

Colonization Factor Receptors

ETEC strains are host-specific. The CFs confer host specificity on the strain. For example, ETEC strains expressing K88 fimbriae cause disease only in pigs, whereas K99-expressing strains cause disease in piglets, calves, and lambs and CFA/I-expressing strains cause disease in humans. For host colonization, an ETEC strain must express a CF type that will bind to a molecule present on the surface of host enterocytes. The host receptors for various CFs have been characterized only partially and are best established for animal-specific CF types. In general, the molecules recognized by ETEC CFs are the oligosaccharide components of glycolipids and glycoproteins. Receptor characterization studies sometimes are performed using nontarget tissues. In such studies, the identified receptor may not fully reflect the receptor *in vivo* due to carbohydrate heterogeneity in and tissue-dependent distribution of glycolipids and glycoproteins.

The characterization of receptors for human-specific CFs is limited. Putative receptors for CFA/I have been identified in human red blood cells. These receptors include the glycoprotein glycophorin A [81], ganglioside GM_2 [82], and sialic acid–containing glycoproteins [83]. Ganglioside GM_1 is a putative receptor for colonization factors CS1, CS2, CS3, and CS4 [84]. The presence of sialic acid on its receptor is important for the binding of CS2 [85]. CS3 binds sialic acid–containing glycoproteins [83]. Additionally, receptor binding by CS3 could be inhibited by GM_1, asialo-GM_1, and GM_2 gangliosides [86]. From these findings, it appears that the specific carbohydrate moiety recognized by CS3 is GalNAcβ1–4Gal [86]. CS6 binds a 55-kDa glycoprotein found in human, bovine, sheep, and pig red blood cells and may interact with both sialic and galactosyl residues on this putative receptor [87]. CS7 and CS17 bind to proteins isolated from rabbit intestinal brush-border membranes [88].

In animal-specific ETEC strains, the receptors for K88, 987P, and K99 fimbriae are the best characterized. The receptors for K88 fimbriae have been investigated extensively (for recent reviews, see refs. 89–91). These fimbriae are associated with disease in piglets and are found in three antigenic variants: K88ab, K88ac, and K88ad, with K88ac encountered most commonly in nature [92]. Several putative receptors have been identified from pig epithelial cells for these variants, including intestinal mucin-type glycoproteins that are bound by K88ab and K88ac [93,94], enterocyte transferrin that is bound by K88ab [95], and a neutral glycosphingolipid that is bound by K88ad [96]. 987P fimbriae also are associated with disease in pigs, and multiple putative receptors have been identified for this fimbrial type, including glycoproteins, ceramide monohexoside, and sulfatide glycolipids [97,98]. K99 fimbriae are associated with disease in piglets, calves, and lambs, and the ganglioside NeuGc-GM_3 has been shown to act as a receptor for this fimbriae [99].

ETEC Enterotoxins

Colonization of the intestinal mucosa allows for the localized delivery of enterotoxins [29,30]. The activities of the enterotoxins lead to net fluid secretion in the jejunum and ileum, with greater fluid loss occurring in the jejunum [31]. This net fluid loss results in the watery diarrhea typically associated with ETEC infection. Two different classes of ETEC enterotoxins have been identified: heat-labile toxin (LT) and heat-stable toxin (ST). About 46% of ETEC isolates express ST alone, 25% express LT alone, and 29% express both ST and LT [74].

Heat-Labile Enterotoxins

The LT toxins are structurally and functionally similar to cholera enterotoxin (CT) produced by *Vibrio cholerae* [100]. Structural similarities include overall toxin architecture and amino acid sequence. Functional similarities include receptor specificity and enzymatic activity. Two antigenically distinct types of LT have been found in ETEC: LT-I and LT-II. Antigenic variants of both type I and II LT toxins have been described [101–105]. LT-I family toxins can be neutralized by cholera toxin antisera, whereas LT-II family toxins cannot [105]. LT-I is produced by ETEC strains that are associated with disease in humans and animals. LT-II is found primarily in animal-specific ETEC strains. However, the production of LT-II is not associated with the diarrheagenic potential of these strains.

LT-I

The LT-I holotoxin is a compound A-B toxin of about 88 kDa [100,106–108]. The A subunit consists of a 30-kDa monomer polypeptide that has enzymatic function. Although synthesized as a single polypeptide, it is proteolytically nicked to form the A_1 and A_2 subunits that remain associated by virtue of a disulfide bond. The A_1 subunit retains the enzymatic function, whereas the A_2 subunit allows the protein to interact with the B subunit. The B subunit mediates receptor binding and is a ring-shaped pentamer of five identical 11.5-kDa polypeptides. The ring-shaped structure forms a central channel into which the A_2 subunit extends. LT is secreted to the E. coli periplasmic space [109,110]. Multiple mechanisms of LT release from the periplasm have been proposed, including intestinal conditions [111], the formation of outer membrane vesicles [112,113], and an LT secretion apparatus [114].

Once secreted, the LT B subunit strongly binds ganglioside GM_1 in the host cell membrane. LT also weakly binds GM_2 and asialo-GM_1, as well as some glycoproteins [101,115–119]. After binding, the receptor-toxin complex undergoes endocytosis, followed by translocation of the A_1 subunit through the vesicle membrane by a mechanism that remains unclear. While associated with the endosome, the toxin is thought to be translocated through the cell via the trans-Golgi vesicular transport apparatus [120,121]. This translocation may allow the toxin to reach its target, which is associated with the basolateral membrane of the cell. Proteolytic cleavage of the A subunit and reduction of the disulfide bond linking the A_1 and A_2 subunits are required for biologic activity of the toxin [122].

The function of LT has been reviewed recently [123,124]. The A_1 peptide functions enzymatically as an ADP-ribosyltransferase. LT transfers the ADP-ribose from NAD to arginine residue 201 on the alpha subunit of the trimeric GTP-binding protein G_s. The function of the $G_{s\alpha}$ regulatory subunit is to transiently activate the basolateral membrane-associated enzyme adenylate cyclase. ADP-ribosylation of $G_{s\alpha}$ inhibits the ability of this protein to spontaneously hydrolyze GTP, as is required for its transient association with adenylate cyclase. Consequently, adenylate cyclase remains permanently activated, leading to increased intracellular concentrations of cyclic AMP (cAMP). LT activity appears to be stimulated by specific low-molecular-weight proteins called *ADP-ribosylation factors*. cAMP is an intracellular messenger that activates the cAMP-dependent protein kinase A kinase. This activation leads to the phosphorylation of ion channels found in the apical membrane of intestine epithelial cells. The cystic fibrosis transmembrane conductance regulator (CFTR) is one of the phosphorylated ion transporters. Owing to this phosphorylation, there is increased secretion of Cl^- from intestinal crypt cells and decreased absorption of Na^+ and Cl^- by villus tip cells. This action reverses the osmotic gradient usually maintained by the intestinal epithelium, resulting in water loss to the gut lumen, which is experienced as diarrhea.

In addition to the ADP-ribosylation of $G_{s\alpha}$ and the subsequent effects on ion transport, LT may stimulate diarrhea through additional mechanisms. Most of the *in vitro* and *in vivo* evidence supporting the existence of additional mechanisms comes from research with CT. However, considering the substantial structural and functional similarity between CT and LT, it is likely that this evidence also applies to LT. A mechanism through which these toxins may act is by stimulating the production and release of prostaglandins (PGE_1 and PGE_2) and platelet-activating factor, which then stimulate intestinal motility and ion transport [124,125]. It remains unclear if the release of these metabolites is in response to ADP-ribosylation of $G_{s\alpha}$ or increased cAMP concentrations resulting from ADP-ribosylation of $G_{s\alpha}$. CT stimulates the release of 5-hydroxytryptamine (serotonin) by intestinal enterochromaffin cells and vasoactive intestinal peptide (VIP) from enteric neurons [124–126]. These hormones then can increase secretion directly by intestine epithelial cells. In addition to the release of VIP, CT also appears to affect the enteric nervous system by influencing smooth muscle contractions and intestinal flow. Release of serotonin and VIP may occur in response to elevated cAMP levels. CT stimulates the release of the proinflammatory cytokine interleukin 6 (IL-6) from cultured rat intestine epithelial cells and ligated mouse intestinal loops [127,128]. IL-6 may promote an inflammatory response that contributes to disease. Finally, CT may increase the rate of villus cell loss, and such a loss may contribute to diarrhea [125]. Some or all of these various alternative mechanisms of CT and LT action may combine with the classic pathway of $G_{s\alpha}$ ADP-ribosylation and altered ion transport to cause diarrhea.

LT-II

LT-II is an enterotoxin that is associated primarily with animal-specific ETEC strains. Like LT-I, LT-II is a compound A-B toxin. While the mass of LT-II is similar to that of LT-I, it is antigenically distinct and is not neutralized by anti-LT-I or anti-CT antisera. Correspondingly, the LT-I and LT-II A subunits have about 57% similarity, whereas the B subunits do not have any significant similarity [129,130]. LT-II enterotoxins are subgrouped into two antigenic variants: LT-IIa and LT-IIb. These variants share 71% and 66% similarity in their A and B subunits, respectively [130]. The A subunit of LT-II acts in the same manner as that of LT-I: ADP-ribosylation of $G_{s\alpha}$ [131,132]. However, the receptor specificity of the LT-II B subunits differs from those of the LT-I B subunits. LT-IIa most efficiently binds ganglioside GD_{1b}, while LT-IIb binds GD_{1a} [101].

As mentioned earlier, LT-II is found primarily in ETEC strains that infect animals and rarely in strains infecting humans [133]. The role of this toxin in the pathogenesis of disease is unclear. *In vitro*, the specific activity of LT-II is 25–50 times greater than that of LT-I in the Y1 mouse adrenal cell assay, and LT-II increases the vascular permeability of rabbit skin [131]. However, LT-II fails to stimulate fluid accumulation in the *in vivo* rabbit ileal loop model [131], a finding that may reflect the absence of appropriate receptors in the rabbit intestine.

Heat-Stable Enterotoxins

Two different types of ETEC heat-stable (ST) enterotoxins have been described: STa and STb. While both these toxins are characterized as low-molecular-mass polypeptides that contain multiple cysteine residues forming disulfide bonds that are essential for their heat stability and biologic activity, the two types of toxins are distinct in their amino acid sequences and modes of action. In contrast to LT, the ST toxins are secreted into the medium [134–136].

STa (STI)

STa is an 18- or 19-amino-acid peptide with a mass of about 2000 [137,138]. STa is synthesized as a 72-amino-acid precursor, and on secretion to the periplasmic space, it is found as a peptide with 53 residues [134]. In the periplasm, the peptide is proteolytically processed to form the mature toxin. Additionally, in the periplasm, three intramolecular disulfide bonds are formed that fold the peptide with three β turns and maintain it in the active conformation [139–143]. The mature toxin is thought to diffuse across the outer membrane from the periplasmic space, although recent work indicates the involvement of the TolC outer membrane protein in this process [144]. Residues 5–17 or 6–18 are sufficient for receptor binding and enterotoxic action of the toxin [145]. Among these 13 amino acids are found the 6 cysteine residues required for toxin folding. Residues 11–14 are thought to be particularly important for toxin activity [140,146]. Two antigenic variants of STa have been described: STaP (porcine STa or ST1a) and STaH (human STa or ST1b). STaP is produced by animal and human strains, whereas STaH is made only by human isolates. The amino acid sequences of the 13 residues sufficient for toxin activity are nearly identical (12 of 13 residues) in these variants [147–149]. Furthermore, the amino acid residues considered important for STa activity (residues 11–14) are identical in these antigenic variants.

STa binds to the extracellular ligand-binding domain of guanylate cyclase C (GC-C) found in the brush-border membrane of intestine epithelial cells, resulting in increased intracellular cGMP concentrations [150–154]. Amino acid residues 387–393 of GC-C are involved in ST binding [155]. GC-C is a transmembrane protein belonging to the atrial natriuretic peptide receptor family and is located in the apical membrane of intestine epithelial cells [156–158]. GC-C consists of an extracellular ligand-binding domain, a transmembrane domain, and a cytoplasmic guanylate cyclase catalytic domain. The natural ligand for GC-C is the peptide hormone guanylin, which is thought to participate in the maintenance of gut homeostasis. STa binding stimulates GC-C activity and chloride secretion more effectively than does guanylin [159–162]; therefore, STa is thought to be a superagonist of GC-C. In humans, the concentration of STa receptors decreases rapidly with age [163], which may contribute to the severity of STa-mediated ETEC diarrhea in children. The density of STa receptors is highest in villus cells and decreases toward the crypt [164]. Whereas GC-C clearly has been identified as an STa receptor, there may be additional STa receptors [157,165,166].

Binding of STa to the GC-C extracellular domain activates the guanylate cyclase activity of this receptor, resulting in elevated levels of cGMP. GC-C activation occurs rapidly, with maximal levels of cGMP observed within 5 minutes [167]. As a consequence of elevated cGMP levels, sodium and chloride adsorption is inhibited in villus epithelial cells, whereas secretion of chloride ions is stimulated in crypt epithelial cells, leading to fluid accumulation in the gut lumen [150,153,168]. The mechanism through which cGMP exerts this effect remains unclear, although proposed pathways include cGMP-dependent kinase and cAMP-dependent kinase [167,169,170]. As with LT, a principal target for these activation pathways appears to be CFTR [171]. Other modes of action for STa on host cells have been proposed but not confirmed. These actions include elevating intracellular calcium levels; stimulating the release of arachidonic acid, prostaglandins, phosphatidylinositol, diacylglycerol, and serotonin; and cytoskeletal rearrangement (reviewed in ref. 124). The inhibition of Na^+/H^+ exchange also has been proposed as a mechanism for STa [172]. Interaction with the enteric nervous system appears to enhance the secretory effect of STa [173].

The gene encoding STa (*estA*) is found on a plasmid as part of a transposon designated Tn1681 [174]. Plasmids coding for STa also may code for colonization factors, antibiotic resistance, and colicin production [175]. STa synthesis is subject to catabolite repression [147,176].

STa is similar to EAST1, a heat-stable enterotoxin that was first identified in enteroaggregative *E. coli* [177]. Subsequently, the gene coding for EAST1 has been found in ETEC strains expressing certain colonization factors [178,179]. EAST1 is a 38-amino-acid peptide (4.1 kDa) with four cysteine residues. EAST1 shows homology with STa but is immunologically distinct from that toxin [177,180]. Similarly to STa, EAST1 appears to act through the stimulation of guanylate cyclase [180]. In addition to EAST1, STa is homologous with heat-stable enterotoxins made by *Yersinia enterocolitica* and *V. cholerae* non-O1 strains [145,181,182]. EAST1 is encoded by the *astA* gene, which has been shown to be embedded within a 1209-bp open reading frame (ORF) of a insertion sequence designated IS1414 [183].

STb (STII)

STb is a 48-amino-acid peptide containing four cysteine residues and two disulfide bonds [184–186]. It is synthesized as a 71-amino-acid precursor containing a 23-residue signal sequence that is proteolytically cleaved to form the mature toxin. STb shares no homology with STa, and anti-STb antiserum fails to neutralize STa [187]. Although the regions of STb participating in receptor binding and endotoxic activity have not been clearly defined, residues Lys22, Lys23, Arg29, and Asp30 are involved in the biologic activity of the toxin [186,188]. The intestinal receptor for STb has not been identified. It has been proposed that STb binds nonspecifically to plasma membrane lipids [189–191]. STb leads to the release of serotonin [192–194], production of prostaglandin E_2 [193–195], and increased free cytosolic calcium ion concentration [196]. The release of serotonin and increased calcium levels appear to result from the activation of the GTP-binding regulatory protein $G_{ai}3$ [192,196]. Histologic damage to intestinal epithelium (loss of villus epithelial cells and villus atrophy) has been associated with STb [197]. Such damage could contribute to the host diarrheal response. STb also increases motility in the mouse intestine through a direct action on ileal muscle cells [198].

Synthesis of STb is restricted primarily to porcine ETEC strains, and STb is the enterotoxin most commonly encountered in porcine ETEC strains [199,200].

However, it is unusual for STb to be the only enterotoxin made by such strains [199]. In studies using neonatal pigs and isogenic adherent STb^+ and STb^- ETEC strains, the presence of STb did not correlate with diarrheal symptoms [201]. Therefore, the contribution of STb to the pathogenesis of ETEC-induced diarrhea in pigs is unclear. The gene encoding STb (*estB*) is part of a 9-kb transposon (Tn4521) that is plasmid-borne [202–204]. *estB*-bearing plasmids may encode other ETEC virulence factors, such as LT, STa, and colonization factors. Additionally, these plasmids may carry mobilization functions as well as antibiotic resistance and colicin production genes [77,78].

Other ETEC Virulence Factors

The importance of CFs and enterotoxins in the pathogenesis of ETEC infections has been clearly established. Although enterotoxin production is believed to be the primary cause of water loss during ETEC infection, studies performed with nontoxigenic strains (i.e., ETEC strains that no longer produce LT or ST) suggest that the production of enterotoxins may not absolutely be required for diarrhea. In these studies, some volunteers developed mild diarrhea after challenge with nontoxigenic ETEC strains [205]. Furthermore, such strains consistently caused fluid accumulation and diarrhea in animal models [206–209] and impaired water and electrolyte absorption by the intestinal mucosa [207]. These observations suggest the presence of uncharacterized virulence factors in ETEC strains. Such virulence factors might consist of novel enterotoxins or other types of virulence mechanisms.

It has been found that ETEC strains are capable of penetrating (invading) cultured human intestine epithelial cell lines [210]. Although epithelial cell invasion has not been thought or shown to play a role in ETEC pathogenesis in humans, ETEC strains can cause septicemia [211] and extraintestinal disease [212] in preweaned and weaned swine. Furthermore, gnotobiotic piglet challenge studies have shown that ETEC can be found within enterocytes [213]. Analysis of the ETEC invasion phenotype has led to identification of the *tia*, *tibA-D*, and *leoA* genes and two pathogenicity islands, all of which may play a role in ETEC pathogenesis.

tia

The *tia* gene codes for a 25-kDa outer membrane protein (Tia) that directs epithelial cell adherence and invasion by ETEC strain H10407 [214]. Tia was purified and shown to bind to HCT8 (human ileocecum epithelial) cells in a dose-dependent and saturable manner, showing its ability to act as an adhesin [215]. Binding of Tia to epithelial cells appears to occur on a specific receptor and leads to host cell cytoskeletal reorganization [216]. Tia was predicted to be an integral outer membrane protein with eight membrane-spanning amphipathic β-sheets and four surface-exposed loops. A peptide corresponding to the second of these loops blocked Tia-mediated invasion, whereas a scrambled peptide had no effect on this phenotype [215]. Tia may participate in the pathogenesis of other organisms: *tia* homologues have been found in other pathogens, such as uropathogenic E. coli, *S. enterica* serovar *typhimurium*, and *Neisseria meningitidis* [217–221]. Additionally, *tia* probes hybridized with some enteropathogenic E. coli (EPEC), enteroaggregative *E. coli* (EAEC), and *Shigella sonnei* strains [214]. The role of Tia in ETEC pathogenesis remains undefined. Tia may function primarily as an adhesin, allowing close association of the bacterium with the enterocyte after CF-mediated colonization. This close adherence may allow for the most efficient delivery of enterotoxins to the host cell. Additionally, Tia-mediated invasion may contribute to the development of diarrheal disease through an uncharacterized mechanism.

DNA sequence analysis revealed that the *tia* gene is part of a large 46-kb genetic element inserted within the *selC* gene on the H10407 chromosome [114]. This element possesses multiple features in common with described pathogenicity islands and thus has been designated *enterotoxigenic pathogenicity-associated island 1* (TPAI1). Considerable TPAI1 heterogeneity was found within ETEC strains: Some strains hybridized with all probes, whereas others failed to hybridize or hybridized with a subset of the probes. EPEC and enterohemorrhagic *E. coli* (EHEC) isolates also hybridized with TPAI1 probes, indicting that this island is found in other enteric pathogens. Several additional ORFs within TPAI1 have been identified. One of these (designated *leoA*) has been shown to participate in the pathogenesis of ETEC infections.

leoA

Analysis of the TPAI1 DNA sequence downstream of *tia* revealed several potential ORFs [114]. Translation of these ORFs showed proteins bearing multiple motifs associated with bacterial secretion apparatuses. An in-frame mutation in one of these candidate ORFs was constructed in ETEC strain H10407. This mutation resulted in a lack of fluid accumulation in the rabbit ileal loop model of disease. Additionally, the mutation resulted in a marked decrease in the secretion of LT to the culture supernatant. Therefore, the mutated ORF was designated *leoA*, for labile enterotoxin output. LeoA may be part of an outer membrane secretion apparatus [114].

tib

The *tib* locus contains four genes (*tibA–D*) and codes for a 104-kDa outer membrane glycoprotein (TibA) that directs epithelial cell adherence and invasion by ETEC strain H10407 [222,223] (Fig. 1). TibA is the first reported glycoprotein

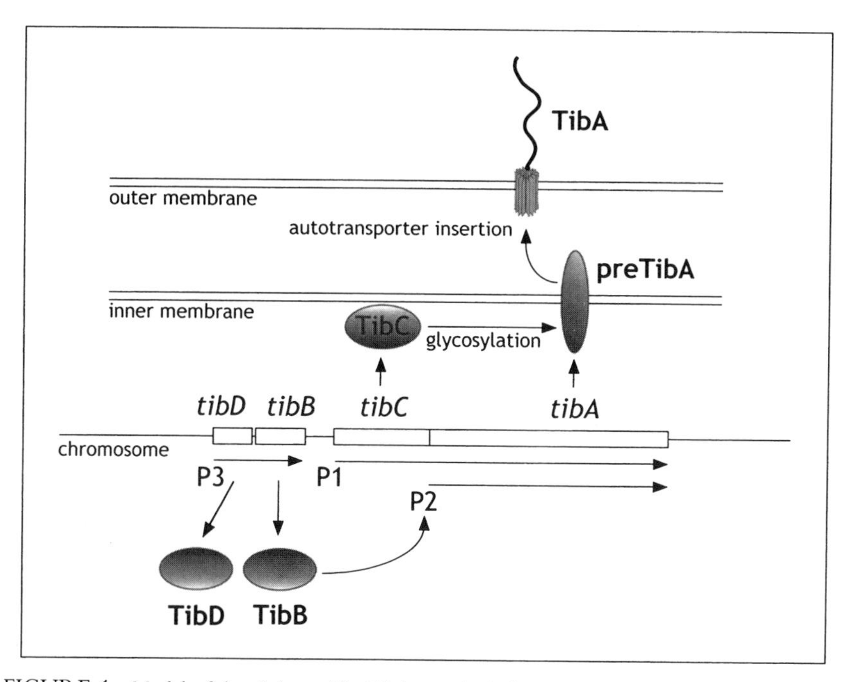

FIGURE 1 Model of the *tib* locus. The TibA protein is the product of the *tib* locus, which contains the *tibA*, *tibB*, *tibC*, and *tibD* genes. Based on T7 promoter/expression and genetic complementation experiments [222] and DNA sequencing, it is proposed that these genes are transcribed from three promoters (P1, P2, and P3). The *tibA* gene codes for a nonglycosylated 100-kDa TibA precursor (pre-TibA) that contains a *sec*-dependent signal sequence [222,223]. The product of the *tibC* gene (TibC) shares homology with RfaQ [223], a protein that is proposed to be a heptosyltransferase III involved in the biosynthesis of the *E. coli* lipopolysaccharide inner core [224]. The cytosolic TibC protein is required for the glycosylation of TibA [225] and therefore is proposed to be a glycosyltransferase that modifies pre-TibA before or during its passage through the inner membrane. Once glycosylated, TibA is 104 kDa [222,223]. TibA is a member of the autotransporter family of outer membrane proteins [223,226]. As an autotransporter, it is thought that the C-terminal third of TibA forms a β-barrel that self-inserts into the outer membrane on reaching the periplasm [223]. The N-terminal two-thirds of TibA (the autotransporter passenger domain) is thought to be translocated to the cell surface through the β-barrel. The TibA passenger domain appears to remain covalently attached to the β-barrel. The TibA passenger domain contains two regions of highly repetitive amino acid sequences that have been proposed to be involved in the biologic activities of the protein (i.e., epithelial cell adherence and invasion) [223]. Based on complementation experiments [222,223], expression of the *tibA* gene from the P2 promoter is activated by the TibD and/or TibB proteins. TibB is similar to RcsA. The RcsA protein is involved in transcription activation of the colanic acid capsule biosynthesis genes in *Escherichia*, *Erwinia*, and *Salmonella* spp. [227]. TibD is similar to CaiF. CaiF activates transcription of the *caiTABCDE* and *fixABCX* operons, which are involved in carnitine metabolism in *E. coli* [228,229].

to be produced by *E. coli*, the only glycoprotein made by H10407, and the only known glycoprotein that appears to act as an invasin [223]. TibA was purified and shown to bind to HCT8 cells in a dose-dependent and saturable manner, showing its ability to act as an adhesin [230]. TibA is similar to AIDA-I, an adhesin from the diarrheagenic diffusely adherent *E. coli*, and to pertactin, an adhesin from *Bordetella pertussis* [223]. The role of TibA in ETEC pathogenesis remains uncharacterized. As with Tia, the primary role of TibA may be to allow close adherence between the bacterium and the host enterocyte, thereby enhancing enterotoxin delivery. However, TibA-mediated invasion may contribute to the development of diarrheal disease through an uncharacterized mechanism.

Analysis of the DNA sequences flanking the *tib* genes suggests that the *tib* locus is part of a second ETEC pathogenicity island, designated TPAI2. TPAI2 is at least 34.4 kb in length, and its point of insertion in the chromosome has yet to be determined. In addition to the *tib* genes, TPAI2 contains other ORFs commonly associated with pathogenicity islands (such as genes involved with DNA recombination). Although the *tib* genes are chromosomally encoded, they appear to be associated with strains that express the plasmid-encoded CF type, CFA/I [222]. Further genetic analysis will be needed to determine if other TPAI2 sequences are present in other ETEC strains and other pathogens.

TREATMENT AND PREVENTION OF ETEC INFECTIONS

Treatment

ETEC infections can be treated using oral rehydration and chemotherapeutic therapies. Maintaining adequate hydration and electrolyte balance is central in dealing with an ETEC infection. Therefore, oral rehydration solutions are recommended. Such therapy can be lifesaving, particularly for infants and children, who are the most prone to diarrhea-induced dehydration. Maintaining hydration is also important for travelers who are experiencing diarrhea. Chemotherapeutic approaches to treating ETEC diarrhea include the use of loperamide or bismuth subsalicylate, both of which can decrease the severity of disease [20]. Loperamide should be administered only in the absence of fever or dysentery or in conjunction with an antibiotic. Antibiotic therapy decreases the duration of symptoms and decreases ETEC shedding in stools [231,232]. Antibiotics often are administered empirically without the determination of susceptibility patterns. However, the emergence of antibiotic resistance in ETEC strains has hampered effective therapy [18,233,234]. Currently, fluoroquinolones (such as such as ciprofloxacin) are the most commonly recommended class of antibiotics for ETEC infections [235]. Trimethoprim-sulfamethoxazole or doxycycline had been recommended for traveler's diarrhea, but resistance to these antibiotics is now common in ETEC strains [236]. It has been suggested that 5-hydroxytryptamine receptor antagonists and other inhibitors of intestinal pathophysiology will be good candidates for antidiarrheal chemotherapy [237].

Prevention

Antibiotic therapy can be effective in the prevention of traveler's diarrhea [18,238]. Owing to the continuing emergence of antibiotic resistance and the potential side effects from such prophylaxis, this approach is not recommended. Instead, it is recommended that antibiotics be used only if diarrhea develops [18]. For travelers, more practical prevention measures include the avoidance of potentially contaminated foods and beverages and the consumption of bismuth subsalicylate four times daily [18,20]. Although not currently developed for routine prevention of traveler's diarrhea, it has been shown that bovine milk immunoglobulin concentrate containing high titers of anti-ETEC antibodies can be protective [239]. While this passive immunization may protect from infection, it does not appear to be a useful therapy to treat ETEC-mediated diarrhea [240].

Vaccines

Owing to the significant morbidity and mortality associated with ETEC infections, considerable effort has been devoted to the development of a vaccine that will prevent ETEC diarrhea in humans. It is clear that natural and experimental ETEC infections do elicit a protective immune response. Volunteer studies have shown that ETEC infection provides protective immunity to challenge with the same strain [7,241]. The decrease in natural infection rates with age in endemic areas also indicates the development of immunity with exposure to various ETEC strains. The immune response to infection includes the production of anti-CF, antienterotoxin, and anti-O-antigen antibodies. Human and animal challenge studies have shown the significance of anti-CF antibodies for protection [242]. However, that protection does not extend to other ETEC strains expressing a different antigenic type of CF. Therefore, the wide antigenic diversity of the ETEC CFs has hampered the development of an effective vaccine. As mentioned earlier, some CFs are encountered more frequently than others, and about 75% of ETEC strains produce CFA/I, CS3, CS6, or CS17 [74]. This finding suggests that a multivalent vaccine that stimulates an immune response to these most common CF types could protect against a majority of ETEC infections.

In addition to anti-CF antibodies, anti-LT antibodies are produced as a result of natural and experimental ETEC infections. The efficacy of antienterotoxin antibodies in disease prevention is unclear. Some studies have shown that anti-LT antibodies do not protect volunteers challenged with heterologous LT-producing ETEC strains (e.g., see ref. 7), whereas other studies indicate that antibodies that recognize LT may be protective (e.g., see refs. 243 and 244) or act synergistically with anti-CF antibodies [245]. While these findings may be contradictory, it is thought that a maximally efficacious vaccine would stimulate immunity to LT. It has been estimated that a vaccine combining the common CFs and LT toxoid would cover at least 85% of 1699 ETEC strains isolated from 18 geographically diverse sites [74].

Natural infections do not elicit anti-ST antibodies. Protein fusion constructs have been developed that allow for the generation of neutralizing anti-ST antibodies (e.g., see refs. 246 and 247). Such constructs may be useful for future vaccine development efforts. ETEC infections do stimulate anti-O-antigen antibodies, and these antibodies may be protective against infections with homologous O groups. However, at least 78 different O serogroups have been associated with ETEC strains [74], decreasing the practicality of basing an ETEC vaccine on O-antigen type.

In consideration of the importance of anti-CF and anti-LT antibodies for protection, multiple approaches have been taken to develop an effective oral ETEC vaccine for humans. These approaches have included inactivated vaccines (e.g., purified CF, toxoids, and killed whole cells) and live vaccines (e.g., attenuated ETEC strains and attenuated *Salmonella*, *Shigella*, or *Vibrio* strains expressing ETEC antigens). Purified CFs have been administered as a vaccine in either their native state or encapsulated within polylactide-polyglycolide microspheres. Administered as a purified protein, CFs are sensitive to gastric acidity and proteolysis and therefore are poorly immunogenic and not protective [248,249]. Purified and encapsulated CFs can stimulate mucosal immunity but may not provide significant levels of protection on challenge with virulent ETEC strains expressing the homologous CF [25]. Further studies will determine if modification of immunization protocols will boost immunogenicity and protective efficacy of encapsulated CF vaccines. Inactivated whole bacteria stimulate anti-CF and anti-LT mucosal immune responses that can protect from subsequent challenge with virulent organisms [250]. A vaccine that consists of formalin-inactivated ETEC strains expressing commonly encountered CF types combined with the B subunit of cholera toxin (CTB) has been studied extensively [251–257]. CTB stimulates the production of antibodies that are cross-protective against ETEC infection [243,244]. This ETEC-CTB vaccine is well tolerated and stimulates significant anti-CF and anti-CTB immune responses in the majority of vaccinees. This vaccine has reached the stage of testing for protective efficacy in the field.

Live-vaccine candidates have included attenuated ETEC strains and other enteric pathogens expressing ETEC antigens [258]. An ETEC-based live oral vaccine would require a mixture of different attenuated strains, with each strain expressing one or more of the most commonly encountered CF types. Attenuated *Salmonella enterica* serovars *typhi* or *typhimurium* [259,260] and *Shigella flexneri* strains [261–263] have been engineered to express ETEC CF antigens as well as mutant LT or the LT B subunit. These strains can stimulate immune responses to *Salmonella* or *Shigella* antigens as well as the heterologous ETEC antigens and may serve as multivalent vaccines.

Alternative ETEC vaccine development strategies are being explored. DNA immunization of BALB/c mice with a CFA/I-expressing plasmid has been shown to stimulate the production of CFA/I-specific serum antibodies [264]. Plant-based vaccines also are being studied. Potato plants were transformed with a construct expressing a hybrid cholera toxin (CT) molecule in which the CT A subunit was fused with CFA/I [265]. CD-1 mice fed the transformed potatoes developed serum and mucosal antibodies against CFA/I and CT, suggesting that food plants might function as a vaccine for ETEC infections.

In addition to its ability to act as an enterotoxin, LT-I also acts as a mucosal adjuvant, a property shared with the closely related cholera toxin. LT mutants lacking ADP-ribosylation activity that retain their mucosal adjuvanticity have been constructed [266–268]. The LT A subunit appears to be important in the ability of the toxin to act as an adjuvant [269]. This adjuvant property is being used for nasal and oral immunization and the development and delivery of novel vaccines for a variety of pathogens.

REFERENCES

1. Institute of Medicine (1986). New vaccine development: Establishing priorities. In *Disease of Importance in Developing Countries*, pp. 159–169. Washington: National Academy Press.
2. World Health Organization (1999). *The World Health Report 1999: Making a Difference*. Geneva: World Health Organization.
3. Wood, L. V., Ferguson, L. E., Hogan, P., *et al.* (1983). Incidence of bacterial enteropathogens in foods from Mexico. *Appl. Environ. Microbiol.* 46:328–332.
4. Black, R. E., Brown, K. H., Becker, S., *et al.* (1982). Contamination of weaning foods and transmission of enterotoxigenic *Escherichia coli* diarrhea in children in rural Bangladesh. *Trans. R. Soc. Trop. Med. Hyg.* 7:259–264.
5. Ryder, R. W., Sack, D. A., Kapikian, A. Z., *et al.* (1976). Enterotoxigenic *Escherichia coli* and reovirus-like agent in rural Bangladesh. *Lancet* 1:659–663.
6. Black, R. E., Brown, K. H., Becker, S., *et al.* (1982). Longitudinal studies of infectious diseases and physical growth of children in rural Bangladesh: II. Incidence of diarrhea and association with known pathogens. *Am. J. Epidemiol.* 115:315–324.
7. Levine, M. M., Nalin, D., Hoover, D. L., *et al.* (1979). Immunity to enterotoxigenic *Escherichia coli*. *Infect. Immun.* 23:729–736.
8. Levine, M. M. (1987). *Escherichia coli* that cause diarrhea: Enterotoxigenic, enteropathogenic, enteroinvasive, enterohemorrhagic, and enteroadherent. *J. Infect. Dis.* 155:377–389.
9. Black, R. E., Merson, M. H., Rowe, B., *et al.* (1981). Entertoxigenic *Escherichia coli* diarrhea: Acquired immunity and transmission in an endemic area. *Bull. WHO* 59:253–258.
10. Levine, M. M., Ferreccio, C., Prado, V., *et al.* (1993). Epidemiological studies of *Escherichia coli* diarrheal infections in a low socioeconomic level peri-urban community in Santiago, Chile. *Am. J. Epidemiol.* 138:849–869.

11. Albert, M. J., Faruque, S. M., Faruque, A. S., *et al.* (1995). Controlled study of *Escherichia coli* diarrheal infections in Bangladeshi children. *J. Clin. Microbiol.* 33:973–977.
12. Rowland, M. G. M., Barrell, R. A. E., and Whitehead, R. G. (1978). Bacterial contamination in traditional Gambian weaning foods. *Lancet* 2:136–138.
13. Levine, M. M., Rennels, M. B., Cisneros, L., *et al.* (1980). Lack of person-to-person transmission of enterotoxigenic *Escherichia coli* despite close contact. *Am. J. Epidemiol.* 111:347–355.
14. Black, R. (1993). Epidemiology of diarrhoeal disease: Implications for control by vaccines. *Vaccine* 11:100–106.
15. Hyams, K. C., Bourgeois, A. L., Merrell, B. R., *et al.* (1991). Diarrheal disease during Operation Desert Shield. *New Engl. J. Med.* 325:1423–1428.
16. Bourgeois, A., Gardiner, C., Thornton, S., *et al.* (1993). Etiology of acute diarrhea among United States military personnel deployed to South America and West Africa. *Am. J. Trop. Med. Hyg.* 48:243–248.
17. Mattila, L. (1994). Clinical features and duration of traveler's diarrhea in relation to its etiology. *Clin. Infect. Dis.* 19:728–734.
18. DuPont, H. L., and Ericsson, C. D. (1993). Prevention and treatment of traveler's diarrhea. *New Engl. J. Med.* 328:1821–1827.
19. Black, R. E. (1990). Epidemiology of traveler's diarrhea and relative importance of various pathogens. *Rev. Infect. Dis.* 12(suppl. 1):S73–S79.
20. Arduino, R. C., and DuPont, H. L. (1993). Traveler's diarrhoea. *Baillieres Clin. Gastroenterol.* 7:365–385.
21. DuPont, H., Olarte, J., and Evans, D. (1976). Comparative susceptibility of Latin America and United States students to enteric pathogens. *New Engl. J. Med.* 295:1520–1521.
22. Bopp, C., Cameron, D., Greene, K., *et al.* (2001). Characterization of enterotoxigenic *Escherichia coli* serotype O6:H16 strains from U.S. outbreaks (abstract Y-7). In *Abstracts of the 101st General Meeting of the American Society for Microbiology 2001*, p. 737. Washington: American Society for Microbiology Press.
23. Dalton C. B., Mintz, E. D., Wells, J. G., *et al.* (1999). Outbreaks of enterotoxigenic *Escherichia coli* infection in American adults: A clinical and epidemiologic profile. *Epidemiol. Infect.* 123:9–16.
24. Sullivan, M. M., Besser, J., and Crouch, N. (2000). High incidence of enterotoxigenic *E. coli* (ETEC) isolated from stools among HMO patients with diarrhea in Minnesota (abstract Y-10). In *Abstracts of the 100th General Meeting of the American Society for Microbiology 2000*, p. 683. Washington: American Society for Microbiology Press.
25. Tacket, C. O., Reid, R. H., Boedeker, E. C., *et al.* (1994). Enteral immunization and challenge of volunteers given enterotoxigenic *E. coli* CFA/II encapsulated in biodegradable microspheres. *Vaccine* 12:1270–1274.
26. Butler, D. G., and Clarke, R. C. (1994). Diarrhoea and dysentery in calves. In C. L. Gyles (ed.), *Escherichia coli in Domestic Animals and Humans*, pp. 91–116. Wallingford, U.K.: CAB International.
27. Gorbach, S. L., Banwell, J. G., Chatterjee, B. D., *et al.* (1967). Studies of intestinal microflora: II. Microorganisms of the small intestine and their relations to oral and fecal flora. *Gastroenterology* 53:856–867.
28. Gorbach, S., Banwell, J., Chattejee, B., *et al.* (1971). Acute undifferentiated diarrhea in the tropics: I. Alteration in the intestinal microflora. *J. Clin. Invest.* 50:881–889.
29. Ofek, I., Zafriri, D., Goldhar, J., and Eisenstein, B. I. (1990). Inability of toxin inhibitors to neutralize enhanced toxicity caused by bacteria adherent to tissue culture cells. *Infect. Immun.* 58:3737–3742.
30. Zafriri, D., Ofon, Y., Eisenstein, B. I., and Ofek, I. (1987). Growth advantage and enhanced toxicity of *Escherichia coli* adherent to tissue culture cells due to restricted diffusion of products by the cells. *J. Clin. Invest.* 79:1210–1216.

31. Banwell, J. G., Gorbach, S. L., Pierce, N. F., *et al.* (1971). Acute undifferentiated human diarrhea in the tropics: II. Alterations in intestinal fluid and electrolyte movements. *J. Clin. Invest.* 50:890–900.
32. Nalin, D. R., McLaughlin, J. C., Rahaman, M., Yunus, M., and Curlin, G. (1975). Enteropathogenic *Escherichia coli* and idiopathic diarrhea in Bangladesh. *Lancet* 2:1116–1119.
33. DuPont, H. L., Formal, S. B., Hornick, R. H., *et al.* (1971). Pathogenesis of *Escherichia coli* diarrhea. *New Engl. J. Med.* 285:1–9.
34. Sack, R., Gorbach, S., Banwell, J., *et al.* (1971). Entertoxigenic *Escherichia coli* isolated from individuals with severe cholera-like disease. *J. Infect. Dis.* 123:378–385.
35. Finkelstein, R., Vasil, M., Jones, J., *et al.* (1976). Clinical cholera caused by enterotoxigenic *Escherichia coli*. *J. Clin. Microbiol.* 3:382–384.
36. Arduino, R., Mosavi, A., Valdez, L., *et al.* (1995). The natural history of enterotoxigenic *Escherichia coli* (ETEC) diarrhea among U.S. travelers in Mexico. Presented at the 33rd Annual Meeting of the Infectious Diseases Society, San Francisco, CA.
37. Guerrant, R. L., Araujo, V., Soares, E., *et al.* (1992). Measurement of fecal lactoferrin as a marker of fecal leukocytes (see comments). *J. Clin. Microbiol.* 30:1238–1242.
38. Rose, R., and Moon, H. W. (1985). Elicitation of enteroluminal neutrophils by enterotoxigenic and nonenterotoxigenic strains of *Escherichia coli* in swine. *Infect. Immun.* 48:818–823.
39. Cassels, F. J., and Wolf, M. K. (1995). Colonization factors of diarrheagenic *E. coli* and their intestinal receptors. *J. Ind. Microbiol.* 15:214–226.
40. Gaastra, W., and Svennerholm, A.-M. (1996). Colonization factors of human enterotoxigenic *Escherichia coli* (ETEC). *Trends Microbiol.* 4:444–452.
41. Jordi, B. J., Willshaw, B. A., van der Zeijst, B. A., and Gaastra, W. (1992). The complete nucleotide sequence of region I of the CFA/I fimbrial operon of human enterotoxigenic *Escherichia coli*. *DNA Seq.* 2:257–263.
42. Perez-Casal, P., Swartley, J. S., and Scott, J. S. (1990). Gene encoding the major subunit of CS1 pili of human enterotoxigenic *Escherichia coli*. *Infect. Immun.* 58:3594–3600.
43. Froehlich, B. J., Karakashian, A., Sakellaris, H., and Scott, J. R. (1995). Genes for CS2 pili of entertoxigenic *Escherichia coli* and their interchangeability with those for CS1 pili. *Infect. Immun.* 63:4849–4856.
44. Wolf, M. K., Andrews, G. P., Tall, B. D., *et al.* (1989). Characterization of CS4 and CS6 antigenic components of PCF8775, a putative colonization factor complex from enterotoxigenic *Escherichia coli* E8775. *Infect. Immun.* 57:164–173.
45. McConnell, M. M., Chart, H., Field, A. M., *et al.* (1989). Characterization of a putative colonization factor (PCFO166) of enterotoxigenic *Escherichia coli* of serogroup O166. *J. Med. Microbiol.* 135:1135–1144.
46. McConnell, M. M., Hibberd, M., Field, A. M., *et al.* (1990). Characterization of a new putative colonization factor (CS17) from a human enterotoxigenic *Escherichia coli* of serotype O114:H21 which produces only heat-labile enterotoxin. *J. Infect. Dis.* 161:343–347.
47. Grewal, H. M., Valvatne, H., Bhan, M. K., *et al.* (1997). A new putative fimbrial colonization factor, CS19, of human enterotoxigenic *Escherichia coli*. *Infect. Immun.* 65:507–513.
48. Clark, C. A., Heuzenroeder, M. W., and Manning, P. A. (1992). Colonization factor antigen CFA/IV (PCF8775) of human enterotoxigenic *Escherichia coli*: Nucleotide sequence of the CS5 determinant. *Infect. Immun.* 60:1254–1257.
49. Hibberd, M. L., McConnell, M. M., Field, A. M., and Rowe, B. (1990). The fimbriae of human enterotoxigenic *Escherichia coli* strain 334 are related to CS5 fimbriae. *J. Gen. Microbiol.* 136:2449–2456.
50. Heuzenroeder, M. W., Elliot, T. R., Thomas, C. J., *et al.* (1990). A new fimbrial type (PCFO9) on enterotoxigenic *Escherichia coli* O9:H-LT$^+$ isolated from a case of infant diarrhea in central Australia. *FEMS Microbiol. Lett.* 66:55–60.

51. Viboud, G. I., Binsztein, N., and Svennerholm, A.-M. (1993). A new fimbrial putative colonization factor, PCFO20, in human enterotoxigenic *Escherichia coli. Infect. Immun.* 61:5190–5197.
52. Valvante, H., Sommerfelt, H., Gaastra, W., *et al.* (1996). Identification and characterization of CS20, a new putative colonization factor of enterotoxigenic *Escherichia coli. Infect. Immun.* 64:2635–2642.
53. Taniguchi, T., Fujino, Y., Yamamoto, K., *et al.* (1995). Sequencing of the gene encoding the major pilin of pilus colonization factor antigen III (CFA/III) of human enterotoxigenic *Escherichia coli* and evidence that CFA/III is related to type IV pili. *Infect. Immun.* 63:724–728.
54. Giron, J. A., Levine, M. M., and Kaper, J. B. (1994). Longus: A long pilus ultrastructure produced by human enterotoxigenic *Escherichia coli. Mol. Microbiol.* 12:71–82.
55. Aubel, D., Darfeuille-Michaud, A., Martin, C., and Joly, B. (1992). Nucleotide sequence of the nfaA gene encoding the antigen 8786 adhesive factor of enterotoxigenic *Escherichia coli. FEMS Microbiol. Lett.* 98:277–284.
56. Pichel, M., Binsztein, N., and Viboud, G. (2000). CS22, a novel human enterotoxigenic *Escherichia coli* adhesin, is related to CS15. *Infect. Immun.* 68:3280–3285.
57. Jalajakumari, M. B., Thomas, C. J., Halter, R., and Manning, P. A. (1989). Genes for biosynthesis and assembly of CS3 pili of CFA/II enterotoxigenic *Escherichia coli*: Novel regulation of pilus production by bypassing an amber codon. *Mol. Microbiol.* 3:1685–1695.
58. Darfeuille-Michaud, A., Forestier, C., Joly, B., and Cluzel, R. (1986). Identification of a nonfimbrial adhesive factor of an enterotoxigenic *Escherichia coli* strain. *Infect. Immun.* 52:468–475.
59. Knutton, S., Lloyd, D. R., and McNeish, A. S. (1987). Identification of a new fimbrial structure in enterotoxigenic *Escherichia coli* (ETEC) serotype O148:H28 which adheres to human intestinal mucosa: a potentially new human ETEC colonization factor. *Infect. Immun.* 58:86–92.
60. Tacket, C. O., Maneval, D. R., and Levine, M. M. (1987). Purification, morphology, and genetics of a new fimbrial putative colonization factor of enterotoxigenic *Escherichia coli* O159:H4. *Infect. Immun.* 55:1063–1069.
61. Sakellaris, H., and Scott, J. R. (1998). New tools in an old trade: CS1 pilus morphogenesis. *Mol. Microbiol.* 30:681–687.
62. Rudin, A., and Svernnerholm, A.-M. (1996). Identification of a cross-reactive continuous B-cell epitope in enterotoxigenic *Escherichia coli* colonization factor antigen I. *Infect. Immun.* 64:4508–4513.
63. Rudin, A., McConnell, M. M., and Svennerholm, A.-M. (1994). Monoclonal antibodies against enterotoxigenic *Escherichia coli* colonization factor antigen I (CFA/I) that cross-react immunologically with heterologous CFAs. *Infect. Immun.* 62:4339–4346.
64. Rudin, A., Olbe, L., and Svennerholm, A.-M. (1996). Monoclonal antibodies against fimbrial subunits of colonization factor antigen I (CFA/I) inhibit binding to human enterocytes and protect against enterotoxigenic *Escherichia coli* expressing heterologous colonization factors. *Microb. Pathog.* 21:35–45.
65. Cassels, F. J., Jarboe, D. L., Reid, R. H., *et al.* (1997). Linear epitopes of colonization factor antigen I and peptide vaccine approach to enterotoxigenic *Escherichia coli. J. Ind. Microbiol. Biotechnol.* 19:66–70.
66. Sakellaris, H., Munson, G. P., and Scott, J. R. (1999). A conserved residue in the tip proteins of CS1 and CFA/I pili of enterotoxigenic *Escherichia coli* that is essential for adherence. *Proc. Natl. Acad. Sci. USA* 96:12828–12832.
67. Viboud, G. I., Jonson, G., Dean-Nystrom, E., and Svennerholm, A.-M. (1996). The structural gene encoding human enterotoxigenic *Escherichia coli* PCFO20 is homologous to that for porcine 987P. *Infect. Immun.* 64:1233–1239.
68. de Graaf, F. K., and Gaastra, W. (1994). Fimbriae of enterotoxigenic *Escherichia coli.* In P. Klemm (ed.), *Fimbriae: Adhesion, Genetics, Biogenesis, and Vaccines*, pp. 53–83. Boca Raton, FL: CRC Press.
69. Knutton, S., McConnell, M. M., Rowe, B., and McNeish, A. S. (1989). Adhesion and ultrastructural properties of human enterotoxigenic *Escherichia coli* producing colonization factor antigens III and IV. *Infect. Immun.* 57:3364–3371.

70. Gomez-Duarte, O. G., Ruiz-Tagle, A., Gomez, D. C., *et al.* (1999). Identification of *lngA*, the structural gene of longus type IV pilus of enterotoxigenic *Escherichia coli. Microbiology* 145:1809–1816.
71. Giron, J. A., Gomez-Duarte, O. G., Jarvis, K. G., and Kaper, J. B. (1997). Longus pilus of enterotoxigenic *Escherichia coli* and its relatedness to other type-4 pili: A minireview. *Gene* 192:39–43.
72. Ogunniyi, A. D., Kotlarski, I., Morona, R., and Manning, P. A. (1997). Role of SefA subunit protein of SEF14 fimbriae in the pathogenesis of *Salmonella enterica* serovar *enteritidis. Infect. Immun.* 65:708–717.
73. Aubel, D., Darfeuille-Michaud, A., and Joly, B. (1991). New adhesive factor (antigen 8786) on a human enterotoxigenic *Escherichia coli* O117:H4 strain isolated in Africa. *Infect. Immun.* 59:1290–1299.
74. Wolf, M. K. (1997). Occurrence, distribution, and associations of O and H serogroups, colonization factor antigens, and toxins of enterotoxigenic *Escherichia coli. Clin. Microbiol. Rev.* 10:569–584.
75. Smyth, C. J. (1982). Two mannose-resistant haemagglutinins of enterotoxigenic *Escherichia coli* of serotype O6:H16 or H$^-$ isolated from travellers' and infantile diarrhoea. *J. Gen. Microbiol.* 128:2081–2096.
76. Thomas, L. V., McConnell, M. M., Rowe, B., and Field, A. M. (1985). The possession of three novel coli surface antigens by entertoxigenic *Escherichia coli* strains positive for the putative colonization factor PCF8775. *J. Gen. Microbiol.* 131:2319–2326.
77. Gyles, C. L., So, M., and Falkow, S. (1974). The enterotoxin plasmids of *Escherichia coli. J. Infect. Dis.* 130:40–49.
78. Harnett, N. M., and Gyles, C. L. (1985). Linkage of genes for heat-stable enterotoxin, drug resistance, K99 antigen, and colicin in bovine and porcine strains of enterotoxigenic *Escherichia coli. Am. J. Vet. Res.* 46:428–433.
79. Wolf, M. K., de Haan, L. A., Cassels, F. J., *et al.* (1997). The CS6 colonization factor of human enterotoxigenic *Escherichia coli* contains two heterologous major subunits. *FEMS Microbiol. Lett.* 148:35–42.
80. Sakellaris, H., Balding, D. P., and Scott, J. R. (1999). Assembly proteins of CS1 pili of enterotoxigenic *Escherichia coli. Mol. Microbiol.* 21:529–541.
81. Pieroni, P., Worobec, E. A., Paranchych, W., and Armstrong, G. D. (1988). Identification of a human erythrocyte receptor for colonization factor antigen I pili expressed by H10407 enterotoxigenic *Escherichia coli. Infect. Immun.* 56:1334–1340.
82. Faris, A., Lindahl, M., and Wadstrom, T. (1980). GM2-like glycoconjugate as possible erythrocyte receptor for the CFA/I and K99 haemagglutinins of enterotoxigenic *Escherichia coli. FEMS Microbiol. Lett.* 7:265–269.
83. Wenneràs, C., Holmgren, J., and Svennerholm, A.-M. (1990). The binding of colonization factor antigens of enterotoxigenic *Escherichia coli* to intestinal cell membrane protein. *FEMS Microbiol. Lett.* 54:107–112.
84. Orø, H. S., Kolst, A. B., Wenneràs, C., and Svennerholm, A.-M. (1990). Identification of asialo GM1 as a binding structure for *Escherichia coli* colonization factor antigens. *FEMS Microbiol. Lett.* 60:289–292.
85. Sjøberg, P. O., Lindahl, M., Porath, J., and Wadstrom, T. (1988). Purification and characterization of CS2, a sialic acid-specific haemagglutinin of enterotoxigenic *Escherichia coli. Biochem. J.* 255:105–111.
86. Wenneràs, C., Neeser, J.-R., and Svennerholm, A.-M. (1995). Binding of the fibrillar CS3 adhesin of enterotoxigenic *Escherichia coli* to rabbit intestinal glycoproteins is competitively provented by Galβ1-4Gal-containing glycoconjugates. *Infect. Immun.* 63:640–646.
87. Grange, P. A., and Cassels, F. J. (2001). Using mammalian erythrocyte membrane glycoprotein extracts to characterize the carbohydrate binding specificity of enterotoxigenic *Escherichia coli* CS6 colonization factor (abstract B-126). In *Abstracts of the 101st General Meeting of the American Society for Microbiology 2001*, p. 68. Washington: American Society for Microbiology Press.

88. Wenneràs, C., Holmgren, J., McConnell, M. M., and Svennerholm, A.-M. (1991). The binding of bacteria carrying CFAs and putative CFAs to rabbit intestinal brush border membranes. In T. Wàdstrom, P. H. Màkelà, A.-M. Svennerholm, and H. Wolf-Watz (eds.), *Molecular Pathogenesis of Gastrointestinal Infections*, pp. 327–330. New York: Plenum Press.
89. Jin, L. Z., and Zhao, X. (2000). Intestinal receptors for adhesive fimbriae of enterotoxigenic *Escherichia coli* (ETEC) K88 in swine: A review. *Appl. Microbiol. Biotechnol.* 54:311–318.
90. Van den Broeck, W., Cox, E., Oudega, B., and Goddeeris, B. M. (2000). The F4 fimbrial antigen of *Escherichia coli* and its receptors. *Vet. Microbiol.* 71:223–244.
91. Francis, D. H., Erickson, A. K., and Grange, P. A. (1999). K88 adhesins of enterotoxigenic *Escherichia coli* and their porcine enterocyte receptors. *Adv. Exp. Med. Biol.* 473:147–154.
92. Westerman, R. B., Mills, K. W., Phillips, R. M., *et al.* (1988). Predominance of the ac variant in K88-positive *Escherichia coli* isolates from swine. *J. Clin. Microbiol.* 26:149–150.
93. Erickson, A. K., Baker, D. R., Bosworth, B. T., *et al.* (1994). Characterization of porcine intestinal receptors for the K88ac fimbrial adhesin of *Escherichia coli* as mucin-type sailoglycoproteins. *Infect. Immun.* 62:5404–5410.
94. Fang, L., Gan, Z., and Marquardt, R. R. (2000). Isolation, affinity purification, and identification of piglet small intestine mucosa receptor for enterotoxigenic *Escherichia coli* K88ac+ fimbriae. *Infect. Immun.* 68:564–569.
95. Grange, P. A., and Mouricout, M. A. (1996). Transferrin associated with the porcine intestinal mucosa is a receptor specific for K88ab fimbriae of *Eshcerichia coli. Infect. Immun.* 64:606–610.
96. Grange, P. A., Erickson, A. K., Levery, S. B., and Francis, D. H. (1999). Identification of an intestinal neutral glycosphingolipid as a phenotype-specific receptor for the K88ad fimbrial adhesin of *Escherichia coli. Infect. Immun.* 67:165–172.
97. Khan, A. S., and Schifferli, D. M. (1994). A minor 987P protein different from the structural fimbrial subunit is the adhesin. *Infect. Immun.* 62:4223–4243.
98. Khan, A. S., Johnston, N. C., Goldfine, H., and Schifferli, D. M. (1996). Porcine 987P glycolipid receptors on intestinal brush borders and their cognate bacterial ligands. *Infect. Immun.* 64:3688–3693.
99. Smit, H., Gaastra, W., Kamerling, J. P., *et al.* (1984). Isolation and structural characterization of the equine erythrocyte receptor for enterotoxigenic *Escherichia coli* K99 fimbrial adhesin. *Infect. Immun.* 46:578–584.
100. Spangler, B. D. (1992). Structure and function of cholera toxin and the related *Escherichia coli* heat-labile enterotoxin. *Microbiol. Rev.* 56:622–647.
101. Fukuta, S., Magnani, J. L., Twiddy, E. M., *et al.* (1988). Comparison of the carbohydrate binding specificities of cholera toxin and *Escherichia coli* heat-labile enterotoxins LTh-1, LT-IIa and LT-IIb. *Infect. Immun.* 56:1748–1753.
102. Finkelstein, R. A., Burks, M. F., Zupan, A., *et al.* (1987). Epitopes of the cholera family of enterotoxins. *Rev. Infect. Dis.* 9:544–561.
103. Holmes, R. K., and Twiddy, E. M. (1983). Characterization of monoclonal antibodies that react with unique and cross-reacting determinants of cholera enterotoxin and its subunits. *Infect. Immun.* 42:914–923.
104. Honda, T., Tsuji, T., Takeda, Y., and Miwatani, T. (1981). Immunological nonidentity of heat-labile enterotoxins from human and procine enterotoxigenic *Escherichia coli. Infect. Immun.* 34:37–340.
105. Guth, B. E. C., Twiddy, E. M., Trabulsi, L. R., and Holmes, R. K. (1986). Variation in chemical properties and antigenic determinants among type II heat-labile enterotoxins of *Escherichia coli. Infect. Immun.* 54:529–536.
106. Sixma, T. K., Kalk, K. H., van Zanten, B. A., *et al.* (1993). Refined structure of *Escherichia coli* heat-labile entertoxin, a close relative of cholera toxin. *J. Mol. Biol.* 230:890–918.
107. Sixma, T. K., Pronk, S. E., Kalk, K. H., *et al.* (1991). Crystal structure of a cholera toxin-related heat-labile enterotoxin from *E. coli. Nature* 351:371–377.

108. Hofstra, H., and Witholy, B. (1985). Heat-labile enterotoxin in *Escherichia coli*: Kinetics of association of subunits into periplasmic holotoxin. *J. Biol. Chem.* 260:16037–16044.
109. Hirst, T. R., Randall, L. L., and Hardy, S. J. S. (1984). Cellular location of heat-labile enterotoxin in *Escherichia coli*. *J. Bacteriol.* 157:637–642.
110. Evans, D. J., Evans, D. G., and Gorbach, S. L. (1974). Polymyxin B—induced release of low-molecular-weight, heat-labile enterotoxin from *Escherichia coli*. *Infect. Immun.* 10:1010–1017.
111. Hunt, P. D., and Hardy, S. J. (1991). Heat-labile enterotoxin can be released from *Escherichia coli* cells by host intestinal factors. *Infect. Immun.* 59:168–171.
112. Horstman, A. L. a. M. J. K. (2000). Enterotoxigenic *Escherichia coli* secretes active heat-labile enterotoxin via outer membrane vesicles. *J. Biol. Chem.* 275:12489–12496.
113. Wai, S. N., Takade, A., and Amako, K. (1995). The release of outer membrane vesicles from the strains of enterotoxigenic *Escherichia coli*. *Microbiol. Immunol.* 39:451–456.
114. Fleckenstein, J. M., Lindler, L. E., Elsinghorst, E. A., and Dale, J. B. (2000). Identification of a gene within a pathogenicity island of enterotoxigenic *Escherichia coli* required for maximal secretion of the heat-labile enterotoxin. *Infect. Immun.* 68:2766–2774.
115. Holmgren, J., Fredman, P., Lindblad, M., *et al.* (1982). Rabbit intestinal glycoprotein receptors for *Escherichia coli* heat-labile enterotoxin lacking affinity for cholera toxin. *Infect. Immun.* 38:424–433.
116. Schengrund, C. L., and Ringler, N. J. (1989). Binding of Vibrio cholera toxin and the heat-labile enterotoxin of *Escherichia coli* to GM_1, derivatives of GM_1, and nonlipid oligosaccharide polyvalent ligands. *J. Biol. Chem.* 264:13233–13237.
117. Griffiths, S. L., Finkelstein, R. A., and Critchley, D. R. (1986). Characterization of the receptor for cholera toxin and *Escherichia coli* heat-labile toxin in rabbit intestinal brush borders. *Biochem. J.* 238:313–322.
118. Orlandi, P. A., Critchley, D. R., and Fishman, P. H. (1994). The heat-labile enterotoxin of *Escherichia coli* binds to polylactosaminoglycan-containing receptors in CaCo-2 human intestine epithelial cells. *Biochemistry* 33:12886–12895.
119. Teneberg, S., Hirst, T. R., Angstrom, J., and Karlsson, K.-A. (1994). Comparison of the glycolipid-binding specificities of cholera toxin and porcine *Escherichia coli* heat-labile enterotoxin: identification of a receptor-active non-ganglioside glycolipid for the heat-labile toxin in infant rabbit small intestine. *Glycoconjugate* 11:533–540.
120. Donta, S. T., Beristan, S., and Tomicic, T. K. (1993). Inhibition of heat-labile cholera and *Escherichia coli* enterotoxins by brefeldin A. *Infect. Immun.* 61:3282–3286.
121. Lencer, W. I., Constable, C., Moe, S., *et al.* (1995). Targeting of cholera toxin and *Escherichia coli* heat-labile toxin in polarized epithelia: role of COOH-terminal KDEL. *J. Cell Biol.* 131:951–962.
122. Mekelanos, J. J., Collier, R. J., and Romig, W. R. (1979). Enzymatic activity of cholera toxin: II. Relationships to proteolytic processing, disulfide bond reduction, and subunit composition. *J. Biol. Chem.* 254:5855–5861.
123. Holmes, R. K., Jobling, M. G., and Connell, T. D. (1995). Cholera toxin and related enterotoxins of gram-negative bacteria. In J. Moss, B. Iglewski, M. Vaughan, and A. T. Tu (eds.), *Bacterial Toxins and Virulence Factors in Disease*, pp. 225–255. New York: Marcel Dekker.
124. Sears, C. L., and Kaper, J. B. (1996). Enteric bacterial toxins: mechanisms of action and linkage to intestinal secretion. *Microbiol. Rev.* 60:167–215.
125. Gyles, C. L. (1994). *Escherichia coli* enterotoxins. In C. L. Gyles (ed.), *Escherichia coli in Domestic Animals and Humans*, pp. 337–364. Wallingford, U.K.: CAB International.
126. Nataro, J. P., and Kaper, J. B. (1998). Diarrheagenic *Escherichia coli*. *Clin. Microbiol. Rev.* 11:142–201.
127. McGee, D. W., Elson, C. O., and McGhee, J. R. (1993). Enhancing effect of cholera toxin on interleukin-6 secretion by IEC-6 intestinal epithelial cells: mode of action and augmenting effect of inflammatory cytokines. *Infect. Immun.* 61:4637–4644.

128. Klimpel, G. R., Asuncion, M., Haithcoat, J., and Niesel, D. W. (1995). Cholera toxin and *Salmonella typhimurium* induce different cytokine profiles in the gastrointestinal tract. *Infect. Immun.* 63:1134–1137.
129. Pickett, C. L., Twiddy, E. M., Coker, C., and Holmes, R. K. (1987). Genetics of type IIa heat-labile enterotoxin of *Escherichia coli*: Operon fusions, nucleotide sequence, and hybridization studies. *J. Bacteriol.* 169:5180–5187.
130. Pickett, C. L., Twiddy, E. M., Coker, C., and Holmes, R. K. (1989). Cloning, nucleotide sequence, and hybridization studies of the type IIb heat-labile enterotoxin gene of *Escherichia coli*. *J. Bacteriol.* 171:4945–4952.
131. Holmes, R. K., Twiddy, E. M., and Pickett, C. L. (1986). Purification and characterization of type II heat-labile enterotoxin of *Escherichia coli*. *Infect. Immun.* 53:464–473.
132. Chang, P. P., Moss, J., Twiddy, E. M., and Holmes, R. K. (1987). Type II heat-labile enterotoxin of *Escherichia coli* activates adenylate cyclase in human fibroblasts by ADP-ribosylation. *Infect. Immun.* 55:1854–1858.
133. Seriwatana, J., Echeverria, P., Taylor, D. N., *et al.* (1988). Type II heat-labile enterotoxin-producing *Escherichia coli* isolated from animals and humans. *Infect. Immun.* 56:1158–1161.
134. Rasheed, J. K., Guzman-Verduzco, L.-M., and Kupersztock, Y. M. (1990). Two precursors of the heat-stable enterotoxin of *Escherichia coli*: Evidence of extracellular processing. *Mol. Microbiol.* 4:265–273.
135. Okomato, K., and Takahara, M. (1990). Synthesis of *Escherichia coli* heat-stable enterotoxin STp as a pre-pro form and role of the pro sequence in secretion. *J. Bacteriol.* 172:5260–5265.
136. Kupersztoch, Y., Tachias, K., Moomaw, C. R., *et al.* (1990). Secretion of methanol-insoluble heat-stable enterotoxin (STb): Energy- and secA-dependent conversion of pre-STb to an intermediate indistinguishable from the extracellular toxin. *J. Bacteriol.* 172:2427–2432.
137. Dreyfus, L. A., Frantz, J. C., and Robertson, D. C. (1983). Chemical properties of heat-stable enterotoxins produced by enterotoxigenic *Escherichia coli* of different origins. *Infect. Immun.* 42:539–548.
138. Aimoto, S., Takao, T., Shimonishi, Y., *et al.* (1982). Amino-acid sequences of a heat-stable enterotoxin produced by human enterotoxigenic *Escherichia coli*. *Eur. J. Biochem.* 129:257–262.
139. Okamoto, K., Yukitake, J., Kawamoto, Y., and Miyama, A. (1987). Substitutions of cysteine residues of *Escherichia coli* heat-stable enterotoxin by oligonucleotide-directed mutagenesis. *Infect. Immun.* 55:2121–2125.
140. Osaki, H., Sato, T., Hubota, H., *et al.* (1991). Molecular structure of the toxic domain of heat-stable enterotoxin produced by a pathogenic strain of *Escherichia coli*. *J. Biol. Chem.* 266:5934–5941.
141. Gariepy, J., Judd, A. K., and Schoolnik, G. K. (1987). Importance of disulfide bridges in the structure and activity of *Escherichia coli* enterotoxin ST1b. *Proc. Natl. Acad. Sci. USA* 84:8907–8911.
142. Gariepy, J., Lane, A., Frayman, F., *et al.* (1986). Structure of the toxic domain of the *Escherichia coli* heat-stable enterotoxin ST I. *Biochemistry* 25:7854–7866.
143. Yamanaka, H., Kameyama, M., Baba, T., *et al.* (1994). Maturation pathway of *Escherichia coli* heat-stable enterotoxin I: Requirement of DsbA for disulfide bond formation. *J. Bacteriol.* 176:2906–2913.
144. Yamanaka, H., Nomura, T., Fujii, Y., and Okamoto, K. (1998). Need for TolC, an *Escherichia coli* outer membrane protein, in the secretion of heat-stable enterotoxin I across the outer membrane. *Microb. Pathog.* 25:111–120.
145. Yoshimura, S., Ikemura, H., Watanabe, H., *et al.* (1985). Essential structure for full enterotoxigenic activity of heat-stable enterotoxin produced by enterotoxigenic *Escherichia coli*. *FEBS Lett.* 181:138–141.
146. Sato, T., Ozaki, H., Hata, Y., *et al.* (1994). Structural characteristics for biological activity of heat-stable enterotoxin produced by enterotoxigenic *Escherichia coli*: X-ray crystallography of weakly toxic and nontoxic analogues. *Biochemistry* 33:8641–8650.

147. Stieglitz, H., Cervantes, L., Robledo, R., *et al.* (1988). Cloning, sequencing, and expression in Ficoll generated minicells of an *Escherichia coli* heat-stable enterotoxin gene. *Plasmid* 20:42–53.
148. Sekizaki, T. H. A., and Terakado, N. (1985). Nucleotide sequences of the genes for *Escherichia coli* heat-stable entertoxin I of bovine, avian, and porcine origins. *Am. J. Vet. Res.* 46:909–912.
149. Mosely, S. L., Samadpour-Motalebi, M., and Falkow, S. (1983). Plasmid association and nucleotide sequence relationships of two genes encoding heat-stable enterotoxin production in *Escherichia coli. J. Bacteriol.* 156:441–443.
150. Field, M. L., Graf, L. H., Laird, W. J., and Smith, P. L. (1978). Heat-stable enterotoxin of *Escherichia coli: In vitro* effects of guanylate cyclase activity, cGMP concentration, and ion transport in small intestine. *Proc. Natl. Acad. Sci. USA* 75:2800–2804.
151. Schulz, S., Green, C. K., Yuen, P. S. T., and Garbers, D. (1990). Guanylyl cyclase is a heat-stable enterotoxin receptor. *Cell* 63:941–948.
152. de Sauvage, F. J., Horuk, R., Bennett, G., *et al.* (1992). Characterization of the recombinant human receptor for *Escherichia coli* heat-stable enterotoxin. *J. Biol. Chem.* 267:6479–6482.
153. Hughes, J. M., Murad, F., Chang, B., and Guerrant, R. L. (1978). Role of cyclic CMP in the activity of heat-stable enterotoxin of *E. coli. Nature* 271:755–756.
154. Guerrant, R. L., Hughes, J. M., Chang, B., *et al.* (1980). Activation of intestinal guanylate cyclase by heat-stable enterotoxin of *Escherichia coli*: Studies of tissue specificity, potential receptors, and intermediates. *J. Infect. Dis.* 142:220–227.
155. Hasegawa, M., Hidaka, Y., Matsumoto, Y., *et al.* (1999). Determination of the binding site on the extracellular domain of guanylyl cyclase C to heat-stable enterotoxin. *J. Biol. Chem.* 274:31713–31718.
156. Vaandrager, A. B., Schulz, S., de Jonge, H. R., and Garbers, D. L. (1993). Guanylyl cyclase C is an N-linked glycoprotein receptor that accounts for multiple heat-stable entertoxin-binding proteins in the intestine. *J. Biol. Chem.* 268:2174–2179.
157. Hirayama, T., Wada, A., Iwata, N., *et al.* (1992). Glycoprotein receptors for a heat-stable enterotoxin (STh) produced by enterotoxigenic *Escherichia coli. Infect. Immun.* 60:4213–4220.
158. Cohen, M. B., Jensen, N. J., Hawkins, J. A., *et al.* (1993). Receptors for *Escherichia coli* heat-stable enterotoxin in human intestine and in a human intestinal cell line (Caco-2). *J. Cell Physiol.* 156:138–144.
159. Carpick, B. W., and Gariepy, J. (1993). The *Escherichia coli* heat-stable enterotoxin is a long-lived superagonist of guanylin. *Infect. Immun.* 61:4710–4715.
160. Currie, M. G., Fok, K. F., Kato, J., *et al.* (1992). Guanylin: An endogenous activator of intestinal guanylate cyclase. *Proc. Natl. Acad. Sci. USA* 89:947–951.
161. Kuhn, M., Adermann, K., Jahne, J., *et al.* (1994). Segmental differences in the effects of guanylin and *Escherichia coli* heat-stable enterotoxin on Cl^- secretion in human gut. *J. Physiol. (Lond.)* 479:433–440.
162. Forte, L. R., Eber, S. L., Turner, J. T., *et al.* (1993). Guanylin stimulation of Cl^- secretion in human intestinal T_{84} cells via guanosine monophosphate. *J. Clin. Invest.* 91:2423–2428.
163. Cohen, M. B., Guarino, a., Shukla, R., and Giannella, R. A. (1988). Age-related differences in receptors for *Escherichia coli* heat-stable enterotoxin in the small and large intestine of children. *Gastroenterology* 94:367–373.
164. Cohen, M. B., Mann, E. A., Lau, C., *et al.* (1992). A gradient in expression of the *Escherichia coli* heat-stable enterotoxin receptor exists along the villus-to-crypt axis of rat small intestine. *Biochem. Biophys. Res. Commun.* 186:483–490.
165. Mann, E. A., Cohen, M. B., and Giannella, R. A. (1993). Comparison of receptors for *Escherichia coli* heat-stable enterotoxin: Novel receptor present in IEC-6 cells. *Am. J. Physiol.* 264:G172–G178.
166. Almenoff, J. S., Williams, S. I., Scheving, L. A., *et al.* (1994). Induction of heat-stable enterotoxin receptor activity by a human Alu repeat. *J. Biol. Chem.* 269:16610–16617.

167. Forte, L. R., Thorne, P. K., Eber, S. L., *et al.* (1992). Stimulation of intestinal Cl^- transport by heat-stable enterotoxin: Activation of cAMP-dependent protein kinase by cGMP. *Am. J. Physiol.* 263:C607–C617.
168. Rao, M. C., Guandalini, S., Smith, P. L., and Field, M. (1980). Mode of action of heat-stable *Escherichia coli* enterotoxin: Tissue and subcellular specificities and role of cyclic GMP. *Biochim. Biophys. Acta* 632:35–46.
169. Lin, M., Nairn, A. C., and Guggino, S. E. (1992). cGMP-dependent protein kinase regulation of a chloride channel in T_{84} cells. *Am. J. Physiol.* 262:C1304–C1312.
170. Vaandrager, A. B., van der Weil, E., and de Jonge, H. R. (1993). Heat-stable enterotoxin activation of immunopurified guanylyl cyclase C. *J. Biol. Chem.* 268:19598–19603.
171. Tien, X., Brasitus, A., Kaetzel, M. A., *et al.* (1994). Activation of the cystic fibrosis transmembrane conductance regulator by cGMP in the human colonic cancer cell line, Caco-2. *J. Biol. Chem.* 269:51–54.
172. Lucas, M. L. (2001). A reconsideration of the evidence for *Escherichia coli* STa (heat-stable) enterotoxin-driven fluid secretion: A new view of STa action and a new paradigm for fluid absorption. *J. Appl. Microbiol.* 90:7–26.
173. Rolfe, V. E., and Levin, R. J. (1999). Vagotomy inhibits the jejunal secretion activated by luminal ileal *Escherichia coli* STa in the rat *in vivo. Gut* 44:615–619.
174. So, M., and McCarthy, B. J. (1980). Nucelotide sequence of the bacterial transposon TN1681 encoding a heat-stable (ST) toxin and its identification in enterotoxigenic *Escherichia coli* strains. *Proc. Natl. Acad. Sci. USA* 77:4011–4015.
175. Echeverria, P., Seriwatana, J., Taylor, D. N., *et al.* (1986). Plasmids coding for colonization factor antigens I and II, heat-labile enterotoxin, and heat-stable enterotoxin A2 in *Escherichia coli. Infect. Immun.* 51:626–630.
176. Alderete, J. F., and Robertson, D. C. (1977). Repression of heat-stable enterotoxin synthesis in enterotoxigenic *Escherichia coli. Infect. Immun.* 17:629–633.
177. Savarino, S. J., Fasano, A., Robertson, D. C., and Levine, M. M. (1991). Enteroaggregative *Escherichia coli* elaborate a heat-stable enterotoxin demonstrable in an in vitro rabbit intestinal model. *J. Clin. Invest.* 87:1450–1455.
178. Savarino, S. J., McVeigh, A., Watson, J., *et al.* (1996). Enteroggregative *Escherichia coli* heat-stable enterotoxin is not restricted to enteroaggregative *E. coli. J. Infect. Dis.* 173:1019–1022.
179. Yamamoto, T., and Echeverria, P. (1996). Detection of the enteroaggregative *Escherichia coli* heat-stable enterotoxin 1 gene sequences in enterotoxigenic *E. coli* strains pathogenic for humans. *Infect. Immun.* 64:1441–1445.
180. Savarino, S. J., Fasano, A., Watson, J., *et al.* (1993). Enteroaggregative *Escherichia coli* heat-stable enterotoxin 1 represents another subfamily of E. coli heat-stable toxin. *Proc. Natl. Acad. Sci. USA* 90:3093–3097.
181. Takao, T., Tominaga, N., Yoshimura, S., *et al.* (1985). Isolation, primary structure and synthesis of heat-labile enterotoxin produced by *Yersinia enterocolitica. Eur. J. Biochem.* 152:199–206.
182. Takao, T., Shimonishi, Y., Kobayashi, M., *et al.* (1985). Amino acid sequence of heat-stable enterotoxin produced by *Vibrio cholerae* non-O1. *FEBS Lett.* 193:250–254.
183. McVeigh, A., Fasano, A., Scott, D. A., *et al.* (2000). IS1414, an *Escherichia coli* insertion sequence with a heat-stable enterotoxin gene embedded in a transposase-like gene. *Infect. Immun.* 68:5710–5715.
184. Sukumar, M., Rizo, J., Wall, M., *et al.* (1980). The structure and *Escherichia coli* heat-stable enterotoxin B by nuclear magnetic resonance and circular dichroism. *Protein Sci.* 4:1718–1729.
185. Fujii, Y. M. H., Hitotsubashi, S., Fuke, Y., *et al.* (1991). Purification and characterization of *Escherichia coli* heat-stable enterotoxin II. *J. Bacteriol.* 173:5516–5522.
186. Dreyfus, L. A., Urban, R. G., Whipp, S. C., *et al.* (1992). Purification of the STb enterotoxin of *Escherichia coli* and the role of selected amino acids on its secretion, stability, and toxicity. *Mol. Microbiol.* 6:2397–2406.

187. Hitotsubashi, S., Fujii, Y., Yamanaka, H., and Okamoto, K. (1992). Some properties of purified *Escherichia coli* heat-stable enterotoxin II. *Infect. Immun.* 60:4468–4474.
188. Fujii, Y., Okamuro, Y., Hitotsubashi, S., *et al.* (1994). Effect of alterations of basic amino acid residues of *Escherichia coli* heat-stable enterotoxin II on enterotoxicity. *Infect. Immun.* 62:2295–2301.
189. Chao, K. L., and Dreyfus, L. A. (1997). Interaction of *Escherichia coli* heat-stable enterotoxin B with cultured human intestinal epithelial cells. *Infect. Immun.* 65:3209–3217.
190. Chao, K. L., and Dreyfus, L. A. (1999). Interaction of *Escherichia coli* heat-stable enterotoxin B with rat intestinal epithelial cells and membrane lipids. *FEMS Microbiol. Lett.* 172:91–97.
191. Rousset, E., and Dubreuil, J. D. (1999). Evidence that *Escherichia coli* STB enterotoxin binds to lipid components extracted from the pig jejunal mucosa. *Toxicon* 37:1529–1537.
192. Harville, B. A., and Dreyfus, L. A. (1996). Release of serotonin from RBL-2H3 cells by the *Escherichia coli* peptide toxin STb. *Peptides* 17:363–366.
193. Harville, B. A., and Dreyfus, L. A. (1995). Involvement of 5-hydroxytryptamine and prostaglandin E_2 in the intestinal secretory action of *Escherichia coli* heat-stable enterotoxin. *Infect. Immun.* 63:745–750.
194. Peterson, J. W., and Whipp, S. C. (1995). Comparison of the mechanisms of action of cholera toxin and the heat-stable enterotoxins of *Escherichia coli*. *Infect. Immun.* 63:1452–1461.
195. Hitosubashi, S., Fujii, Y., Yamanaka, H., and Okamoto, K. (1992). Some properties of purified *Escherichia coli* heat-stable enterotoxin II. *Infect. Immun.* 60:4468–4474.
196. Dreyfus, L. A., Harville, B., Howard, D. E., *et al.* (1993). Calcium influx mediated by the *Escherichia coli* heat-stable enterotoxin B (STb). *Proc. Natl. Acad. Sci. USA* 90:3202–3206.
197. Whipp, S. C., Kokue, E., Morgan, R. W., *et al.* (1987). Functional significance of histological alterations induced by *Escherichia coli* pig-specific, mouse-negative, heat-labile enterotoxin (ST_b). *Vet. Res. Commun.* 11:41–55.
198. Hitotsubashi, S., Akagi, M., Saitou, A., *et al.* (1992). Action of *Escherichia coli* heat-stable enterotoxin II on isolated sections of mouse ileum. *FEMS Micobiol. Lett.* 90:249–252.
199. Moon, H. W., Schneider, R. A., and Moseley, S. L. (1986). Comparative prevalence of four enterotoxin genes among *Escherichia coli* isolated from swine. *Am. J. Vet. Res.* 47:210–212.
200. Handl, C. E., Olsson, E., and Flock, J. I. (1992). Evaluation of three different STb assays and comparison of enterotoxin pattern over a 5-year period in Swedish porcine *Escherichia coli*. *Diagn. Microbiol. Infect. Dis.* 15:505–510.
201. Casey, T. A. C. J. H., Schneider, R. A., *et al.* (1998). Expression of heat-stable enterotoxin STb by adherent *Escherichia coli* is not sufficient to cause severe diarrhea in neonatal pigs. *Infect. Immun.* 66:1270–1272.
202. Hu, S. T., and Lee, C. H. (1988). Characterization of the transposon carrying the STII gene of enterotoxigenic *Escherichia coli*. *Mol. Gen. Genet.* 214:490–495.
203. Hu, S. T., Yang, M. K., Spandau, D. F., and Lee, C. H. (1987). Characterization of the terminal sequences flanking the transposon that carries the *Escherichia coli* enterotoxin STII gene. *Gene* 55:157–167.
204. Lee, C. H., Hu, S. T., Swiatek, P. J., *et al.* (1985). Isolation of a novel transposon which carries the *Escherichia coli* enterotoxin STII gene. *J. Bacteriol.* 162:615–620.
205. Levine, M. M., Kaper, J. B., Black, R. E., and Clements, M. L. (1983). New knowledge on pathogenesis of bacterial enteric infections as applied to vaccine development. *Microbiol. Rev.* 47:510–550.
206. Sack, R. B., Kline, R. L., and Spira, W. M. (1988). Oral immunization of rabbits with enterotoxigenic *Escherichia coli* protects against intraintestinal challenge. *Infect. Immun.* 56:387–394.
207. Schlager, T. A., Wanke, C. A., and Guerrant, R. L. (1990). Net fluid secretion and impaired villous function induced by colonization of the small intestine by nontoxigenic colonizing *Escherichia coli*. *Infect. Immun.* 58:1337–1343.

208. Smith, H. W., and Linggood, M. A. (1971). Observations of the pathogenic properties of K88, Hly, and Ent plasmids of *Escherichia coli* with special reference to porcine diarrhea. *J. Med. Microbiol.* 4:467–485.
209. Wanke, C., and Guerrant, R. (1987). Small-bowel colonization alone is cause of diarrhea. *Infect. Immun.* 55:1924–1926.
210. Elsinghorst, E. A., and Kopecko, D. J. (1992). Molecular cloning of epithelial cell invasion determinants from enterotoxigenic *Escherichia coli. Infect. Immunity* 60:2409–2417.
211. Fairbrother, J. M., and Ngeleka, M. (1994). Extraintestinal *Escherichia coli* infections in pigs. In C. L. Gyles (ed.), *Escherichia coli in Domestic Animals and Humans*, pp. 221–236. Wallingford, U.K.: CAB International.
212. Moxley, R. A., Erickson, E. D., and Breisch, S. (1988). Shock associated with enteric colibacillosis in suckling and weaned swine. In *Proceedings of the George A. Young Swine Conference and Annual Nebraska SPF Swine Conference.*
213. Moxley, R. A., Berberov, E. M., Francis, D. H., *et al.* (1998). Pathogenicity of an enterotoxigenic *Escherichia coli* hemolysin (*hlyA*) mutant in gnotobiotic piglets. *Infect. Immun.* 66:5031–5035.
214. Fleckenstein, J. M., Kopecko, D. J., Warren, R. L., and Elsinghorst, E. A. (1996). Molecular characterization of the tia invasion locus from enterotoxigenic *Escherichia coli. Infect. Immun.* 64:2256–2265.
215. Mammarappallil, J. G., and Elsinghorst, E. A. (2000). Epithelial cell adherence mediated by the enterotoxigenic *Escherichia coli* Tia protein. *Infect. Immun.* 68:6595–6601.
216. Mammarappallil, J. G., and Elsinghorst, E. A. (2000). The enterotoxigenic *E. coli* Tia protein binds a specific host cell membrane protein (abstract B-286). In *Abstracts of the 100th General Meeting of the American Society for Microbiology 2000*, p. 106. Washington: American Society for Microbiology Press.
217. Conner, C., Heithoff, D., Julio, S., *et al.* (1998). Differential patterns of acquired virulence genes distinguish *Salmonella* strains. *Proc. Natl. Acad. Sci. USA* 95:4641–4645.
218. Heithoff, D. M., Conner, C. P., Hanna, P. C., *et al.* (1997). Bacterial infection as assessed by *in vivo* gene expression. *Proc. Natl. Acad. Sci. USA* 94:934–939.
219. Heithoff, D. M., Conner, C. P., Hentschel, U., *et al.* (1999). Coordinate intracellular expression of *Salmonella* genes induced during infection. *J. Bacteriol* 181:799–807.
220. Morelli, G., Malorny, B., Mueller, K., *et al.* (1997). Clonal descent and microevolution of *Neisseria meningitidis* during 30 years of epidemic spread. *Mol. Microbiol.* 25:1047–1064.
221. Swenson, D. L., Bukanov, N., Berg, D., and Welch, R. (1996). Two pathogenicity islands in uropathogenic *Escherichia coli* J96: Cosmid cloning and sample sequencing. *Infect. Immun.* 64:3736–3743.
222. Elsinghorst, E. A., and Weitz, J. A. (1994). Epithelial cell invasion and adherence directed by the enterotoxigenic *Escherichia coli tib* locus is associated with a 104-kilodalton outer membrane protein. *Infect. Immun.* 62:3463–3471.
223. Lindenthal, C., and Elsinghorst, E. A. (1999). Identification of a glycoprotein produced by enterotoxigenic *Escherichia coli. Infect. Immun.* 67:4084–4091.
224. Parker, C. T., Pradel, E., and Schnaitman, C. A. (1992). Identification and sequences of the lipopolysaccharide core biosynthetic genes *rfaQ*, *rfaP*, and *rfaG* of *Escherichia coli* K-12. *J. Bacteriol* 174:930–934.
225. Mammarappallil, J. G., Hronek, B. W., Mettenburg, L. D., and Elsinghorst, E. A. (2001). Glycosylation of the enterotoxigenic *Escherichia coli* TibA adhesin requires the TibC protein (abstract B-57). In *Abstracts of the 101st General Meeting of the American Society for Microbiology 2001*, p. 53. Washington: American Society for Microbiology Press.
226. Hendrickson, I. R. a. J. P. N. (2001). Virulence functions of autotransporter proteins. *Infect. Immun.* 69:1231–1243.

227. Kelm, O., Kiecker, C., Geider, K., and Bernhard, F. (1997). Interaction of the regulator proteins RcsA and RcsB with the promoter of the operon for amylovoran biosynthesis in *Erwinia amylovora. Mol. Gen. Genet.* 256:72–83.
228. Eichler, K., Buchet, A., Lemke, R., *et al.* (1996). Identification and characterization of the *caiF* gene encoding a potential transcriptional activator of carnitine metabolism in *Escherichia coli. J. Bacteriol.* 178:1248–1257.
229. Buchet, A., Nasser, W., Eichler, K., and Mandrand-Berthelot, M. A. (1999). Positive co-regulation of the *Escherichia coli* carnitine pathway *cai* and *fix* operons by CRP and the CaiF activator. *Mol. Micrbiol.* 34:562–575.
230. Lindenthal, C., and Elsinghorst, E. A. (2001). The enterotoxigenic *Escherichia coli* TibA glycoprotein adheres to human intestine epithelial cells. *Infect. Immun.* 69:52–57.
231. DuPont, H. L. (1995). Pathogenesis of traveler's diarrhea. *Chemotherapy* 41(suppl.1):33–39.
232. Black, R. E., Levine, M. M., Clements, M. L., *et al.* (1982). Treatment of experimentally induced enterotoxigenic *Escherichia coli* diarrhea with trimethoprim, trimethoprim-sulfamethoxazole, or placebo. *Rev. Infect. Dis.* 4:540–545.
233. Jiang, Z.-D., Mathewson, J. J., Ericsson, C. D., *et al.* (2000). Characterization of enterotoxigenic *Escherichia coli* strains in patients with travelers' diarrhea acquired in Guadalajara, Mexico, 1992–1997. *J. Infect. Dis.* 181:779–782.
234. Vila, J., Vargas, M., Casals, C., *et al.* (1999). Antimicrobial resistance of diarrheagenic *Escherichia coli* isolated from children under the age of 5 years from Ifakara, Tanzania. *Antimicrob. Agents Chemother.* 43:3022–3024.
235. Juckett, G. (1999). Prevention and treatment of traveler's diarrhea. *Am. Fam. Phys.* 60:119–124, 135–136.
236. Vila, J., Vargas, M., Ruiz, J., *et al.* (2000). Quinolone resistance in enterotoxigenic *Escherichia coli* causing diarrhea in travelers to India in comparison with other geographic areas. *Antimicrob. Agents Chemother.* 44:1731–1733.
237. Farthing, M. J. (2000). Novel targets for the pharmacotherapy of diarrhoea: A view for the millennium. *J. Gastroenterol. Hepatol.* 15(suppl.):G38–G45.
238. Heck, J. E., Staneck, J. L., Cohen, M. B., *et al.* (1994). Prevention of traveler's diarrhea: ciprofloxacin verses trimethoprim/sulfamethoxazole in adult volunteers working in Latin America and the Carribean. *J. Travel Med.* 1:136–142.
239. Tacket, C. O., Losonsky, G., Link, H., *et al.* (1988). Protection by milk immunoglobulin concentrate against oral challenge with enterotoxigenic *Escherichia coli. New Engl. J. Med.* 318:1240–1243.
240. Casswall, T. H., Sarker, S. A., Faruque, S. M., *et al.* (2000). Treatment of enterotoxigenic and enteropathogenic *Escherichia coli*—induced diarrhoea in children with bovine immunoglobulin milk concentrate from hyperimmunized cows: A double-blind, placebo-controlled, clinical trial. *Scand. J. Gastroenterol.* 35:711–718.
241. Levine, M. M., Black, R. E., Clements, M. L., *et al.* (1981). Volunteer studies in development of vaccines against cholera and enterotoxigenic *Escherichia coli*: A review. In T. Holme, J. Holmgren, M. H. Merson, and R. Møllby (eds.), *Acute Enteric Infections in Children: New Prospects for Treatment and Prevention*, pp. 443–459. Amsterdam: Elsevier North-Holland Biomedical Press.
242. Levine, M. M. (1990). Vaccines against enterotoxigenic *Escherichia coli* infections. In G. C. Woodrow, and M. M. Levine (eds.), *New Generation Vaccines*, pp. 649–660. New York: Marcell Dekker.
243. Peltola, H., Siitonen, A., Kyronseppa, H., *et al.* (1991). Prevention of traveller's diarrhoea by oral B-subunit/whole-cell cholera vaccine. *Lancet* 338:1285–1289.
244. Clemens, J., Sack, D., Harris, J. R., *et al.* (1988). Cross-protection by B-subunit-whole cell cholera vaccine against diarrhea associated with heat-labile toxin-producing enterotoxigenic *Escherichia coli*: Results of a large-scale field trial. *J. Infect. Dis.* 158:372–377.

245. à6hren, C. M., and Svennerholm, A.-M. L. (1982). Synergistic protective effect of antibodies against *Escherichia coli* enterotoxin and colonization factor antigens. *Infect. Immun.* 38:74–79.
246. Dubreuil, J. D., Letellier, A., and Harel, J. (1996). A recombinant *Escherichia coli* heat-stable enterotoxin B (STb) fusion protein eliciting nuetralizing antibodies. *FEMS Immunol. Med. Microbiol.* 13:317–323.
247. Pereira, C. M., Guth, B. E., Sbrogio-Almeida, M. E., and Castilho, B. A. (2001). Antibody response against *Escherichia coli* heat-stable enterotoxin expressed as fusions to flagellin. *Microbiology* 147:861–867.
248. Levine, M. M., Morris, J. G., Losonsky, G., *et al.* (1986). Fimbriae (pili) adhesins as vaccines. In D. L. Lark, S. Normark, B. E. Uhlin, and H. Wolf-Watz (eds.), *Protein-Carbohydrate Interactions in Biological Systems: The Molecular Biology of Microbial Pathogenicity*, pp. 143–145. London: Academic Press.
249. Evans, D. G., Graham, D. Y., Evans, D. J., Jr., and Opekun, A. (1984). Administration of purified colonization factor antigens (CFA/I, CFA/II) of enterotoxigenic *Escherichia coli* to volunteers. *Gastroenterology* 87:934–940.
250. Evans, D. G., Evans, D. J., Jr., Opekun, A., and Graham, D. Y. (1988). Non-replicating whole cell vaccine protective against enterotoxigenic *Escherichia coli* (ETEC) diarrhea: stimulation of anti-CFA (CFA/I) and anti-enterotoxin (anti-LT) intestinal IgA and protection against challenge with ETEC belonging to heterologous serotypes. *FEMS Microbiol. Lett.* 47:117–125.
251. Savarino, S. J., Brown, F. M., Hall, E., *et al.* (1998). Safety and immunogenicity of an oral, killed enterotoxigenic *Escherichia coli*–cholera toxin B subunit vaccine in Egyptian adults. *J. Infect. Dis.* 177:796–799.
252. Savarino, S. J., Hall, E. R., Bassily, S., *et al.* (1999). Oral, inactivated, whole cell enterotoxigenic *Escherichia coli* plus cholera toxin B subunit vaccine: results of the initial evaluation in children. PRIDE Study Group. *J. Infect. Dis.* 179:107–114.
253. Jertborn, M., à6hren, C., Holmgren, J., and Svennerholm, A.-M. (1998). Safety and immunogenicity of an oral, inactivated enterotoxigenic *Escherichia coli* vaccine. *Vaccine* 16:255–260.
254. Wennerà6s, C., Qadri, F., Bardhan, P. K., *et al.* (1999). Intestinal immune responses in patients infected with enterotoxigenic *Escherichia coli* and in vaccinees. *Infect. Immun.* 67:6234–6241.
255. Åhren, C., Jertborn, M., and Svennerholm, A.-M. (1998). Intestinal immune responses to an inactivated oral enterotoxigenic *Escherichia coli* vaccine and associated immunoglobulin A responses in blood. *Infect. Immun.* 66:3311–3316.
256. Svennerholm, A.-M., à6hren, C., and Jertborn, M. (1997). Vaccines against enterotoxigenic *Escherichia coli* infections: I. Oral inactivated vaccines against enterotoxigenic *Escherichia coli.* In M. M. Levine, G. C. Woodrow, J. B. Kaper, and G. S. Cobon (eds.), *New Generation Vaccines*, 2d ed., pp. 865–873. New York: Marcel Dekker.
257. Qadri, F., Wennerà6s, C., Ahmed, F., *et al.* (2000). Safety and immunogenicity of an oral, inactivated enterotoxigenic *Escherichia coli* plus cholera toxin B subunit vaccine in Bangladeshi adults and children. *Vaccine* 18:2704–2712.
258. Tacket, C. O., and Levine, M. M. (1997). Vaccines against enterotoxigenic *Escherichia coli* infections: II. Live oral vaccines and subunit (purified fimbriae and toxin subunit) vaccines. In M. M. Levine, G. C Woodrow, J. B. Kaper, and G. S. Cobon (eds.), *New Generation Vaccines*, 2d ed., pp. 875–883. New York: Marcel Dekker.
259. Levine, M. M., Giron, J. A., and Noriega, F. (1994). Fimbrial vaccines. In P. Klemm (ed.), *Fimbriae: Adhesion, Biogenics, Genetics, and Vaccines*, pp. 255–270. Boca Raton, FL: CRC Press.
260. Giron, J. A., Xu, J.-G., Gonzalez, C. R., *et al.* (1995). Simultaneous constitutive expression of CFA/I and CS3 colonization factors of enterotoxigenic *Escherichia coli* by *aroC, aroD Salmonella typhi* vaccine strain CVD908. *Vaccine* 10:939–946.
261. Noriega, F. R., Losonsky, G., Wang, Y. Y., *et al.* (1996). Further characterization of *aroA, virG Shigella flexneri* 2a strain CVD 1203 as a mucosal *Shigella* vaccine and as a live vector vaccine for delivering antigens of enterotoxigenic *Escherichia coli. Infect. Immun.* 64:23–27.

262. Koprowski, H., II, Levine, M. M., Anderson, R. J., *et al.* (2000). Attenuated *Shigella flexneri* 2a strain CVD1204 expressing colonization factor antigen I and mutant heat-labile enterotoxin of enterotoxigenic *Escherichia coli. Infect. Immun.* 68:4884–4892.
263. Altboum, Z., Barry, E. M., Losonsky, G., *et al.* (2001). Attenuated *Shigella flexneri* 2a ΔguaBA strain CVD 1204 expressing enterotoxigenic *Escherichia coli* (ETEC) CS2 and CS3 fimbriae as a live mucosal vaccine against *Shigella* and ETEC infection. *Infect. Immun.* 69:3150–3158.
264. Alves, A. M. B., Lasaro, M. O., Almeida, D. F., and Ferreira, L. C. S. (2001). DNA immunisation against the CFA/I fimbriae of enterotoxigenic *Escherichia coli* (ETEC). *Vaccine* 19:788–795.
265. Yu, J., and Langridge, H. R. (2001). A plant-based multicomponent vaccine protects mice from enteric diseases. *Nature Biotechnol.* 19:548–552.
266. Nasher, T. O., Webb, H. M., Eaglestone, S., *et al.* (1996). Potent immunogenicity of the B subunits of *Escherichia coli* heat-labile enterotoxin: receptor binding is essential and induces differential modulation of lymphocyte subsets. *Proc. Natl. Acad. Sci. USA* 93:226–230.
267. Douce, G., Turcotte, C., Cropley, I., *et al.* (1995). Mutants of *Escherichia coli* heat-labile toxin lacking ADP-ribosyltransferase activity act as nontoxic, mucosal adjuvants. *Proc. Natl. Acad. Sci. USA* 92:1644–1648.
268. Dickinson, B. L., and Clements, J. D. (1995). Dissociation of *Escherichia coli* heat-labile enterotoxin adjuvanticity from ADP-ribosyltransferase activity. *Infect. Immun.* 63:1617–1623.
269. Bowman, C., and Clements, J. D. (2001). Differential biological and adjuvant activities of cholera toxin and *Escherichia coli* heat-labile enterotoxin hybrids. *Infect. Immun.* 69:1528–1535.

CHAPTER 6

Enteroaggregative and Diffusely Adherent Escherichia Coli

James P. Nataro
Center for Vaccine Development, University of Maryland School of Medicine, Baltimore, Maryland

Theodore Steiner
Division of Infectious Diseases, University of British Columbia, Vancouver, British Columbia, Canada

In 1979, Cravioto and colleagues [1] reported that *Escherchia coli* isolates from the stools of diarrhea patients frequently adhered to HEp-2 cells in culture. In this study, HEp-2 adherence was most common among strains of traditional EPEC serotypes. Later, however, diarrhea-associated *E. coli* of other serotypes also were shown to adhere to HEp-2 cells; interestingly, however, the adherence patterns for the latter *E. coli* were clearly distinguishable from those of EPEC [2,3]. This simple observation gave rise to two new enteric pathotypes: enteroaggregative *E. coli* (EAEC) and diffusely adherent *E. coli* (DAEC). We now know that these pathotypes exhibit distinctive clinical, epidemiologic, and pathogenetic features.

ENTEROAGGREGATIVE *E. COLI*

EAEC currently are defined as *E. coli* that do not secrete the enterotoxigenic *E. coli* (ETEC) heat-labile or heat-stable enterotoxins and which adhere to HEp-2 cells in an aggregative (AA) pattern (Fig. 1). The AA pattern is recognized by the distinctive "stacked brick" autoagglutination of the bacteria either on the

Escherichia coli: Virulence Mechanisms of a Versatile Pathogen
ISBN 0-12-220751-3

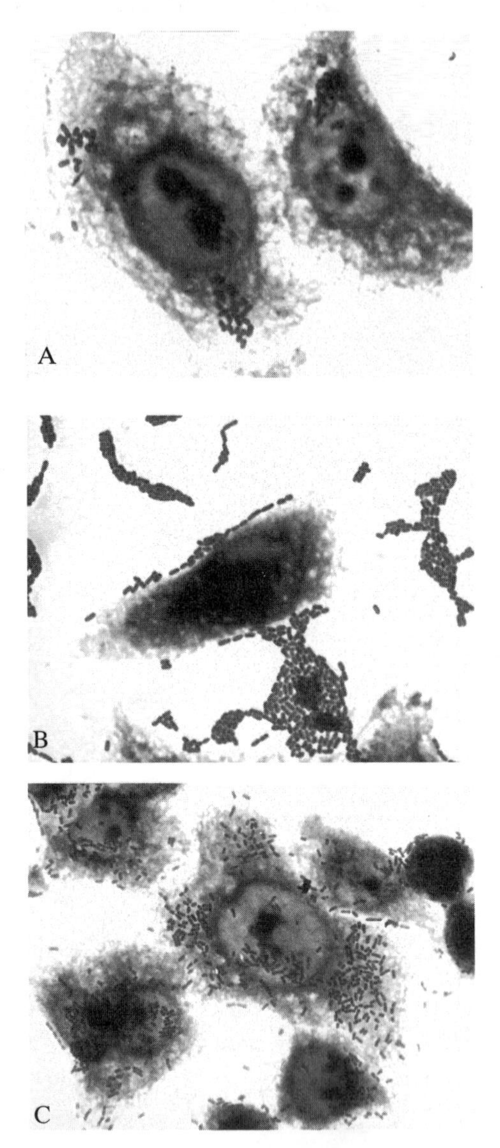

FIGURE 1 HEp-2 adherence patterns of diarrheagenic *E. coli*. Light photomicrographs of EAEC strain 042, EPEC strain E2348/69, and DAEC strain C1845 incubated with HEp-2 cells for 3 hours in the standard adherence assay [59]. (*A*) Localized adherence. Note clusters or microcolonies of bacteria on the surface of the HEp-2 cells. This pattern is typical of enteropathogenic *E. coli*. (*B*) Aggregative adherence. Note aggregation of bacteria in typical "stacked brick" pattern on the surface of the cell as well as free from the cell. (*C*) Diffuse adherence. Bacteria are scattered over the surface of the cell with little aggregation or adherence to the glass background.

surface of the HEp-2 cells or on the glass substratum. This definition could encompass both pathogenic and nonpathogenic bacteria, and therefore, much EAEC research focuses on identification of factors that will define truly virulent strains.

Microbiology

The serotypes characteristic of EAEC are not well characterized, but several serotypes are found commonly. Most characteristic are the flagellar antigens H18, H2, and H33. EAEC are commonly found to be O-antigen nontypeable, although Vial and colleagues [4] demonstated that at least some of these strains are resistant to rough-specific phages, and Weintraub and colleagues [5] have indeed characterized an O polysaccharide from one roughlike EAEC strain.

EAEC Clinical Illness

The best description of illness due to EAEC is derived from outbreaks and volunteer studies. Mathewson and colleagues [6] reported that strain JM221, isolated from the stool of a diarrhea patient in Mexico, induced mild watery diarrhea in three of eight volunteers; two additional subjects developed other enteric complaints, including abdominal cramps and borborygmi. No volunteer developed fever.

Nataro and colleagues [7] fed four EAEC strains of different serotypes to groups of 5 volunteers at a dose of 10^{10} colony-forming units (cfu). Peruvian EAEC strain 042 (O44:H18) induced clinically significant diarrhea in 3 volunteers, and one other recipient of this strain reported low-volume liquid stools and borborygmi. Stools typically were mucoid, without blood or polymorphonuclear cells. No volunteer reported fever. Notably, of the 15 volunteers fed other EAEC strains (including JM221), none met the study definition of diarrhea, and only 1 of the 15 volunteers had stools looser than normal. These data support strongly the pathogenicity of EAEC in humans but suggest that the virulence of these strains is not uniform. The basis for this difference in virulence is unknown.

Outbreaks of EAEC diarrhea (see below) suggest an illness similar to that seen in volunteers. Notably, in natural outbreaks a subset of patients typically will develop a prolonged watery diarrhea lasting several weeks [8,9].

Despite the lack of frank signs and symptoms suggesting inflammation, such as fever and the presence of fecal leukocytes, there is some evidence that EAEC may induce a mild inflammatory enteritis. Steiner and colleagues [10] demonstrated elevated lactoferrin titers and supranormal concentrations of interleukin-8 (IL-8) and interleukin-1β (IL-1β) and elevated ratios of IL-1β/IL-1RA (receptor antagonist) in stool samples from children with EAEC persistent diarrhea. This study also found that even asymptomatic children excreting EAEC had significantly higher levels of fecal cytokines than control children. Dramatically, EAEC shedding (with or without diarrhea) was associated with shortfalls in height and weight for age. A similar observation linking EAEC and growth faltering has been found in Australian aboriginal children (S. Elliott and J. Nataro, unpublished observations), and studies in adult travelers with EAEC infection have corroborated the presence of elevated fecal cytokines (H. Dupont, unpublished observations).

EAEC Epidemiology

Outbreaks

The first reported EAEC outbreak occurred in a Serbian nursery, and 19 infected infants developed watery diarhea [8]. Three affected infants developed persistent diarrhea; none died. Two additional nosocomial outbreaks among infants were reported from Mexico City [11]. This scenario is typical of enteropathogenic *E. coli* (EPEC), and indeed, several classic outbreaks presumed due to EPEC were later shown to be due to EAEC strains [12].

EAEC also has been implicated in food-borne outbreaks among older children and adults. The largest such outbreak involved nearly 2700 Japanese children, whose infection was linked to consumption of a contaminated school lunch [9]. Four food-borne outbreaks were reported in the United Kingdom; infection occurred in persons of all ages [13]. A large outbreak in an Indian village was attributed to contamination of the village well water with more than one EAEC strain [14].

Endemic Sporadic Diarrhea

EAEC was implicated initially as a pathogen in studies of infants and children in Santiago, Chile [3]. Since then, the organism has been reported with increasing frequency. As with most enteric pathogens, there appears to be geographic variation in EAEC epidemiology; areas of high incidence include Mexico [15], Brazil [16,17], and India [18,19]. EAEC may be recognized increasingly but also may be increasing in incidence. In a recent study outside São Paulo, Brazil, EAEC was implicated as the predominant diarrheal pathogen, whereas previous studies of this community had implicated EPEC as the principal pathogen (I. Skaletsky, personal communication). A similar observation was reported among persistent diarrhea patients in Bangladesh [20].

EAEC infection is not limited to developing countries; the pathogen has been implicated recently as an important diarrheal pathogen in the United Kingdom

[21], Germany [22], Austria [23], and Switzerland [24]. Tompkins and colleagues [21] reported that EAEC was associated with diarrhea among all age groups in a large prospective cohort study in England. In that study, EAEC was second only to *Campylobacter* among the bacterial pathogens. EAEC has been implicated in human immunodeficiency virus (HIV)–associated diarrhea in the United States [25].

Pathogenesis

The study of EAEC has been hampered by the fact that strain virulence is apparently heterogeneous and that observations made in nonpathogenic isolates may not be relevant to virulent EAEC. However, the identification of virulent isolates from outbreaks and volunteer studies has provided a small but invaluable collection of prototype strains.

Models of EAEC Infection

Although flawed, animal models have suggested several features of EAEC pathogenesis. In the rabbit ligated ileal loop, EAEC adheres abundantly to the mucosa and elicits shortening of the villi, hemorrhagic necrosis of the villous tips, and a mild inflammatory response, manifested as edema and mononuclear infiltration of the submucosa [4]. Notably, transmission electron microscopy (TEM) reveals normal microvillar architecture without invasion.

The only good whole-animal model for EAEC infection is the gnotobiotic piglet. Tzipori and colleagues [26] observed that of six piglets fed EAEC strain 17-2, four developed severe enteric signs, leading to death in two. Necropsy revealed a histopathologic lesion of the ileum similar to that observed in the rabbit ileal loop but with more prominent edema and less necrosis of the villous tips. Bacteria adhered to the mucosa in a thick biofilm. Strains JM221 and 042 elicited significant but less severe illness.

Histopathology of EAEC-infected humans has not been reported. However, Hicks and colleagues [27] have infected normal human intesinal biopsy specimens *in vitro* (the *in vitro* organ culture, or IVOC, model) and have shown that EAEC strains induce exfoliation of intestinal epithelial cells within 6–8 hours. Once again, EAEC adhere to the intestinal mucosa in a thick biofilm similar to that observed in animal models.

Adherence

AA to HEp-2 cells, aggregation in liquid culture, and adherence to the intestinal mucosa by EAEC require a 60- to 65-MDa plasmid designated pAA. In most EAEC strains, these properties are mediated by the plasmid-encoded aggregative

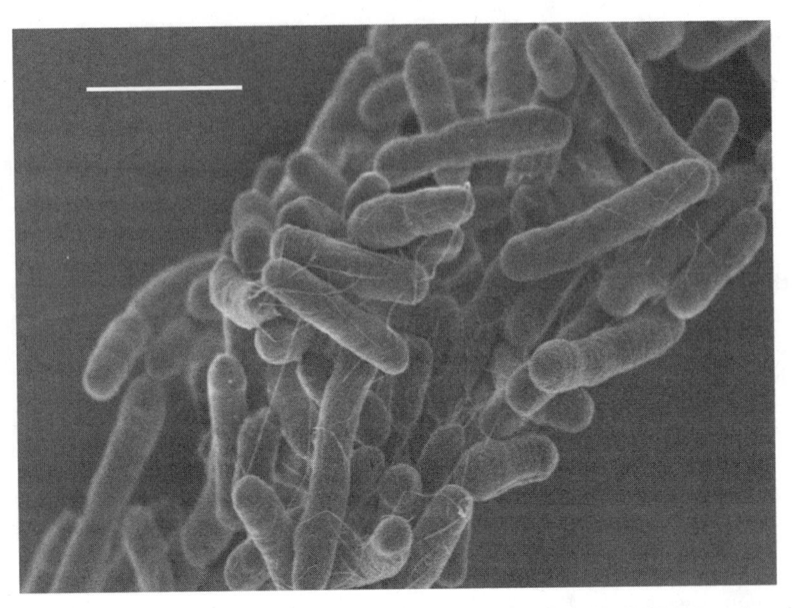

FIGURE 2 EAEC adheres by virtue of aggregative adherence fimbriae (AAFs). Scanning electron microscopy of strain 042 grown under AAF/II-inducing conditions. AAF structures wrap around adjoining bacteria (bar = 3 μm).

adherence fimbriae (AAFs), of which four allelic varieties have been described [28–31]. AAFs are semiflexible and form small bundles; they frequently extend great distances (>10 μm) and wrap around neighboring bacteria (Fig. 2). AAF biogenesis requires a cluster of genes including a fimbrial usher and chaperone as well as the pilin subunit [32,33]. A fourth gene in this cluster encodes a 14-kDa secreted protein that is related to the Dr family invasin AfaD [34]. Indeed, all four AAF biogenesis genes display significant homology with gene clusters encoding Dr adhesins. However, all other members of the Dr family mediate diffuse adherence to HEp-2 cells rather than the AA characteristic of EAEC [30,35].

The four AAFs thus far described display a high level of conservation of the accessory genes but with much greater divergence of the pilin genes. The conservation of accessory proteins extends to the function level because AAF/II subunits can be assembled by AAF/I accessory proteins [29]. Expression of AAF biogenesis genes requires the action of AggR, a transcriptional regulator of the AraC/XylS family [36]. AggR is most similar to the Rns activator of CS1 expression in ETEC, and the two proteins are capable of heterologous activation.

Only a minority of EAEC strains express any of the characterized AAF adhesins [29,31,37]. In contrast, the majority of EAEC carry genes homologous to AggR, suggesting that as yet uncharacterized AAFs also may require AggR or that AggR is a more general activator of EAEC virulence genes.

AAFs may not be the only EAEC adhesins. Knutton and colleagues [38] studied 44 EAEC strains and characterized four different fimbrial morphologies: hollow rod, rod, fibrillar, and fibrillar bundles. All the strains expressed the fibrillar bundle morphology but expressed at least one of the other morphologies as well. In the same study, the investigators showed that adherence of EAEC to cells in culture and to IVOC revealed a gap between the bacterium and the cell, as reported in animal models by Vial and colleagues [4] and in contrast to the intimate adherence seen with EPEC.

Sheikh and colleagues [38a] showed that EAEC forms a biofilm on plastic or glass surfaces when grown in cell culture medium with added glucose (0.4%). Moreover, these investigators found not only that AAFs were necessary for biofilm formation by EAEC strains but also that the pAA plasmid of strain 042 was sufficient to confer abundant biofilm formation on commensal *E. coli* HS.

Invasion of cultured epithelial cells by EAEC has been reported [39,40]. The AafD/AggD protein, encoded by the AAF gene clusters, may act as an invasin because the proteins have been shown to complement mutations in AfaD in uropathogenic *E. coli* [34]. Notably, however, invasion has not been demonstrated in the IVOC model. The role of invasiveness in EAEC pathogenesis remains to be elucidated.

Many EAEC strains, particularly those carrying the *aggR* gene, also express an approximately 10-kDa secreted protein called Aap (antiaggregation protein; formerly designated AspU [37]). EAEC aggregates strongly on the intestinal mucosa, but single bacteria are also observed dispersed across the epithelial surface; *aap* mutants, in contrast, aggregate to a greater extent, yielding virtually no individual bacteria on the mucosal surface (J. Czeculin, S. Hicks, A. Philips, I. Henderson, and J. Nataro, unpublished data). The manner in which Aap mediates this phenomenon is not known, but it does not appear to affect the number of AAFs assembled on the surface of the bacterium. The high prevalence of Aap (approximately 80% of strains) and its demonstrated immunogenicity [30] make this a promising candidate immunogen.

Toxins

Multiple studies suggest that the pathogenetic strategy of EAEC consists of mucosal adherence with elaboration of enterotoxins, the result of which is cellular damage (Fig. 3*A*). Savarino and colleagues [41,42] described a homologue of the ETEC ST designated EAST. EAST is a 38-amino-acid protein that induces net increases in short-circuit current in the Ussing chamber model. The role of EAST in disease has not been established with certainty, and many supposed nonpathogenic strains of *E. coli* appear to harbor the EAST gene [43].

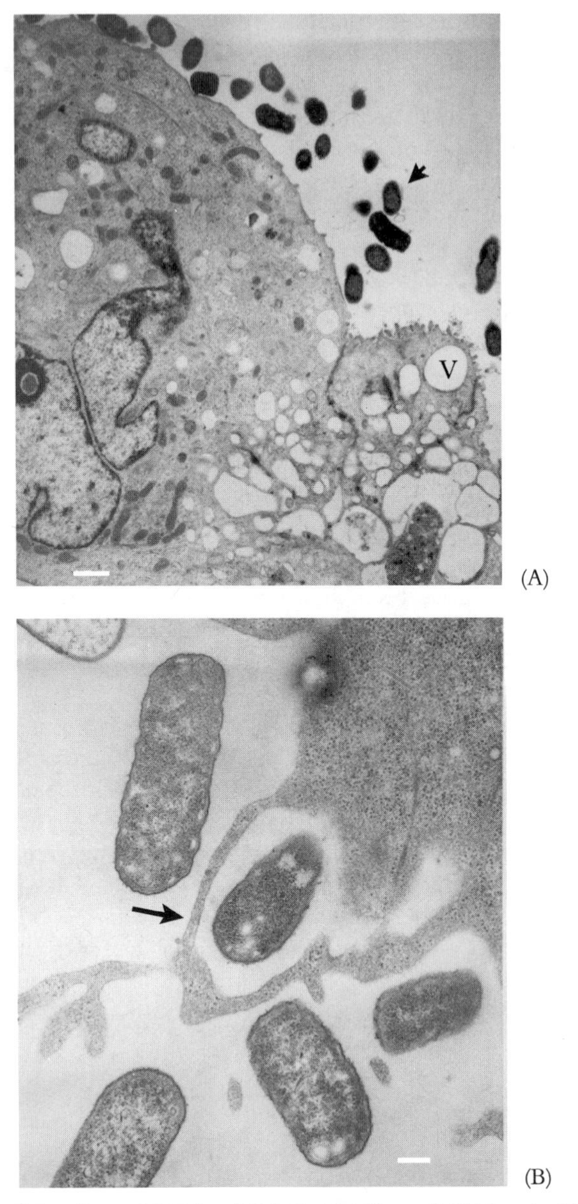

FIGURE 3 Cellular effects of EAEC and DAEC. (*A*) Effects of EAEC strain 042 on T84 cells. Bacteria were incubated with polarized cells for 6 hours, with change of cell culture medium at 3 hours. Note adherence of bacteria (*arrow*), loss of microvilli, and vacuolation (V) of cells (bar = 1 μm). (*B*) Epithelial cell effects induced by DAEC strain C1845. Bacteria were incubated with Hep-2 cells for 3 hours. Long, thin processes (*arrow*) wrap around bacteria, although complete internalization is unusual at this time point (bar = 200 μm). (From ref. 82 with permission.)

Eslava and colleagues [44] and Navarro-Garcia and colleagues [45,46] have characterized a high-molecular-weight enterotoxin called *Pet* that induces net secretion in the Ussing chamber. The Pet toxin is a member of the autotransporter family of secreted proteins, so named because they mediate their own translocation through the outer membrane via a dedicated C-terminal domain [47]. Henderson and colleagues [48] demonstrated that 042 carrying a mutation in the *pet* gene did not induce exfoliation of epithelial cells in the colonic IVOC model [48].

The mechanism of action of the Pet toxin is partially elucidated. Navarro-Garcia and colleagues [46,49] have shown that the Pet toxin is a serine protease and that mutation of the catalytic serine residue abolishes its toxic effects. Moreover, Pet has been shown to enter target epithelial cells [50]. Pet's activity can be inhibited with Brefeldin A [49], suggesting that entry and trafficking through the cell's vesicular system are necessary steps in the toxin's mode of action.

Early events following Pet intoxication include redistribution of the spectrin membrane cytoskeleton and blebbing [49] (R. Cappello and J. P. Nataro, unpublished data), followed by dissolution of actin stress fibers. Villaseca and colleagues [51] have shown that intoxication of epithelial cells in culture is associated with degradation of spectrin α and β and that the purified toxin cleaves both spectrins *in vitro.* Thus spectrin may be the target of the Pet protease *in vivo,* although it has not been proven that cleavage of spectrin is essential to cytopathic effects. Moreover, although Pet contributes to histopathology in the explant model, it is notable that the majority of EAEC clinical isolates do not harbor the *pet* gene [37]. Therefore, Pet may be an accessory virulence factor in some EAEC strains, but it does not represent the critical factor that links all pathogenic EAEC.

A third EAEC toxin is the *Shigella* enteroxin 1 (ShET1), a 55-kDa oligomeric toxin also present in *Shigella flexneri* 2a strains [52,53]. ShET1 induces fluid secretion in the Ussing chamber model but does not induce cytotoxic effects [53]. The mechanism of action of ShET1 is not known, and the toxin does not appear to act via the classic cAMP and cGMP pathways [53]. Some studies have suggested that most EAEC strains from diarrhea patients express the ShET1 toxin [37,54].

Inflammation

Steiner and colleagues [55] demonstrated that culture filtrates of 042, 17-2, and several other EAEC strains induce release of IL-8 from cultured intestinal epithelial cells. This effect was linked to expression of a novel EAEC flagellin protein, which is homologous to a flagellin encoded by *S. dysenteriae.* Several other EAEC strains expressing H18 antigen also induce IL-8 release, as do several H10 isolates. The clinical importance of proinflammatory flagellar expression remains to be determined. However, neutrophil transmigration across the epithelium in response to IL-8 can itself lead to tissue disruption and fluid secretion, both of which could contribute to the clinical features of EAEC infections.

Persistent Diarrhea

EAEC has been associated with persistent diarrhea of infants in several studies, but the mechanism underlying such an association remains obscure. One potential mechanism is delayed repair of epithelial surfaces after damage caused by Pet and other toxins; intestinal repair is delayed in malnourished children and some normal infants [56]. In addition, the presence of a thick mucosal biofilm may thwart host attempts at eradication of EAEC infection, although this has yet to be confirmed in any model.

Genetic Diversity and Phylogenetics

The AA phenotype could be a property of both pathogenic and nonpathogenic bacteria; indeed, aggregation of bacteria is known to be mediated by many hydrophobic surface structures [57]. Volunteer studies and outbreak investigations suggest that there are at least some human pathogens among the EAEC, but identification of pathogenic strains in prospective fashion cannot yet be done.

A related question asks whether or not EAEC strains display clonality similar to that of other pathogens, or whether they instead represent a wide diversity of bacteria. The first suggestion that EAEC may represent a related group of organisms was the discovery of the EAEC probe (see below) by Buadry and colleagues [58]. These investigators identified a conserved DNA fragment from the pAA plasmids of strains 042 and 17-2 and showed that this fragment was 89% sensitive and 99% specific in the detection of EAEC. This report thus suggested that many EAEC strains share at least the large pAA plasmid. Subsequent studies have confirmed that many genes on the pAA plasmid are shared among EAEC strains and that, therefore, like *Shigella*, *Salmonella*, *Yersinia*, enterohemorrhagic *E. coli* (EHEC), and enteropathogenic *E. coli* (EPEC), most EAEC strains harbor a conserved virulence plasmid.

Czeczulin and colleagues [37] described the phylogeny of EAEC by characterizing both plasmid and chromosomal loci in a collection of EAEC and DAEC strains. By multilocus enzyme electrophoresis analysis of 20 enzymes, EAEC strains were shown to cluster in three well-defined groups (Fig. 4). Moreover, conserved pAA plasmids as well as the chromosomal loci Pic/Set and yersiniabactin were found with high frequency in all three EAEC clusters. These data suggest that like other enteric pathogens, EAEC share conserved chromosomal as well as plasmid-borne loci and that the characteristic combinations have arisen on multiple occasions. Figure 4 also suggests that although EAEC and DAEC segregate into different clusters, there is significant overlap, with some EAEC and DAEC differing at but a single enzyme locus.

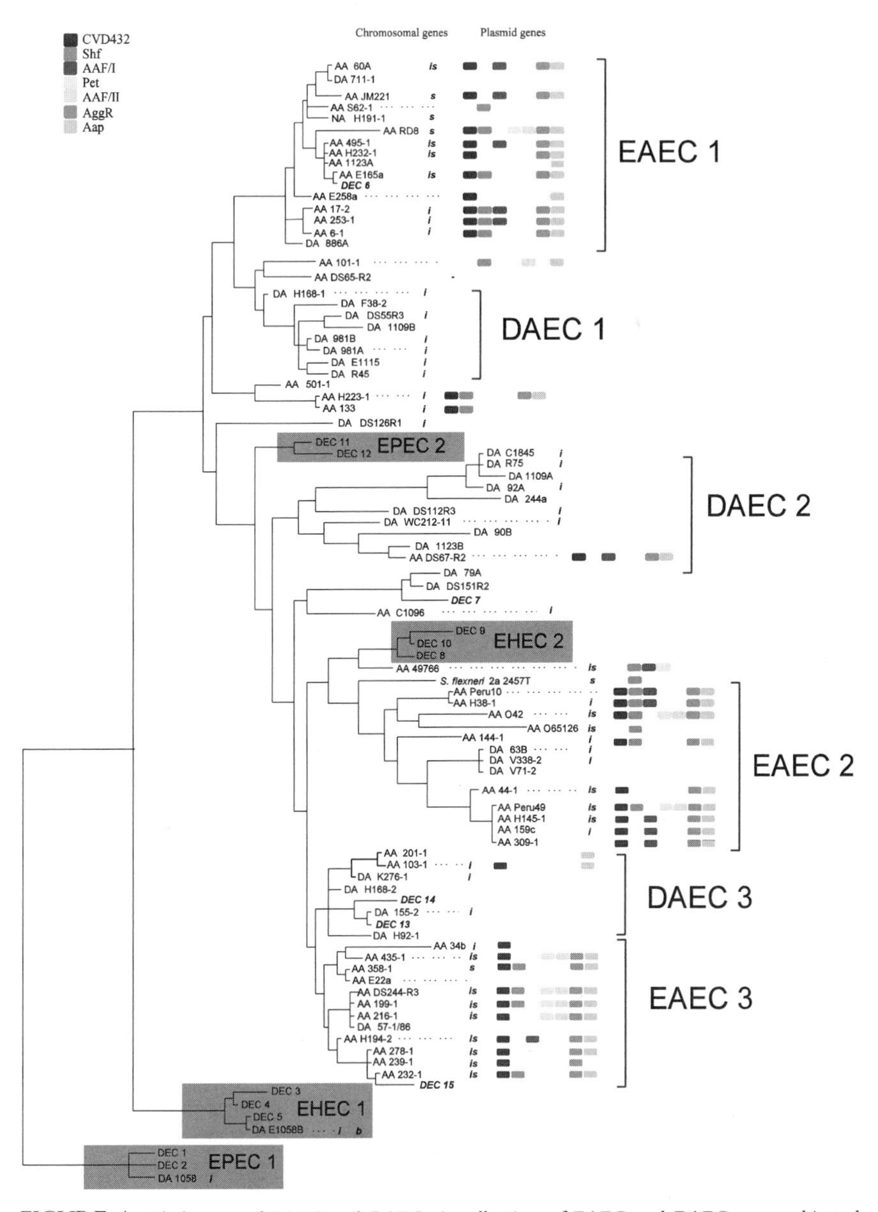

FIGURE 4 Phylogeny of EAEC and DAEC. A collection of EAEC and DAEC were subjected to multilocus enzyme electrophoresis and hybridized with DNA fragment probes derived from either the EAEC plasmid or from sequenced chromosomal loci. Chromosomal loci: i, yersiniabactin; s, ShET1/Pic locus. (Adapted from ref. 37 with permission.)

Detection of EAEC

EAEC infection is diagnosed by the isolation of *E. coli* displaying AA from the stools of patients. The AA phenotype in the HEp-2 assay continues to best define EAEC because molecular methods are not as sensitive and no defining virulence genes have been elucidated. The original HEp-2 assay protocol (for 3 hours of incubation) still provides the best differentiation of localized, diffuse, and aggregative adherence patterns [59]. Since AAFs are expressed in static L-broth cultures at 37°C, we incubate all HEp-2 assay inocula under these conditions.

It is notable that the AA probe [58] was derived empirically from the AA plasmid (see above) and does not correspond to any known virulence factor. The percentage of EAEC strains that hybridize with the AA probe varies among studies, and whether probe-positive strains are more virulent than probe-negative strains is still controversial, although some data suggest that this may be the case [60]. Implication of specific virulence factors in diarrhea is also unresolved, although one study suggested that AAF/II-expressing strains may be more virulent than strains that do not express this adhesin [61]. A recent study from Mexico similarly has implicated the *pic* locus [62].

EAEC detection methods other than HEp-2 adherence and DNA probe assays have been proposed. Albert and colleagues [63] reported that AA probe-positive organisms form pellicles in Mueller-Hinton broth. Similarly, EAEC strains produce an abundant biofilm on polystyrene culture tubes [64] (J. Sheikh and J. Nataro, in press). Either of these techniques is a convenient substitute for the DNA probe, but further confirmation of their sensitivity and specificity is required. It should be emphasized, moreover, that until epidemiologic studies establish greater pathogenicity of probe-positive strains over probe-negative strains, the HEp-2 assay should remain the gold standard for EAEC detection.

Treatment

Glandt and colleagues [65] have suggested that treatment with ciprofloxacin may ameliorate EAEC diarrhea. However, Yamamoto and colleagues [66] and Okeke and colleagues [61] have suggested that EAEC generally are resistant to many of the antibiotics commonly used for the treatment of diarrhea. Okeke and colleagues [61] suggested that multiple antibiotic resistance genes may be encoded on the pAA plasmid.

DIFFUSELY ADHERING *E. COLI*

Microbiology

DAEC is defined by its ability to adhere to HEp-2 cells in a diffuse pattern, with little adherence to the substratum. One impediment to understanding of DAEC is the lack of prototype strains of proven pathogenicity. No DAEC outbreak has yet been documented, and volunteer studies using two different DAEC strains did not result in diarrhea [67]. Little is known therefore about the pathogenetic and clinical characteristics of DAEC, assuming that there are indeed at least some true enteric pathogens within this group. No published reports allow characterization of characteristic serotypes, although our experience suggests that nontypeable and roughlike strains are characteristic of DAEC, as they are for EAEC (J. Nataro, unpublished data).

Epidemiology

Although early studies failed to find an association between DAEC and diarrheal disease, some more recent studies have documented such an association [68–71]. Generally, in the latter studies, DAEC are linked to diarrhea only outside infancy.

Gunzberg and colleagues [71] studied the adherence characteristics of 138 *E. coli* samples from the stools of aboriginal children from whom no recognized pathogen had been isolated. Twenty-five (36.8%) of 68 children with diarrhea and 32 (45.7%) of 70 without diarrhea had diffusely adherent (DA) isolates. After age stratification, however, DAEC strains were found to be significantly associated with diarrhea in children older than 18 months of age ($p < 0.05$). Levine and colleagues [70] performed a prospective cohort study of 360 Chilean children under 5 years of age; the relative risk of diarrhea with DAEC excretion increased with each year of age, maximizing at a relative risk of 2.1 at 48–60 months.

Baqui and colleagues [69] have suggested that DAEC may cause persistent diarrhea. Following a cohort of 705 children in rural Bangladesh, these investigators implicated DAEC in 16.4% of 177 patients with persistent diarrhea, 10.3% of patients with matched acute diarrhea, and 8.2% of asymptomatic matched controls ($p < 0.05$ for persistent diarrhea versus both acute and no diarrhea). These data suggest not only that DAEC may cause persistent diarrhea but also that studies in which patients with diarrhea are not stratified by duration may be less likely to link DAEC with diarrhea.

Like EAEC, DAEC may be a pathogen in industrialized countries. Jallat and colleagues [72] characterized 262 strains of *E. coli* isolated from diarrhea patients of all ages hospitalized in Clermont-Ferrand, France. One-hundred strains (38.2%) exhibited the DA phenotype compared with 8.9% (8 of 90) of *E. coli* isolated from asymptomatic patients ($p < 0.0001$).

Pathogenesis

Several features of DAEC pathogenesis are beginning to emerge. Bilge and colleagues [73] have characterized surface fimbriae (designated F1845) that are responsible for the DA phenotype in strain C1845. Approximately 75% of DAEC strains may carry F1845 or a related adhesin (J. Nataro, unpublished data). The genes encoding F1845 can be found on either the bacterial chromosome or a plasmid [73,74].

Like AAFs, F1845 fimbriae are homologous at the nucleotide and amino acid level with members of the Dr family of adhesins [73]. The receptor for F1845 and other members of the Dr family is the decay-accelerating factor (DAF, or CD55), a glycosylphosphatidylinositol-anchored protein that normally protects cells from damage by the complement system [75]. Deletion of the DAF short consensus repeat 3 (SCR3) domain or coincubation with antibodies to this region prevents adherence by DAEC [76]. It also has been suggested that adherence to SCR3 results in DAF clustering and endocytosis of the adhering bacteria [76].

DAEC strains induce a cytopathic effect characterized by the development of long cellular extensions that wrap around the adherent bacteria [77–82] (see Fig. 3*B*). This characteristic effect requires binding and clustering of the DAF receptor by the Dr family fimbriae [79]. All members of the Dr family (including uropathogenic *E. coli* as well as DAEC strain C1845) elicit this effect; AAFs producing EAEC do not and do not bind DAF [83]. Binding of Dr adhesins is accompanied by the activation of several signal-transduction cascades, including activation of PI-3 kinase [77].

LeBouguenec and colleagues [84] have shown that all Dr adhesins express a protein that mediates binding to and internalization by epithelial cells (see EAEC above). This protein, designated DaaD in the F1845 system, could promote invasion of the mucosa by DAEC strains, although this has not yet been demonstrated in clinical specimens. The related AfaD protein has been localized to the surface of the bacterium [84,85] in uropathogenic Dr producers, but its location and role in DAEC are not fully characterized.

Benz and colleagues [86] have described AIDA-I, a 100-kDa outer membrane autotransporter protein that mediates DA in a strain of serotype O126:H27. Use of an AIDA-I DNA probe suggests that this factor is expressed by a small minority of DAEC isolates (J. Nataro, unpublished data).

It is noteworthy that DAEC may express either F1845 or an afimbrial adhesin, the latter presumed to be a virulence factor for the urinary tract [83]. Although molecular approaches can distinguish among members of the Dr family, it is not yet clear which diffusely adhering strains are enteric pathogens (if indeed any). Peiffer and colleagues [81] have reported that infection of an intestinal cell line by DAEC strains impairs the activities and reduces the abundance of brush-border-associated sucrase isomaltase and dipeptidylpeptidase IV. This effect is independent of the DAF-associated pathway described earlier. This report therefore provides a feasible mechanism for DAEC-induced enteric disease and also suggests the presence of virulence factors in DAEC other than Dr adhesins. Identification of the bacterial factors responsible for this effect could permit improved definition and detection of DAEC.

Diagnosis

Presence of the DA pattern in the HEp-2 assay defines DAEC. A DAEC DNA probe was derived from the *daaC* gene [87] that encodes the fimbrial usher of the F1845 fimbriae. Due to the genetic relatedness of the F1845 apparatus with those of other members of the Dr family, false-positive results with the DA probe occur. Since the sequence of the F1845 pilin subunit is available (GenBank Accession Number M27725), specific detection of F1845 could provide additional benefit in detection of DAEC.

REFERENCES

1. Cravioto, A., Gross, R. J., Scotland, S. M., and Rowe, B. (1979). An adhesive factor found in strains of *Escherichia coli* belonging to the traditional infantile enteropathogenic serotypes. *Curr. Microbiol.* 3:95–99.
2. Scaletsky, I. C., Silva, M. L., and Trabulsi, L. R. (1984). Distinctive patterns of adherence of enteropathogenic *Escherichia coli* to HeLa cells. *Infect. Immun.* 45:534–536.
3. Nataro, J. P., Kaper, J. B., Robins Browne, R., *et al.* (1987). Patterns of adherence of diarrheagenic *Escherichia coli* to HEp-2 cells. *Pediatr. Infect. Dis. J.* 6:829–831.
4. Vial, P. A., Robins Browne, R., Lior, H., *et al.* (1988). Characterization of enteroadherent-aggregative *Escherichia coli,* a putative agent of diarrheal disease. *J. Infect. Dis.* 158:70–79.
5. Weintraub, A., Leontein, K., Widmalm, G., *et al.* (1993). Structural studies of the O-antigenic polysaccharide of an enteroaggregative *Escherichia coli* strain. *Eur. J. Biochem.* 213:859–864.
6. Mathewson, J. J., Johnson, P. C., DuPont, H. L., *et al.* (1986). Pathogenicity of enteroadherent *Escherichia coli* in adult volunteers. *J. Infect. Dis.* 154:524–527.
7. Nataro, J. P., Yikang, D., Cookson, S., *et al.* (1995). Heterogeneity of enteroaggregative *Escherichia coli* virulence demonstrated in volunteers. *J. Infect. Dis.* 171:465–468.
8. Cobeljic, M., Miljkovic-Selimovic, B., Paunovic-Todosijevic, D., *et al.* (1996). Enteroaggregative *Escherichia coli* associated with an outbreak of diarrhoea in a neonatal nursery ward. *Epidemiol. Infect.* 117:11–16.
9. Itoh, Y., Nagano, I., Kunishima, M., and Ezaki, T. (1997). Laboratory investigation of enteroaggregative *Escherichia coli* O untypeable:H10 associated with a massive outbreak of gastrointestinal illness. *J. Clin. Microbiol.* 35:2546–2550.
10. Steiner, T. S., Lima, A. A., Nataro, J. P., and Guerrant, R. L. (1998). Enteroaggregative *Escherichia coli* produce intestinal inflammation and growth impairment and cause interleukin-8 release from intestinal epithelial cells. *J. Infect. Dis.* 177:88–96.
11. Eslava, C., Villaseca, J., Morales, R., *et al.* (1993). Identification of a protein with toxigenic activity produced by enteroaggregative *Escherichia coli.* In *Proceedings of the 93rd General Meeting of the American Society for Microbiology*, p. 44. Atlanta: American Society for Microbiology.
12. Smith, H. R., Scotland, S. M., Willshaw, G. A., *et al.* (1994). Isolates of *Escherichia coli* O44:H18 of diverse origin are enteroaggregative. *J. Infect. Dis.* 170:1610–1613.
13. Spencer, J., Smith, H. R., and Chart, H. (1999). Characterization of enteroaggregative *Escherichia coli* isolated from outbreaks of diarrhoeal disease in England. *Epidemiol. Infect.* 123:413–421.
14. Pai, M., Kang, G., Ramakrishna, B. S., *et al.* (1997). An epidemic of diarrhoea in south India caused by enteroaggregative *Escherichia coli. Ind. J. Med. Res.* 106:7–12.
15. Cravioto, A., Tello, A., Navarro, A., *et al.* (1991). Association of *Escherichia coli* HEp-2 adherence patterns with type and duration of diarrhoea. *Lancet* 337:262–264.

16. Wanke, C. A., Schorling, J. B., Barrett, L. J., *et al.* (1991). Potential role of adherence traits of *Escherichia coli* in persistent diarrhea in an urban Brazilian slum. *Pediatr. Infect. Dis. J.* 10:746–751.
17. Lima, A. A., Fang, G., Schorling, J. B., *et al.* (1992). Persistent diarrhea in northeast Brazil: etiologies and interactions with malnutrition. *Acta Paediatr. Suppl.* 381:39–44.
18. Bhan, M. K., Khoshoo, V., Sommerfelt, H., *et al.* (1989). Enteroaggregative *Escherichia coli* and *Salmonella* associated with nondysenteric persistent diarrhea. *Pediatr. Infect. Dis. J.* 8:499–502.
19. Bhan, M. K., Raj, P., Levine, M. M., *et al.* (1989). Enteroaggregative *Escherichia coli* associated with persistent diarrhea in a cohort of rural children in India. *J. Infect. Dis.* 159:1061–1064.
20. Bardhan, P. K., Albert, M. J., Alam, N. H., *et al.* (1998). Small bowel and fecal microbiology in children suffering from persistent diarrhea in Bangladesh. *J. Pediatr. Gastroenterol. Nutr.* 26:9–15.
21. Tompkins, D. S., Hudson, M. J., Smith, H. R., *et al.* (1999). A study of infectious intestinal disease in England: Microbiological findings in cases and controls (see comments). *Commun. Dis. Public Health* 2:108–113.
22. Huppertz, H. I., Rutkowski, S., Aleksic, S., and Karch, H. (1997). Acute and chronic diarrhoea and abdominal colic associated with enteroaggregative *Escherichia coli* in young children living in western Europe. *Lancet* 349:1660–1662.
23. Presterl, E., Nadrchal, R., Wolf, D., *et al.* (1999). Enteroaggregative and enterotoxigenic *Escherichia coli* among isolates from patients with diarrhea in Austria. *Eur. J. Clin. Microbiol. Infect. Dis.* 18:209–212.
24. Nadal, D., Pabst, W., Kind, C., *et al.* (2001). Prevalence of enteropathogenic (EPEC) and enteroaggregative (EaggEC) *E. coli* in children with and without diarrhea in switzerland. Presented at the American Society of Microbiology 101st General Meeting, Orlando, FL.
25. Wanke, C. A., Gerrior, J., Blais, V., *et al.* (1998). Successful treatment of diarrheal disease associated with enteroaggregative *Escherichia coli* in adults infected with human immunodeficiency virus. *J. Infect. Dis.* 178:1369–1372.
26. Tzipori, S., Montanaro, J., Robins-Browne, R. M., *et al.* (1992). Studies with enteroaggregative *Escherichia coli* in the gnotobiotic piglet gastroenteritis model. *Infect. Immun.* 60:5302–5306.
27. Hicks, S., Candy, D. C., and Phillips, A. D. (1996). Adhesion of enteroaggregative *Escherichia coli* to formalin-fixed intestinal and ureteric epithelia from children. *J. Med. Microbiol.* 44:362–371.
28. Rich, C., Favre-Bonte, S., Sapena, F., *et al.* (1999). Characterization of enteroaggregative *Escherichia coli* isolates. *FEMS Microbiol. Lett.* 173:55–61.
29. Czeczulin, J. R., Balepur, S., Hicks, S., *et al.* (1997). Aggregative adherence fimbria II, a second fimbrial antigen mediating aggregative adherence in enteroaggregative *Escherichia coli. Infect. Immun.* 65:4135–4145.
30. Nataro, J. P., Deng, Y., Maneval, D. R., *et al.* (1992). Aggregative adherence fimbriae I of enteroaggregative *Escherichia coli* mediate adherence to HEp-2 cells and hemagglutination of human erythrocytes. *Infect. Immun.* 60:2297–2304.
31. Bernier, C., Labigne, A. F., and Le Bouguenec, C. (2001). Enteroaggregative *Escherichia coli* isolate 55989 produces the AAF-III adhesion system and induces cellular damage. Presented at the 101st General Meeting of the American Society for Microbiology, Orlando, FL (Abstract B-3).
32. Savarino, S. J., Fox, P., Deng, Y., and Nataro, J. P. (1994). Identification and characterization of a gene cluster mediating enteroaggregative *Escherichia coli* aggregative adherence fimbria I biogenesis. *J. Bacteriol.* 176:4949–4957.
33. Elias, W. P., Jr., Czeczulin, J. R., Henderson, I. R., *et al.* (1999). Organization of biogenesis genes for aggregative adherence fimbria II defines a virulence gene cluster in enteroaggregative *Escherichia coli. J. Bacteriol.* 181:1779–1785.
34. Garcia, M. I., Jouve, M., Nataro, J. P., *et al.* (2000). Characterization of the AfaD-like family of invasins encoded by pathogenic *Escherichia coli* associated with intestinal and extra-intestinal infections. *FEBS Lett.* 479:111–117.

35. Nataro, J. P., Yikang, D., Giron, J. A., *et al.* (1993). Aggregative adherence fimbria I expression in enteroaggregative *Escherichia coli* requires two unlinked plasmid regions. *Infect. Immun.* 61:1126–1131.
36. Nataro, J. P., Yikang, D., Yingkang, D., and Walker, K. (1994). AggR, a transcriptional activator of aggregative adherence fimbria I expression in enteroaggregative *Escherichia coli. J. Bacteriol.* 176:4691–4699.
37. Czeczulin, J. R., Whittam, T. S., Henderson, I. R., *et al.* (1999). Phylogenetic analysis of enteroaggregative and diffusely adherent *Escherichia coli. Infect. Immun.* 67:2692–2699.
38. Knutton, S., Shaw, R. K., Bhan, M. K., *et al.* (1992). Ability of enteroaggregative *Escherichia coli* strains to adhere *in vitro* to human intestinal mucosa. *Infect. Immun.* 60:2083–2091.
38a. Sheikh, J., Hicks, S., Dall'Agnol, M., Phillips, A. D., and Nataro, J. P. (2001). Roles for Fis and Yafk in biofilm formation by enteroaggregative *Escherichia coli. Mol. Microbiol.* 41:983–997.
39. Wanke, C. A., Cronan, S., Goss, C., *et al.* (1990). Characterization of binding of *Escherichia coli* strains which are enteropathogens to small-bowel mucin. *Infect. Immun.* 58:794–800.
40. Abe, C. M., Knutton, S., Pedroso, M. Z., *et al.* (2001). Ultrastructural studies on the interaction of enteroaggregative *Escherichia coli* strains of EPEC serogroups with Caco-2 and T84 cells and presence of putative virulence gene sequence. Presented at the 101st General Meeting of the American Society for Microbiology, Orlando, FL (Abstract B-123).
41. Savarino, S., Fasano, A., Robertson, D., and Levine, M. (1991). Enteroaggregative *Escherichia coli* elaborate a heat-stable enterotoxin demonstrable in an *in vitro* intestinal model. *J. Clin. Invest.* 87:1450–1455.
42. Savarino, S., Fasano, A., Watson, J., *et al.* (1993). Enteroaggregative *Escherichia coli* heat-stable enterotoxin 1 represents another subfamily of *E. coli* heat-stable toxin. *Proc. Natl. Acad. Sci. USA* 90:3093–3097.
43. Savarino, S., McVeigh, A., Watson, J., *et al.* (1996). Enteroaggregative *Escherichia coli* heat-stable enterotoxin is not restricted to enteroaggregative *E. coli. J. Infect. Dis.* 173:1019–1022.
44. Eslava, C. E., Navarro-Garcia, F., Czeczulin, J. R., *et al.* (1998). Pet, an autotransporter enterotoxin from enteroaggregative *Escherichia coli. Infect. Immun.* 66:3155–3163.
45. Navarro-Garcia, F., Eslava, C., Villaseca, J. M., *et al.* (1998). *In vitro* effects of a high-molecular-weight heat-labile enterotoxin from enteroaggregative *Escherichia coli. Infect. Immun.* 66:3149–3154.
46. Navarro-Garcia, F., Sears, C., Eslava, C., *et al.* (1999). Cytoskeletal effects induced by pet, the serine protease enterotoxin of enteroaggregative *Escherichia coli. Infect. Immun.* 67:2184–2192.
47. Henderson, I. R., Navarro-Garcia, F., and Nataro, J. P. (1998). The great escape: structure and function of the autotransporter proteins. *Trends Microbiol.* 6:370–378.
48. Henderson, I. R., Hicks, S., Navarro-Garcia, F., *et al.* (1999). Involvement of the Enteroaggregative *Escherichia coli* plasmid-encoded toxin in causing human intestinal damage. *Infect. Immun.* 67:5338–5344.
49. Navarro-Garcia, F., Eslava, C. E., Villaseca, J. M., *et al.* (1998). In vitro effects of a high-molecular weight heat-labile enterotoxin from enteroaggregative *Escherichia coli. Infect. Immun.* 66:349–354.
50. Navarro-Garcia, F., Canizalez-Roman, A., Luna, J., *et al.* (2001). Plasmid-encoded toxin of enteroaggregative *Escherichia coli* is internalized by epithelial cells. *Infect. Immun.* 69:1053–1060.
51. Villaseca, J. M., Navarro-Garcia, F., Mendoza-Hernandez, G., *et al.* (2000). Pet toxin from enteroaggregative *Escherichia coli* produces cellular damage associated with fodrin disruption. *Infect. Immun.* 68:5920–5927.
52. Fasano, A., Noriega, F. R., Maneval, D. R., *et al.* (1995). *Shigella* enterotoxin 1: An enterotoxin of *Shigella flexneri* 2a active in rabbit small intestine *in vivo* and *in vitro. J. Clin. Invest.* 95:2853–2861.
53. Fasano, A., Noriega, F., Liao, J., *et al.* (1997). Effect of *Shigella* enterotoxin 1 (ShET1) on rabbit intestine *in vitro* and *in vivo. Gut* 40:505–511.

54. Vila, J., Vargas, M., Henderson, I. R., *et al.* (2000). Enteroaggregative *Escherichia coli* virulence factors in traveler's diarrhea strains. *J. Infect. Dis.* 182:1780–1783.
55. Steiner, T. S., Nataro, J. P., Poteet-Smith, C. E., *et al.* (2000). Enteroaggregative *Escherichia coli* expresses a novel flagellin that causes IL-8 release from intestinal epithelial cells. *J. Clin. Invest.* 105:1769–1777.
56. Guerrant, R., Schorling, J., McAuliffe, J., and Souza, M. D. (1992). Diarrhea as a cause and effect of malnutrition: Diarrhea prevents catch-up growth and malnutrition increases diarrhea frequency and duration. *Am. J. Trop. Med. Hyg.* 47:28–35.
57. Irvin, R. (1990). Hydrophobicity of proteins and bacterial fimbriae. In R. J. and M. R. Doyle (eds.), *Microbial Cell Surface Hydrophobicity,* pp. 137–177. Washington: American Society for Microbiology.
58. Baudry, B., Savarino, S. J., Vial, P., *et al.* (1990). A sensitive and specific DNA probe to identify enteroaggregative *Escherichia coli,* a recently discovered diarrheal pathogen. *J. Infect. Dis.* 161:1249–1251.
59. Vial, P. A., Mathewson, J. J., DuPont, H. L., *et al.* (1990). Comparison of two assay methods for patterns of adherence to HEp-2 cells of *Escherichia coli* from patients with diarrhea. *J. Clin. Microbiol.* 28:882–885.
60. Durrer, P., Zbinden, R., Fleisch, F., *et al.* (2000). Intestinal infection due to enteroaggregative *Escherichia coli* among human immunodeficiency virus-infected persons. *J. Infect. Dis.* 182:1540–1544.
61. Okeke, I. N., Lamikanra, A., Czeczulin, J., *et al.* (2000). Heterogeneous virulence of enteroaggregative *Escherichia coli* strains isolated from children in Southwest Nigeria. *J. Infect. Dis.* 181:252–260.
62. Hernandez, U., Villaseca, J., Navarro, A., *et al.* (2001). Expression of Pet and Pic the serine proteases from enteroaggregative *Escherichia coli,* in strains isolated from Mexican children with and without diarrhea. Presented at the 101st General Meeting of the American Society for Microbiology, Orlando, FL (Abstract B-226).
63. Albert, M. J., Qadri, F., Haque, A., and Bhuiyan, N. A. (1993). Bacterial clump formation at the surface of liquid culture as a rapid test for identification of enteroaggregative *Escherichia coli. J. Clin. Microbiol.* 31:1397–1399.
64. Yamamoto, T., Koyama, Y., Matsumoto, M., *et al.* (1992). Localized, aggregative, and diffuse adherence to HeLa cells, plastic, and human small intestines by *Escherichia coli* isolated from patients with diarrhea. *J. Infect. Dis.* 166:1295–1310.
65. Glandt, M., Adachi, J. A., Mathewson, J. J., *et al.* (1999). Enteroaggregative *Escherichia coli* as a cause of traveler's diarrhea: Clinical response to ciprofloxacin. *Clin. Infect. Dis.* 29:335–338.
66. Yamamoto, T., Echeverria, P., and Yokota, T. (1992). Drug resistance and adherence to human intestines of enteroaggregative *Escherichia coli. J. Infect. Dis.* 165:744–749.
67. Tacket, C. O., Moseley, S. L., Kay, B., *et al.* (1990). Challenge studies in volunteers using *Escherichia coli* strains with diffuse adherence to HEp-2 cells. *J. Infect. Dis.* 162:550–552.
68. Giron, J. A., Jones, T., Millan-Velasco, F., *et al.* (1991). Diffuse-adhering *Escherichia coli* (DAEC) as a putative cause of diarrhea in Mayan children in Mexico. *J. Infect. Dis.* 163:507–513.
69. Baqui, A. H., Sack, R. B., Black, R. E., *et al.* (1992). Enteropathogens associated with acute and persistent diarrhea in Bangladeshi children less than 5 years of age. *J. Infect. Dis.* 166:792–796.
70. Levine, M. M., Ferreccio, C., Prado, V., *et al.* (1993). Epidemiologic studies of *Escherichia coli* diarrheal infections in a low socioeconomic level periurban community in Santiago, Chile. *Am. J. Epidemiol.* 138:849–869.
71. Gunzburg, S. T., Chang, B. J., Elliott, S. J., *et al.* (1993). Diffuse and enteroaggregative patterns of adherence of enteric *Escherichia coli* isolated from aboriginal children from the Kimberley region of Western Australia. *J. Infect. Dis.* 167:755–758.
72. Jallat, C., Livrelli, V., Darfeuille-Michaud, A., *et al.* (1993). *Escherichia coli* strains involved in diarrhea in France: High prevalence and heterogeneity of diffusely adhering strains. *J. Clin. Microbiol.* 31:2031–2037.

73. Bilge, S. S., Clausen, C. R., Lau, W., and Moseley, S. L. (1989). Molecular characterization of a fimbrial adhesin, F1845, mediating diffuse adherence of diarrhea-associated *Escherichia coli* to HEp-2 cells. *J. Bacteriol.* 171:4281–4289.
74. Carnoy, C., and Moseley, S. L. (1997). Mutational analysis of receptor binding mediated by the Dr family of *Escherichia coli* adhesins. *Mol. Microbiol.* 23:365–379.
75. Nicholson-Weller, A., and Wang, C. E. (1994). Structure and function of decay accelerating factor CD55. *J. Lab. Clin. Med.* 123:485–491.
76. Selvarangan, R., Goluszko, P., Popov, V., *et al.* (2000). Role of decay-accelerating factor domains and anchorage in internalization of Dr-fimbriated *Escherichia coli*. *Infect. Immun.* 68:1391–1399.
77. Peiffer, I., Servin, A. L., and Bernet-Camard, M. F. (1998). Piracy of decay-accelerating factor (CD55) signal transduction by the diffusely adhering strain *Escherichia coli* C1845 promotes cytoskeletal F-actin rearrangements in cultured human intestinal INT407 cells. *Infect. Immun.* 66:4036–4042.
78. Guignot, J., Bernet-Camard, M. F., Pous, C., *et al.* (2001). Polarized entry of uropathogenic Afa/Dr diffusely adhering *Escherichia coli* strain IH11128 into human epithelial cells: evidence for alpha5beta1 integrin recognition and subsequent internalization through a pathway involving caveolae and dynamic unstable microtubules. *Infect. Immun.* 69:1856–1868.
79. Bernet-Camard, M. F., Coconnier, M. H., Hudault, S., and Servin, A. L. (1996). Pathogenicity of the diffusely adhering strain *Escherichia coli* C1845: F1845 adhesin–decay-accelerating factor interaction, brush border microvillus injury, and actin disassembly in cultured human intestinal epithelial cells. *Infect. Immun.* 64:1918–1928.
80. Peiffer, I., Guignot, J., Barbat, A., *et al.* (2000). Structural and functional lesions in brush border of human polarized intestinal Caco-2/TC7 cells infected by members of the Afa/Dr diffusely adhering family of *Escherichia coli*. *Infect. Immun.* 68:5979–5990.
81. Peiffer, I., Bernet-Camard, M. F., Rousset, M., and Servin, A. L. (2001). Impairments in enzyme activity and biosynthesis of brush border-associated hydrolases in human intestinal Caco-2/TC7 cells infected by members of the Afa/Dr family of diffusely adhering *Escherichia coli*. *Cell. Microbiol.* 3:341–357.
82. Cookson, S. T., and Nataro, J. P. (1996). Characterization of HEp-2 cell projection formation induced by diffusely adherent *Escherichia coli*. *Microb. Pathog.* 21:421–434.
83. Nowicki, B., Selvarangan, R., and Nowicki, S. (2001). Family of *Escherichia coli* Dr adhesins: decay-accelerating factor receptor recognition and invasiveness. *J. Infect. Dis.* 183(suppl 1):S24–27.
84. Garcia, M. I., Jouve, M., Nataro, J. P., *et al.* (2000). Characterization of the AfaD-like family of invasins encoded by pathogenic *Escherichia coli* associated with intestinal and extra-intestinal infections. *FEBS Lett.* 479:111–117.
85. Gounon, P., Jouve, M., and Le Bouguenec, C. (2000). Immunocytochemistry of the AfaE adhesin and AfaD invasin produced by pathogenic *Escherichia coli* strains during interaction of the bacteria with HeLa cells by high-resolution scanning electron microscopy. *Microb. Infect.* 2:359–365.
86. Benz, I., and Schmidt, M. A. (1990). Diffuse adherence of enteropathogenic *Escherichia coli* strains. *Res. Microbiol.* 141:785–786.
87. Germani, Y., Begaud, E., Duval, P., and Le Bouguenec, C. (1996). Prevalence of enteropathogenic, enteroaggregative, and diffusely adherent *Escherichia coli* among isolates from children with diarrhea in New Caledonia. *J. Infect. Dis.* 174:1124–1126.

CHAPTER 7

Shigella and Enteroinvasive *Escherichia coli*: Paradigms for Pathogen Evolution and Host-Parasite Interactions

William A. Day*

Department of Microbiology and Immunology, Uniformed Services University of the Health Sciences, F. Edward Hebert School of Medicine, Bethesda, Maryland

Anthony T. Maurelli

Department of Microbiology and Immunology, Uniformed Services University of the Health Sciences, F. Edward Hebert School of Medicine, Bethesda, Maryland

INTRODUCTION

Bacillary dysentery or shigellosis is caused by members of the *Shigella* species and a group of pathogenic strains of *Escherichia coli* known as enteroinvasive *E. coli* (EIEC). *Dysentery* was the term used by Hippocrates to describe an illness characterized by frequent passage of stools containing blood and mucus accompanied by painful abdominal cramps. The disease has had a tremendous impact on human society over the centuries, with perhaps one of the greatest impacts being its powerful influence in military operations. Long, protracted military campaigns and sieges almost always spawned epidemics of dysentery and caused large numbers of military and civilian casualties. The Napoleonic campaigns and the American Civil War were nineteenth-century examples of the devastation caused by dysentery. This capacity of dysentery to factor in the outcome of nation-shaping events continues in the modern era. Soldiers were stricken with the

The opinions and assertions contained herein are the private ones of A.T.M. and are not to be construed as official or reflecting the views of the Department of Defense or the Uniformed Services University of the Health Sciences.

*Present address: Bacteriology Division, U.S. Army Medical Research Institute of Infectious Diseases, Fort Detrick, Maryland 21702.

Escherichia coli: Virulence Mechanisms of a Versatile Pathogen
ISBN 0-12-220751-3

disease in essentially all major conflicts of the twentieth century, including World War I, World War II, the Korean Conflict, the Vietnam War, and the Gulf War. Indeed, this ancient disease goes hand in hand with political and socioeconomic strife because dysentery continues to claim the lives of hundreds of thousands of people living in the unsanitary conditions brought on by war and poverty. The low infectious dose required to cause disease coupled with oral transmission of the bacteria via fecally contaminated food and water accounts for the spread of dysentery caused by *Shigella* spp. in the wake of many natural (e.g., earthquakes, floods, and famine) as well as human-made (e.g., war) disasters. For example, refugees of a recent conflict in Zaire suffered high morbidity and mortality rates brought on by bacillary dysentery [1]. Even apart from these special circumstances, shigellosis remains an important disease in developed countries as well as in underdeveloped countries. The global impact of shigellosis has been estimated at over 163 million cases per year with nearly 1 million deaths, the majority occurring in children younder than 5 years of age [2].

The appropriateness of a chapter on *Shigella* in a book devoted to pathogenic *E. coli* derives from the fact that EIEC cause dysentery that is clinically indistinguishable from that caused by members of the *Shigella* species. Indeed, the strains that comprise the EIEC are so taxonomically similar to *Shigella* that some scientists have argued that they should be placed in the same genus. In this chapter we will consider these pathogens together. Since most of the research on the pathogenesis of bacillary dysentery has been done on *Shigella*, we will focus on these organisms. No single review can be completely comprehensive. Therefore, the reader is encouraged to refer to several excellent recent reviews for additional information [3–7].

CLASSIFICATION AND BIOCHEMICAL CHARACTERISTICS

The Japanese microbiologist Shiga isolated an organism (*Shigella dysenteriae*) from the dysenteric stool of a sticken individual in 1898. Three more *Shigella* species (*S. boydii*, *S. flexneri*, and *S. sonnei*) were identified subsequently and grouped by serotype and metabolic activities. The four species of the genus *Shigella* are grouped serologically (41 serotypes) based on their somatic O-antigens: *S. dysenteriae* (group A), *S. flexneri* (group B), *S. boydii* (group C), and *S. sonnei* (group D). As members of the family Enterobacteriaceae, they are nearly genetically identical to the Escherichieae and closely related to the Salmonelleae [8]. *Shigella* are nonmotile gram-negative rods. Some important biochemical characteristics that distinguish these bacteria from other enterics are their inability to use citric acid as a sole carbon source and their inability to ferment lactose, although some strains of *S. sonnei* may ferment lactose slowly. They are oxidase-negative, do not produce H_2S (except for *S. flexneri* serotype 6 and *S. boydii* serotypes 13 and 14), and do not produce gas from glucose. *Shigella* spp. are inhibited by potassium cyanide and do not synthesize lysine decarboxylase [9].

EIEC isolates that cause a diarrheal illness identical to *Shigella* dysentery were identified in the 1970s [10,11]. The pathogenic and biochemical properties that EIEC share with *Shigella* pose a problem in distinguishing these pathogens. For example, unlike normal-flora *E. coli*, EIEC are nonmotile, and 70% of isolates are unable to ferment lactose [12]. These are also features of *Shigella*. More striking is the observation that strains of EIEC are almost universally negative for lysine decarboxylase (LDC) activity, whereas almost 90% of normal-flora *E. coli* are positive. In this respect, EIEC also resemble *Shigella*, which uniformly lack LDC activity. Some serotypes of EIEC even share identical O-antigens with Shigella [13]. By contrast, EIEC resemble *E. coli* in their ability to ferment xylose and to produce gas from glucose, both traits for which *Shigella* are negative [12].

CHARACTERISTICS OF DISEASE

Cell Biology

Shigellosis, or bacillary dysentery, is an acute inflammatory disease of the colonic mucosa that results from invasion and intercellular spread of the bacteria. Infection is via the oral route by ingestion of the bacteria in contaminated food or water. Shigella and EIEC display a remarkable ability to survive the acidity of the stomach [14], which may account for the low infectious dose required to cause disease (see below). After passing through the stomach, the bacteria transit through the small intestine to reach the colon, where the infection is established. Studies using several animal models demonstrate that the initial portals of entry into the human host are M cells. These specialized cells, which do not produce microvilli and are not covered with mucous or glycocalyx, normally sample intestinal antigens that are transcytosed unmodified to underlying lymphoid follicles. *Shigella* gain access to the colonic subepithelium and establish an infection through these cells. In the subepithelium, the bacteria are ingested rapidly by resident macrophages. However, rather than being killed, *Shigella* escape the phagosome and rapidly induce apoptosis of the infected macrophage [15]. Death of the phagocyte has dual effects because the bacteria are released unharmed along with the proinflammatory cytokine interleukin 1β (IL-1β) that initiates an inflammatory cascade [16]. Following release from the macrophage, *Shigella* penetrate colonic epithelial cells through the basolateral surface via interaction with α5-β1 integrins [17] (the virulence systems and molecular interactions required for invasion will be reviewed in a later section). As in the macrophage, internalized bacteria escape the endosomal vacuole and spread to adjacent cells. In contrast to the fate of the macrophage, there is no evidence that *Shigella* infection of epithelial cells induces apoptosis. Infection of the colonocytes induces the release of still more proinflammatory cytokines, namely, IL-6 and IL-8 [18,19]. The high levels of proinflammatory cytokines produced by epithelial cells and macrophages in response to *Shigella* infection signal the massive recruitment of polymorphonuclear neutrophils (PMNs) that are able to kill the pathogens. Thus a paradox of *Shigella*-host interaction exists: How can *Shigella* be successful pathogens (able to infect new hosts) when the organisms are readily killed and eventually cleared by the massive innate host immune response? One possible answer has been proposed recently.

The severe signs and symptoms of shigellosis are due to extensive tissue destruction by the intense inflammatory response. The major effectors of this damage are PMNs. Inhibition of PMN recruitment to the site of infection using anti-IL-8 antibodies significantly reduces the severity of symptoms [19]. PMN degranulation mediates tissue destruction and destabilizes tight junctions between colonocytes [20]. *Shigella* present in the intestinal lumen exploit this window of opportunity to gain access to the subepithelium, where the pathogens can begin another round of infection by invasion of colonocytes through the basolateral surface [21]. Intracellular bacteria likely are protected from the PMN assault. Thus the initial focus of infection is amplified through successive rounds of infection made possible by the activity of PMNs. Amplification of the infection increases the number of *Shigella* in the infected host and increases the likelihood of successful transmission to another host. Consistent with this model, recruitment of PMNs across the epithelium to the apical side (which faces the intestinal lumen) requires virulent invasive *Shigella* [22]. Recruitment of PMNs further requires that the bacteria interact with the basolateral membrane of the epithelial cells, where IL-8 is secreted. Bacteria present on the apical side do not invade colonocytes efficiently, nor do they recruit PMNs. Presently, it is unknown how bacteria residing within cells or in the subepithelium mediate PMN recruitment across the epithelium to the apical surface in model monolayers. Collectively, these observations suggest that *Shigella* play dangerous games with effectors of the host innate immune response by first inducing and then exploiting the intense inflammatory response to ensure their own survival and evolutionary success (passage to another host).

Clinical Presentation

The incubation period for shigellosis is 1 to 7 days, but the illness usually begins within 3 days. Signs and symptoms associated with dysentery include fever, severe abdominal cramps, tenesmus, and diarrhea composed of watery stools containing mucous and traces of blood. While nearly all patients with shigellosis experience abdominal pain and diarrhea, fever occurs in about one-third and gross blood in the stools in about 40% of patients [23]. The feces also contain high numbers of viable infectious *Shigella* bacteria and PMNs.

The clinical picture of shigellosis ranges from a mild watery diarrhea to severe dysentery. The dysentery stage of the disease may or may not be preceded by watery diarrhea. This stage probably reflects the transient multiplication of the bacteria as they pass through the small bowel. Enterotoxins produced by *Shigella* and EIEC are believed to induce fluid secretion in the small intestine. In addition, as bacterial invasion and destruction of the colonic mucosa commence, jejunal secretions probably are not reabsorbed effectively in the colon due to transport abnormalities caused by the tissue destruction [24]. The dysentery stage of disease correlates with extensive bacterial colonization of the colonic mucosa. The bacteria invade the epithelial cells of the colon, spread from cell to cell, but penetrate only as far as the lamina propria. Foci of individually infected cells produce microabscesses that coalesce, forming large abscesses and mucosal ulcerations. As the infection progresses, dead cells of the mucosal surface slough off, thus leading to the presence of blood, pus, and mucus in the stools. Bacillary dysentery is also characterized by a massive concentration of PMNs in the subepithelium that migrate into the intestinal lumen. The intense inflammatory response of the host is initiated by *Shigella* and sustained by release of proinflammatory mediators.

S. dysenteriae 1 causes the most severe disease, whereas *S. sonnei* produces the mildest disease. *S. flexneri* and *S. boydii* infections can be either mild or severe. Volunteer studies and clinical reports show that EIEC produce dysentery with a clinical presentation indistinguishable from that produced by *Shigella* with symptoms ranging from mild to severe [10,11]. Despite the severity of the disease, shigellosis is self-limiting. If left untreated, clinical illness usually persists for 1 to 2 weeks (although it may be as long as a month), and then the patient recovers.

Complications

Dysentery can be a very painful and incapacitating disease and is more likely to require hospitalization than other bacterial diarrheas. It is not usually life-threatening, and mortality is rare, except in malnourished children, immunocompromised individuals, and the elderly [25]. However, serious complications can arise from the disease and include severe dehydration, intestinal perforation, toxic megacolon, septicemia, seizures, hemolytic-uremic syndrome (HUS), and Reiter's syndrome [26]. These last two syndromes have received increased research attention recently. HUS is a rare but potentially fatal complication associated with infection by *S. dysenteriae* 1 [27]. The syndrome is characterized by hemolytic anemia, thrombocytopenia, and acute renal failure. Epidemiologic studies suggest that Shiga toxin produced by *S. dysenteriae* 1 is the cause of HUS [28]. This hypothesis is supported by the fact that HUS is also caused by strains of enterohemorrahagic *E. coli* (EHEC) that produce high levels of Shiga-like toxin [29]. It is believed that Shiga toxin (and the Shiga-like toxins) causes HUS by entering the bloodstream and damaging vascular endothelial cells such as those in the kidney [28–30]. Reiter's syndrome, a form of reactive arthritis, is a postinfectious sequela to shigellosis that is strongly associated with individuals of the HLA-B27 histocompatibility group [31]. The syndrome is comprised of three symptoms, urethritis, conjunctivitis, and arthritis, with the latter being the most dominant symptom. Infections caused by several other gram-negative enteric pathogens also can lead to this type of sterile inflammatory polyarthropathy [32]. A PubMed search of the literature failed to reveal any reports of Reiter's syndrome as a complication of infection with EIEC.

Infectious Dose, Transmission, and Epidemiology

An important aspect of *Shigella* pathogenesis is the extremely low ID_{50}, i.e., the experimentally determined oral dose required to cause disease in 50% of volunteers challenged with a virulent strain of the organism. The ID_{50} for *S. flexneri*, *S. sonnei*, and *S. dysenteriae* is approximately 5000 organisms. Volunteers become ill when doses as low as 200 organisms are given [33]. The low infectious dose of *Shigella* underlies the high communicability of bacillary dysentery and gives the disease great explosive potential for person-to-person spread as well as food- and waterborne outbreaks of diarrhea. In contrast, at least 10^8 EIEC must be ingested to produce disease [10]. The reason for the significantly higher infectious dose for EIEC is unknown.

Person-to-person transmission of *Shigella* is by the fecal-oral route, with most cases of shigellosis being caused by the ingestion of fecally contaminated food or water. In the case of foods, the major factor for contamination is the poor personal hygiene of food handlers. *Shigella* can spread from infected carriers by several routes, including food, fingers, feces, and flies. The latter usually transmit the bacteria from fecal matter to foods. The highest incidence of shigellosis occurs during the warmer months of the year. Improper storage of contaminated foods is the second most common factor that accounted for food-borne outbreaks due to *Shigella* [34].

Shigella and EIEC are frank pathogens capable of causing disease in otherwise healthy individuals. Certain populations, however, may be predisposed to infection and disease due to the nature of transmission of the organisms. The greatest frequency of illness due to *Shigella* occurs among children younger than 6 years of age [2]. In the United States, outbreaks of shigellosis and other diarrheal diseases are increasing in day-care centers as more single-parent and two-parent working families turn to these facilities to care for their children [35,36]. Inadequate personal hygiene habits coupled with typical toddler behavior such as oral exploration of the environment create conditions ideally suited to transmission of bacterial, protozoan, and viral pathogens that are spread by fecal contamination. Transmission of *Shigella* in this population is very efficient, and the low infectious dose for causing disease increases the risk for shigellosis. Family contacts of day-care attendees are also at increased risk [37].

Shigellosis can be endemic in other institutional settings as well. Prisons, mental hospitals, and nursing homes, where crowding and/or insufficient hygienic conditions may exist, can provide an environment for direct fecal-oral contamination. Crowded conditions and poor sanitation contribute to dysentery being endemic in developing countries as well.

When natural or human-caused disasters destroy a region's sanitary waste treatment and water purification infrastructure, developed countries take on the conditions of a developing country. These conditions place a population at increased risk for diarrheal diseases such as cholera and dysentery. Recent examples include famine and political upheaval in Somalia and the war in Bosnia [35]. Massive population displacement (e.g., refugees fleeing from Rwanda into Zaire in 1994) also has led to explosive epidemics of diarrheal disease caused by *Vibrio cholerae* and *S. dysenteriae* 1 [1].

Reservoirs and Vehicles of Infection

Shigella spp. are highly host-adapted pathogens. Humans are the natural reservoir of *Shigella* infections, and nonhuman primates such as the rhesus monkeys are the only animals in which *Shigella* cause disease. Interestingly, several cases of transmission from monkeys to humans have been reported. In one instance, three animal caretakers at a monkey house complained of having diarrhea. *S. flexneri* 1b was isolated from stool samples of these employees, and further investigation showed that four monkeys were shedding the identical serotype. The disease apparently was spread by direct contact of the caretakers with excrement from the infected monkeys [38].

Asymptomatic carriers of *Shigella* may contribute to the maintenance and spread of this pathogen in developing countries. Two studies, one in Bangladesh [39] and the other in Mexico [40], showed that *Shigella* could be isolated from stool samples from asymptomatic children younger than 5 years of age. *Shigella* rarely were found in infants younger than 6 months of age.

Treatment and Prevention

Stool fluid losses with dysentery are not as massive as with other bacterial diarrheas. However, the diarrhea associated with dysentery combined with water loss due to fever and decreased water intake due to anorexia may result in severe dehydration [25]. Although intravenous rehydration may be required in the treatment of very young and elderly patients, oral intake generally can replace these fluid losses.

The antibiotic of choice for treatment of dysentery is trimethoprim-sulfamethoxazole [23]. However, there is some controversy regarding the use of antibiotics in treating dysentery. Since the infection is self-limited in normally healthy patients and full recovery occurs without the use of antibiotics, drug therapy usually is not indicated. In addition, multiple-antibiotic resistance among isolates of *Shigella* is becoming more common. Clinical isolates resistant to sulfonamides, ampicillin, trimethoprim-sulfamethoxazole, tetracycline, chloramphenicol, and streptomycin have been found [41,42]. Extensive use of antibiotics selects for drug-resistant organisms, and therefore, many practitioners believe that antimicrobial therapy for shigellosis should be reserved only for the most severely ill patients. On the other hand, there are persuasive public health arguments for the use of antibiotics to manage shigellosis. Antibiotic treatment limits the duration of disease and shortens the period of fecal excretion of bacteria [43]. Since an infected person or asymptomatic carrier can be an index case for person-to-person and food- and waterborne spread, antibiotic treatment of these individuals can be an effective public health tool to contain the spread of dysentery. However, antibiotics are not a substitute for improved hygienic conditions to contain secondary spread of dysentery. The single most effective means of preventing secondary transmission is handwashing.

Despite many years of intensive effort, an effective vaccine against shigellosis still has not been developed. Attenuated oral vaccine strains of *S. flexneri* are being tested currently. One such strain is SC602. This strain has mutations that block iron uptake and abolish the intracellular and intercellular motility phenotypes [44]. A limited-challenge study showed protection among vaccinees. However, one major drawback still to be resolved is the transient fever and mild diarrhea associated with administration of the vaccine [45]. This study highlights one of the persistent problems impeding the development of a safe *Shigella* vaccine: designing a strain that can induce a protective immune response without producing unacceptable side effects. A second consideration is whether a live, attenuated vaccine strain can afford cross-protection against all species and serotypes of *Shigella* and against EIEC as well.

VIRULENCE FACTORS AND GENETICS

Hallmarks of Virulence

Shigella spp. and EIEC cause disease by overt invasion of epithelial cells in the large intestine (Fig. 1). The clinical symptoms of dysentery can be attributed directly to the following hallmarks of virulence: the ability to induce diarrhea, invasion of epithelial cells of the intestine, multiplication inside these cells, and spread from cell to cell.

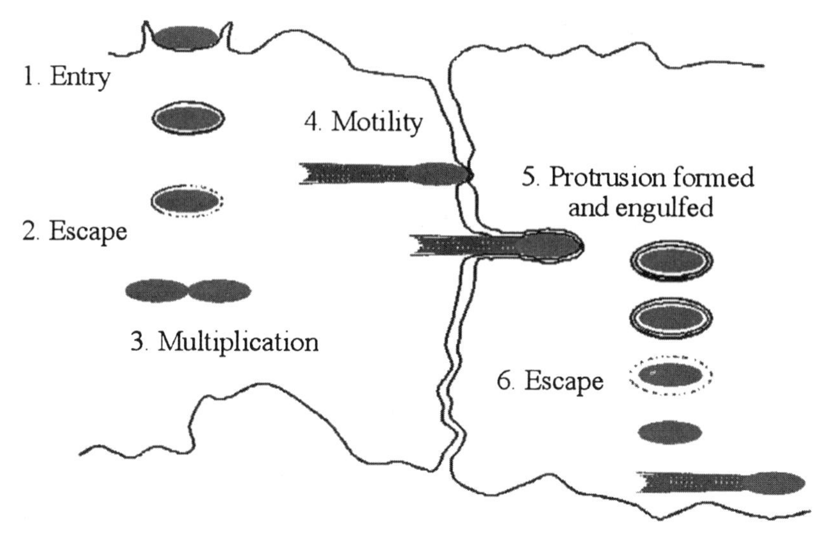

FIGURE 1 Stages in the invasion and intercellular spread of *Shigella*.

Shigella colonize the small intestine only transiently and cause little tissue damage [46]. Production of enterotoxins by *Shigella* and EIEC while they are in the small bowel probably results in the diarrhea that generally precedes onset of dysentery [47,48]. It is believed that the jejunal secretions elicited by these toxins facilitate passage of the bacteria through the small intestine and into the colon, where they colonize and invade the epithelium.

Formal and colleagues [49] established the essential role of epithelial cell invasion in *Shigella* pathogenesis in a landmark study that employed both *in vitro* tissue culture assays for invasion and animal models. These authors showed that spontaneous colonial variants of *S. flexneri* 2a that are unable to invade mammalian cells in tissue culture do not cause disease in monkeys. Further, they showed the presence of wild-type bacteria within the epithelial cells of the large intestine in the experimentally infected animals.

Gene transfer studies using *E. coli* K-12 donors and *S. flexneri* 2a recipients established the third hallmark of *Shigella* virulence. An *S. flexneri* 2a recipient that inherits the *xyl-rha* region of the *E. coli* K-12 chromosome retains the ability to invade epithelial cells but has a reduced ability to multiply within these cells [50]. This hybrid strain fails to cause a fatal infection in the opium-treated guinea pig model and is unable to cause disease when fed to rhesus monkeys [51]. The high frequency of recombinants in these conjugation experiments also served to confirm the close genetic relatedness of *E. coli* and *Shigella*.

While it is necessary for *Shigella* to be able to multiply within the host epithelial cell after invasion, intracellular multiplication is not sufficient to cause disease.

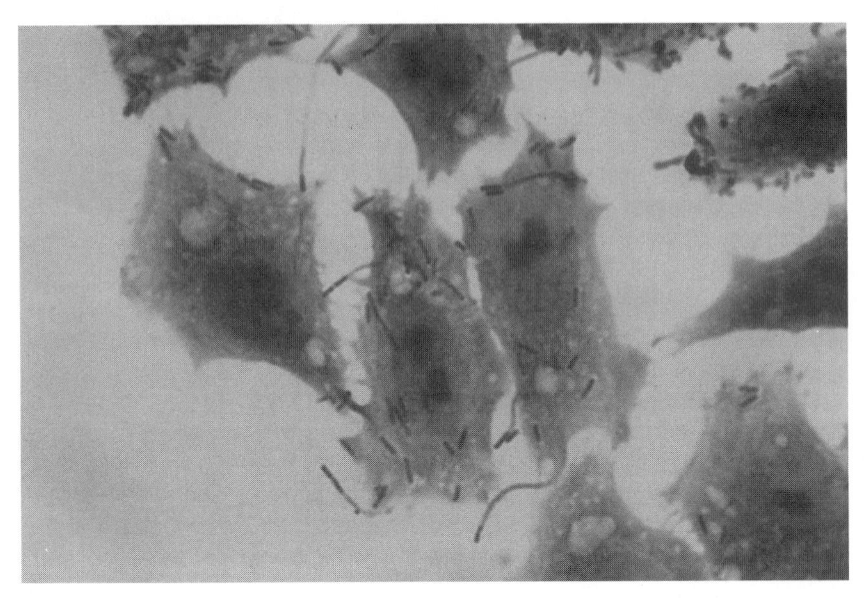

FIGURE 2 Invasion of mammalian cells by *Shigella*. A semiconfluent monolayer of mouse L2 fibroblasts was infected with *S. flexneri* 2a for 30 min. The monolayer was washed and tissue culture medium with gentamicin was added to kill extracellular bacteria. The monolayer was fixed and stained 2 hours later. Bacteria can be seen as darkly staining rods within the cytoplasm of the eucaryotic cells. Intracellular movement of the bacteria and contact with the plasma membrane causes formation of "fireworks," cytoplasmic protrusions that contain bacteria at the ends. See Color Plate 6.

The bacterium also must be able to spread through the epithelial lining of the colon by cell-to-cell spread in a manner that does not require the bacterium to leave the intracellular environment and be reexposed to the intestinal lumen (Fig. 2). Mutants of *Shigella* that are competent for invasion and multiplication but are unable to spread between cells in this fashion have been isolated. These mutants established intracellular spread as the fourth hallmark of *Shigella* virulence and will be discussed in a later section.

Along with the ability to colonize and cause disease, an intrinsic part of any bacterium's pathogenicity is its machinery for regulating expression of virulence genes. Virulence in *Shigella* spp. and EIEC is regulated by growth temperature. After growth at 37°C, virulent strains of *Shigella* are able to invade mammalian cells, but when cultivated at 30°C, they are noninvasive. This noninvasive phenotype is reversible by shifting the growth temperature to 37°C, where the bacteria reexpress their virulence properties [52]. Similarly, the invasive phenotype of EIEC is also temperature-regulated [53]. Temperature regulation of virulence gene expression is a characteristic of other human pathogens such as pathogenic *E. coli*, *Salmonella typhimurium*, *Bordetella pertussis*, *Yersinia* spp., and *Listeria monocytogenes*. Regulation of gene expression in response to environmental temperature is a useful strategy for bacteria. By sensing the ambient temperature of the mammalian host (e.g., 37°C for humans) to trigger gene expression, this strategy permits *Shigella* and EIEC to economize energy that otherwise would be expended on the synthesis of virulence products when the bacteria are outside the host. The system also permits the bacteria to coordinately regulate expression of multiple unlinked genes that are required for the full virulence phenotype.

Genetics

Virulence Plasmid Genes

Given the complexity of the interactions between host and pathogen, it is not surprising that the virulence of *Shigella* and EIEC is multigenic, involving both chromosomal and plasmid-encoded genes. Another landmark study on the pathogenicity of *Shigella* was the demonstration of the indispensable role for a large plasmid in invasion. A 180-kb plasmid in *S. sonnei* and a 220-kb plasmid in *S. flexneri* were first shown to be required for invasion [54,55]. Other *Shigella* spp., as well as strains of EIEC, also contain large plasmids that are functionally interchangeable and share significant degrees of DNA homology with the plasmid described in *S. flexneri* [13,56,57]. Thus it is probable that the plasmids of *Shigella* and EIEC derive from a common ancestral plasmid [58]. Analysis of the complete DNA sequence of the 220-kb plasmid in *S. flenxeri* 5a reveals a complex history [59,60]. The plasmid is essentially a genetic mosaic composed of blocks of genes from different origins as well as the remnants of four different ancestral plasmids. For instance, the genes required for the invasive phenotype have a G + C content of 31–37%. This value is significantly lower than the composition of loci required for intercellular spread, and the composition of these loci is significantly different from that of loci encoding a surface protease. These genes are interspersed with close to 100 insertion sequences comprising one-third of the virulence plasmid. Evidence exists of active transposition as well as rearrangements and deletions in the plasmid mediated by homologous recombination between these elements. These events likely led to the size and organizational differences observed between the virulence plasmids of different *Shigella* species and lineages.

A 37-kb region of the invasion plasmid of *S. flexneri* 2a contains all the genes necessary to permit the bacteria to penetrate into tissue culture cells (Fig. 3). This DNA segment was identified as the minimal region of the virulence plasmid necessary to allow a plasmid-cured derivative of *S. flexneri* (and *E. coli* K-12) to invade tissue culture cells [61]. The 34 virulence genes encoded in this region are organized into three large operons (*ipa*, *mxi*, and *spa*) that are transcribed in opposite directions. The essential role of these genes in expression of the inva-

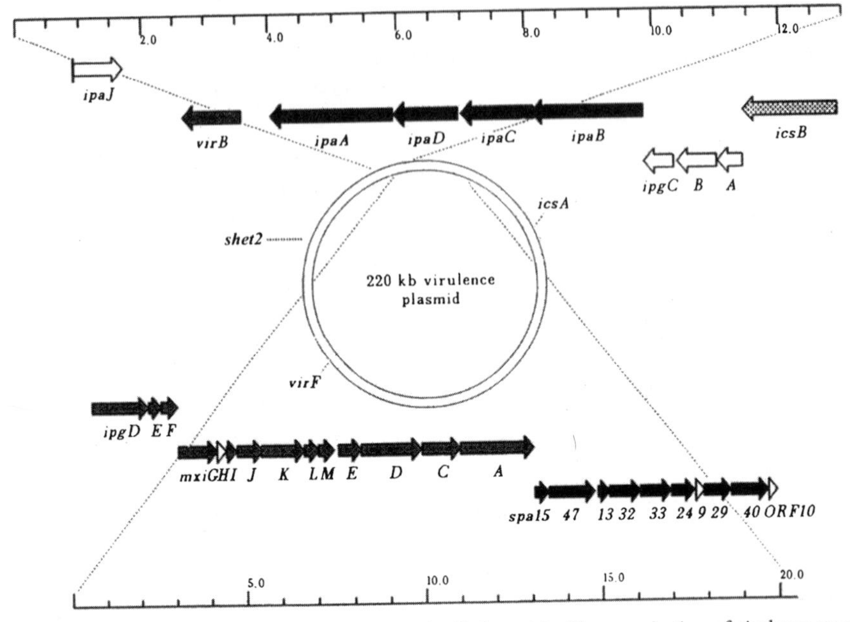

FIGURE 3 Map of virulence plasmid form *Shigella flexneri* 2a. The organization of virulence genes and their approximate location on the circular plasmid map is shown. Sections of the plasmid are expanded to illustrate the virulence loci encoded in these regions. The expanded *ipa* and *mxi-spa* regions are contiguous and cover 32 kb. The open reading frame for *icsB* is separated from that of *ipgD* by 314 base pairs. Three additional virulence genes (*virF*, *icsA*, and *shet2*) that map outside this gene cluster are also shown.

sive *Shigella* phenotype and virulence has been demonstrated using mutation analyses. Organization of these loci into large transcription units suggests that the genes are coregulated and encode factors that function together in a multicomponent virulence (invasion) system. The *mxi* and *spa* operons encode subunits of a type III secretion system (TTSS), and the *ipa* operon encodes proteins (effectors) secreted by this system. TTSS are used widely by several animal and plant pathogens (enterohemorrhagic and enteropathogenic *E. coli*, *Salmonella* spp., *Yersinia* spp., *Pseudomonas* spp., *Erwinia caratovara*, and *Xanthomonas campestris*, to name a few) for delivery of virulence factors to the surface or interior of host cells [62]. These secreted effectors of virulence allow the pathogens to reprogram host cells to serve the pathogen's particular needs. The TTSS delivery machines are conserved structurally, functionally, and genetically and share several core characteristics.

The genes comprising the *ipaBCDA* (invasion plasmid antigens) cluster encode the immunodominant antigens of *Shigella* and EIEC that are detected with sera from convalescent patients and experimentally challenged monkeys [63,64]. *ipaBCD* have been demonstrated experimentally to be absolutely required for invasion of mammalian cells [65]. An *ipaA* mutant is still invasive but has a 10-fold reduced ability compared with wild type [66]. This operon also encodes a molecular chaperone, IpgC, that binds IpaB and IpaC in the bacterial cytosol prior to secretion and prevents their interaction and proteolytic degradation [67]. Following secretion, IpaB and IpaC form a complex that binds to molecules such as α5-β1 integrins and CD44 that are present on the basolateral surfaces of host cells [17,68]. After binding, the Ipa complex inserts into the host cell membrane and causes a massive recruitment of host cell actin and other cytoskeletal factors beneath the site of attachment. This exploitation of the host cell cytoskeleton is observed with IpaBC-coated beads or whole *Shigella* bacteria and leads to the formation of localized membrane projections that encompass and internalize either beads or bacteria in roughly 5 minutes [69,70]. Purified IpaC induces cytoskeletal reorganization, including formation of filopodia and lamellipodial extensions on permeabilized cells [71]. By analogy to homologues in other pathogens such as YopB of *Yersinia* spp., the IpaBC complex may form a channel through which IpaA and other secreted factors gain access to the host cell cytoplasm [62]. Inside the host cell, IpaA interacts with vinculin to promote F-actin depolymerizaton. Through this action, IpaA may promote formation of a pseudo-focal adhesion complex that organizes the actin assembly complex beneath the bacterium and directs optimal uptake of the pathogen [66,72].

IpaB serves several other roles in *Shigella* virulence that may be accomplished by separate domains of the protein. IpaB, along with IpaD, forms a plug in the *Shigella* TTSS and prevents secretion of as many as 25 target molecules [73]. *ipaB* and *ipaD* mutants constitutively secrete TTSS effectors. IpaB is also required for escape of *Shigella* from both the phagosomal (following uptake by macrophages) and endosomal (following uptake by epithelial cells) compartments [74]. This activity requires formation of a hydrophobic N-terminal alpha-helix and is linked to the ability of IpaB to insert into host cell membranes and lyse erythrocytes. In addition, IpaB binds to caspase 1 *in vitro* and *in vivo* and triggers hydrolysis of the cysteine protease [75]. In macrophages, this activity leads to secretion of IL-1β (which initiates inflammation) and induction of apoptosis of the phagocyte. Evidence is emerging that this activation of caspase 1 cleavage also requires the N-terminal hydrophobic alpha-helix domain of IpaB [76].

The *icsA* gene (also known as *virG*), which is encoded outside the 31-kb invasion region on the *Shigella* virulence plasmid, is an essential virulence determinant. *icsA* encodes a 120-kDa outer membrane protein that is required for motility of *Shigella* inside cells and spread of the bacteria to adjacent cells [77,78]. *icsA* mutants do not form plaques in confluent tissue culture cell monolayers and are significantly attenuated in the macaque monkey challenge model [79] (Fig. 4).

The IcsA protein is unusual in that it is expressed asymmetrically on the bacterial surface, being found only at one pole [80]. Unipolar localization of IcsA

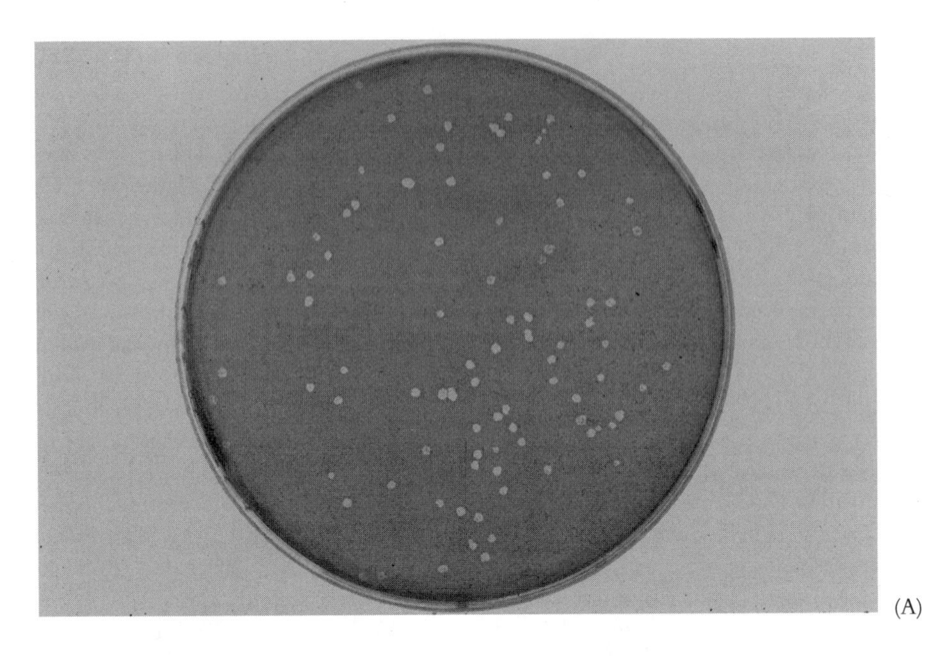

(A)

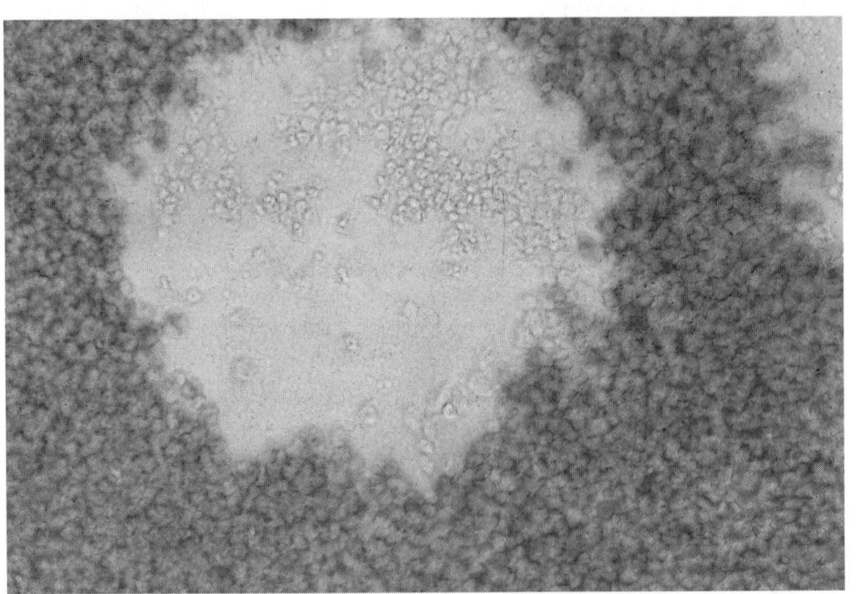

(B)

FIGURE 4 Intercellular spread of *Shigella* measured in the plaque assay. Bacteria are allowed to invade a confluent monolayer of cells in tissue culture for 90 min. The monolayer is washed and an overlay of 0.5% agarose-containing tissue culture medium with gentamicin is added and the monolayers incubated for 48–72 hr. The gentamicin kills extracellular bacteria. Intracellular bacteria that move from cell to cell by forming protrusions that "invade" surrounding cells are protected from the antibiotic and propagate the foci of infection. Panel A shows plaques formed by *S. flexneri* 2a in a monolayer of L2 fibroblasts after 48 hr. Panel B shows the appearance of a plaque at higher magnification. The monolayer is stained with neutral red, which stains living cells. Dead (unstained) cells can be seen in the center of the plaque. See Color Plate 7.

imparts directionality of movement to the organism. IcsA mediates motility by binding host cytoplasmic N-WASP and Arp2/Arp3 complex, which catalyze polymerization of actin at that end of the bacterium [81,82]. Polymerization of actin monomers at one end of the bacterium provides the force that drives the organism through the host cell cytoplasm and into adjacent cells.

Correct unipolar localization of IcsA in *Shigella* is essential for virulence. Although the mechanism of unipolar localization of IcsA is unknown, it depends on synthesis of a complete lipopolysaccharide (LPS) [83,84]. LPS mutants of *Shigella* express IcsA but fail to confine IcsA to the pole, and IcsA is found uniformly over the cell surface in these mutants. The protein is still capable of polymerizing actin, but actin polymerizes around the entire cell. As a result, movement is restricted as the bacterium becomes encased in a shell of actin. A plasmid-encoded protease, SopA/IcsP, has been shown to cleave IcsA and is proposed to play a role in unipolar localization of IcsA [85,86]. However, *E. coli* and plasmid-cured derivatives of *S. flexneri* transformed with a cloned *icsA* gene localize IcsA normally [87]. These results suggest that IcsA localization does not require any other virulence plasmid–encoded gene and that motifs within IcsA itself contain the information that directs the protein to the pole.

The products of the *ipa* genes are actively secreted into the extracellular medium even though they contain no signal sequence for recognition by the usual general secretory pathway of gram-negative bacteria. Ipa secretion is mediated by a TTSS composed of gene products from the *mxi/spa* loci [7]. A characteristic feature of TTSS is a contact-dependent secretion mechanism. Delivery of effectors to their targets (surface and cytoplasmic) occurs only after outer membrane elements of the TTSS complex contact the host cell [73,88]. This feature is observed in the *Shigella* TTSS as the Ipa proteins are deposited on the host cell surface after contact with host cells. Other compounds mimic the host secretion signal or short circuit regulation of secretion in the *Shigella* system. These compounds include fibronectin, laminin, collagen type IV, Congo red stain, bile salts, and serum [89].

The *mxi* (membrane expression of invasion plasmid antigens) genes comprise an operon that encodes several lipoproteins (MxiJ and MxiM), a transmembrane protein (MxiA), and proteins containing signal sequences (MxiD, MxiJ, and MxiM) [90–93]. MxiH, MxiJ, MxiD, and MxiA share homology with proteins involved in secretion of virulence proteins (Yops) in *Yersinia* spp. The *spa* (surface presentation of Ipa antigens) genes encode proteins that share significant homologies with proteins involved in flagellar synthesis in *E. coli*, *Salmonella typhimurium*, *Bacillus subtilis*, and *Caulobacter crescentus* [94,95]. Included among these genes is *spa47*, which encodes a protein that has sequence similarities with ATPases of the flagellar assembly machinery of other bacteria [94]. It has been proposed that Spa47 is the energy-generating component of the secretion apparatus. Nonpolar null mutations in all the *mxi/spa* genes tested so far result in loss of the ability to secrete the Ipas and concomitant loss of invasive capacity for cultured cells. Thus the TTSS is an essential component of *Shigella* virulence.

Recently, a role for the *Shigella* TTSS in intercellular spread was demonstrated [96]. These studies showed that expression of the Ipa proteins (and the secretion machinery required for export) is essential for spread of *Shigella* to adjacent cells. Bacteria that shut off their expression of Ipa, Mxi, and Spa inside infected cells do not form plaques in confluent cultured cell monolayers. Scanning electron microscopic analysis revealed that the bacterium is unable to lyse the double membrane that forms following penetration of the bacterium into adjacent cells. These observations suggest that intracellular secretion of Ipa proteins is required for *Shigella* to escape into the cytosol of an adjacent cell after IcsA-mediated movement of the bacterium into the neighboring cell.

Analysis of the culture supernatants from a Δ*ipaBCDA Shigella* strain (constitutive TTSS activation) revealed the presence of roughly 15 proteins that are not secreted by a TTSS-deficient strain [73]. Among these factors are the VirA and IpaH(7.8) proteins. VirA, encoded on the virulence plasmid near *icsA*, is a 44.7-kDa protein required for optimal *Shigella* invasion and intercellular spread [97]. IpaH(7.8) is encoded by one of five *ipaH* alleles located on both the chromosome and the virulence plasmid that have arisen by gene duplication that likely was mediated by adjacent insertion sequences [98]. Of these alleles, only *ipaH(7.8)* is implicated in virulence because mutants are impaired in phagosomal escape [99]. Considering the dedicated role of TTSS in virulence, the remaining newly identified TTSS substrates are strong candidates for additional effectors of virulence used by *Shigella* to reprogram host cells. N-terminal sequencing of these proteins and BLAST searches of the completed *Shigella* virulence plasmid sequence identified the genes encoding these proteins (Table I). These genes are designated as *osps* for outer *Shigella* proteins. The percentage G + C content of these genes is very similar to that of genes in the entry region, suggesting that these putative effector molecules were acquired at the same time as genes encoding the TTSS and Ipas. Several *osps* have paralogues in the virulence plasmid that likely arose through gene duplication. One of these is the *Shigella* enterotoxin gene *ospD3/senA*. The *ospD3/senA* gene product is a 60-kDa protein that induces fluid secretion in an animal model and alters the ion potential of cultured cells lines by an unknown mechanism [48]. Two *ospD3/senA* paralogues in the virulence plasmid are *ospD1* (a truncated allele of *ospD3*) and *ospD2*, which encodes a protein of unknown function 38% identical to OspD3 [59].

Chromosomal Virulence Genes

In contrast to the genes of the virulence plasmid that are responsible for invasion of mammalian tissues, most of the chromosomal loci associated with *Shigella* virulence are involved in regulation or survival within the host. Mutations that alter O-antigen and core synthesis or assembly lead to a so-called rough pheno-

TABLE I *Shigella* Virulence-Associated Loci

Locus	Gene Product	Role in Virulence
Plasmid		
icsA	120-kDa outer membrane protein	Catalyzes actin polymerization
ipaA	70-kDa TTSS[a] secreted protein	Invasion; binds vinculin; promotes F-actin depolymerization
ipaB	62-kDa TTSS secreted protein	Invasion; lysis of vacuole; apoptosis induction; antisecretion plug
ipaC	43-kDa TTSS secreted protein	Invasion; induces cytoskeletal reorganization
ipaD	38-kDa TTSS secreted protein	Invasion; antisecretion plug
ipaH(7.8)	62-kDa TTSS secreted protein	Escape from vacuole
ipgB1	25-kDa TTSS secreted protein	Unknown
ipgC	17-kDa cytosolic protein	Chaperone for IpaB and IpaC
ipgD	60-kDa TTSS secreted protein	Unknown
ipgE	14-kDa protein	Chaperone for IpgD
mxi/spa	20 proteins	TTS for secretion of Ipa and Osp proteins
ospB	31-kDa TTSS secreted protein	Unknown
ospC1	58-kDa TTSS secreted protein	Unknown
ospD1	25-kDa TTSS secreted protein	Unknown
ospE1	10-kDa TTSS secreted protein	Unknown
ospF	28-kDa TTSS secreted protein	Unknown
ospG	24-kDa TTSS secreted protein	Unknown
sen	ShET2	Enterotoxin
virA	44.7-kDa TTSS secreted protein	Invasion and intercellular spread
virB	Transcriptional activator	Temperature regulation of *ipa, mxi*, and *spa* virulence operons
virF	Transcriptional activator	Temperature regulation of *virB*
Chromosomal		
hns	Histone-like DNA-binding protein	Repressor of virulence gene expression
iuc	Siderophore and cognate receptor synthesis	Iron acquisition in the host
rfa; rfb	Synthesis of LPS core and O-antigen	Polar localization of IcsA; promote inflammatory response
Set[b]	ShET1	Enterotoxin
Stx[c]	*N*-glycosidase	Disruption of host protein synthesis

[a]TTSS = type III secretion system.
[b]Contained on *S. flexneri* 2a SHI-1 island.
[c]*stx* locus and Shiga toxin production are only observed in *S. dysenteriae* 1.

type and render *Shigella* avirulent. Synthesis of a complete LPS, which is crucial for correct unipolar localization of IcsA (see above), requires chromosomal loci such as *rfa* and *rfb*. In the case of *S. sonnei* and *S. dysenteriae* 1, plasmid-encoded genes are also necessary for synthesis of the LPS O-sidechain [100]. LPS also has been implicated as an important indirect chemoattractant because the molecule induces IL-8 expression in epithelial cells to recruit PMNs to the intestinal lumen, thereby promoting tissue injury and amplification of the *Shigella* infectious cycle.

The *stx* locus encodes Shiga toxin in *S. dysenteriae* 1. Shiga toxin inhibits protein synthesis by cleaving the *N*-glycosidic bond at adenine 4324 in the 28S ribosomal RNA of mammalian cells (for a review, see ref. 101). Animal studies suggest that Shiga toxin is responsible for the more severe vascular damage in the colonic tissue observed during *S. dysenteriae* 1 infection. Shiga toxin is also produced, at even higher levels, by enterohemorrhagic *E. coli* (EHEC) and is associated with glomerular damage and kidney failure. The resulting HUS may lead to death directly linked to Shiga toxin–producing pathogens [28]. A mutation in the *stx* locus of *S. dysenteriae* 1 does not alter the ability of the organism to invade epithelial cells or cause keratoconjunctivitis in the Sereny test. However, when administered orally to macaque monkeys, the mutant strain causes less vascular damage in the colonic tissue than the toxin-producing parent [102]. Thus production of Shiga toxin may account for the generally more severe infections caused by *S. dysenteriae* 1 as compared with the other species of *Shigella*.

Although *Shigella* and *E. coli* are very closely related at the genetic level, there are significant differences beyond the presence of the virulence plasmid in *Shigella* and EIEC. Two pathogenicity islands have been identified in the chromosome of *S. flexneri*, and a third has been described recently in *S. boydii*. The *Shigella* pathogenicity island 1 (SHI-1), which is present only in *S. flexneri* 2a, contains the *set* gene that encodes a second enterotoxin, ShET-1 [48]. The *set* gene is contained within the open reading frame of another gene, *she*, that encodes a protein with putative hemagglutinin and mucinase activity [103]. The SHI-2 pathogenicity island is broadly distributed in the *Shigella* and contains the *iuc* locus, which encodes the genes for synthesis of aerobactin and its outer membrane receptor [104,105]. Aerobactin is a hydroxamate siderophore that *S. flexneri* uses to scavenge iron. When the *iuc* locus is inactivated in *S. flexneri*, the aerobactin-deficient mutants retain their capacity to invade host cells but are altered in virulence, as measured in animal models. These results suggest that aerobactin synthesis is important for bacterial growth within the mammalian host [106]. SHI-3 of *S. boydii* contains an aerobactin locus that is 97% identical at the DNA level to that of SHI-2 [107]. This island shares other genes found on SHI-2, but its location on the genome is different. SHI-2 is found inserted downstream of *selC*, whereas SHI-3 is located between *lysU* and *pheU*. Although EIEC synthesize aerobactin, the *iuc* genes are not contained on a SHI-2-like pathogenicity island near *selC*. The chromosomal location of the *iuc* genes in EIEC remains to be determined.

In addition to extra genes in the *Shigella* chromosome, there are genes that are present in the closely related *E. coli* but that are missing from the chromosome of *Shigella*. *ompT* is part of a cryptic prophage in the *E. coli* K-12 chromosome and encodes an outer membrane protease. This prophage is not present in the genome of *Shigella* spp. When the *ompT* gene is introduced into *S. flexneri* by conjugation, the strain loses the ability to spread from cell to cell and is attenuated. This phenotype is due to degradation of IcsA by OmpT protease [108]. Another example of a missing genetic locus or "black hole" in the *Shigella* genome is *cadA*, the gene for lysine decarboxylase. Although lysine decarboxylase activity is present in more than 85% of *E. coli* strains, it is missing in all strains of *Shigella* spp. and EIEC. When *cadA* is introduced into *S. flexneri*, the production of cadaverine (by the decarboxylation of lysine) inhibits the action of the *Shigella* enterotoxins [109]. Thus these examples of antivirulence genes illustrate another way that pathogens evolve from their nonpathogenic commensal relatives by both acquiring genes (e.g., the virulence plasmid) that contribute to virulence and deleting genes that are incompatible with expression of these new virulence traits (see section below on evolution).

Virulence Gene Regulation

Overcoming multiple host barriers often requires coordinated expression of multiple virulence genes encoding complex virulence systems such as the *Shigella* TTSS. In many cases, genes encoding these systems are unlinked. Furthermore, it is beneficial to the invading pathogen to express virulence factors required for access to and fitness in host tissues only when these tissues are encountered. Like other bacterial pathogens, *Shigella* and EIEC have evolved regulatory cascades that respond to an environmental stimulus (temperature). This signal coordinates virulence gene expression with transit from the environment into the human gut. Temperature regulation of *Shigella* and EIEC operates at the level of gene transcription, and virulence is governed by both plasmid and chromosomal loci.

Expression of the *ipa*, *mxi*, and *spa* operons is induced 50- to 100-fold when the bacteria are shifted from 30 to 37°C [110]. Genes governing this thermal regulation include *virR/hns*, located on the chromosome, and two virulence plasmid genes, *virB* and *virF*. The histone-like protein (H-NS) encoded by *virR/hns* acts as a repressor of virulence gene expression. In *hns* mutants, the *ipa*, *mxi*, and *spa* operons are expressed at the normally repressive temperature (30°C) [111]. The *virR/hns* locus is allelic with regulatory loci in other enteric bacteria, and like *virR/hns*, these alleles act as repressors of their respective regulons (references cited in ref. 4). Several different models to explain how VirR/H-NS acts as a transcriptional repressor have been proposed and are analyzed in a recent review [4]. However, because VirR/H-NS is involved in gene regulation in response to diverse environmental stimuli such as osmolarity, pH, and temperature, a comprehensive model to explain its activity has been elusive.

One of the targets of H-NS regulation is *virB* [112]. Expression of genes in the *ipa* and *mxi/spa* clusters depends on activation by VirB, and mutations in *virB* abolish the bacteria's ability to invade tissue culture cells [113,114]. Transcription of *virB* depends on growth temperature and VirF [115]. VirB is probably a DNA-binding protein, as suggested by the homology it shares with the plasmid-partitioning proteins ParB of bacteriophage P1 and SopB of plasmid F. However, targets for VirB binding in the promoter regions of the *ipa* and *mxi/spa* genes have not yet been identified, and the mechanism by which VirB activates virulence gene expression is unknown.

The product of the *virF* locus is a key element in temperature regulation of the *Shigella* virulence regulon. A helix-turn-helix motif in the C-terminal portion of VirF is characteristic of members of the AraC family of transcriptional activators. Consistent with its predicted role as a DNA-binding protein, VirF binds to sequences upstream of *virB* [112]. Current models suggest that H-NS binds the *virB* operator at 30°C, precluding VirF binding and impairing *virB* transcription [4]. On a shift to 37°C, H-NS is displaced from the *virB* operator, allowing VirF to bind and induce the expression of *virB*, which in turn induces the expression of the *ipa*, *mxi*, and *spa* operons. Thus, binding of VirF may act as an antagonist to binding by H-NS and thereby provides a mechanism for responding to temperature. However, expression of *virF* itself is apparently not subject to temperature regulation, leaving open the possibility that an unknown regulator may be involved in activating VirF. Alternatively, temperature may induce conformational changes in VirF that influence its binding affinity and thus its ability to activate promoters.

In addition to temperature regulation of genes required for invasion, *Shigella* relies on postinvasion signals following contact with host cells to cue expression of virulence genes required for intracellular growth and survival [116]. Growth at 37°C is not sufficient for expression of *ipaH* and *virA*. Expression of these genes is induced after contact with host cells and induction of secretion through the *Shigella* TTSS. Similar regulatory strategies have been described in *Yersinia* because secretion of the LcrQ repressor through the organism's TTSS allows expression of several TTSS substrates required for virulence [117]. Since IpaH may facilitate escape from the endosome, and since VirA is required for intracellular spread, this pattern of expression is consistent with the intracellular roles of these factors in *Shigella* virulence. Thus this second mechanism of virulence gene regulation ensures expression of factors where they are most effective. It is expected that this strategy will emerge as an important means by which *Shigella* maximizes its fitness in host tissues and reprograms host cells to serve its own needs.

EVOLUTION OF *SHIGELLA* SPECIES AND EIEC

The four species of *Shigella* and EIEC are invasive organisms that penetrate the colonic mucosa and induce an intense inflammatory response. Early genetic studies demonstrated that essentially all the virulence factors required for expression of the invasive phenotype are encoded on a large plasmid and that this genetic element not only is required for virulence but also is present in all *Shigella* and EIEC isolates examined [13,54,55]. Therefore, from the pathogenesis perspective, EIEC appear to be closely related to the *Shigella*. However, unlike *Shigella*, EIEC strains express traits that prevented inclusion in any of the *Shigella* species (see above). The EIEC were recognized as a heterogeneous group of pathogens that resembled *Shigella* in their pathogenic potential (and in certain metabolic traits) but were clearly related to *E. coli*. These observations led many investigators to propose, incorrectly, that the EIEC represented a missing link in the evolution from *E. coli* to the *Shigella*.

Several studies have established that, like EIEC, the four species of *Shigella* are so closely related to *E. coli* that they should all be included in a single species. The chromosomes of these organisms are largely colinear and are more than 90% homologous [118]. Therefore, *Shigella* spp., like EIEC, are a group of pathogenic *E. coli*. In fact, several investigators using different approaches have established that the four species of *Shigella* evolved from commensal *E. coli* [8,119–121]. Bacteria expressing the *Shigella* phenotype were generated through horizontal transfer of the virulence plasmid from an unknown donor bacterium to commensal *E. coli* [120]. Thus horizontal transfer of the virulence plasmid provided an early example of the role of the gain-of-function pathway in the generation of a pathogenic species, i.e., *Shigella*. Through horizontal gene transfer, other commensal *E. coli* acquired many genes encoding whole virulence systems that permit the new pathogen access to and survival in novel host niches (resulting in host disease). These pathogenic *E. coli* are discussed in other chapters in this book.

Studies that employed a variety of methods demonstrated that *Shigella* spp. do not form a single subgroup of *E. coli* but instead are derived from separate *E. coli* strains [8,119–121]. Seven different *Shigella* lineages have been identified through sequence analysis of multiple chromosomal loci. Thus horizontal transfer of the *Shigella* virulence plasmid to commensal *E. coli* has occurred multiple times, each time giving rise to new *Shigella* clones. These findings suggest that traits unique to and shared by *Shigella* spp. and EIEC are the result of convergent evolution either through gain-of-function mutations (e.g., horizontal transfer of the virulence plasmid) or loss-of-function mutations (deletion of genes encoding traits expressed by ancestral *E. coli* strains). These insights provide a possible answer to lingering questions about the relationship of EIEC to *Shigella* and the evolution of these pathogens. The heterogeneous characteristics of EIEC are consistent with a polyphyletic history and reflect the nonclonal origins of these isolates. Moreover, these findings strongly suggest that the EIEC are not *Shigella* ancestors or "missing links." Rather, the EIEC clones may be recently evolved pathogens that have not yet completely adapted to a pathogenic lifestyle and fully developed the *Shigella* phenotype [120]. This proposal suggests that expression of the full *Shigella* virulence phenotype requires not only the virulence plasmid but also additional modifications to the *E. coli* genome for optimal transmissibility, fitness in host tissues, and virulence.

Several recent studies have provided evidence of a new pathway of evolution, termed *antagonistic pleiotrophy*, that fine-tunes pathogen genomes for maximal fitness and virulence in host tissues. This pathway and the selective forces that drive its function can be observed in the transformation of commensal *E. coli* to virulent *Shigella* pathogens. Following acquisition of the virulence plasmid and expression of its virulence genes, commensal *E. coli* gained access to new host tissues. However, the newly evolved pathogen, which expresses the full complement of ancestral traits as well as virulence factors, may not be optimally suited for this new pathogenic lifestyle. Selective pressures encountered in the new environment (within colonocytes) are very different from those in the ancestral niche (lumen of the colon). In fact, some ancestral traits may interfere with the expression or function of factors required for survival within host tissues. Loci encoding these interfering factors are designated *antivirulence genes.* Thus the newly evolved pathogen must inactivate ancestral antivirulence genes for optimal fitness and virulence. These modifications to the new pathogen's genome are termed *pathoadaptive mutations* [122].

The convergent evolution of the seven *Shigella* lineages and EIEC presents a unique opportunity for the identification and study of antivirulence genes and pathoadaptive mutations as well as the study of pathogen evolution. Ancestral traits that interfere with virulence are lost from the newly evolved pathogen genome early on as the increased fitness of the adapted clones fixes these beneficial mutations in the newly or recently evolved pathogen population. Therefore, traits that are absent in all *Shigella* and EIEC but commonly expressed in *E. coli* are strong indicators of pathoadaptive mutations that have arisen by convergent evolution. Not surprisingly, evidence supporting this new pathogen evolution pathway was first provided in *Shigella*. Lysine decarboxylase (LDC) activity, which is encoded by the *cadA* gene, is expressed in over 85% of *E. coli* isolates. In contrast, none of the *Shigella* or EIEC clones expresses LDC activity [12]. Lack of LDC activity in the *Shigella* and EIEC is consistent with a pathoadaptive mutation. Experimental evidence of the antivirulence nature of the *cadA* gene has been demonstrated because the product of LDC activity, cadaverine, blocks the action of the virulence plasmid–encoded *Shigella* enterotoxin [109]. Sequence analysis of the *cadA* region of four *Shigella* lineages revealed novel genetic arrangements that were distinct in each strain examined [123]. Thus the pathoadaptive mutation of *cadA* was accomplished by different mechanisms in each *Shigella* strain, indicating that each newly evolved *Shigella* (or EIEC) clone

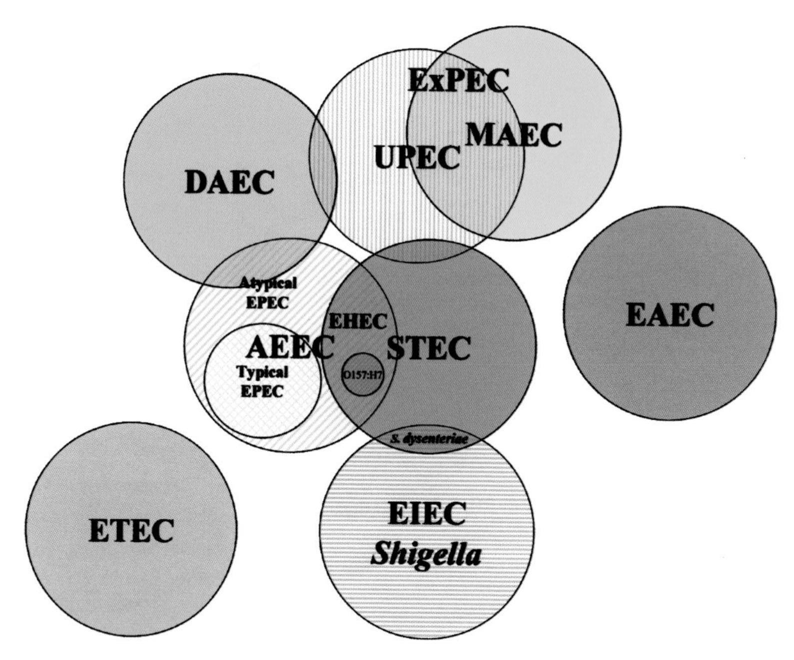

COLOR PLATE 1 Venn diagram illustrating the complex relationships among different pathotypes of *E. coli* that cause disease in humans. Extraintestinal pathogenic *E. coli* (ExPEC) strains include meningitis-associated *E. coli* (MAEC, *yellow*) and uropathogenic *E. coli* (UPEC, *vertical stripes*). These strains share many virulence factors, and it is clear that single clones can cause both types of infections [8]. It is less clear whether or not strains exist that are capable of causing one syndrome and not the other. Among the UPEC, some strains exhibit diffuse adherence to tissue culture cells and share with diffuse adhering *E. coli* (DAEC, *orange*) the same adhesins. DAEC is a heterogeneous pathotype that has been epidemiologically linked to diarrhea. There are reports of DAEC strains recovered from individuals with both urinary tract infections (UTIs) and diarrhea [9]. There are also reports of Shiga toxin–producing *E. coli* (STEC, *green*) strains causing UTI [10]. STEC are defined by production of Shiga toxins, usually encoded by bacteriophages. Among STEC, some strains are also capable of attaching intimately to epithelial cells, effacing microvilli, and eliciting the formation of adhesion pedestals composed of cytoskeletal proteins, a property that defines the attaching and effacing *E. coli* (AEEC, *diagonal stripes*). Such strains, which are both STEC and AEEC, are known as enterohemorrhagic *E. coli* (EHEC). The most important serotype found within the EHEC pathotype is O157:H7. AEEC strains that do not produce Shiga toxins are referred to as enteropathogenic *E. coli* (EPEC). Among EPEC, many strains produce a bundle-forming pilus and attach to tissue culture cells in a localized adherence pattern. These are referred to as typical EPEC (*checkered*), whereas those which do not are known as atypical EPEC. There are some strains of atypical EPEC that exhibit diffuse adherence. Enteroinvasive *E. coli* (EIEC, *horizontal stripes*) invade tissue culture cells with high efficiency, multiply in the cytoplasm, and spread from cell to cell. These strains include the organisms commonly classified in the genus *Shigella*, which in fact is a subset of the species *E. coli*. Strains classified as *S. dysenteriae* produce Shiga toxins and therefore are members of both the EIEC and STEC pathotypes. Enteroaggregative *E. coli* (EAEC, *blue*) is a heterogeneous pathotype that causes acute and persistent diarrhea and is defined by its pattern of adherence. Enterotoxigenic *E. coli* (ETEC, *violet*) strains cause acute diarrhea and are defined by production of heat-labile and/or heat-stable enterotoxins.

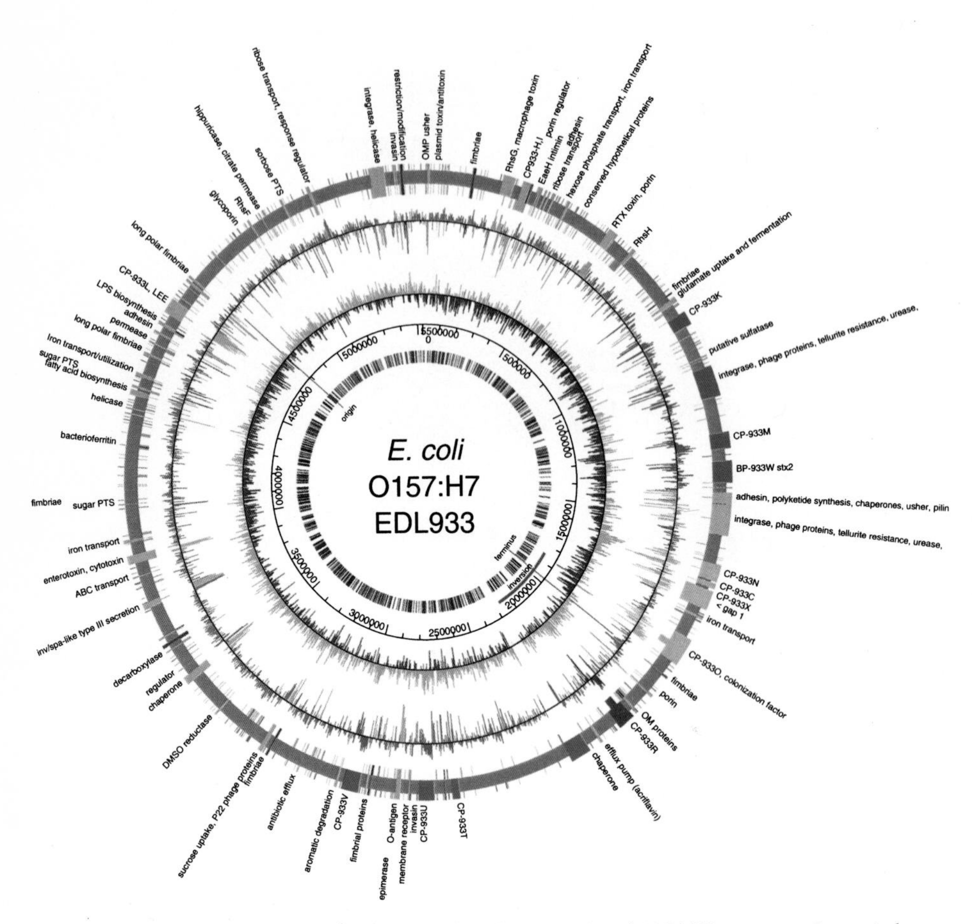

COLOR PLATE 2 Circular map comparing the EDL933 and MG1655 genomes. Outer circle shows the distribution of islands: shared colinear backbone (*blue*); positions of EDL933-specific sequences (O-islands) (*red*); MG1655-specific sequences (K-islands) (*green*); O-islands and K-islands at the same locations in the backbone (*tan*); hypervariable (*purple*). Second circle: G + C content calculated for each gene >100 aa, plotted around the mean value for the whole genome, color coded like outer circle. Third circle: GC skew for codon third position, calculated for each gene >100 aa; positive values (*lime*), negative (*dark green*). Fourth circle: scale in bp. Fifth circle: distribution of the highly skewed octamer Chi (GCTGGTGG), *bright blue* and *purple* indicating the two DNA strands. The origin and terminus of replication and the chromosomal inversion are shown. The gap at ~2,300,000 is now closed, and the GenBank file and Web site are updated accordingly. [Reprinted with permission from Perna et al., 2001, *Nature* 409(6819):529–533; map figures created by Genvision (DNASTAR).]

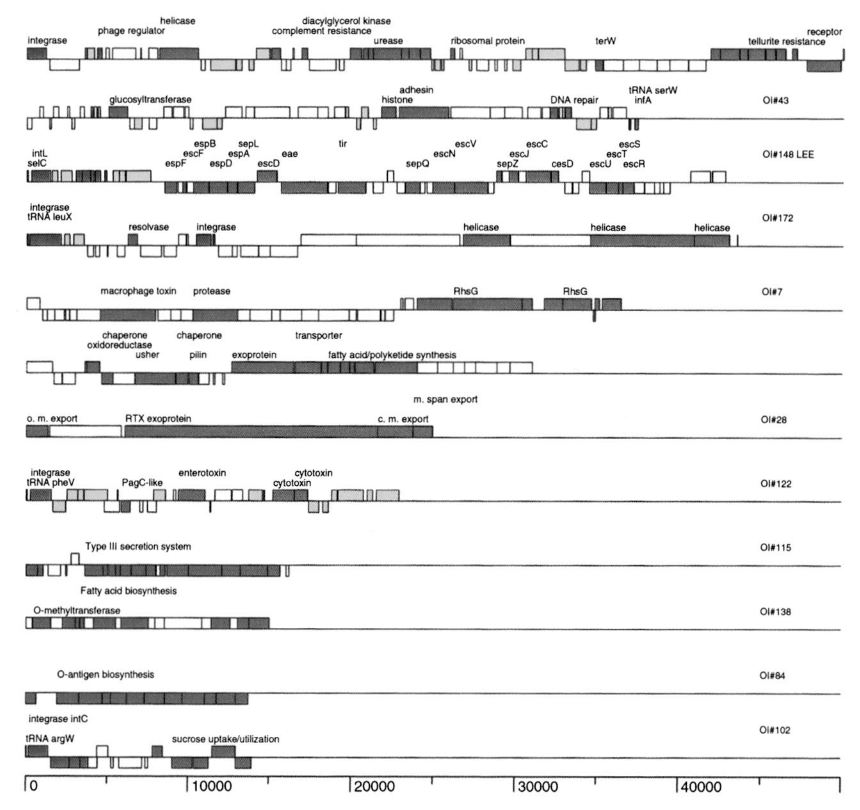

COLOR PLATE 3 Twelve largest O-islands. Boxes represent ORFs and their orientation by position above or below the axis line. O-island identification numbers are shown at the right of each map. *Red* ORFs, function or putative function assigned; *open boxes*, unknown function; *turquoise*, tRNA; *green*, phage; *yellow*, IS elements. OI#43 is 87 kb and occupies the first two panels. OI#48 (not shown) is identical to OI#43. Abbreviations: m, membrane; o. m., outer membrane; c. m., cytoplasmic membrane. Labels are aligned with the left border of the corresponding ORF.

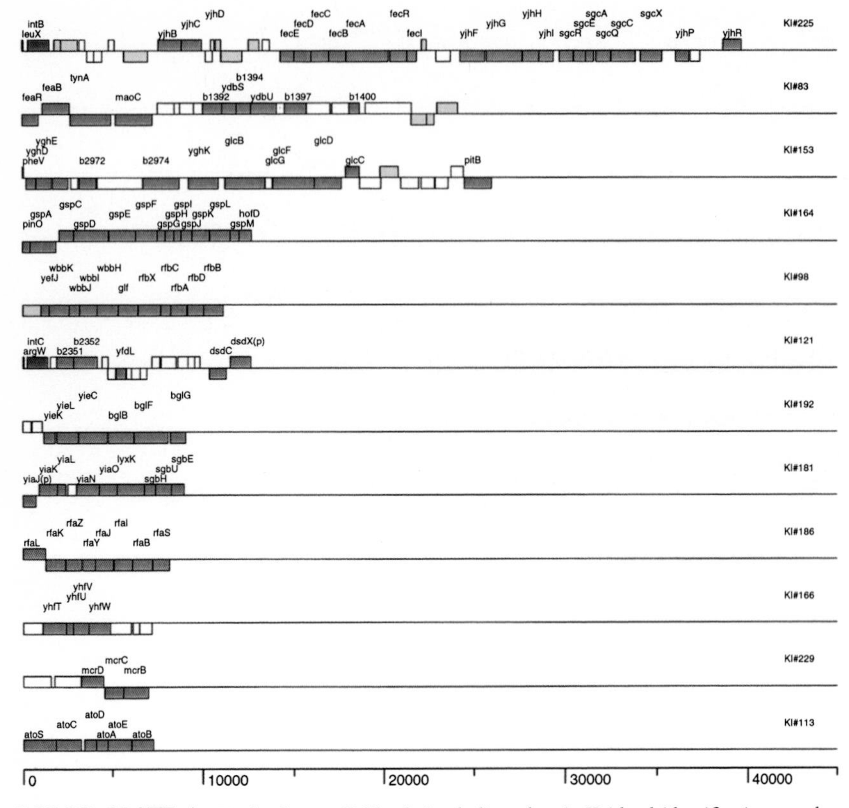

COLOR PLATE 4 Twelve largest K-islands (excluding phage). K-island identification numbers are shown at the right. *Mauve* ORFs, function or putative function assigned; *open boxes*, unknown function; *turquoise*, tRNA; *green*, phage; *yellow*, IS elements. (p) indicates that the ORF spans the island boundary.

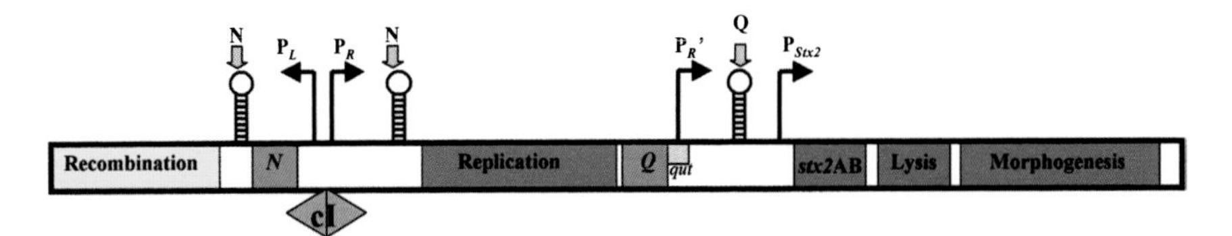

COLOR PLATE 5 Diagram of Stx2-encoding phages. Shown are some of the relevant genes, promoters, terminators, and operators, not drawn to scale. Induction inactivates the repressor cI, resulting in transcription initiating at the early promoters P_L and P_R and expression of the N antiterminator. N modification of RNA polymerase allows read-through of N antiterminator sites, resulting in production of proteins catalyzing excision and replication of the phage genome and synthesis of Q. Q is an antiterminator that binds to *qut* at the site of the late $P_{R'}$ promoter, allowing read-through of the *stx* genes as well as the genes of the phage lysis cassette. Stx1 phages are similarly arranged, except that the Stx1 promoter replaces the Stx2 promoter. The Stx1 promoter is under the control of the *fur* gene and therefore is regulated by iron, as described in the text. (Courtesy of Drs. P. L. Wagner and M. K. Waldor.)

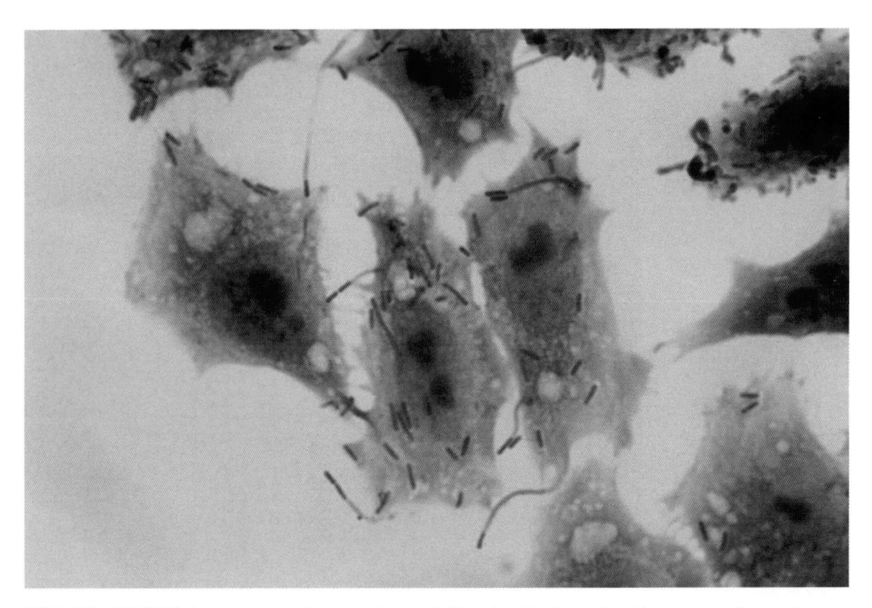

COLOR PLATE 6 Invasion of mammalian cells by *Shigella*. A semiconfluent monolayer of mouse L2 fibroblasts was infected with *S. flexneri* 2a for 30 min. The monolayer was washed and tissue culture medium with gentamicin was added to kill extracellular bacteria. The monolayer was fixed and stained 2 hours later. Bacteria can be seen as darkly staining rods within the cytoplasm of the eucaryotic cells. Intracellular movement of the bacteria and contact with the plasma membrane causes formation of "fireworks," cytoplasmic protrusions that contain bacteria at the ends.

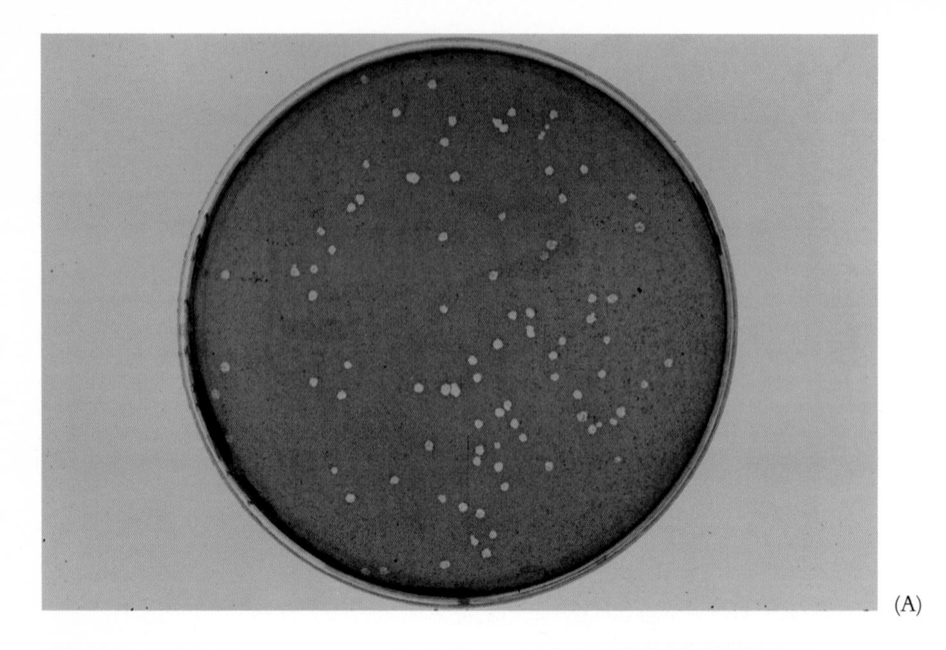

(A)

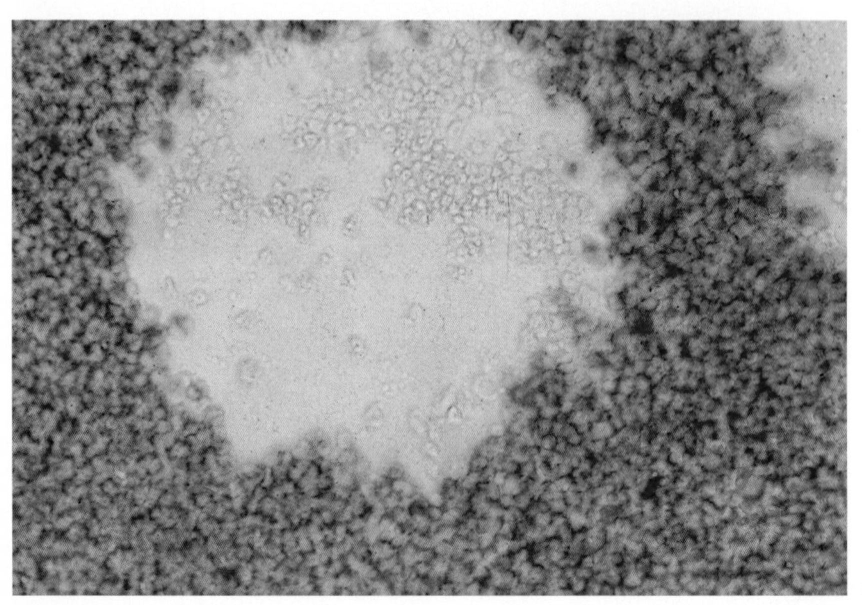

(B)

COLOR PLATE 7 Intercellular spread of *Shigella* measured in the plaque assay. Bacteria are allowed to invade a confluent monolayer of cells in tissue culture for 90 min. The monolayer is washed and an overlay of 0.5% agarose-containing tissue culture medium with gentamicin is added and the monolayers incubated for 48–72 hr. The gentamicin kills extracellular bacteria. Intracellular bacteria that move from cell to cell by forming protrusions that "invade" surrounding cells are protected from the antibiotic and propagate the foci of infection. Panel A shows plaques formed by *S. flexneri* 2a in a monolayer of L2 fibroblasts after 48 hr. Panel B shows the appearance of a plaque at higher magnification. The monolayer is stained with neutral red, which stains living cells. Dead (unstained) cells can be seen in the center of the plaque.

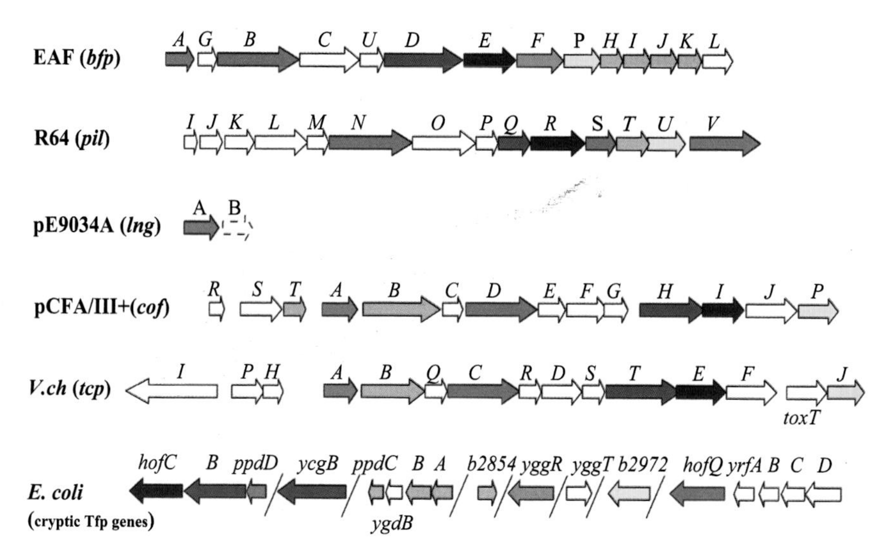

COLOR PLATE 8 Scheme of the BFP gene cluster of EPEC in comparison with the gene organization of other type IVB biogenesis clusters. Lower part shows genes within the *E. coli* K-12 genome with homologies to other Tfp genes. The putative major pilin gene *ppdD* belongs into the class A of Tfp. Arrows: red (prepilin); magenta (secretin); dark green (putative NTP-binding protein); light green (putative NTP-binding protein/retraction), dark blue (polytopic inner membrane protein); yellow (prepilin peptidase); light blue (lytic transglycosylase); orange (prepilin-like proteins).

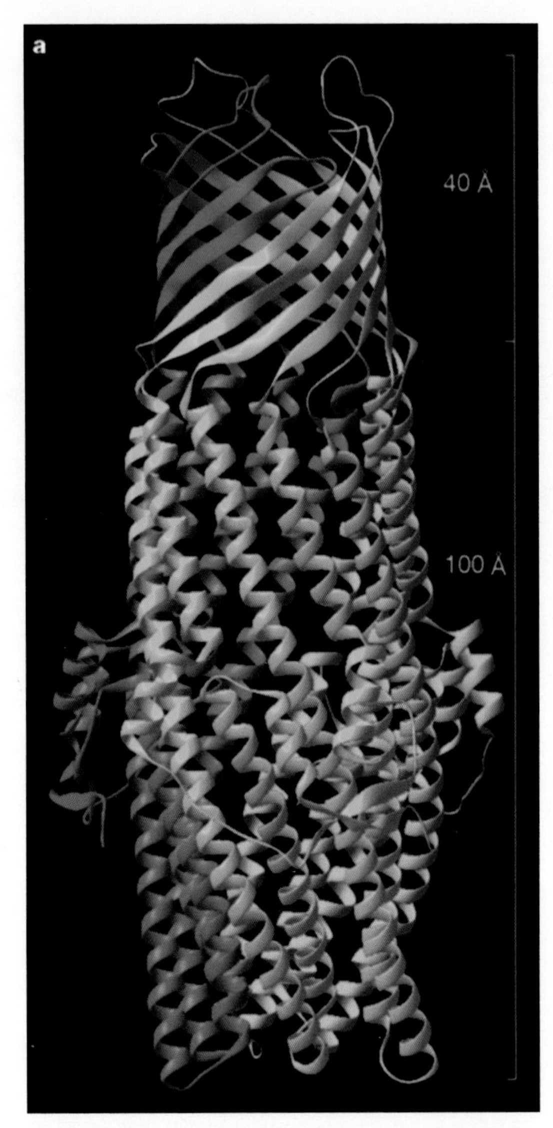

COLOR PLATE 9 The structure of trimeric TolC. The protomers are individually colored. The β-barrel channel (outer membrane) domain is at the top, and the α-helical tunnel (periplasmic) domain is at the bottom. Reprinted with permission from Nature 405:914–919, Fig 1a.

increased virulence through the loss of an ancestral trait, LDC. Similar studies, when extended to other pathogens, should provide new insights into bacterial pathogen evolution. These insights also may identify novel antipathogen agents that are produced by ancestral nonpathogenic species that interfere with virulence functions.

CONCLUSIONS

A century of research has yielded significant insights into the mechanisms used by *Shigella* and EIEC to interact with host tissues. These insights have revealed a family of pathogens that are master cell biologists that reprogram host cells and exploit the actions of host phagocytes to serve the needs of the parasite. Furthermore, the unique evolution of these bacteria has promoted understanding of pathogen evolution and driven discovery of novel evolutionary pathways. Thus *Shigella* and EIEC have served as models of several aspects of pathogen biology and provided paradigms that have been applied to other disease-causing bacteria. Yet, despite the advances in our knowledge of *Shigella* and EIEC pathogenesis, there remain many mysteries at the genetic and cell biology levels. Furthermore, in the absence of an effective vaccine against dysentery, these pathogens will continue to remain a formidable global threat to public health.

ACKNOWLEDGMENTS

Research on the genetics of *Shigella* virulence in the laboratory of A.T.M. is supported by USUHS Protocol RO-7385 and Public Health Service Grant AI24656 from the National Institute of Allergy and Infectious Diseases.

REFERENCES

1. Goma Epidemiology Group (1995). Public health impact of Rwandan refugee crisis: What happened in Goma, Zaire, in July 1994? *Lancet* 345:339–344.
2. Kotloff, K. L., Winickoff, J. P., Ivanoff, B., *et al.* (1999). Global burden of *Shigella* infections: Implications for vaccine development and implementation of control strategies. *Bull. WHO* 77:651–666.
3. Parsot, C., and Sansonetti, P. J. (1996). Invasion and the pathogenesis of *Shigella* infections. *Curr. Top. Microbiol. Immunol.* 209:25–42.
4. Dorman, C. J., and Porter, M. E. (1998). The *Shigella* virulence gene regulatory cascade: A paradigm of bacterial gene control mechanisms. *Mol. Microbiol.* 29:677–684.
5. Tran Van Nhieu, G., Bourdet-Sicard, R., Dumenil, G., *et al.* (2000). Bacterial signals and cell responses during *Shigella* entry into epithelial cells. *Cell. Microbiol.* 2:187–193.
6. Maurelli, A. T., and Lampel, K. A. (2001). *Shigella.* In Y. H. Hui, M. D. Pierson, and J. R. Gorham (eds.), *Foodborne Disease Handbook,* Vol. I, 2d ed., pp. 323–343. New York: Marcel Dekker.

7. Schuch, R., and Maurelli, A. T. (2000). The type III secretion pathway: dictating the outcome of bacterial-host interactions. In K. A. Brogden, J. A. Roth, T. B. Stanton, *et al.* (ed.), *Virulence Mechanisms of Bacterial Pathogens*, 3d ed., pp. 203–223. Washington: American Society for Microbiology Press.
8. Ochman, H., Whittam, T. S., Caugant, D. A., and Selander, R. K. (1983). Enzyme polymorphism and genetic population structure in *Escherichia coli* and *Shigella*. *J. Gen. Microbiol.* 129:2715–2726.
9. Edwards, P. R., and Ewing, W. H. (1972). *Identification of Enterobacteriaceae.* Minneapolis: Burgess Publishing.
10. DuPont, H. L., Formal, S. B., Hornick, R. B., *et al.* (1971). Pathogenesis of *Escherichia coli* diarrhea. *New Engl. J. Med.* 285:1–9.
11. Tulloch, E. F., Jr., Ryan, K. J., Formal, S. B., and Franklin, F. A. (1973). Invasive enteropathic *Escherichia coli* dysentery: An outbreak in 28 adults. *Ann. Intern. Med.* 79:13–17.
12. Silva, R. M., Regina, M., Toledo, F., and Trabulsi, L. R. (1980). Biochemical and cultural characteristics of invasive *Escherichia coli*. *J. Clin. Microbiol.* 11:441–444.
13. Sansonetti, P. J., Hale, T. L., and Oaks, E. V. (1985). Genetics of virulence in enteroinvasive *Escherichia coli*. In D. Schlessinger (ed.), *Microbiology—1985*, pp. 74–77. Washington: American Society for Microbiology Press.
14. Gorden, J., and Small, P. L. C. (1993). Acid resistance in enteric bacteria. *Infect. Immun.* 61:364–367.
15. Zychlinsky, A., Prevost, M. C., and Sansonetti, P. J. (1992). *Shigella flexneri* induces apoptosis in infected macrophages. *Nature* 358:167–169.
16. Zychlinsky, A., and Sansonetti, P. J. (1997). Apoptosis as a proinflammatory event: what can we learn from bacteria-induced cell death? *Trends Microbiol.* 5:201–204.
17. Watarai, M., Funato, S., and Sasakawa, C. (1996). Interaction of Ipa proteins of *Shigella flexneri* with α5-β1 integrin promotes entry of the bacteria into mammalian cells. *J. Exp. Med.* 183:991–999.
18. Raqib, R., Lindberg, A. A., Wretlind, B., *et al.* (1995). Persistence of local cytokine production in shigellosis in acute and convalescent stages. *Infect. Immun.* 63:289–296.
19. Sansonetti, P. J., Arondel, J., Huerre, M., *et al.* (1999). Interleukin-8 controls bacterial transepithelial translocation at the cost of epithelial destruction in experimental shigellosis. *Infect. Immun.* 67:1471–1480.
20. Perdomo, O. J., Cavaillon, J. M., Huerre, M., *et al.* (1994). Acute inflammation causes epithelial invasion and mucosal destruction in experimental shigellosis. *J. Exp. Med.* 180:1307–1319.
21. Perdomo, J. J., Gounon, P., and Sansonetti, P. J. (1994). Polymorphonuclear leukocyte transmigration promotes invasion of colonic epithelial monolayer by *Shigella flexneri*. *J. Clin. Invest.* 93:633–643.
22. McCormick, B. A., Siber, A. M., and Maurelli, A. T. (1998). Requirement of the *Shigella flexneri* virulence plasmid in the ability to induce trafficking of neutrophils across polarized monolayers of the intestinal epithelium. *Infect. Immun.* 66:4237–4243.
23. DuPont, H. L. (1995). *Shigella* species (bacillary dysentery). In G. L. Mandell, J. E. Bennett, and R. Dolin (eds.), *Principles and Practice of Infectious Diseases*, 4th ed., pp. 2033–2039. New York: Churchill Livingstone.
24. Kinsey, M. D., Formal, S. B., Dammin, G. J., and Giannella, R. A. (1976). Fluid and electrolyte transport in rhesus monkeys challenged intracecally with *Shigella flexneri* 2a. *Infect. Immun.* 14:368–371.
25. Bennish, M. L., Harris, J. R., Wojtynaik, B. J., and Struelens, M. (1990). Death in shigellosis: Incidence and risk factors in hospitalized patients. *J. Infect. Dis.* 161:500–506.
26. Bennish, M. L. (1991). Potentially lethal complications of shigellosis. *Rev. Infect. Dis.* 13(suppl. 4):S319–S324.
27. Raghupathy, P., Date, A., Shastry, J. C. M., *et al.* (1978). Haemolytic-uraemic syndrome complicating *Shigella* dysentery in south Indian children. *Br. Med. J.* 1:1518–1521.

28. Lopez, E. L., Diaz, M., Grinstein, S., *et al.* (1989). Hemolytic uremic syndrome and diarrhea in Argentine children: The role of Shiga-like toxins. *J. Infect. Dis.* 160:469–475.
29. Karmali, M. A., Petric, M., Lim, C., *et al.* (1985). The association between idiopathic hemolytic syndrome and infection by Verotoxin-producing *Escherichia coli*. *J. Infect. Dis.* 151:775–782.
30. Tesh, V. L., and O'Brien, A. D. (1991). The pathogenic mechanisms of Shiga toxin and the Shiga-like toxins. *Mol. Microbiol.* 5:1817–1822.
31. Simon, D. G., Kaslow, R. A., Rosenbaum, J., *et al.* (1981). Reiter's syndrome following epidemic shigellosis. *J. Rheumatol.* 8:969–973.
32. Bunning, V. K., Raybourne, R. B., and Archer, D. L. (1988). Foodborne enterobacterial pathogens and rheumatoid disease. *J. Appl. Bacteriol. Symp. Suppl.* 65:87S–107S.
33. DuPont, H. L., Levine, M. M., Hornick, R. B., and Formal, S. B. (1989). Inoculum size in shigellosis and implications for expected mode of transmission. *J. Infect. Dis.* 159:1126–1128.
34. Smith, J. L. (1987). *Shigella* as a food-borne pathogen. *J. Food Protect.* 50:788–801.
35. Levine, M. M., and Levine, O. S. (1994). Changes in human ecology and behavior in relation to the emergence of diarrheal diseases, including cholera. *Proc. Natl. Acad. Sci. USA* 91:2390–2394.
36. Pickering, L. K., Bartlett, A. V., and Woodward, W. E. (1986). Acute infectious diarrhea among children in day-care: epidemiology and control. *Rev. Infect. Dis.* 8:539–547.
37. Weissman, J. B., Schmerler, A., Weiler, P., *et al.* (1974). The role of preschool children and day-care centers in the spread of shigellosis in urban communities. *J. Pediatr.* 84:797–802.
38. Kennedy, F. M., Astbury, J., Needham, J. R., and Cheasty, T. (1993). Shigellosis due to occupational contact with non-human primates. *Epidemiol. Infect.* 110:247–257.
39. Hossain, M. A., Hasan, K. Z., and Albert, M. J. (1994). *Shigella* carriers among non-diarrhoeal children in an endemic area of shigellosis in Bangladesh. *Trop. Geogr. Med.* 46:40–42.
40. Guerrero, L., Calva, J. J., Morrow, A. L., *et al.* (1994). Asymptomatic *Shigella* infections in a cohort of Mexican children younger than 2 years of age. *Pediatr. Infect. Dis. J.* 13:597–602.
41. Bennish, M. L., and Salam, M. A. (1992). Rethinking options for the treatment of shigellosis. *J. Antimicrob. Chemother.* 30:243–247.
42. Hoge, C. W., Gambel, J. M., Srijan, A., *et al.* (1998). Trends in antibiotic resistance among diarrheal pathogens isolated in Thailand over 15 years. *Clin. Infect. Dis.* 26:341–345.
43. Haltalin, K., Nelson, J., and Ring, R. (1967). Double-blind treatment study of shigellosis comparing ampicillin, sulfadiazone and placebo. *J. Pediatr.* 70:970–981.
44. Barzu, S., Fontaine, A., Sansonetti, P., and Phalipon, A. (1996). Induction of a local anti-IpaC antibody response in mice by use of a *Shigella flexneri* 2a vaccine candidate: Implications for use of IpaC as a protein carrier. *Infect. Immun.* 64:1190–1196.
45. Coster, T. S., Hoge, C. W., VanDeVerg, L. L., *et al.* (1999). Vaccination against shigellosis with attenuated *Shigella flexneri* 2a strain SC602. *Infect. Immun.* 67:3437–3443.
46. Rout, W. R., Formal, S. B., Giannella, R. A., and Dammin, G. J. (1975). Pathophysiology of *Shigella* diarrhea in the rhesus monkey: Intestinal transport, morphological, and bacteriological studies. *Gastroenterology* 68:270–278.
47. Fasano, A., Noriega, F. R., Maneval, D. R., Jr., *et al.* (1995). *Shigella* enterotoxin 1: An enterotoxin of *Shigella flexneri* 2a active in rabbit small intestine *in vivo* and *in vitro*. *J. Clin. Invest.* 95:2853–2861.
48. Nataro, J. P., Seriwatana, J., Fasano, A., *et al.* (1995). Identification and cloning of a novel plasmid-encoded enterotoxin of enteroinvasive *Escherichia coli* and *Shigella* strains. *Infect. Immun.* 63:4721–4728.
49. LaBrec, E. H., Schneider, H., Magnani, T. J., and Formal, S. B. (1964). Epithelial cell penetration as an essential step in the pathogenesis of bacillary dysentery. *J. Bacteriol.* 88:1503–1518.
50. Falkow, S., Schneider, H., Baron, L., and Formal, S. B. (1963). Virulence of *Escherichia-Shigella* genetic hybrids for the guinea pig. *J. Bacteriol.* 86:1251–1258.

51. Formal, S. B., LaBrec, E. H., Kent, T. H., and Falkow, S. (1965). Abortive intestinal infection with an *Escherichia coli–Shigella flexneri* hybrid strain. *J. Bacteriol.* 89:1374–1382.
52. Maurelli, A. T., Blackmon, B., and Curtiss, R., III. (1984). Temperature-dependent expression of virulence genes in *Shigella* species. *Infect. Immun.* 43:195–201.
53. Small, P. C., and Falkow, S. (1988). Identification of regions on a 230-kilobase plasmid from enteroinvasive *Escherichia coli* that are required for entry into Hep-2 cells. *Infect. Immun.* 56:225–229.
54. Sansonetti, P. J., Kopecko, D. J., and Formal, S. B. (1981). *Shigella sonnei* plasmids: Evidence that a large plasmid is necessary for virulence. *Infect. Immun.* 34:75–83.
55. Sansonetti, P. J., Kopecko, D. J., and Formal, S. B. (1982). Involvement of a plasmid in the invasive ability of *Shigella flexneri. Infect. Immun.* 35:852–860.
56. Harris, J. R., Wachsmuth, I. K., Davis, B. R., and Cohen, M. L. (1982). High-molecular-weight plasmid correlates with *Escherichia coli* enteroinvasiveness. *Infect. Immun.* 37:1295–1298.
57. Sansonetti, P. J., d'Hauteville, H., Formal, S. B., and Toucas, M. (1982). Plasmid-mediated invasiveness of "*Shigella*-like" *Escherichia coli. Ann. Microbiol. (Inst. Pasteur)* 132A:351–355.
58. Sansonetti, P. J., d'Hauteville, H., Ecobichon, C., and Pourcel, C. (1983). Molecular comparison of virulence plasmids in *Shigella* and enteroinvasive *Escherichia coli. Ann. Microbiol. (Inst. Pasteur)* 134A:295–318.
59. Buchrieser, C., Glaser, P., Rusniok, C., *et al.* (2000). The virulence plasmid pWR100 and the repertoire of proteins secreted by the type III secretion apparatus of *Shigella flexneri. Mol. Microbiol.* 38:760–71.
60. Venkatesan, M. M., Goldberg, M. B., Rose, D. J., *et al.* (2001). Complete DNA sequence and analysis of the large virulence plasmid of *Shigella flexneri. Infect. Immun.* 69:3271–3285.
61. Maurelli, A. T., Baudry, B., d'Hauteville, H., *et al.* (1985). Cloning of virulence plasmid DNA sequences involved in invasion of HeLa cells by *Shigella flexneri. Infect. Immun.* 49:164–171.
62. Hueck, C. J. (1998). Type III protein secretion systems in bacterial pathogens of animals and plants. *Microbiol. Mol. Biol. Rev.* 62:379–433.
63. Hale, T. L., Oaks, E. V., and Formal, S. B. (1985). Identification and antigenic characterization of virulence-assiciated, plasmid-coded proteins of *Shigella* spp. and enteroinvasive *Escherichia coli. Infect. Immun.* 50:620–629.
64. Oaks, E. V., Hale, T. L., and Formal, S. B. (1986). Serum immune response to *Shigella* protein antigens in rhesus monkeys and humans infected with *Shigella* spp. *Infect. Immun.* 53:57–63.
65. Ménard, R., Sansonetti, P. J., and Parsot, C. (1993). Nonpolar mutagenesis of the *ipa* genes defines IpaB, IpaC, and IpaD as effectors of *Shigella flexneri* entry into epithelial cells. *J. Bacteriol.* 175:5899–5906.
66. Tran Van Nhieu, G., Ben-Ze'ev, A., and Sansonetti, P. J. (1997). Modulation of bacterial entry into epithelial cells by association between vinculin and the *Shigella* IpaA invasin. *EMBO J.* 16:2717–2729.
67. Ménard, R., Sansonetti, P., Parsot, C., and Vasselon, T. (1994). Extracellular association and cytoplasmic partitioning of the IpaB and IpaC invasins of *S. flexneri. Cell.* 79:515–525.
68. Skoudy, A., Mounier, J., Aruffo, A., *et al.* (2000). CD44 binds to the *Shigella* IpaB protein and participates in bacterial invasion of epithelial cells. *Cell. Microbiol.* 2:19–33.
69. Ménard, R., Prevost, M. C., Gounon, P., *et al.* (1996). The secreted Ipa complex of *Shigella flexneri* promotes entry into mammalian cells. *Proc. Natl. Acad. Sci. USA* 93:1254–1258.
70. Sansonetti, P. J., Tran Van Nhieu, G., and Egile, C. (1999). Rupture of the intestinal epithelial barrier and mucosal invasion by *Shigella flexneri. Clin. Infect. Dis.* 28:466–475.
71. Tran Van Nhieu, G., Caron, E., Hall, A., and Sansonetti, P. J. (1999). IpaC induces actin polymerization and filopodia formation during *Shigella* entry into epithelial cells. *EMBO J.* 18:3249–3262.
72. Bourdet-Sicard, R., Rudiger, M., Jockusch, B. M., *et al.* (1999). Binding of the *Shigella* protein IpaA to vinculin induces F-actin depolymerization. *EMBO J.* 18:5853–5862.

73. Ménard, R., Sansonetti, P. J., and Parsot, C. (1994). The secretion of the *Shigella flexneri* Ipa invasins is induced by the epithelial cell and controlled by IpaB and IpaD. *EMBO J.* 13:5293–5302.
74. High, N., Mounier, J., Prevost, M. C., and Sansonetti, P. J. (1992). IpaB of *Shigella flexneri* causes entry into epithelial cells and escape from the phagocytic vacuole. *EMBO J.* 11:1991–1999.
75. Zychlinsky, A., Kenny, B., Ménard, R., *et al.* (1994). IpaB mediates macrophage apoptosis induced by *Shigella flexneri. Mol. Microbiol.* 11:619–627.
76. Guichon, A., Hersh, D., Smith, M. R., and Zychlinsky, A. (2001). Structure-function analysis of the *Shigella* virulence factor IpaB. *J. Bacteriol.* 183:269–1276.
77. Makino, S., Sasakawa, C., Kamata, K., *et al.* (1986). A genetic determinant required for continuous reinfection of adjacent cells on large plasmid in *Shigella flexneri* 2a. *Cell.* 46:551–555.
78. Bernardini, M. L., Mounier, J., d'Hauteville, H., *et al.* (1989). Identification of *icsA*, a plasmid locus in *Shigella flexneri* that governs bacterial intra- and intercellular spread through interaction with F-actin. *Proc. Natl. Acad. Sci. USA* 86:3867–3871.
79. Sansonetti, P. J., Arondel, J., Fontaine, A., *et al.* (1991). *OmpB* (osmo-regulation) and *icsA* (cell-to-cell spread) mutants of *Shigella flexneri:* Vaccine candidates and probes to study the pathogenesis of shigellosis. *Vaccine* 9:416–422.
80. Goldberg, M. B., Barzu, O., Parsot, C., and Sansonetti, P. J. (1993). Unipolar localization and ATPase activity of IcsA, a *Shigella flexneri* protein involved in intracellular movement. *J. Bacteriol.* 175:2189–2196.
81. Egile, C., Loisel, T. P., Laurent, V., *et al.* (1999). Activation of the CDC42 effector N-WASP by the *Shigella flexneri* IcsA protein promotes actin nucleation by Arp2/3 complex and bacterial actin-based motility. *J. Cell. Biol.* 146:1319–1332.
82. Suzuki, T., Saga, S., and Sasakawa, C. (1996). Functional analysis of *Shigella* VirG domains essential for interaction with vinculin and actin-based motility. *J. Biol. Chem.* 271:21878–21885.
83. Sandlin, R. C., Lampel, K. A., Keasler, S. P., *et al.* (1995). Avirulence of rough mutants of *Shigella flexneri*: Requirement of O-antigen for correct unipolar localization of IcsA in bacterial outer membrane. *Infect. Immun.* 63:229–237.
84. Sandlin, R. C., Goldberg, M. B., and Maurelli, A. T. (1996). Effect of O side-chain length and composition on the virulence of *Shigella flexneri* 2a. *Mol Microbiol.* 22:63–73.
85. Egile, C., d'Hauteville, H., Parsot, C., and Sansonetti, P. J. (1997). SopA, the outer membrane protease responsible for polar localization of IcsA in *Shigella flexneri. Mol. Microbiol.* 23:1063–1073.
86. Shere, K. D., Sallustio, S., Manessis, A., *et al.* (1997). Disruption of IcsP, the major *Shigella* protease that cleaves IcsA, accelerates actin-based motility. *Mol. Microbiol.* 25:451–462.
87. Sandlin, R. C., and Maurelli, A. T. (1999). Establishment of unipolar localization of IcsA in *Shigella flexneri* 2a is not dependent on virulence plasmid determinants. *Infect. Immun.* 67:350–356.
88. Watarai, M., Tobe, T., Yoshikawa, M., and Sasakawa, C. (1995). Contact of *Shigella* with host cells triggers release of Ipa invasins and is an essential function of invasiveness. *EMBO J.* 14:2461–2470.
89. Bahrani, F. K., Sansonetti, P. J., and Parsot, C. (1997). Secretion of Ipa proteins by *Shigella flexneri:* Inducer molecules and kinetics of activation. *Infect. Immun.* 65:4005–4010.
90. Allaoui, A., Sansonetti, P. J., and Parsot, C. (1992). MxiJ, a lipoprotein involved in secretion of *Shigella* Ipa invasins, is homologous to YscJ, a secretion factor of the *Yersinia* Yop proteins. *J. Bacteriol.* 174:7661–7669.
91. Allaoui, A., Sansonetti, P. J., and Parsot, C. (1993). MxiD, an outer membrane protein necessary for the secretion of the *Shigella flexneri* Ipa invasins. *Mol. Microbiol.* 7:59–68.
92. Andrews, G. P., Hromockyj, A. E., Coker, C., and Maurelli, A. T. (1991). Two novel virulence loci, *mxiA* and *mxiB*, in *Shigella flexneri* 2a facilitate excretion of invasion plasmid antigen. *Infect. Immun.* 59:1997–2005.

93. Andrews, G. P., and Maurelli, A. T. (1992). *mxiA* of *Shigella flexneri* 2a, which facilitates export of invasion plasmid antigens, encodes a homologue of the low-calcium response protein, LcrD, of *Yersinia pestis. Infect. Immun.* 60:3287–3295.
94. Venkatesan, M. M., Buysse, J. M., and Oaks, E. V. (1992). Surface presentation of *Shigella flexneri* invasion plasmid antigens requires the products of the *spa* locus. *J. Bacteriol.* 174:1990–2001.
95. Sasakawa, C., Komatsu, K., Tobe, T., *et al.* (1993). Eight genes in region 5 that form an operon are essential for invasion of epithelial cells by *Shigella flexneri* 2a. *J. Bacteriol.* 175:2334–2346.
96. Schuch, R., Sandlin, R. C., and Maurelli, A. T. (1999). A system for identifying post-invasion functions of invasion genes: requirements for the Mxi-Spa type III secretion pathway of *Shigella flexneri* in intercellular dissemination. *Mol. Microbiol.* 34:675–689.
97. Uchiya, K., Tobe, T., Komatsu, K., *et al.* (1995). Identification of a novel virulence gene, *virA*, on the large plasmid of *Shigella*, involved in invasion and intercellular spreading. *Mol. Microbiol.* 17:241–250.
98. Hartman, A. B., Venkatesan, M., Oaks, E. V., and Buysse, J. M. (1990). Sequence and molecular characterization of a multicopy invasion plasmid antigen gene, *ipaH,* of *Shigella flexneri. J. Bacteriol.* 172:1905–1915.
99. Fernandez-Prada, C. M., Hoover, D. L., Tall, B. D., *et al.* (2000). *Shigella flexneri* IpaH(7.8) facilitates escape of virulent bacteria from the endocytic vacuoles of mouse and human macrophages. *Infect. Immun.* 68:3608–3619.
100. Brahmbhatt, H. N., Lindberg, A. A., and Timmis, K. N. (1992). *Shigella* lipopolysaccharide: Structure, genetics, and vaccine development. *Curr. Top. Microbiol. Immunol.* 180:45–64.
101. O'Brien, A. D., and Holmes R. K. (1995). Protein toxins of *Escherichia coli* and *Salmonella.* In F. D. Neidhardt, R. Curtiss III, C. A. Gross, *et al.* (eds.), *Escherichia coli and Salmonella typhimurium: Cellular and Molecular Biology*, 2d ed., pp. 2788–2802. Washington: American Society for Microbiology Press.
102. Fontaine, A., Arondel, J., and Sansonetti, P. J. (1988). Role of the Shiga toxin in the pathogenesis of bacillary dysentery studied by using a Tox$^-$ mutant of *Shigella dysenteriae* 1. *Infect. Immun.* 56:3099–3109.
103. Rajakumar, K., Sasakawa, C., and Adler, B. (1997). Use of a novel approach, termed island probing, identifies the *Shigella flexneri she* pathogenicity island which encodes a homologue of the immunoglobulin A protease-like family of proteins. *Infect. Immun.* 65:4606–4614.
104. Vokes, S. A., Reeves, S. A., Torres, A. G., and Payne, S. M. (1999). The aerobactin iron transport system genes in *Shigella flexneri* are present within a pathogenicity island. *Mol. Microbiol.* 33:63–73.
105. Moss, J. E., Cardozo, T. J., Zychlinsky, A., and Groisman, E. A. (1999). The *selC*-associated SHI-2 pathogenicity island of *Shigella flexneri. Mol. Microbiol.* 33:74–83.
106. Nassif, X., Mazert, M. C., Mounier, J., and Sansonetti, P. J. (1987). Evaluation with an *iuc::Tn10* mutant of the role of aerobactin production in the virulence of *Shigella flexneri. Infect. Immun.* 55:1963–1969.
107. Purdy, G. E., and Payne, S. M. (2001). The SHI-3 iron transport island of *Shigella boydii* 0-1392 carries the genes for aerobactin synthesis and transport. *J. Bacteriol.* 183:4176–4182.
108. Nakata, N., Tobe, T., Fukuda, I., *et al.* (1993). The absence of a surface protease, OmpT, determines the intercellular spreading ability of *Shigella*: The relationship between the *ompT* and *kcpA* loci. *Mol. Microbiol.* 9:459–468.
109. Maurelli, A. T., Fernández, R. E., Bloch, C. A., *et al.* (1998). "Black holes" and bacterial pathogenicity: A large genomic deletion that enhances the virulence of *Shigella* spp. and enteroinvasive *Escherichia coli. Proc. Natl. Acad. Sci. USA* 95:3943–3948.
110. Hromockyj, A. E., and Maurelli, A. T. (1989). Identification of *Shigella* invasion genes by isolation of temperature-regulated *inv::lacZ* operon fusions. *Infect. Immun.* 57:2963–2970.
111. Maurelli, A. T., and Sansonetti, P. J. (1988). Identification of a chromosomal gene controlling temperature-regulated expression of *Shigella* virulence. *Proc. Natl. Acad. Sci. USA* 85:2820–2824.

112. Tobe, T., Yoshikawa, M., Mizuno, T., and Sasakawa, C. (1993). Transcriptional control of the invasion regulatory gene *virB* of *Shigella flexneri*: Activation by VirF and repression by H-NS. *J. Bacteriol.* 175:6142–6149.
113. Adler, B., Sasakawa, C., Tobe, T., *et al.* (1989). A dual transcriptional activation system for the 230-kb plasmid genes coding for virulence-associated antigens of *Shigella flexneri*. *Mol. Microbiol.* 3:627–635.
114. Buysse, J. M., Venkatesan, M. M., Mills, J., and Oaks, E. V. (1990). Molecular characterization of a transacting, positive effector (*ipaR*) of invasion plasmid antigen synthesis in *Shigella flexneri* serotype 5. *Microb. Pathog.* 8:197–211.
115. Tobe, T., Nagai, S., Okada, N., *et al.* (1991). Temperature-regulated expression of invasion genes in *Shigella flexneri* is controlled through the transcriptional activation of the *virB* gene on the large plasmid. *Mol. Microbiol.* 5:887–893.
116. Demers, B., Sansonetti, P. J., and Parsot, C. (1998). Induction of type III secretion in *Shigella flexneri* is associated with differential control of transcription of genes encoding secreted proteins. *EMBO J.* 17:2894–2903.
117. Pettersson, J., Nordfelth, R., Dubinina, E., *et al.* (1996). Modulation of virulence factor expression by pathogen target cell contact. *Science* 273:1231–1233.
118. Brenner, D. J., Fanning, G. R., Johnson, K. E., *et al.* (1969). Polynucleotide sequence relationships among members of Enterobacteriaceae. *J. Bacteriol.* 98:637–650.
119. Pupo, G. M., Karaolis, D. K., Lan, R., and Reeves, P. R. (1997). Evolutionary relationships among pathogenic and nonpathogenic *Escherichia coli* strains inferred from multilocus enzyme electrophoresis and *mdh* sequence studies. *Infect. Immun.* 65:2685–2692.
120. Pupo, G. M., Lan, R., and Reeves, P. R. (2000). Multiple independent origins of *Shigella* clones of *Escherichia coli* and convergent evolution of many of their characteristics. *Proc. Natl. Acad. Sci. USA* 97:10567–10572.
121. Rolland, K., Lambert-Zechovsky, N., Picard, B., and Denamur, E. (1998). *Shigella* and enteroinvasive *Escherichia coli* strains are derived from distinct ancestral strains of *E. coli*. *Microbiology* 144:2667–2672.
122. Sokurenko, E. V., Hasty, D. L., and Dykhuizen, D. E. (1999). Pathoadaptive mutations: gene loss and variation in bacterial pathogens. *Trends Microbiol.* 7:191–195.
123. Day, W. A., Fernández, R. E., and Maurelli, A. T. (2001). Pathoadaptive mutations that enhance virulence: Genetic organization of the *cadA* regions of *Shigella* spp. *Infect. Immun.* 69:7471–7480.

CHAPTER 8

Uropathogenic Escherichia coli

Farah Bahrani-Mougeot
Division of Infectious Diseases, Department of Medicine, University of Maryland School of Medicine, Baltimore, Maryland

Nereus W. Gunther IV
Department of Microbiology and Immunology, University of Maryland School of Medicine, Baltimore, Maryland

Michael S. Donnenberg
Division of Infectious Diseases, Department of Medicine, University of Maryland School of Medicine, Baltimore, Maryland

Harry L.T. Mobley
Department of Microbiology and Immunology, University of Maryland School of Medicine, Baltimore, Maryland

EPIDEMIOLOGY AND CLINICAL SYMPTOMS

The urinary tract is among the most common sites of bacterial infection, and *Escherichia coli* is by far the most common infecting agent at this site [1]. Individuals at high risk for symptomatic urinary tract infection (UTI) include neonates, preschool girls, sexually active women, and elderly women and men. In 1991, the last year for which pertinent information is available, UTIs were the cause of 9.6 million physician visits [2] and were noted in 1.5 million hospital discharges [3]. These frequencies placed UTIs first among kidney and urologic diseases in terms of total cost, beyond that of chronic renal failure with its attendant dialysis and transplant expenses.

E. coli is the predominant cause of uncomplicated cystitis and acute pyelonephritis. Of women going to physicians for a UTI, 95% do so for symptoms of cystitis [4]. It is estimated that 40% of adult women will suffer symptoms of cystitis during their lifetimes, and *E. coli* will be identified as the etiologic agent in 75–80% of these women. Although less frequent, the serious clinical syndrome of acute pyelonephritis is also caused most commonly by strains of *E. coli* [5].

Escherichia coli: Virulence Mechanisms of a Versatile Pathogen
ISBN 0-12-220751-3

Uncomplicated UTIs include cystitis infections in adult women who are not pregnant and do not suffer from structural or neurologic dysfunction [6]. Cystitis is a clinical diagnosis presumed to represent an infection of the bladder. Cystitis often is defined by the presence of 10^3 bacteria per milliliter or more in a midstream clean-catch urine sample from a patient with symptoms including dysuria, urinary urgency, and frequency [7,8].

The most serious UTI is acute pyelonephritis. This infection in one or both kidneys usually results from ascent of organisms from the bladder via the ureter and is distinguished from other UTIs clinically, pathologically, and by characteristics of the causative organisms [9]. The patient with acute pyelonephritis classically presents with the triad of fever, flank pain, and bacteriuria with or without diaphoresis, rigors, abdominal or groin pain, and nausea and vomiting. Laboratory studies show leukocytosis, bacteriuria, pyuria, diminished renal concentrating ability, and elevated C-reactive protein [1]. On pathologic examination, wedge-shaped areas of inflammation containing predominantly polymorphonuclear leukocytes (PMNs) extend from papillae to cortex, tubules are filled with PMNs, and necrosis of proximal tubular epithelial cells is evident. Glomeruli tend to be spared, even in areas of intense inflammation [10]. Localized inflammation may coalesce to form a renal abscess, and perinephric abscess and pyonephrosis may result, particularly in the presence of urinary stones or outflow obstruction [11]. The infection may spread beyond the urinary tract if bacteria enter the bloodstream directly. Indeed, 12% of patients with pyleonephritis have bacteremia, which can accentuate the severity of pyelonephritis infections and, if left untreated, may be lethal [7,12].

Studies suggest that up to 95% of all UTIs develop by an ascending route of infection [6], meaning that infection begins by colonizing the periurethral area, followed by an upward progression to infect the bladder and, in some cases, continued progression of the bacteria through the ureters to infect the kidneys if conditions of infection allow. The *E. coli* that colonize the periurethral area arise from the host's intestinal flora. Such colonizing bacteria, which normally are confined to the host's colon, thus have access to a new environment [6].

UPEC AS A DEFINED GROUP OF STRAINS

The subset of *E. coli* causing UTIs is thought to be distinct from the commensal *E. coli* strains that comprise a majority of the *E. coli* populating the lower colon of humans. *E. coli* strains that cause UTI are termed *uropathogenic E. coli* (UPEC) and may be considered a subgroup of extraintestinal pathogenic *E. coli* (ExPEC). Generally, UPEC strains differ from commensal *E. coli* strains in that they possess extra genetic material that codes for gene products that may contribute to bacterial pathogenesis. UPEC express a number of factors that are proposed to play roles in the process of causing disease. These factors include

Factor	**Epidemiology**[1]	**Attenuated Mutant**[2]	**Complementation**[3]	**Volunteers**[4]
Type-1 Fimbriae	✓	✓	✓	
P Fimbriae	✓	✓		✓
Dr adhesins[5]	✓	✓	✓	
TonB	✓	✓	✓	
Hemolysin	✓			
CNF	✓			
Capsule				
LPS O-Antigen	✓	✓		
proP, guaA, argC		✓		

[1]Gene or expression more frequent in UPEC than control strains.
[2]Mutant in gene less able to cause UTI in animal model.
[3]Ability to cause UTI restored to mutant by re-introduction of gene.
[4]Evidence of role in UTI from experiments in humans.
[5]Attenuation and complementation confirmed in murine model of chronic pyelonephritis.

FIGURE 1 Virulence determinants of UPEC. Features of *E. coli* that have been implicated as contributing to virulence in the urinary tract are shown. These factors have been demonstrated experimentally to contribute to virulence in animal models of infection or have been associated epidemiologically with strains cultured from the urine of patients with cystitis or pyelonephritis. Such phenotypes are found significantly less frequently in fecal strains.

hemolysins, membrane-bound and secreted proteins, lipopolysaccharides, capsule, iron-acquisition systems, and fimbriae (Fig. 1).

Virulent strains of *E. coli* that cause these infections are capable of causing outbreaks of UTI and display specific phenotypic traits. Such isolates typically carry large blocks of genes, called *pathogenicity-associated islands* (PAIs), not found in fecal isolates. Thus UPEC can be defined as a virulent group of related strains that carry a set of genes beyond that of fecal *E. coli* that allow them to infect the urinary tract of an immunocompetent host.

Molecular Epidemiology

Certain *E. coli* phenotypes have been found more frequently in the urine of patients with acute pyelonephritis than in patients with cystitis or asymptomatic bacteriuria or in feces of normal individuals (see Fig. 1). *E. coli* from a small number of O-serogroups (six O-groups cause three-fourths of UTIs) have phenotypes epidemiologically associated with acute pyelonephritis in the normal urinary tract that include expression of P fimbriae, hemolysin, aerobactin, serum resistance, and encapsulation. For example, about 80% of *E. coli* strains from patients with pyelonephritis produce P fimbriae, a prevalence that is six times that of fecal strains. These observations have been interpreted as evidence for the existence of a set of virulence factors that allow specific clonally related (uropathogenic) strains of *E. coli* to cause pyelonephritis. Therefore, from studying the phenotypes of virulent uropathogenic strains and commensal fecal strains, it generally has been concluded that *E. coli* strains causing cystitis and acute pyelonephritis have a distinct genetic profile beyond the "base model" fecal strain represented by *E. coli* K-12.

Urovirulent Clones

Some of the most persuasive evidence that there exist *E. coli* strains that are particularly suited to infect the urinary tract comes from reports of *E. coli* clones that have caused either UTI outbreaks or geographically dispersed UTIs. In the 1980s, there occurred a sudden increase in community-acquired *E. coli* urosepsis in southeastern London. A single clone of *E. coli* serotype O15:K52:H1 was responsible for the excess cases [13]. More recently, a clone of *E. coli*, many members of which are resistant to trimethoprim-sulfamethoxasole, was found to cause a significant proportion of episodes of cystitis in young women in California, Michigan, and Minnisota. This clone appears to be related to the one that caused the outbreak in London [14]. An epidemic of pyelonephritis among infants in Sweden was traced to nosocomial acquisition during prior hospitalization of a serotype O6:H5 strain also found in the feces of hospital personnel [15]. Analysis of cystitis isolates has shown that closely related strains often cause a high proportion of these infections [16,17]. In addition, a clone of *E. coli* serotype O4:H5 has been isolated from patients with cystitis and urosepsis from diverse geographic areas [18]. These reports provide convincing evidence that all *E. coli* strains do not share the same propensity to cause UTI.

Pathogenicity Islands

Uropathogens synthesize potential virulence factors such as hemolysin, adhesins, iron-acquisition systems, and toxins. Virulence-associated genes that encode these proteins are commonly clustered within regions of DNA termed *pathogenicity islands* (PAIs). PAIs were first defined by Jorg Hacker and colleagues [19] to include DNA regions greater than 30 kb that are associated with pathogenic organisms and that harbor virulence-associated genes not commonly found in the genome of fecal *E. coli*. In addition, PAIs frequently are associated with tRNA genes, have a G + C content different from the host DNA, and include the presence of mobility genes including transposons and insertion elements. All these characteristics suggest that PAIs have been inserted in the chromosome of UPEC via a lateral-transfer mechanism [19].

Multiple PAIs have been described in each of three uropathogenic *E. coli* strains: 536, J96, and CFT073 [19–23]. For the most part, PAIs have been associated with pyelonephritis strains. For example, two PAIs have been identified within the chromosome of *E. coli* CFT073. The first 57,998-bp PAI (PAI I) carries the *pap* and *hly* operons, which encode P fimbriae and hemolysin, respectively [20,23]. PAI I sequences of strain CFT073 were shown by hybridization to be associated significantly more often with both pyelonephritis and cystitis isolates in comparison with fecal isolates [23]. In addition, PAI I has a G + C content of 42.9% (significantly lower than the 50.8% seen in the K12 chromosome), is flanked by 9-bp direct repeats, and contains numerous transposases and insertion elements; these features are indicative of a lateral transfer event. The second PAI of *E. coli* CFT073 (PAI II), exceeding 80 kb, includes a second copy of the *pap* operon. In addition, putative iron-transport genes as well as the gene encoding the secreted autotransporter toxin (Sat) reside within CFT073 PAI II.

A hallmark of PAIs, as seen in *E. coli* CFT073, is the presence of genes encoding hemolysin and/or P fimbriae [19]. These genes frequently are found in strains that colonize the urinary tract. Numerous studies, however, have shown that hemolysin and P fimbriae are not the sole virulence factors involved in the pathogenesis of UPEC. For example, it was shown in our laboratory that neither a double *pap* mutant nor a hemolysin mutant of *E. coli* CFT073 resulted in the attenuation of virulence in a CBA mouse model of UTI [24,25]. In contrast, the spontaneous loss of PAIs I and II of another *E. coli* strain, 536, led not only to the reduction of hemolytic activity and cell binding but also to reduced serum resistance, mannose-resistant hemagglutination, and virulence in mice [21]. These results suggest that other PAI-encoded genes may play a role in the virulence of uropathogenic organisms.

VIRULENCE FACTORS

Introduction and Models

The finding that certain phenotypic traits are expressed more commonly by strains isolated from the urine of patients with UTI than from the feces of healthy individuals suggests that these factors may be involved in the pathogenesis of UTI. However, these factors merely may be genetically linked to true virulence determinants and irrelevant to UTI. One of the best-accepted methods to distinguish between these alternatives is to create a strain that has a specific mutation in a gene encoding a putative virulence factor and test this strain, along with the wild-type strain from which it was derived and the mutant strain complemented with the wild-type copy of the gene, in a suitable model of infection. Factors that are proven to play a role in infection by fulfilling these so-called molecular Koch's postulates are considered to be *bona fide* virulence factors [26].

Several models have been used to study uropathogens [27]. Arguably the most relevant is that which employs the adult volunteer. Several studies have used individuals, usually those who already suffer from recurrent UTI, to test various aspects of the pathogenesis of UTI [28,29]. Other studies have used nonhuman primates to study UTI pathogenesis, arguing that these animals share many of the same putative adhesin receptors with humans [30,31]. The murine model of ascending UTI offers a much less costly alternative to primates, which has additional advantages. First, the urinary tract of mice has many of the same adhesin receptors as that of humans (reviewed in ref. 27). Second, mice are easily manipulated and, if carefully inoculated, can be infected by the ascending route by direct transurethral inoculation without causing reflux to the kidneys. Third and most important, the murine model of ascending UTI can accurately distinguish UPEC strains from control strains and thus can be used to identify the virulence factors that allow the former to cause UTI [24,32]. Recently, a murine model was shown to distinguish between strains isolated from patients with cystitis and strains isolated from patients with pyelonephritis, offering promise that factors required for tissue tropism within the urinary tract also may be identified using this model [33].

Adhesins

The current dogma of bacterial pathogenesis identifies adherence, colonization, avoidance of host defenses, and damage to host tissues as events vital for achieving bacterial virulence [34]. As the first step in this process, adherence plays a prominent role. Bacteria employ many different surface structures by which they bind to targets on the host cell surface. Fimbriae are common elements used by bacteria to adhere. The words *fimbriae* (derived from the Latin term meaning "threads" or "fringe") and *pili* (Latin for "hair") are used to describe these organelles. Fimbriae project from and are distributed over the entire surface of the bacteria, giving the impression that the bacteria have a fringe of threads or hair covering their exterior.

UPEC strains have been shown to express a number of different fimbriae, including P, F1C, S, M, Dr, and type 1 fimbriae [35–39]. The fimbriae bind different host receptors and subsequently are believed to target different areas within the urinary tract [35,38]. Originally, fimbriae were subdivided into two different groups based on a phenotypic assay for receptor specificity and categorized by their ability to hemagglutinate erythrocytes isolated from a variety of sources, including humans, sheep, and guinea pigs [40–42]. The ability of some fimbrial types to hemagglutinate erythrocytes was blocked by the presence of mannose, whereas other fimbriae remained capable of hemagglutination in the presence of mannose [41,43,44]. Therefore, fimbriae can be grouped as mannose-sensitive (MSHA) or mannose-resistant (MRHA) based on the ability to hemagglutinate erythrocytes. P fimbriae, S fimbriae, F1C fimbriae, and Dr fimbriae all demonstrate mannose-resistant phenotype of hemagglutination of erythrocytes, whereas type 1 fimbriae do not hemagglutinate erythrocytes in the presence of mannose and therefore are defined as mannose-sensitive [38].

Type 1

Type 1 fimbriae are mannose-binding hairlike projections expressed on the surface of *E. coli* and other members of the *Enterobacteriaceae*. The fimbriae function as a virulence factor in the pathogenesis of *E. coli* UTI. However, the genes responsible for type 1 fimbriae are found in almost all subgroups of *E. coli*, not just in UPEC strains. The ubiquitous presence of the type 1 fimbrial genes may be explained by evidence that suggests that the fimbriae are required for colonization of the oropharynx as a prelude to intestinal colonization [45]. Type 1 fimbrial expression is primarily under the control of an invertible element that contains the promoter responsible for transcription of the major structural subunit. Cystitis and pyelonephritis isolates have been observed to differ in the control of the invertible element orientation and the expression of type 1 fimbriae during the course of an infection. Recent research has focused on the role of type 1 fimbriae in the invasion of host cells and host immune responses, the potential crosstalk between the contol of type 1 fimbriae and other fimbriae types, and the development of a vaccine directed against type 1 fimbriae in UTIs.

Brinton and colleagues [40] originally described type 1 fimbria on the basis of mannose-sensitive agglutination of erythrocytes. These fimbriae are, on average, 0.5–2 μm in length and have a diameter of 7 nm and a center axial hole 0.2–0.25 nm in diameter [46]. At the distal end of the fimbria, a fibrillar structure projects roughly an additional 16 nm in length that contains the adhesin responsible for binding and for hemagglutination.

Type 1 fimbriae are encoded by an operon consisting of nine separate genes present on the chromosome in the following order: *fimB*, *fimE*, *fimA*, *fimI*, *fimC*, *fimD*, *fimF*, and *fimH* [47–49] (Fig. 2). The operon is transcribed as at least three separate transcripts, one for encoding the main structural subunit of the fimbriae (*fimA*) and two others responsible for the adhesin (*fimH*) and accessory proteins necessary for assembly of the fimbriae [38].

The FimA polypeptides form the shaft of the fimbriae by polymerizing into a helix formation with $3\frac{1}{8}$ FimA proteins constituting one full turn [46]. The FimH protein is the adhesin of the fimbriae [50–53]. The type 1 fimbriae bind, by means of the FimH adhesin, to mannosides that are widely distributed on epithelial surfaces of humans, including uroplakin receptors that coat transitional epithelial cells of the bladder [54]. The collective products of the *fimG* and *fimF* genes with the FimH adhesin combine to form the fimbrillar tip found at the end of the fimbriae [55]. FimC is a periplasmic chaperone protein with an

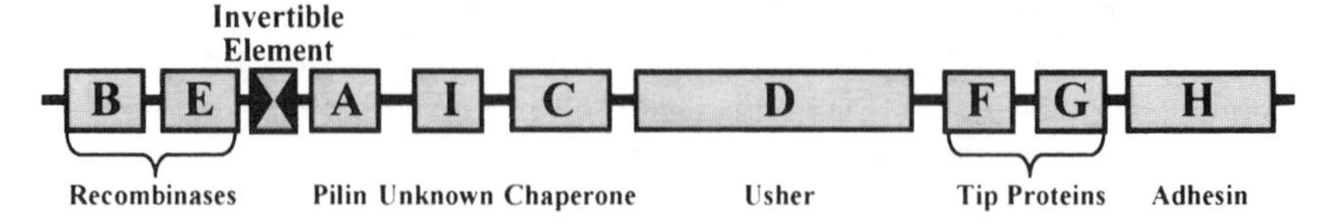

FIGURE 2 The type 1 fimbriae operon. Nine genes contribute to the expression of the type 1 fimbriae and the invertible element region responsible for control of fimbriae expression. Of note, *fimB* and *fimE* encode recombinases that switch the orientation of the invertible element from on to off and from off to on. *fimA* encodes the major structural pilin that comprises the majority of the fimbrial structure. The mannose-binding adhesin is encoded by *fimH*.

immunoglobulin-like fold responsible for stabilizing the fimbrial components as they are transported to the site of fimbriae construction on the outer membrane [56]. This chaperone prevents premature interactions between fimbrial subunits. The FimD protein has been shown to function as an outer membrane usher that recognizes the chaperone–fimbrial subunit complexes and assists in localizing them to the proper area for fimbriae construction [57]. Further details of the chaperone-usher pathway of fimbrial assembly are provided elsewhere (see Chap. 10). FimB and FimE are recombinases that control the expression of type 1 fimbriae by acting on a specific region of DNA in the fimbrial operon that is capable of being inverted into two different orientations by the recombinases [58]. A type 1 fimbriated bacterium typically expresses approximately 500 fimbriae per cell, which accounts for about 8% of the total protein content of the bacterium at any given time.

Expression of type 1 fimbriae undergoes phase variation. The individual bacterial cells either express the fimbriae over their entire surface or do not express any fimbriae at all. This phase variability of type 1 fimbriae is controlled at the transcriptional level by the previously mentioned invertible stretch of DNA that is acted on by the FimB and FimE recombinases [59]. The σ70 promoter, responsible for transcription of the major fimbrial subunit FimA, is located within this invertible DNA element [60,61]. The 314-bp invertible element is flanked on both ends by inverted DNA repeats of 9 bp in length; these repeat sequences are essential for proper inversion [62]. The FimB and FimE recombinases also have been shown to bind to specific sites on the DNA that flank either side of the inverted repeats.

The orientation of the invertible element and subsequent expression of type 1 fimbriae are believed to be controlled by a variety of environmental factors. Environmental conditions can affect the orientation of the invertible element by acting via the FimB and FimE recombinases, by affecting one of the many elements that the recombinases require for normal switching, or by selecting for bacteria that express or fail to express the fimbriae. The invertible element in the on orientation and subsequent expression of type 1 fimbriae occur in a majority of strains when *E. coli* are grown statically in nutrient broth. Alternatively,

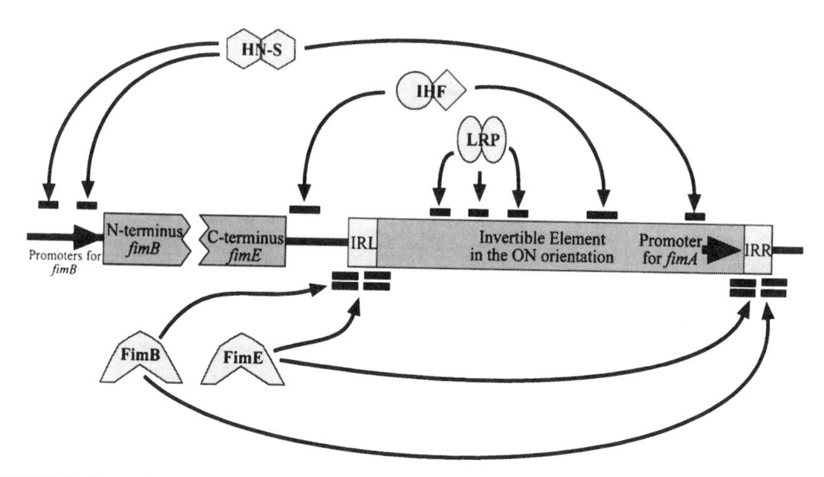

FIGURE 3 Regulatory mechanisms that influence type 1 fimbriae expression. The molecular elements that affect switching of type 1 fimbriae's invertible elements and their known binding sites are depicted. (LRP, leucine-responsive protein; IHF, integration host factor; HN-S, histone-like nucleoid structuring protein; IRR, inverted repeat right; IRL, inverted repeat left; see text for discussion of these elements.)

when the bacteria are cultured on solid nutrient agar, the invertible elements are in the off orientation, and type 1 fimbriae are not expressed [63,64].

The switching of the invertible element and subsequent expression of the type 1 fimbriae ultimately are achieved through the actions of the FimB and FimE recombinases. However, different molecular elements influence the two recombinases and contribute to the dynamic switching behavior demonstrated by *E. coli*, especially UPEC strains (Fig. 3). Three molecular factors, leucine-responsive protein (LRP), integration host factor (IHF), and the histone-like protein H-NS, have been shown to affect the switching of the invertible element and type 1 fimbrial expression by binding to DNA sequences around and within the invertible element region, thus assisting or blocking the switching actions of the recombinases [65].

Type 1 fimbriae are a proven virulence factor for *E. coli* in the urinary tract. In a series of experiments conducted by Connell and colleagues [66], molecular Koch's postulates were satisfied, defining a role for type 1 fimbriae as a virulence factor in *E. coli* infecting the urinary tract. In this study, an *E. coli* strain was selected that was isolated previously from a patient with a UTI and was shown to be capable of causing infections in the murine model of UTI. An isogenic mutation was introduced into this strain within the *fimH* adhesin gene. The *fimH* mutant was found to infect the mouse model in significantly lower numbers (1–3 logs less) when compared with mice similarly infected with the wild-type parent strain. Finally, the *fimH* defect was complemented in the mutant strain by placing a functional copy of the *fimH* gene on a plasmid and transforming the mutant with this plasmid construct. The complemented strain was found to infect the mouse urinary tract in numbers determined to be statistically similar to the original wild-type strain, thus demonstrating successful complementation of the former defect in type 1 fimbriae.

Recently, a novel role has been proposed for type 1 fimbriae in the pathogenesis of UTI. Mulvey and colleagues [67] used high-resolution electron microscopy to demonstrate binding of the tips of type 1 fimbriae to the hexagonal arrays of the bladder cell surface glycoprotein uroplakin *in vivo*. They also noted apoptosis and loss of surface umbrella cells, whereas the bacteria seemed to resist being shed by invading into underlying cells. They proposed that UPEC use type 1 pili to attach to the bladder mucosa, induce apoptosis of host cells, and facilitate invasion and colonization of the damaged tissue.

A quantitative assay has been developed to measure the orientation of the type 1 fimbriae invertible element under *in vivo* conditions [68]. The expression of type 1 fimbriae by *E. coli* can be predicted by determining the percentage of invertible elements in the on orientation per sample. The percentage correlates with the number of bacteria capable of expressing type 1 fimbriae. The assay was used originally to measure the orientation of the switch of fecal, cystitis, and pyelonephritis isolates under various *in vitro* growth conditions that selected for or against the invertible element switching into the on orientation [69]. In these experiments, it was observed that fecal and cystitis isolates were, under defined conditions, able to switch a greater percentage of their invertible elements into the on orientation when compared with pyelonephritis isolates. The differences observed in the invertible element orientation between cystitis and pyelonephritis isolates under *in vitro* conditions suggested that the two groups of isolates may use type 1 fimbriae differently during *in vivo* infections.

Indeed, when representative strains isolated from patients with pyelonephritis and cystitis were compared as to the orientation of their invertible elements during experimental infections of CBA mice, significant differences were observed [68]. Cystitis isolate F11, expelled in the urine of mice, had a larger percentage (median value 84.5%) of its invertible elements switched to the on orientation 24 hours after inoculation of the bladder when compared with pyelonephritis strain CFT073 (median value 33.6%) at the same time point. The infecting population of strain F11 also maintained a majority of its invertible elements in the on orientation as far out as 96 hours after bladder inoculation, whereas the infecting population of strain CFT073 had the vast majority of its invertible elements in the off orientation 48 hours after infection. When these observations were expanded to include three additional cystitis strains and three additional pyelonephritis strains, the switch position of the respective isolates continued to differ significantly 24 hours after experimental inoculation of the bladder. The median on value for cystitis strains equaled 81%, whereas the median on value for pyelonephritis strains equaled only 2%. These data suggest that cystitis and pyelonephritis isolates control the expression of type 1 fimbriae differently during the course of an infection.

A vaccine using the type 1 fimbrial adhesin protein (FimH), which binds to mannosylated receptors in the urinary tract, has been developed by Soloman Langermann and colleagues (MedImmune, Inc., Gaithersburg, Maryland) for protection against the development of UTI [70]. Mice vaccinated with the fimbrial adhesin had significantly lower concentrations of bacteria in the bladder compared with control mice after challenge with *E. coli*. Similar results were obtained in a primate model, although the numbers of animals used precluded meaningful statistical analysis [71]. Phase 1 clinical trials in humans are underway.

P Fimbriae

P fimbria was the first virulence factor described for UPEC. In 1976, Svanborg-Eden and Hansson discovered that *E. coli* isolated from patients with pyelonephritis adhered in higher numbers to exfoliated uroepithelial cells than did fecal strains of *E. coli* [72]. This adherence soon was associated with the presence of fimbriae [73], which when isolated also could specifically adhere to the surface of uroepithelial cells [74]. Bacteria expressing this fimbria could agglutinate human type O erythrocytes, and the hemagglutination was not inhibitable by mannose (i.e., MRHA); this pattern of agglutination differentiated this new fimbrial type from type 1 fimbriae, which display mannose-sensitive hemagglutination. Indeed, Kallenius and colleagues [75] used erythrocytes to identify the eucaryotic receptor for P fimbriae. They identified the very common P blood group antigen (present in over 99.9% of the world's population) as the receptor for P fimbriae. The antigen was found to be a glycosphingolipid with a lipid moiety anchored in the cell membrane and a carbohydrate chain exposed on the surface of the erythrocytes. The key epitope turned out to be a digalactoside, α-D-Gal-(1 → 4)-β-D-Gal. This digalactoside itself was capable of inhibiting P fimbrial attachment to its receptor.

An understanding of fimbrial structure and biogenesis was advanced by the cloning of the genes encoding P fimbriae by Hull and colleagues [76]. A cosmid clone carrying the *pap* (pyelonephritis-associated pili) operon conferred the ability to produce P fimbriae on a laboratory strain of *E. coli*. Sequencing of the smallest DNA fragment capable of encoding fimbriae revealed 11 genes. Normark and Hultgren advanced these studies to elucidate a model of fimbrial biogenesis that is currently applicable to a wide range of enterobacterial fimbriae [77–85]. Details of the chaperone-usher pathway can be found in Chap. 10.

Following the initial correlation of the presence of P fimbriae and virulent uropathogenic strains, a number of investigators confirmed and extended this work [38,86–95]. In an analysis of published, controlled studies, Donnenberg and Welch reported that *E. coli* from otherwise healthy patients with pyelonephritis are six times as likely to possess P fimbriae as strains from the feces of controls [96]. Approximately 80% of such strains have P fimbriae. There is also a strong relationship between the severity of the infection and the prevalence of P fimbriae. Studies of humans with UTI demonstrating expression of P fimbriae by *E. coli* in urine [97] and antibodies to P fimbriae in serum [98] indicate that *E. coli* produce and display P fimbriae *in vivo* in infected patients.

The expression of P fimbriae by pyelonephritogenic strains was long thought to explain the virulence of these strains in the urinary tract [91,99]. The adherence phenotype conferred by this cell surface structure logically explained the affinity of these strains for uroepithelium. Therefore, intense investigation of the role of P fimbriae in the pathogenesis of acute pyelonephritis was undertaken. Although the genetics of the *pap* operon are now well understood [78,81,100,101], and numerous studies have implicated P fimbriae and hemolysin in the pathogenesis of acute pyelonephritis [38,96], only two reports have appeared in which true isogenic P fimbrial mutants of virulent clinical isolates have been constructed for testing in animal models of infection [25,102]. In a study by Mobley and colleagues [25], CBA mice (N = 100) were challenged transurethrally with 10^5, 10^6, 10^7, or 10^9 colony-forming units (cfu) of strains CFT073 (parent strain) and UPEC76 (P fimbria–negative mutant). After 1 week, no substantive differences in organism concentration or histologic findings between parent and mutant were detected in urine, bladder, or kidney at any challenge concentration. It was concluded that adherence by P fimbriae of uropathogenic *E. coli* strain CFT073 plays at best a subtle role in the development of acute pyelonephritis in the CBA mouse model. Roberts and colleagues [102] studied the effect of a *pap*G mutation (does not bind the digalactoside receptor) on colonization of the bladder and on the development of pyelonephritis in cynomolgus monkeys. While they found no difference in colonization between the wild-type and mutant strain after bladder inoculation, they found that the wild-type strain colonized the urinary tract for significantly longer periods than the mutant strain after instillation into the ureter. Furthermore, there was evidence of loss of kidney function and mass on the inoculated side only in recipients of the wild-type strain [102]. Although no mutant complemented with a copy of the wild-type allele was tested, this study provides the best evidence available that P fimbriae contribute to the pathogenesis of pyelonephritis. However, the fact that the recipients of the mutant strain also had renal colonization indicates that these organelles are not indispensable for infection. Clearly, P fimbrial genes (*pap*) are preferentially found in urovirulent strains, may serve as a marker for the presence of additional genes that confer virulence, and themselves contribute to virulence.

Strains that carry P fimbriae can be categorized by which of three PapG adhesins are expressed at the tip of the fimbria. PapG I was the first for which the gene was cloned but is not common in human isolates, PapG II is the most common of the adhesins, and PapG III is found in some cystitis isolates. The three adhesins recognize slightly different portions of the Gal-Gal disaccharide-containing glycosphingolipids [103].

Other Fimbriae

Dr Adhesin

The Dr adhesin family is composed of fimbiral and afimbrial structures on the surface of *E. coli* that bind to the Dr blood group antigen [104], a portion of the decay-accelerationg factor, which is a membrane protein that prevents cell lysis by complement [105,106]. Within the urinary tract, Dr adhesins bind to bladder epithelium and type IV collagen on basement membranes [107]. At present, the role that Dr adhesins play in the pathogenesis of UTI is unclear. These adhesins are present in a minority of UTI strains; however, pooled data indicate that the genes for members of the Dr family are more prevalent among pyelonephritis and cystitis strains than among fecal control strains of *E. coli* [96]. Some other data are intriguing: Intravenously administered purified Dr adhesins bound for months to the mesangium of rat kidneys [108], and a Dr-positive *E. coli* strain but not its Dr-negative isogenic mutant caused a disease pathologically similar to chronic tubulointerstitial nephritis in a mouse model of ascending UTI [109]. This difference was reversed on complementation of the mutant with the cloned gene.

F1C Fimbriae and S Fimbriae Family

A family of *E. coli* fimbriae that includes S and F1C fimbriae has highly related biogenesis genes but different adhesin alleles. The fimbriae have been linked to UTI and other extraintestinal infections, particularly neonatal meningitis. Although the epidemiology of the fimbriae remains poorly studied, pooled data from several reports indicate that F1C fimbriae in particular are more common among pyelonephritis and cystitis strains than among fecal control strains of *E. coli* [96]. F1C fimbriae are found commonly in the O-antigen serotypes that are overrepresented among pyelonephritis isolates and in strains with P fimbriae and hemolysin [110–112]. F1C fimbriae adhere to human distal tubular and collecting tubular epithelium and vascular endothelium on kidney cross sections [113]. They are expressed *in vivo*, as determined by immunofluorescence of organisms in the urine of patients with UTIs [114]. They do not, however, agglutinate erythrocytes, and identification of the F1C receptor has been delayed. Results of experiments using true isogenic mutants in an attempt to determine the role of these in UTI have not been reported.

Toxins

UPEC produce three proteins that may be regarded as toxins: hemolysin, cytotoxic necrotizing factor 1 (CNF1), and the recently discovered secreted autotransporter toxin (Sat).

Hemolysin

E. coli hemolysin is the prototypic member of the family of RTX (repeats-in-toxin) toxins. The *hly* operon (*hlyCABD*) is often located adjacent to the P fimbrial genes on the same pathogenicity island on the chromosome of UPEC strains [22,23,115,116]. Therefore, it is not surprinsing that UPEC are more likely to have the *hly* genes than are fecal strains [97]. For example, Dudgeon and colleagues [117] reported as early as 1921 that 50% of UTI *E. coli* isolates were hemolytic compared with only 13% of fecal isolates.

The *hly* genes encode the proteins required for synthesis and secretion of hemolysin [118]. The *hlyA* gene encodes the prohemolysin protein. Generation of the mature and active form of the toxin is followed by fatty acylation of the prohemolysin protein by HlyC and an acyl carrier protein [119,120]. Hemolysin is secreted to the extracellular milieu by type I secretion system. The export system consists of an ATP-binding cytoplasmic membrane protein, HlyB; a cytoplasmic membrane protein with a large periplasmic domain, HlyD; and an outer membrane protein not encoded within the hemolysin operon, TolC [121]. Thanabalu and colleagues [122] have shown that HlyB and HlyD form a complex to which hemolysin binds. Hemolysin binding then triggers a comformational change that brings in the TolC that forms a channel through the outer membrane. Only when the hemolysin is in does the channel open. Recently, the crystal structure of TolC revealed that the protein is a unique trimer that spans both the outer membrane and the periplasmc space to form the channel [123].

Expression of the hemolysin operon is positively regulated by RfaH, which is also a positive regulator of lipopolysaccharide synthesis [124]. Inactivation of the *rfaH* gene resulted in a 10-fold decrease in the hemolytic activity of UPEC strain 536 [125]. The PAI-associated tRNA gene *leuX*, which encodes leucyl-tRNA, also has been shown to modulate the expression of hemolysin in strain 536 [126].

Once secreted, hemolysin requires calcium to assume its functional tertiary structure and to bind to the cell membrane [127–129]. Hemolysin inserts into the outer leaflet of the host cell membrane rather than forming a pore through the membrane, as originally proposed [130]. The receptor for the toxin is as yet unidentified.

At high doses, hemolysin is cytotoxic not only to erythrocytes but also to a variety of nucleated cell types, including leukocytes [131], fibroblasts [132], and more relevant to UTI, uroepithelial cells [24,133]. Hemolysin is also suggested to play a role in invasion of the renal parenchyma by destruction of the epithelial barrier. In a study by Trifillis and colleagues [134], a hemolytic strain penetrated a monolayer of proximal renal tubular cells in higher numbers than a nonhemolytic mutant of the same strain. Sublytic concentrations of hemolysin have been shown to induce oscillations of the intracellular concentration of Ca^{2+} in renal epithelial cells. This oscillation stimulated production of interleukin 6 (IL-6) and IL-8 [135]. *In vivo*, hemolysin can stimulate production of cytokines by this mechanism, which can lead to inflammatory responses. Antihemolysin antibodies increase during UTI, indicating that hemolysin is expressed *in vivo* [136].

The genetic evidence for the role of hemolysin in UTI rests on studies that did not use true isogenic mutant strains. In one such study, the addition of the *hly* operon on a plasmid rendered a nonhemolytic UPEC able to cause more kidney damage and deaths than the parental strain in a murine model of ascending UTI [91]. In another study, reintroduction of the *hly* operon on a high-copy-number plasmid into a strain deleted of both known pathogenicity islands contributed to restoration of virulence in a rat model of ascending UTI [137]. In a more recent study, deletion of both pathogenicity islands of UPEC strain 536 carrying hemolysin genes resulted in complete loss of virulence in the kidneys of intravenously challenged mice [138]. Since deletion of two pathogenicity islands eliminates many other genes besides hemolysin genes, the results of this study are difficult to interpret. Therefore, the role of hemolysin in UTI remains unproven until the true isogenic mutants of *hlyA* are established. More details on hemolysin and type I secretion can be found in Chap. 13.

Cytotoxic Necrotizing Factor 1 (CNF1)

CNF1 is found commonly in strains of *E. coli* isolated from patients with UTI but nearly always in association with hemolysin, with which it is genetically linked [116,138]. CNF1 is a 110-kDa protein that is encoded by a single gene [139]. CNF1 induces the formation of actin stress fibers and membrane ruffling in HEp-2 cells [140]. The cytoskeleton rearrangement results from constitutive activation of members of the Rho family of small GTP-binding proteins due to deamidation of glutamine 63 of Rho protein into glutamic acid by CNF1 [141,142].

Numerous studies suggest a potential role for CNF1 in the pathogenesis of UPEC. Treatment of HEp-2 cells with CNF1 enabled them to internalize latex beads and noninvasive bacteria [143]. In addition, CNF1 induced effacement of intestinal microvilli, reduced PMN transmigration in intestinal T84 epithelial cells [144], and increased permeability in polarized intestinal cell monolayers [145]. In a recent study, CNF1 was shown to increase adherence of PMNs to T84 monolayers and decrease their phagocytic effect [146]. CNF1 also causes apoptosis in the 5637 bladder cell line, a phenomenon that might explain the exfoliation of bladder epithelial cells after infection with UPEC [147]. Although these effects together suggest that CNF1 may play a role in UTI, an isogenic mutant of *cnf1* was no different from its wild-type parent UPEC strain CFT073 in its ability to colonize or cause inflammation in a murine model of UTI, and thus the role of CNF1 in UTI remains unconfirmed [148].

Secreted Autotransporter Toxin (Sat)

Another mechanism by which uropathogens may promote infection is by targeting proteins to extracytoplasmic sites using secretion systems. Most recently, putative virulence determinants have been identified that are secreted via the autotransporter system. Studies on a number of autotransporters [149] have demonstrated that translocation across the inner membrane occurs via the *sec*-dependent pathway and across the outer membrane through a β-barrel porin structure formed by the C-terminal autotransporter domain. Once presented to the outside environment, the passenger domain may remain attached to the outer membrane or be released by proteolytic cleavage [149].

A 107-kDa secreted protein, designated *Sat* (secreted autotransporter toxin), is expressed significantly more often by *E. coli* strains associated with the clinical syndromes of acute pyelonephritis (86% of strains) than by fecal strains (14% of strains) [150]. The polypeptide, isolated from *E. coli* CFT073, shares highest similarity with the subcategory of autotransporters termed SPATE (serine protease autotransporters of *Enterobacteriaceae*) proteins, which are produced by diarrheagenic *E. coli* and *Shigella* spp. isolates [149]. The native Sat protein includes the three characteristic domains of SPATE proteins: an unusually long signal sequence, a secreted passenger domain (the mature protein) to which the phenotypes of such proteins are attributed, and an autotransporter C-terminal domain. The mature Sat protein was shown to have a cytopathic effect on various kidney and bladder cell lines and to elicit an immune response by mice infected with *E. coli* CFT073 (Guyer, Radulovich, and Mobley, unpublished observations), evidence that may suggest a role of Sat in bacterial pathogenesis. In addition, the *sat* gene was shown to reside within PAI II of *E. coli* CFT073, indicating Sat as another possible PAI-encoded virulence determinant. To date, Sat is the first autotransporter protein to be associated preferentially with uropathogenic *E. coli*.

Extracellular Polysaccharides

E. coli may produce a variety of extracellular polysaccharides, including the O-antigen and core polysaccharides of LPS, colanic acid or group I capsule, group II capsules, and eubacterial common antigen (see Chap. 14 for more details). The potential roles of these surface molecules in the pathogenesis of UTI and other infections are not clear. It has been proposed that certain of these surface carbohydrates may help the bacteria resist phagocytosis, survive in human serum, or adhere to cells. The epidemiologic data concerning the occurrence of capsule in strains isolated from patients with UTIs is quite confusing and contradictory. Some studies have suggested that certain capsular types are overrepresented in strain isolated from UTIs. However, our review of the published data did not confirm that certain capsules are more common in UPEC than in other *E. coli*, nor did it confirm that such strains were more likely to be resistant to human serum [96].

The importance of capsule and lipopolysaccharide O sidechains in UPEC pathogenesis has been controversial. Russo and colleagues [151,152] found no difference in the ability of a wild-type O4:K54:H5 blood isolate and an isogenic mutant unable to synthesize K54 capsule to cause UTI in a mouse model. In addition, there is no experimental evidence to support the hypothesis that having a particular O-antigen type per se increases the pathogenic potential for an *E. coli* strain. However, an isogenic mutant of an O4:K54:H5 strain that was no longer able to synthesize the O4-antigen was significantly more impaired in its ability to colonize the mouse bladder, kidney, and urine than was the wild-type strain from which it was derived [152]. This suggests that the inability to produce any O-antigen may be detrimental to *E. coli* in the urinary tract. Schilling and colleagues [153] also studied the role of LPS in the pathogenesis of UPEC. These authors found that UPEC LPS is required for invasion of the bladder epithelial cells, which leads to production of IL-6 both in cultured bladder epithelial cells and in mice.

Lipopolysaccharide and capsular carbohydrates seem to play a role in the ability of certain strains of *E. coli* to grow in the presence of human serum. Cross and colleagues [154], demonstrated that certain O-antigens confer resistance to human serum, but for some of these O-antigens, the presence of the capsule also was required. By using a defined LPS mutant and an acapsular mutant of UPEC 075:K5, Burns and Hall [155] showed that LPS was more important for serum resistance than the capsule and that the length of the O-antigen is crucial for exerting such effect. Capsules are also known to play a role in resistance to phagocytosis in the absence of anticapsular opsonizing antibody [156]. Indeed, in a study by Burns and Hull [157], both O- and K-antigens were found to play roles in resistance to phagocytosis by UPEC O75:K5.

In a recent large-scale screen of signature-tagged transposon (STM) mutants of UPEC strain CFT073 in a murine model of UTI, several survival-attenuated mutants were found to have insertions in genes involved in the biogenesis of various extracellular polysaccharides [158]. Three mutants that were deficient in colonization of the murine urinary tract had disruptions in loci linked to capsule production. Each of these mutants did not make detectable capsule when grown *in vivo*. In addition, a *manB* mutant was found to be defective in survival in the bladder but not in the kidneys. Phosphomannomutase, the product of *manB*, is required for the production of colanic acid, the exopolysaccharide associated with group I capsules. Previously, colanic acid had been thought to play no role in the pathogenesis of sepsis by extraintestinal *E. coli* isolates [151]. STM studies also showed one attenuated mutant with an insertion in *rffA*, a gene encoding a transaminase required for the synthesis of enterobacterial common antigen [158]. This mutant was confirmed to be unable to produce this extracellular polysaccharide, ubiquitous among members of the family Eubacteriaceae but heretofore unimplicated in pathogenesis.

Iron Acquisition

Iron, an essential cofactor for enzymes found in all organisms, is sequestered in the human body by a variety of iron-binding proteins, and therefore, bacteria that invade host tissues must have systems for procuring this element. It has long been appreciated that the ability to produce aerobactin, an iron-chelating siderophore, is more common among strains of *E. coli* isolated from the urine of patients with UTIs than among control strains isolated from the feces of healthy individuals [159,160]. Recently, however, definitive genetic evidence of the role of iron acquisition in the pathogenesis of *E. coli* UTI was obtained. Torres and colleagues [161] isolated a spontaneous *tonB* mutant of a virulent UPEC strain. TonB is an inner membrane transport protein required for iron uptake via several systems. These authors found that the *tonB* mutant was significantly attenuated in its ability infect the kidney in a murine model of ascending UTI [161]. Furthermore, they showed that this defect was corrected when the *tonB* gene was restored to the mutant on a plasmid, thereby fulfilling molecular Koch's postulates. To determine which of the several iron-uptake systems was involved, they constructed mutants with defects in aerobactin-, enterobactin-, and heme-mediated iron uptake. They found that a strain with mutations in both the aerobactin and eneterobactin systems was slightly deficient in kidney colonization but that strains with either system alone still intact were not (they did not complement these mutants). Furthermore, using the more sensitive coinfection assay, they found that the wild-type strain could outcompete either a mutant with a defective aerobactin receptor or a mutant with a defective heme receptor when inoculated together in a mixed infection. Thus they suggest that multiple iron-uptake systems work in concert to allow growth of UPEC strains in the urinary tract. Not surprisingly, they did not find a defect in the ability of any of these mutants to grow in human urine, where iron is not limiting.

Growth in Urine

Flushing by urine is a primary defense mechanism of the host against microbial colonization of the urinary system. In addition, the low pH, high osmolality, and high concentration of urea in urine are inhibiting factors for bacterial growth [162]. UPEC strains, however, are able to grow in urine, and this may contribute to the pathogenesis of UTI [163]. Given the high osmolarity of urine, a number of investigators have studied osmoregulatory proteins in UTI isolates. In one such study it was reported that a pyelonephritis isolate had an approximately three-fold higher capacity to transport proline betaine than did *E. coli* K-12. Furthermore, deletion of the *proP* locus encoding the proline betaine transporter resulted in a 100-fold decrease in the ability of a pyelonephritis isolate to colonize the bladder, but the mutation had no effect on kidney colonization [164]. In a screen for mutants of a prototrophic pyelonephritis isolate that are deficient for growth in urine, Hull and Hall [165] found that guanine, arginine, and glutamine auxotrophs were severely defective for growth in urine and that growth in urine was reduced in serine, proline, leucine, phenylalanine, and methionine auxotrophs. In contrast, in an *in vivo* screen of 20 uncharacterized auxotrophic mutants generated by signature-tagged mutagenesis, only a pyrimidine auxotroph was slightly attenuated for survival in mice [158], suggesting that perhaps the close association of the organisms with the urinary tract epithelium provides sufficient nutrients for survival of many UPEC auxotrophic mutants. In another study, Russo and colleagues [166] screened mutants of an *E. coli* blood isolate for diminished growth in urine or for induction of *lacZ* expression in urine. They found that mutations in *guaA* caused the organism to be sensitive to urine. They also found that the *argC* locus was induced in urine and that *argC* mutants exhibited reduced growth in urine. The *guaA* mutant was significantly impaired in its ability to colonize the urine, bladder, and kidneys of mice, whereas the *argC* mutant was impaired in its ability to colonize only the kidneys [166].

Other Potential Virulence Determinants

Advances in genomic and functional genomic studies have facilitated the discovery of new virulence determinants of the uropathogens. A screen of 2000 signature-tagged mutants of UPEC strain CFT073 in a murine model of UTI revealed new uropathogenic genes in addition to the previously described ones [158]. Among these genes was *phoU*, the last gene in the *pst-phoU* operon, which encodes for a phosphate transport system and negatively regulates Pho regulon [167]. Since Pho regulon regulates many other genes, the effect of *phoU* gene mutation on bacterial survival may have been indirect.

A subtractive hybridization strategy also has revealed a number of genes on the chromosome of the prototypic UPEC strain CFT073 that are absent from the genome of *E. coli* K-12, thus representing potential virulence-associated genes [168]. Genes found by this subtractive procedure include iron-utilization genes, genes involved in the cleavage of the glycosidic linkages, bacteriophage-encoded toxin genes, and insertion sequences.

In a more directed approach, pathogenicity islands were identified by examining sequences of strain CFT073 that flanked the two functional copies of the *pap* operon. The first was a 60-kb stretch of DNA that contained 44 open-reading frames (ORFs) including genes encoding P fimbriae, hemolysin, and an iron-acquisition system [20,23]. Nineteen ORFs showed no homology with *E. coli* K-12 sequences. A second PAI was identified that carries more than 72 kb with 89 ORFs including *pap* genes [169]. DNA probes isolated from both these PAIs hybridized significantly more frequently with DNA from pyelonephritis and cystitis isolates than with DNA from fecal strains [20].

In addition, many of the genes identified by these genomic approaches were found to be unknown genes, some of which having a G + C content different from *E. coli*, suggesting an origin in a foreign species. Soon the complete genome of *E. coli* CFT073 will be assembled, and we will know for the first time the number of additional genes, beyond those found in the genome of *E. coli* K-12, that are present in a uropathogenic strain. Based on annotation, it will be possible to identify homologues of known virulence genes. These genes then can be mutated individually and tested for virulence in an appropriate animal model of infection. This may be a monumental task because the uropathogenic strains may contain as much as a megabase of extra DNA located within pathogenicity islands. Nevertheless, the complete genome will allow sequential testing of mutants for virulence.

MODEL

Despite the fact that the complete nucleotide sequence of the first uropathogenic strain is nearing completion and that we do not fully appreciate the role of many ORFs in the development of UTI, much prior research allows us to construct a model of pathogenesis. Infection likely begins with colonization of the bowel with a uropathogenic strain in addition to the commensal flora. This strain, which by virtue of factors encoded within pathogenicity islands is capable of infecting an immunocompetent host, colonizes the periurethral area and ascends the urethra to the bladder. Little is understood about these early steps in UTI pathogenesis, but it is clear that certain risk factors (such as the use of spermicidal gel in the case of periurethral colonization and sexual activity in the case of introduction into the bladder) facilitate these processes. Between 4 and 24 hours later, the new environment in the bladder selects for the expression of type 1 fimbriae, which clearly play a critical role early in the development of a UTI. Type 1 fimbriated *E. coli* attach to mannose moieties of the uroplakin receptors that coat transitional epithelial cells. Attachment triggers apoptosis and exfoliation; for at least one strain, invasion of bladder epithelium has been observed, and it is argued that invaded epithelial cells may act as a reservoir for recurrent infection.

In cystitis strains, type 1 fimbriae are expressed continually, and the infection is confined to the bladder. In pyelonephritis strains, the invertible element that controls type 1 fimbriae expression turns to the off position, and type 1 fimbriae expression ceases. One could argue that this releases the *E. coli* strain from bladder epithelial cell receptors and allows the organism to ascend the ureters to the kidneys where the organism can attach via P fimbriae to digalactoside receptors expressed on the kidney epithelium. At this stage, hemolysin could damage epithelium and together with other bacterial products, including LPS, an acute inflammatory response, recruit PMNs to the site. Secretion of Sat, a vacuolating cytotoxin, damages glomeruli and is cytopathic for surrounding epithelium. In some cases the barrier provided by the one-cell-thick proximal tubules can be breached, and bacteria can penetrate the endothelial cell to enter the bloodstream, leading to bacteremia.

SUMMARY

UPEC represent a heterogeneous group of isolates, restricted to a small number of O-serogroups, with different phenotypes. Although many UTI isolates appear to be clonal, there is clearly not a single phenotypic profile that causes UTI. Therefore, there may be several subclasses of UPEC. Specific adhesins including P fimbriae and type 1 fimbriae appear to aid in colonization. Several toxins are produced, including hemolysin, cytotoxic necrotizing factor, and an autotransported protease, Sat. Uropathogenic strains possess large and small pathogenicity islands containing blocks of genes not found in the chromosome of fecal strains. Completion of the sequencing of the chromosome of *E. coli* CFT073 by Fred Blattner, Rodney Welch, and colleagues (University Wisconsin Medical School, Madison) and efforts by other investigators to identify virulence genes by signature-tagged mutagenesis and other methods should advance the field significantly and allow the development of a comprehensive model of pathogenesis for UPEC.

REFERENCES

1. Kunin, C. M. (1987). *Detection, Prevention and Management of Urinary Tract Infections*, 4th ed. Philadelphia: Lea & Febiger.
2. Unpublished data from National Ambulatory Medical Care Survey (1991). Washington: National Center for Health Statistics.
3. Detailed Diagnosis and Procedures, National Hospital Discharge Survey (1991). Washington: National Center for Health Statistics, Centers for Disease Control and Prevention, U.S. Department of Health and Human Services, February 1994.
4. Ferry, S., Burman, L. G., and Holm, S. E. (1988). Clinical and bacteriological effects of therapy of urinary tract infection in primary health care: relation to *in vitro* sensitivity testing. *Scand. J. Infect. Dis.* 20:535–544.

5. Warren, J. W. (1996). In H. L. T. Mobley and J. W. Warren (eds.), *Urinary Tract Infections: Molecular Pathogenesis and Clinical Management*, pp. 3–27. Washington: ASM Press.
6. Bacheller, C. D., and Bernstein, J. M. (1997) Urinary tract infections. *Med. Clin. North Am.* 81:719–730.
7. Faro, S., and Fenner, D. E. (1998). Urinary tract infections. *Clin. Obstet. Gynecol.* 41:744–754.
8. Warren, J. W., Abrutyn, E., Hebel, J. R., *et al.* (1999). Guidelines for antimicrobial treatment of uncomplicated acute bacterial cystitis and acute pyelonephritis in women. Infectious Diseases Society of America (IDSA). *Clin. Infect. Dis.* 29:745–758.
9. Warren, J., Harry, W., Mobley, L. T., and Donnenberg, M. S. (2001). Host-parasite interactions and host defense mechanisms. In R. W. Schrier (ed.), *Diseases of the Kidney and Urinary Tract*, pp. 903–921. Philadelphia: Lippincott Williams & Wilkins.
10. Hepstinstall, R. H. (1983). Pyelonephritis: Pathologic features. In R. H. Hepstinstall (ed.), *Pathology of the Kidney*, 3d ed., Vol. III, pp. 1323–1396. Boston: Little, Brown.
11. Roberts, J. A. (1986). Pyelonephritis, cortical abscess, and perinephric abscess. *Urol. Clin. North Am.* 13:637–645.
12. Ikaheimo, R., Siitonen, A., Karkkainen, U., *et al.* (1994). Community-acquired pyelonephritis in adults: Characteristics of *E. coli* isolates in bacteremic and nonbacteremic patients. *Scand. J. Infect. Dis.* 26:289–296.
13. Phillips, I., Eykyn, S., King, A., *et al.* (1988). Epidemic multiresistant *Escherichia coli* infection in West Lambeth health district. *Lancet* 1:1038–1041.
14. Manges, A. R., Johnson, J. R., Foxman, B., *et al.* (2001). Widespread distribution of urinary tract infections caused by a multidrug-resistant *Escherichia coli* clonal group. *New Engl. J. Med.* 345(14):1007–1013.
15. Tullus, K., Hörlin, K., Svenson, S. B., and Källenius, G. (1984). Epidemic outbreaks of acute pyelonephritis caused by nosocomial spread of P fimbriated *Escherichia coli* in children. *J. Infect. Dis.* 150:728–736.
16. Kunin, C. M., Hua, T. H., Krishnan, C., *et al.* (1993). Isolation of a nicotinamide-requiring clone of *Escherichia coli* O18:K1:H7 from women with acute cystitis: Resemblance to strains found in neonatal meningitis. *Clin. Infect. Dis.* 16:412–416.
17. Zhang, L. X., Foxman, B., Tallman, P., *et al.* (1997). Distribution of *drb* genes coding for Dr binding adhesins among uropathogenic and fecal *Escherichia coli* isolates and identification of new subtypes. *Infect. Immun.* 65(6):2011–2018.
18. Johnson, J. R., Russo, T. A., Scheutz, F., *et al.* (1997). Discovery of disseminated J96-like strains of uropathogenic *Escherichia coli* O4:H5 containing genes for both $papG_{J96}$ (class I) and $prsG_{J96}$ (class III) Gal(α-1–4)Gal-binding adhesins. *J. Infect. Dis.* 175(4):983–988.
19. Hacker, J., Bender, L., Ott, M., *et al.* (1990). Deletions of chromosomal regions coding for fimbriae and hemolysins occur *in vivo* and *in vitro* in various extraintestinal *Escherichia coli* isolates. *Microb. Pathog.* 8:213–225.
20. Guyer, D. M., Kao, J. S., and Mobley, H. L. T. (1998). Genomic analysis of a pathogenicity island in uropathogenic *Escherichia coli* CFT073: Distribution of homologous sequences among isolates from patients with pyelonephritis, cystitis, and catheter-associated bacteriuria and from fecal samples. *Infect. Immun.* 66:4411–4417.
21. Knapp, S., Hacker, J., Jarchau, T., and Goebel, W. (1986). Large, unstable inserts in the chromosome affect virulence properties of uropathogenic *Escherichia coli* O6 strain 536. *J. Bacteriol.* 168:22–30.
22. Swenson, D. L., Bukanov, N. O., Berg, D. E., and Welch, R. A. (1996). Two pathogenicity islands in uropathogenic *Escherichia coli* J96: Cosmid cloning and sample sequencing. *Infect. Immun.* 64:3736–3743.
23. Kao, J. S., Stucker, D. M., Warren, J. W., and Mobley, H. L. T. (1997). Pathogenicity island sequences of pyelonephritogenic *Escherichia coli* CFT073 are associated with virulent uropathogenic strains. *Infect. Immun.* 65:2812–2820.

24. Mobley, H. L. T., Green, D. M., Trifillis, A. L., *et al.* (1990). Pyelonephritogenic *Escherichia coli* and killing of cultured human renal proximal tubular epithelial cells: Role of hemolysin in some strains. *Infect. Immun.* 58:1281–1289.
25. Mobley, H. L. T., Jarvis, K. G., Elwood, J. P., *et al.* (1993). Isogenic P-fimbrial deletion mutants of pyelonephritogenic *Escherichia coli*: The role of alpha Gal-(1–4)-β-Gal binding in virulence of a wild-type strain. *Mol. Microbiol.* 10:143–155.
26. Falkow, S. (1988). Molecular Koch's postulates applied to microbial pathogenicity. *Rev. Infect. Dis.* 10(suppl. 2):S274–S276.
27. Johnson, D. E., and Russell, R. G. (1996). Animal models of urinary tract infection. In H. L. T. Mobley and J. W. Warren (eds.), *Urinary Tract Infections: Molecular Pathogenesis and Clinical Management*, pp. 377–403. Washington: ASM Press.
28. Hedges, S., Anderson, P., Lidin-Janson, G., *et al.* (1991). Interleukin-6 response to deliberate colonization of the human urinary tract with gram-negative bacteria. *Infect. Immun.* 59:421–427.
29. Andersson, P., Engberg, I., Lidin-Janson, G., *et al.* (1991). Persistence of *Escherichia coli* bacteriuria is not determined by bacterial adherence. *Infect. Immun.* 59:2915–2921.
30. Roberts, J. A., Kaack, B., Källenius, G., *et al.* (1984). Receptors for pyelonephritogenic *Escherichia coli* in primates. *J. Urol.* 131:163–168.
31. Langermann, S., Mollby, R., Burlein, J. E., *et al.* (2000). Vaccination with FimH adhesin protects cynomolgus monkeys from colonization and infection by uropathogenic *Escherichia coli*. *J. Infect. Dis.* 181(2):774–778.
32. Hagberg, L., Engberg, I., Freter, R., *et al.* (1983). Ascending, unobstructed urinary tract infection in mice caused by pyelonephritogenic *Escherichia coli* of human origin. *Infect. Immun.* 40:273–283.
33. Johnson, D. E., Lockatell, C. V., Russell, R. G., *et al.* (1988). Comparison of *Escherichia coli* strains recovered from human cystitis and pyelonephritis infections in transurethrally challenged mice. *Infect. Immun.* 66(7):3059–3065.
34. Finlay, B. B., and Falkow, S. (1989). Common themes in microbial pathogenicity. *Microbiol. Rev.* 53:210–230.
35. Virkola, R., Westerlund, B., Holthofer, H., *et al.* (1988). Binding characteristics of *Escherichia coli* adhesins in human urinary bladder. *Infect. Immun.* 56:2615–2622.
36. Nowicki, B., Svanborg-Eden, C., Hull, R., and Hull, S. (1989). Molecular analysis and epidemiology of the Dr hemagglutinin of uropathogenic *Escherichia coli*. *Infect. Immun.* 57:446–451.
37. Pere, A., Nowicki, B., Saxen, H., *et al.* (1987). Expression of P, type-1, and type-1C fimbriae of *Escherichia coli* in the urine of patients with acute urinary tract infection. *J. Infect. Dis.* 156:567–574.
38. Johnson, J. R. (1991). Virulence factors in *Escherichia coli* urinary tract infection. *Clin. Microbiol. Rev.* 4:80–128.
39. Nowicki, B., Rhen, M., Vaisanen-Rhen, V., *et al.* (1984). Immunofluorescence study of fimbrial phase variation in Escherichia coli KS71. *J. Bacteriol.* 160:691–695.
40. Brinton, C. C. (1959) Non-flagellar appendages of bacteria. *Nature* 183:782–786.
41. Duguid, J. P., Clegg, S., and Wilson, M. I. (1979). The fimbrial and non-fimbrial haemagglutinins of *Escherichia coli*. *J. Med. Microbiol.* 12:213–227.
42. Duguid, J. P., Smith, I .W., Dempster, G., and Edmunds, P. N. (1955). Nonflagellar filamentous appendages (fimbriae) and haemagglutinating activity in *Bacterium coli*. *J. Pathol. Bacteriol.* 70:335–349.
43. Evans, D. J., Jr., Evans, D. G., Hohne, C., *et al.* (1981). Hemolysin and K antigens in relation to serotype and hemagglutination type of *Escherichia coli* isolated from extraintestinal infections. *J. Clin. Microbiol.* 13:171–178.
44. Green, C. P., and Thomas, V. L. (1981). Hemagglutination of human type O erythrocytes, hemolysin production, and serogrouping of *Escherichia coli* isolates from patients with acute pyelonephritis, cystitis, and asymptomatic bacteriuria. *Infect. Immun.* 31:309–315.

45. Orndorff, P. E., and Bloch, C. A. (1990). The role of type 1 pili in the pathogenesis of *Escherichia coli* infections: A short review and some new ideas. *Microb. Pathog.* 9:75–79.
46. Brinton, C. C., Jr. (1965). The structure, function, synthesis and genetic control of bacterial pili and a molecular model for DNA and RNA transport in gram negative bacteria. *Trans. N.Y. Acad. Sci.* 27:1003–1054.
47. Schilling, J. D., Mulvey, M. A., and Hultgren, S. J. (2001). Structure and function of *Escherichia coli* type 1 pili: New insight into the pathogenesis of urinary tract infections. *J. Infect. Dis.* 183(suppl. 1):S36–S40.
48. Hull, R. A., Gill, R. E., Hsu, P., *et al.* (1981). Construction and expression of recombinant plasmids encoding type 1 or D-mannose-resistant pili from a urinary tract infection *Escherichia coli* isolate. *Infect. Immun.* 33:933–938.
49. Hultgren, S. J., Normark, S., and Abraham, S. N. (1991). Chaperone-assisted assembly and molecular architecture of adhesive pili. *Annu. Rev. Microbiol.* 45:383–415.
50. Abraham, S. N., and Beachey, E. H. (1987). Assembly of a chemically synthesized peptide of *Escherichia coli* type 1 fimbriae into fimbria-like antigenic structures. *J. Bacteriol.* 169:2460–2465.
51. Abraham, S. N., Goguen, J. D., and Beachey, E. H. (1988). Hyperadhesive mutant of type 1-fimbriated *Escherichia coli* associated with formation of FimH organelles (fimbriosomes). *Infect. Immun.* 56:1023–1029.
52. Hacker, J. (1990). Genetic determinants coding for fimbriae and adhesins of extraintestinal *Escherichia coli. Curr. Top. Microbiol. Immunol.* 151:1–27.
53. Krogfelt, K. A., Bergmans, H., and Klemm, P. (1990). Direct evidence that the FimH protein is the mannose-specific adhesin of *Escherichia coli* type 1 fimbriae. *Infect. Immun.* 58:1995–1998.
54. Wu, X. R., Sun, T. T., and Medina, J. J. (1996). *In vitro* binding of type 1 fimbriated *Escherichia coli* to uroplakins Ia and Ib: Relation to urinary tract infections. *Proc. Natl. Acad. Sci. USA* 93:9630–9635.
55. Jones, C. H., Pinkner, J. S., Roth, R., *et al.* (1995). FimH adhesin of type 1 pili is assembled into a fibrillar tip structure in the Enterobacteriaceae. *Proc. Natl. Acad. Sci. USA* 92:2081–2085.
56. Jones, C. H., Pinkner, J. S., Nicholes, A. V., *et al.* (1993). FimC is a periplasmic PapD-like chaperone that directs assembly of type 1 pili in bacteria. *Proc. Natl. Acad. Sci. USA* 90:8397–8401.
57. Klemm, P., and Christiansen, G. (1990). The *fimD* gene required for cell surface localization of *Escherichia coli* type 1 fimbriae. *Mol. Gen. Genet.* 220:334–338.
58. Klemm, P. (1986). Two regulatory *fim* genes, *fimB* and *fimE*, control the phase variation of type 1 fimbriae in *Escherichia coli. EMBO J.* 5:1389–1393.
59. Abraham, J. M., Freitag, C. S., Clements, J. R., and Eisenstein, B. I. (1985). An invertible element of DNA controls phase variation of type 1 fimbriae of *Escherichia coli. Proc. Natl. Acad. Sci. USA* 82:5724–5727.
60. Eisenstein, B. I. (1981). Phase variation of type 1 fimbriae in *Escherichia coli* is under transcriptional control. *Science* 214:337–339.
61. Olsen, P. B., and Klemm, P. (1994). Localization of promoters in the *fim* gene cluster and the effect of H-NS on the transcription of *fimB* and *fimE. FEMS Microbiol. Lett.* 116:95–100.
62. McClain, M. S., Blomfield, I .C., Eberhardt, K. J., and Eisenstein, B. I. (1993). Inversion-independent phase variation of type 1 fimbriae in *Escherichia coli. J. Bacteriol.* 175:4335–4344.
63. Hultgren, S. J., Schwan, W. R., Schaeffer, A. J., and Duncan, J. L. (1986). Regulation of production of type 1 pili among urinary tract isolates of *Escherichia coli. Infect. Immun.* 54:613–620.
64. Klemm, P. (1985). Fimbrial adhesions of *Escherichia coli. Rev. Infect. Dis.* 7:321–340.
65. Gally, D. L., Bogan, J. A., Eisenstein, B. I., and Blomfield, I. C. (1993). Environmental regulation of the *fim* switch controlling type 1 fimbrial phase variation in *Escherichia coli* K-12: Effects of temperature and media. *J. Bacteriol.* 175:6186–6193.
66. Connell, I., Agace, W., Klemm, P., *et al.* (1996). Type 1 fimbrial expression enhances *Escherichia coli* virulence for the urinary tract. *Proc. Natl. Acad. Sci. USA* 93:9827–9832.

67. Mulvey, M. A., Lopez-Boado, Y. S., Wilson, C. L., *et al.* (1998). Induction and evasion of host defenses by type 1-piliated uropathogenic *Escherichia coli. Science* 282:1494–1497.
68. Gunther, N. W., Lockatell, V., Johnson, D. E., and Mobley, H. L. (2001). *In vivo* dynamics of type 1 fimbria regulation in uropathogenic *Escherichia coli* during experimental urinary tract infection. *Infect. Immun.* 69:2838–2846.
69. Lim, J. K., Gunther, N. W., Zhao, H., *et al.* (1998). *In vivo* phase variation of *Escherichia coli* type 1 fimbrial genes in women with urinary tract infection. *Infect. Immun.* 66:3303–3310.
70. Langermann, S., Palaszynski, S., Barnhart, M., *et al.* (1997). Prevention of mucosal *Escherichia coli* infection by FimH-adhesin-based systemic vaccination (see comments). *Science* 276:607–611.
71. Langermann, S., Mollby, R., Burlein, J. E., *et al.* (2000). Vaccination with FimH adhesin protects cynomolgus monkeys from colonization and infection by uropathogenic *Escherichia coli. J. Infect. Dis.* 181(2):774–778.
72. Svanborg-Eden, C. S., Hanson, L. A., Jodal, U., *et al.* (1976). Variable adherence to normal human urinary-tract epithelial cells of *Escherichia coli* strains associated with various forms of urinary-tract infection. *Lancet* 1(7984):490–492.
73. Svanborg-Eden, C. S., and Hansson, H. A. (1978). *Escherichia coli* pili as possible mediators of attachment to human urinary tract epithelial cells. *Infect. Immun.* 21:229–237.
74. Korhonen, T. K., Virkola, R., and Holthofer, H. (1986). Localization of binding sites for purified *Escherichia coli* P fimbriae in the human kidney. *Infect. Immun.* 54(2):328–332.
75. Kallenius, G., and Mollby, R. (1979). Adhesion of *Escherichia coli* to human periurethral cells correlated to mannose-resistant agglutination of human erythrocytes. *FEMS Microbiol. Lett.* 5:295.
76. Hull, R. A., Gill, R. E., Hsu, P., *et al.* (1981). Construction and expression of recombinant plasmids encoding type 1 or D-mannose-resistant pili from a urinary tract infection *Escherichia coli* isolate. *Infect. Immun.* 33:933–938.
77. Normark, S., Lark, D., and Hull, R. (1983). Genetics of digalactoside-binding adhesin from a uropathogenic *Escherichia coli* strain. *Infect. Immun.* 41:942–949.
78. Norgren, M., Normark, S., Lark, D., *et al.* (1984). Mutations in *E. coli* cistrons affecting adhesion to human cells do not abolish Pap pili fiber formation. *EMBO J.* 3:1159–1165.
79. Uhlin, B. E., Norgren, M., Baga, M., and Normark, S. (1985). Adhesion to human cells by *Escherichia coli* lacking the major subunit of a digalactoside-specific pilus-adhesin. *Proc. Natl. Acad. Sci. USA* 82:1800–1804.
80. Lund, B., Lindberg, F., Marklund, B. I., and Normark, S. (1987). The PapG protein is the alpha-D-galactopyranosyl-(1–4)-β-D-galactopyranose-binding adhesin of uropathogenic *Escherichia coli. Proc. Natl. Acad. Sci. USA* 84:5898–5902.
81. Kuehn, M. J., Heuser, J., Normark, S., and Hultgren, S. J. (1992). P pili in uropathogenic *E. coli* are composite fibres with distinct fibrillar adhesive tips. *Nature* 356:252–255.
82. Hultgren, S. J., Abraham, S., Caparon, M., *et al.* (1993). Pilus and nonpilus bacterial adhesins: Assembly and function in cell recognition. *Cell* 73:887–901.
83. Dodson, K. W., Jacob-Dubuisson, F., Striker, R. T., and Hultgren, S. J. (1993). Outer-membrane PapC molecular usher discriminately recognizes periplasmic chaperone-pilus subunit complexes. *Proc. Natl. Acad. Sci. USA* 90:3670–3674.
84. Kuehn, M. J., Normark, S., and Hultgren, S. J. (1991). Immunoglobulin-like PapD chaperone caps and uncaps interactive surfaces of nascently translocated pilus subunits. *Proc. Natl. Acad. Sci. USA* 88:10586–10590.
85. Vaisanen, V., Elo, J., Tallgren, L. G., *et al.* (1981). Mannose-resistant haemagglutination and P antigen recognition are characteristic of *Escherichia coli* causing primary pyelonephritis. *Lancet* 2:1366–1369.
86. Kallenius, G., Mollby, R., Svenson, S. B., *et al.* (1981). Occurrence of P-fimbriated *Escherichia coli* in urinary tract infections. *Lancet* 2:1369–1372.

87. Hagberg, L., Jodal, U., Korhonen, T. K., *et al.* (1981). Adhesion, hemagglutination, and virulence of *Escherichia coli* causing urinary tract infections. *Infect. Immun.* 31:564–570.
88. Latham, R. H., and Stamm, W. E. (1984). Role of fimbriated *Escherichia coli* in urinary tract infections in adult women: Correlation with localization studies. *J. Infect. Dis.* 149:835–840.
89. Jacobson, S. H., *et al.* (1985). P fimbriated *Escherichia coli* in adults with acute pyelonephritis. *J. Infect. Dis.* 152:426.
90. O'Hanley, P., Low, D., Romero, I., *et al.* (1985). Gal-Gal binding and hemolysin phenotypes and genotypes associated with uropathogenic *Escherichia coli*. *New Engl. J. Med.* 313:414–420.
91. Dowling, K. J., Roberts, J. A., and Kaack, M. B. (1987). P-fimbriated *Escherichia coli* urinary tract infection: A clinical correlation. *South. Med. J.* 80:1533–1566.
92. Sandberg, T., Kaijser, B., Lidin-Jason, G., *et al.* (1988). Virulence of *Escherichia coli* in relation to host factors in women with symptomatic urinary tract infection. *J. Clin. Microbiol.* 26:1471–1476.
93. Ulleryd, P., Lincoln, K., and Scheutz, F. (1994). Virulence characteristics of *Escherichia coli* in relation to host response in men with symptomatic urinary tract infection. *Clin. Infect. Dis.* 18:579–584.
94. Westerlund, B., Siitonen, A., Elo, J., *et al.* (1988). Properties of *Escherichia coli* isolates from urinary tract infections in boys. *J. Infect. Dis.* 158:996–1002.
95. Johnson, J. R., Roberts, P. L., and Stamm, W. E. (1987). P fimbriae and other virulence factors in *Escherichia coli* urosepsis: Association with patients' characteristics. *J. Infect. Dis.* 156:225–229.
96. Donnenberg, M. S., and Welch, R. A. (1996). Virulence determinants of uropathogenic *Escherichia coli*. In H. L. T. Mobley and J. W. Warren (eds.), *Urinary Tract Infections: Molecular Pathogenesis and Clinical Management*, pp. 135–174. Washington: ASM Press.
97. Kisielius, P. V., Schwan, W. R., Amundsen, S. K., *et al.* (1989). *In vivo* expression and variation of *Escherichia coli* type 1 and P pili in the urine of adults with acute urinary tract infections. *Infect. Immun.* 57:1656–1662.
98. de Ree, J. M., and van den Bosch, J. F. (1987). Serological response to the P fimbriae of uropathogenic *Escherichia coli* in pyelonephritis. *Infect. Immun.* 55:2204–2207.
99. Lomberg, H., Hanson, L. A., Jacobsson, B., *et al.* (1983). Correlation of P blood group vesicoureteral reflux, and bacterial attachment in patients with recurrent pyelonephritis. *New Engl. J. Med.* 308:1189–1192.
100. Baga, M., Normark, S., Hardy, J., *et al.* (1984). Nucleotide sequence of the *papA* gene encoding the *pap* pilus subunit of human uropathogenic *Escherichia coli*. *J. Bacteriol.* 157:330–333.
101. Lindberg, F., Lund, B., Johansson, L., and Normark, S. (1987). Localization of the receptor-binding protein adhesin at the tip of the bacterial pilus. *Nature* 328:84–87.
102. Roberts, J. A., Marklund, B. I., Ilver, D., *et al.* (1994). The Gal(α-1–4)Gal-specific tip adhesin of *Escherichia coli* P-fimbriae is needed for pyelonephritis to occur in the normal urinary tract. *Proc. Natl. Acad. Sci. USA* 91:11889–11893.
103. Stromberg, N., *et al.* (1991). Saccharide orientation at the cell surface affects glycolipid receptor function. *Proc. Natl. Acad. Sci. USA* 88:9340.
104. Nowicki, B., *et al.* (1990). The Dr hemagglutinin, afimbrial adhesins AFA-I and AFA-III, and F1845 fimbriae of uropathogenic and diarrhea-associated *Escherichia coli* belong to a family of hemagglutinins with Dr receptor recognition. *Infect. Immun.* 58:279.
105. Nowicki, B., *et al.* (1988). A hemagglutinin of uropathogenic *Escherichia coli* recognizes the Dr blood group antigen. *Infect. Immun.* 56:1057.
106. Nowicki, B., *et al.* (1993). Short consensus repeat-3 domain of recombinant decay-accelerating factor is recognized by *Escherichia coli* recombinant Dr adhesin in a model of a cell-cell interaction. *J. Exp. Med.* 178:2115.
107. Westerlund, B., *et al.* (1989). The O75X adhesin of uropathogenic *Escherichia coli* is a type IV collagen-binding protein. *Mol. Microbiol.* 3:329.

108. Miettinen, A., *et al.* (1993). Binding of bacterial adhesins to rat glomerular mesangium in vivo. *Kidney Int.* 43:592.
109. Goluszko, P., Moseley, S. L., Truong, L. D., *et al.* (1997). Development of experimental model of chronic pyelonephritis with *Escherichia coli* O75:K5:H-bearing Dr fimbriae: Mutation in the *dra* region prevented tubulointerstitial nephritis. *J. Clin. Invest.* 99(7):1662–1672.
110. Pere, A., *et al.* (1985). Occurrence of type-1C fimbriae on *Escherichia coli* strains isolated from human extraintestinal infections. *J. Gen. Microbiol.* 131:1705.
111. Zingler, G., *et al.* (1992). Clonal analysis of *Escherichia coli* serotype O6 strains from urinary tract infections. *Microb. Pathog.* 12:299.
112. Zingler, G., *et al.* (1990). Clonal differentiation of uropathogenic *Escherichia coli* isolates of serotype O6:K5 by fimbrial antigen typing and DNA long-range mapping techniques. *Med. Microbiol. Immunol.* 182:13.
113. Korhonen, T. K., *et al.* (1990). Tissue tropism of *Escherichia coli* adhesins in human extraintestinal infections. *Curr. Top. Microbiol. Immunol.* 151:115.
114. Pere, A., *et al.* (1987). Expression of P, type-1, and type-1C fimbriae of *Escherichia coli* in the urine of patients with acute urinary tract infection. *J. Infect. Dis.* 156:567.
115. Blum, G., Falbo, V., Caprioli, A., and Hacker, J. (1995). Gene clusters encoding the cytotoxic necrotizing factor type 1, Prs-fimbriae and α-hemolysin from the pathogenicity island II of the uropathogenic *Escherichia coli* strain J96. *FEMS Microbiol. Lett.* 126:189–196.
116. High, N. J., Hales, B. A., Jann, K., and Boulnois, G. J. (1988). A block of urovirulence genes encoding multiple fimbriae and hemolysin in *Escherichia coli* O4:K12:H^-. *Infect. Immun.* 56:513–517.
117. Dudgeon, L. S., Worldley, E., and Bawtree, F. (1921). On *Bacillus coli* infections of the urinary tract, especially in relation to hemolytic organisms. *J. Hygiene* 10:137.
118. Femlee, T., Pellett, S., and Welch, R. A. (1985). Nucleotide sequence of an *Escherichia coli* chromosomal hemolysin. *J. Bacteriol.* 163:94–105.
119. Issartel, J. P., Koronakis, V., and Hughes, C. (1991). Activation of *Escherichia coli* prohemolysin to the mature toxin by acyl carrier protein-dependent fatty acylation. *Nature* 351:759–761.
120. Lim, K. B., Walker, C. R. B., Guo, L., *et al.* (2000). *Escherichia coli* α-hemolysin is heterogeneously acylated *in vivo* with 14-, 15-, and 17-carbon fatty acids. *J. Biol. Chem.* 275(47): 36698–36702.
121. Koronakis, V., and Hughes, C. (1996). Synthesis, maturation and export of the *E. coli* hemolysin. *Med. Microbial. Immunol.* 185:65–71.
122. Thanabalu, T., Koronakis, E., Hughes, C., and Koronakis, V. (1998). Substrate-induced assembly of a contiguous channel for protein export from *E. coli*: Reversible bridging of an inner-membrane translocase to an outer membrane exit pore. *EMBO J.* 17:6487–6496.
123. Koronakis, V., Sharff, A., Koronakis, E., *et al.* (2000). Crystal structure of the bacterial membrane protein TolC central to multidrug efflux and protein export. *Nature* 405:914–919.
124. Bailey, M. J. A., Koronakis, V., Schmoll, T., and Hughes, C. (1992). *Escherichia coli* HlyT protein, a transcriptional activator of hemolysin synthesis and secretion, is encoded by the *rfaH* (*sfrB*) locus required for expression of sex factor and lipopolysaccharide genes. *Mol. Microbiol.* 6:1003–1012.
125. Nagy, G., Dobrindt, U., Blum-Oehler, G., *et al.* (2000). In Emody *et al.* (eds.), *Genes and Proteins Underlying Microbial Urinary Tract Virulence*, pp. 57–61. New York: Kluwer Academic/Plenum Publishers.
126. Dobrindt, U., Janke, B., Piechaczek, K., *et al.* (2000). Toxin genes on pathogenicity islands: Impact for microbial evolution. *Int. J. Med. Microbiol.* 290:307–311.
127. Ludwig, A, Jarchau, T., Benz, R., and Goebel, W. (1988). The repeat domain of *Escherichia coli* hemolysin (HlyA) is responsible for its Ca^{2+}-dependent binding to erythrocytes. *Mol. Gen. Genet.* 214:553–561.

128. Boehm, D. F., Welch, R. A., and Snyder, I. S. (1990). Calcium is required for binding of *Escherichia coli* hemolysin (HlyA) to erythrocyte membrane. *Infect. Immun.* 58:1951–1958.
129. Boehm, D. F., Welch, R. A., and Snyder, I. S. (1990). Domains of *Escherichia coli* hemolysin (HlyA) involved in binding of calcium and erythrocyte membranes. *Infect. Immun.* 58:1959–1964.
130. Soloaga, A., Veiga, M. P., Garcia-Segura, L. M., *et al.* (1999). Insertion of *Escherichia coli* α-hemolysin in lipid bilayers as a nontransmembrane integral protein: prediction and experiment. *Mol. Microbiol.* 31:1013–1024.
131. Cavalieri, S. J., and Snyder, I. S. (1982). Effect of *Escherichia coli* α-hemolysin on human peripheral leukocyte viability *in vitro*. *Infect. Immun.* 36:455–461.
132. Cavalieri, S. J., and Snyder, I. S. (1982). Cytotoxic activity of partially purified *Escherichia coli* α-hemolysin. *J. Med. Microbiol.* 15:11–21.
133. Island, M. D., Cui, X. L., Foxman, B., *et al.* (1998). Cytotoxicity of hemolytic, cytotoxic necrotizing factor 1–positive and –negative *Escherichia coli* to human T24 bladder cells. *Infect. Immun.* 66:3384–3389.
134. Triffilis, A. L., Donnenberg, M. S., Cui, X., *et al.* (1994). Binding to and killing of human renal epithelial cells by hemolytic P-fimbriated *E. coli*. *Kidney Int.* 46:1083–1091.
135. Uhlen, P., Laestadius, A., Jahnukainen, T., *et al.* (2000). α-Haemolysin of uropathogenic *E. coli* induces Ca^{2+} oscillations in renal epithelial cells. *Nature* 405:694–697.
136. Seetharama, S., Cavalieri, S. J., and Snyder, I. S. (1988). Immune response to *Escherichia coli* α-hemolysin in patients. *J. Clin. Microbiol.* 26:850–856.
137. Marre R., Hacker, J., Henkel, W., and Gobel, W. (1986). Contribution of cloned virulence factors from uropathogenic *Escherichia coli* strains to nephropathogenicity in an experimental rat pyelonephritis model. *Infect. Immun.* 54:761–767.
138. Falbo, V., Famiglietti, M., and Caprioli, A. (1992). Gene block encoding production of cytotoxic necrotizing factor 1 and hemolysin in *Escherichia coli* isolates from extraintestinal infections. *Infect. Immun.* 60:2182–2187.
139. Falbo, V., Pace, T., Picci, L., *et al.* (1993). Isolation and nucleotide sequence of the gene encoding cytotoxic necrotizing factor 1 of *Escherichia coli*. *Infect. Immun.* 61:4909–4914.
140. Fiorentini, C., Arancia, G., Caprioli, A., *et al.* (1988). Cytoskeletal changes induced in Hep-2 cells by the cytotoxic necrotizing factor of *Escherichia coli*. *Toxicon* 26:1047–1056.
141. Flatau, G., Lemichez, E., Gauthler, M., *et al.* (1997). Toxin-induced activation of the G protein p21 Rho by deamidation of glutamine. *Nature* 387:729–733.
142. Schmidt, G., Sehr, P., Wilm, M., *et al.* (1997). Gln 63 of Rho is deamidated by *Escherichia coli* cytotoxic necrotizing factor-1. *Nature* 387:725–729.
143. Falzano, L., Fiorentini, C., Donelli, G., *et al.* (1993). Induction of phagocytic behavior in human epithelial cells by *Escherichia coli* cytotoxic necrotizing factor type 1. *Mol. Microbiol.* 9(6):1247–1254.
144. Hofman, P., Flatau, G., Selva, E., *et al.* (1998). *Escherichia coli* cytotoxic necrotizing factor 1 effaces microvilli and decreases transmigration of polymorphonuclear leukocytes in intestinal T84 epithelial cell monolayers. *Infect. Immun.* 66:2494–2500.
145. Gerhard, R., Schmidt, G., Hofmann, F., and Aktories, K. (1998). Activation of Rho GTPases by *Escherichia coli* cytotoxic necrotizing factor 1 increases intestinal permeability in Caco-2 cells. *Infect. Immun.* 66:5125–5131.
146. Hofman, P., Negrate, G. L., Mograbi, B., *et al.* (2000). *Escherichia coli* cytotoxic necrotizing factor-1 (CNF-1) increases the adherence to epithelia and the oxidative burst of human polymorphonuclear leukocytes but decreases bacteria phogocytosis. *J. Leukocyte Biol.* 68:522–528.
147. Mills, M., Meysick, K. C., and O'Brien, A. D. (2000). Cytotoxic necrotizing factor type 1 of uropathogenic *Escherichia coli* kills cultured human uroepithelial 5637 cells by an apoptotic mechanism. *Infect. Immun.* 68(10):5869–5880.

148. Johnson, D. E., Drachenberg, C., Lockatell, C. V., *et al.* (2000). The role of cytotoxic necrotizing factor-1 in colonization and tissue injury in a murine model of urinary tract infection. *FEMS Immunol. Med. Microbiol.* 28:37–41.
149. Henderson, I. R., Navarro-Garcia, F., and Nataro, J. P. (1998). The great escape: structure and function of the autotransporter proteins. *Trends Microbial.* 6:370–378.
150. Guyer, D. M., Henderson, I. R., Nataro, J. P., and Mobley, H. L. T. (2000). Identification of Sat, an autotransporter toxin produced by uropathogenic *E. coli*. *Mol. Microbiol.* 38(1):53–66.
151. Russo, T. A., Sharma, G., Weiss, J., and Brown, C. (1995). The construction and characterization of colanic acid deficient mutants in an extraintestinal isolate of *Escherichia coli* (O4/K54/H5). *Microb. Pathog.* 18:269–278.
152. Russo, T. A., Brown, J. J., Jodush, S. T., and Johnson, J. R. (1996). The O4 specific antigen moiety of lipopolysaccharide but not the K54 group 2 capsule is important for urovirulence of an extraintestinal isolate of *Escherichia coli*. *Infect. Immun.* 64(6):2343–2348.
153. Schilling, J. D., Mulvey, M. A., Vincent, C. D., *et al.* (2001). Bacterial invasion augments epithelial cytokine responses to *Escherichia coli* through a lipopolysaccharide-dependent mechanism. *J. Immunol.* 166(2):1148–1155.
154. Cross, A. S., Kim, K. S., Wright, D. C., *et al.* (1986). Role of lipopolysaccharide and capsule in the serum resistance of bacteremic strains of *Escherichia coli*. *J. Infect. Dis.* 154:497–503.
155. Burns, S. M., and Hull, S. I. (1998). Comparison of loss of serum resistance by defined lipopolysaccharide mutants and an acapsular mutant of uropathogenic *Escherichia coli* O75:K5. *Infect. Immun.* 66:4244–4253.
156. Horwitz, M. A., and Silverstein, S. C. (1980). Influence of the *Escherichia coli* capsule on complement fixation and on phagocytosis and killing by human phagocytes. *J. Clin. Invest.* 65:82–94.
157. Burns, S. M., and Hull, S. I. (1999). Loss of resisitance to ingestion and phagocytic killing by O^- and K^- mutants of a uropathogenic *Escherichia coli* O75:K5 strain. *Infect. Immun.* 67:3757–3762.
158. Bahrani-Mougeot, F. K., Buckles E., Lockatell, C. V., *et al.* (2001). Type-1 fimbriae and extracellular polysaccharides are preeminent uropathogenic *Escherichia coli* virulence determinants in the murine urinary tract. (submitted.)
159. Carbonetti, N. H., Boonchai, S., Parry, S. H., *et al.* (1986). Aerobactin-mediated iron uptake by *Escherichia coli* isolates from human extraintestinal infections. *Infect. Immun.* 51:966–968.
160. Johnson, J. R., Moseley, S. L., Roberts, P. L., and Stamm, W. E. (1988). Aerobactin and other virulence factor genes among strains of *Escherichia coli* causing urosepsis: association with patient characteristics. *Infect. Immun.* 56:405–412.
161. Torres, A. G., Redford, P., Welch, R. A., and Payne, S. M. (2001). TonB-dependent systems of uropathogenic *Escherichia coli*: aerobactin and heme transport and TonB are required for virulence in the mouse. *Infect. Immun.* 69:6179–6185.
162. Kaye, D. (1968). Antibacterial activity of human urine. *J. Clin. Invest.* 47:2374–2390.
163. Gordon, D. M., and Riley, M. A. (1992). A theoretical and experimental analysis of bacterial growth in the bladder. *Mol. Microbiol.* 6:555–562.
164. Culham, D. E., Dalgado, C., Gyles, C. L., *et al.* (1998). Osmoregulatory transporter ProP influences colonization of the urinary tract by *Escherichia coli*. *Microbiology* 144:91–102.
165. Hull, R. A., and Hall, S. I. (1997). Nutritional requirements for growth of uropathogenic *Escherichia coli* in human urine. *Infect. Immun.* 65:1960–1961.
166. Russo, T. A., Jodush, S. T., Brown, J. J., and Johnson, J. R. (1996). Identification of two previously unrecognized genes (*guaA* and *argC*) important for uropathogenesis. *Mol. Microbiol.* 22(2):217–229.
167. Surin, B .P., Rosenberg, H., and Cox, G. B. (1985). Phosphate-specific transport system of *Escherichia coli*: Nucleotide sequence and gene-polypeptide relationships. *J. Bacteriol.* 161:189–198.

168. Bahrani-Mougeot, F. K., Pancholi, S., Daoust, M., and Donnenberg, M. S. (2001). Identification of putative urovirulence genes by subtractive cloning. *J. Infect. Dis.* 183(suppl. 1):S21–23.
169. Rasko, D. A., Phillips, J. A., Li, X., and Mobley, H. L. T. (2001). Identification of pyelonephritis and cystitis DNA sequences from a putative second pathogenicity island of uropathogenic *Escherichia coli* CFT073: Probes specific for uropathogenic sequences. *J. Infec. Dis.* 184:1041–1049.

CHAPTER 9

Meningitis-Associated Escherichia coli

Kwang Sik Kim
Pediatric Infectious Diseases Division, Johns Hopkins University School of Medicine, Baltimore, Maryland

INTRODUCTION

The mortality and morbidity associated with neonatal gram-negative bacillary meningitis have remained significant despite advances in antimicrobial chemotherapy and supportive care. Case-fatality rates have ranged between 15 and 40%, and approximately 50% of the survivors sustain neurologic sequelae [1,2]. A major contributing factor is the incomplete understanding of the pathogenesis of this disease.

Escherichia coli is the most common gram-negative organism that causes meningitis during the neonatal period [1,2]. Give the plethora of *E. coli* serotypes (e.g., 103 capsular of K, 170 LPS, and 56 flagella or H-antigens), it is striking that *E. coli* strains possessing the K1 capsular polysaccharide are predominant (approximately 80%) among isolates from neonatal *E. coli* meningitis and that most of these K1 isolates are associated with a limited number of O types (e.g., O-18, O-7, O-1) [3–6]. Most cases of bacterial meningitis develop as a result of hematogenous spread, but it is unclear how circulating *E. coli* crosses the blood-brain barrier [7–9]. *E. coli* K1 has been shown to translocate from blood to the central nervous system (CNS) without altering the integrity of the blood-brain barrier [9–11]. These studies of bacterial translocation from blood to the CNS

Escherichia coli: Virulence Mechanisms of a Versatile Pathogen
ISBN 0-12-220751-3

have become feasible by the availability of both *in vitro* and *in vivo* models of the blood-brain barrier.

MODELS OF BLOOD-BRAIN BARRIER

In vitro models of the blood-brain barrier have been developed by isolation and cultivation of brain microvascular endothelial cells (BMECs) from humans [11–13]. On cultivation on collagen-coated Transwell inserts, these human BMECs form a continuous lining of endothelial cells (largely a single monolayer with occasionally two or three endothelial cells overlying one another) and exhibit transendothelial electrical resistance of $200–600\,\Omega\text{-cm}^2$ [11], a unique property of the brain microvascular endothelial monolayer compared with systemic vascular endothelium.

The *in vivo* model of the blood-brain barrier was established by induction of hematogenous meningitis in 5-day-old rats [14–19]. In this experimental meningitis model, *E. coli* is injected via intracardiac or subcutaneous injection, resulting in bacteremia and subsequent entry into the CNS, which most likely occurs at the sites of the blood-brain barrier. The development of techniques for atraumatic collection of blood and cerebrospinal fluid (CSF) specimens allows for use of this *in vivo* animal model to examine the pathogenetic mechanisms involved in crossing of the blood-brain barrier by circulating *E. coli*.

A THRESHOLD LEVEL OF BACTEREMIA

Several studies in humans and experimental animals suggest a relationship between the magnitude of *E. coli* bacteremia and the development of meningitis. For example, Dietzman and colleagues [20] reported a significantly higher incidence of *E. coli* meningitis in neonates who had bacterial counts in blood of greater than 10^3 colony-forming units (cfu) per milliliter (6 of 11, or 55%) compared with those with bacterial counts in blood of less than 10^3 cfu/ml (1 of 19, or 5%). Consistent with these clinical findings, a high degree of bacteremia was shown to be a primary determinant of meningeal invasion by *E. coli* K1 in an experimental hematogenous meningitis model [14]. As stated earlier, *E. coli* is commonly associated with neonatal meningitis, and *E. coli* strains possessing the K1 capsular polysaccharide are predominant (approximately 80% among isolates from *E. coli* meningitis) [3–6]. Thus one of the reasons for the close association of *E. coli* K1 with meningitis is their ability to escape from host defenses and to achieve the threshold level of bacteremia necessary for meningeal invasion. Of interest, rates of *E. coli* meningitis (defined as positive CSF cultures) were found to be similar between neonatal and adult animals developing a high degree of bacteremia (e.g., $>10^5$ cfu/ml of blood); however, an approximately 10^6-fold greater inoculum of *E. coli* K1 was required in adult animals to induce a similar high-level bacteremia compared with neonatal animals [14]. These findings indicate that the age dependency of *E. coli* meningitis is due to relative resistance of adults to high-level bacteremia, which precedes the development of meningitis, less likely due to greater invasion of neonatal BMECs compared with adults BMECs [13].

BACTERIAL STRUCTURES CONTRIBUTING TO BINDING TO BMECs

As described previously, we presently do not know how circulating *E. coli* cross the blood-brain barrier. Parkkinen and colleagues [21] reported that purified S fimbriae or a recombinant *E. coli* HB101 strain expressing S fimbriae bind to the luminal surfaces of the brain vascular endothelium in neonatal rat brain tissues. S fimbriae are shown to be associated more frequently with *E. coli* isolates from CSF then from blood or feces (64 versus 22 or 4%, respectively) [22]. S fimbriae, a functional protein complex, have been shown to consist of the major subunit SfaA (16 kDa) and minor subunits SfaG (17 kDa), SfaS (14 kDa), and SfaH (29 kDa) [23]. The two binding domains of S fimbriae have been identified, i.e., SfaS adhesin for binding to BMEC sialoglycoprotein and SfaA for binding to BMEC sulfated glycolipid (sulfatide) [24,25]. Since SfaS is shown to be preferentially located at the tip of S fimbriae [26], it is likely that the critical attachment of S-fimbriated *E. coli* to BMECs occurs via SfaS-BMEC sialoglycoprotein interaction, which is followed by an association of the bacteria with sulfatide for a more intimate contact of organism to BMECs to withstand blood flow *in vivo*, and that such a binding probably is required for subsequent crossing of the blood-brain barrier. However, at present, there is no *in vivo* information available to support the role of S fimbriae in the penetration of *E. coli* K1 across the blood-brain barrier. It is also unclear whether other types of fimbriae such as type 1 and P fimbriae contribute to *E. coli* K1 binding to BMECs and subsequent invasion into the CNS.

BACTERIAL STRUCTURES CONTRIBUTING TO INVASION OF BMECs

Recent studies indicate that a high degree of bacteremia is necessary but not sufficient for the development of meningitis by *E. coli* K1 and that invasion of BMECs is a prerequisite for *E. coli* K1 penetration of the blood-brain barrier *in vivo* [9]. This was shown by the demonstration in infant rats with experimental hematogenous meningitis that several isogenic mutants of *E. coli* K1 strain RS 218 (018:K1:H7) were significantly less able to induce meningitis (defined as

TABLE I Development of Bacteremia and Meningitis (Defined as a Positive CSF Culture) in Newborn Rats Receiving *E. coli* K1 Strain E44[b] (RS 218 Rif, 018:K1:H7) or Its Isogenic Mutants

E. coli strain (no. of animals)	Bacteremia (log cfu/ml blood)	No. (%) of animals with positive CSF	Reference
E44 (19)	7.18 ± 0.63	12 (63%)	36
Δ*ompA* (22)	7.05 ± 0.49	6 (27%)[a]	
E44 (24)	7.51 ± 1.25	16 (67%)	15
Δ*ibeA* (25)	6.97 ± 1.21	4 (16%)[a]	
E44 (27)	7.01 ± 1.17	15 (56%)	16
Δ*ibeB* (25)	7.06 ± 1.29	4 (16%)[a]	
E44 (24)	7.53 ± 0.40	18 (75%)	17
Δ*ibeC* (24)	7.80 ± 0.67	10 (42%)[a]	
E44 (17)	7.50 ± 0.32	14 (82%)	18
Δ*aslA* (22)	7.60 ± 0.49	7 (32%)[a]	
E44 (51)	7.22 ± 0.59	34 (67%)	19
Δ*traJ* (50)	7.10 ± 0.44	23 (46%)[a]	
E44 (26)	6.06 ± 1.49	10 (38%)	32
Δ*cnf1* (28)	6.07 ± 1.21	3 (11%)[a]	

[a]Reprinted with permission from *Infection and Immunity* [47].
[b]Significantly less than E44.

positive CSF cultures) compared with the parent strain despite having similar levels of bacteremia [15–19,32,36], indicating that these *E. coli* K1 structures are necessary for crossing the blood-brain barrier *in vivo* (Table I). The following is a summary regarding how these *E. coli* K1 structures contributing to traversal of the blood-brain barrier have been identified.

Outer membrane protein A (OmpA) was identified as a potential contributor to *E. coli* K1 invasion of BMECs on the basis of its homology with *Neisseria* Opa proteins, which have been shown to be involved in invasion of eukaryotic cells [27]. The isogenic *ompA* deletion mutant of *E. coli* K1 strain RS218 was less invasive in BMECs and less able to cross the blood-brain barrier in the hematogenous *E. coli* meningitis model compared with its parent *E. coli* K1 strain (see Table I). OmpA is a major outer membrane protein in *E. coli*, and its N-terminal domain crosses the membrane eight times in antiparallel β-strands with four relatively large and hydrophilic surface-exposed loops. The N-terminal surface-exposed loops of OmpA have been shown to contribute to *E. coli* K1 invasion of BMECs [27]. For example, the synthetic peptides representing a part of the first loop and the tip of the second loop of OmpA have been shown to inhibit *E. coli* K1 invasion of BMECs [27]. OmpA interacts with the

GlcNAcβ1–4 GlcNAc epitope of the BMEC receptor glycoprotein [28]. Of interest, our recent studies using atomic-level simulations to predict the binding sites and energy of OmpA interactions with GlcNAcβ1–4 GlcNAc and other sugars indicate that the first and second loops of OmpA are important for GlcNAcβ1–4 GlcNAc, which are consistent with the results obtained with the synthetic peptides [27]. The receptor glycoprotein is found to be present on BMECs but undetectable on systemic vascular endothelial cells [28]. These findings indicate that OmpA contributes to BMEC invasion via a ligand-receptor interaction. *E. coli* K1 OmpA is highly homologous with *E. coli* K12 OmpA, and only 3 of the 325 deduced amino acid residues differ between the K1 and K12 OmpA. The function of OmpA in *E. coli* invasion of BMECs also was found to be similar between *E. coli* K1 and K12 OmpA, as shown by successful complementation of the noninvasive *ompA* deletion mutant of *E. coli* K1 to invade BMECs with the *E. coli* K12 *ompA* gene [36].

Invasion of brain endothelial cells (Ibe) proteins A, B, and C and AslA were identified by cloning and characterizing the Tn*phoA* insertion sites from strains 10A-23, 7A-33, 23A-20, and 27A-6, respectively, which are the noninvasive mutants of *E. coli* K1 strain RS218 (018:K1:H7) [15–18]. *ibeB*, *ibeC*, and *aslA* were found to have K12 homologues *p77211*, *yijp*, and *aslA*, respectively, whereas *ibeA* was unique to *E. coli* K1. The *ibeB* sequence is found to be 97% identical to a gene encoding a 50-kDa hypothelical protein (*p77211*), and the *ibeC* sequence is almost identical to an open reading frame (ORF) *yijp*, whose function is currently unknown. The *E. coli* K1 *aslA* sequence is highly homologous with the *E. coli* K12 *aslA*, a putative arylsulfatase-like gene. This *E. coli* K12 gene was named because the deduced protein sequence contains sulfatase consensus motifs I and II, which are homologous (55 and 70% identity) with those of *Klebsiella pneumoniae* AtsA, an arylsulfatase involved in sulfate metabolism. The roles of IbeA, IbeB, IbeC, and AslA in *E. coli* K1 invasion of BMECs were verified by deletion and complementation experiments; i.e., isogenic deletion mutants were less invasive in BMEC *in vitro* and less able to cross the blood-brain barrier *in vivo* (see Table I), and their invasion abilities were restored by complementation in *trans* with individual genes. Sequence analysis indicates that IbeA and IbeB encode outer membrane proteins with three and two transmembrane domains, respectively, whereas IbeC has some features of outer membrane protein, including a signal peptide–like sequence and five or six transmembrane segments at its N terminus. The *E. coli* K1 *aslA* encodes a 52-kDa protein with two transmembrane domains and an N-terminal signal sequence [15–18]. Of interest, recombinant Ibe proteins inhibited *E. coli* K1 invasion of BMECs, suggesting that Ibe proteins contribute to BMEC invasion by ligand-receptor interactions. This concept was supported by the demonstration of a 45-kDa BMEC surface protein interactive with IbeA, and a polyclonal antibody raised against this receptor protein inhibited *E.coli* K1 invasion of BMECs [29]. Partial characterization by N-terminal and internal amino acid sequencing of this receptor protein reveals that it represents a novel albumin-like protein present on BMECs (29). Studies are in progress to identify whether BMECs possess receptor proteins for IbeB and IbeC. The *E. coli* K1 AslA, based on its protein sequence, indicates a member of the arylsulfatase family of enzymes that contains highly conserved sulfatase motifs. In bacteria, these genes are expressed under conditions of sulfur starvation. Of interest, unlike *Klebsiella*, both *E. coli* K1 and K12 AslA failed to exhibit *in vitro* arylsulfatase activity [18]. It remains unclear how AslA contributes to *E. coli* K1 invasion of BMECs *in vitro* and traversal of the blood-brain barrier *in vivo*.

A recent study has shown that certain environmental conditions positively and negatively affect *E. coli* K1 invasion of BMECs *in vitro* and traversal of the blood-brain barrier *in vivo* [30]. For example, the bacterial growth conditions, such as microaerophilic growth and medium supplemented with newborn bovine serum or iron, have been shown to enhance *E. coli* K1 invasion of BMECs. Growth conditions that significantly repressed invasion include iron chelation and high osmolarity [30]. Using differential fluorescence induction and screening of gfp-fusion library, TraJ was identified as a contributor to *E. coli* K1 invasion of BMECs [31]. TraJ was, as expected, found to be differentially expressed at the transcriptional level; e.g., an increase in transcript levels of *traJ* was demonstrated when *E. coli* K1 was grown in the presence of serum compared with medium alone. A *traJ* mutant was less invasive in BMECs *in vitro* and less able to cross the blood-brain barrier *in vivo* [19]. More important, when wild-type DNA was supplied in *trans*, the invasion phenotype was fully restored for the *traJ* mutant. *traJ* belongs to a cluster of genes within the F-like plasmid R1-19 transfer region called the *tra* operon. TraJ has homology with a component of the bacterial conjugation system, suggesting that type IV secretion system may be relevant to TraJ, but it remains to be determined whether *E. coli* K1 possesses a type IV secretion system and that this secretion system participates in the pathogenesis of *E. coli* K1 meningitis. Studies with signature-tagged mutagenesis also identified TraJ and cytotoxic necrotizing factor 1 (CNF1) as contributors to *E. coli* K1 invasion of BMECs [19]. An isogenic *cnf1* deletion mutant of *E. coli* K1 strain RS 218 is less invasive in BMECs *in vitro* and less able to penetrate the blood-brain barrier *in vivo* (see Table I). CNF1 has been shown to activate Rho GTPases, resulting in polymerization of F-actin and increased formation of stress fibers [33,34]. Actin cytoskeletal rearrangements are required for *E. coli* K1 invasion of BMECs, as shown by invasive *E. coli* K1–associated F-actin condensation and blockade of invasion by the microfilament-disrupting agents cytochalasin D and latrunculin A [35]. CNF1 has been shown to activate RhoA GTPase as well as enhance *E. coli* K1 invasion in BMECs [32]. Taken together, these findings indicate that CNF1 contributes to *E. coli* K1 invasion of BMECs most likely via RhoA activation.

At present, it is unclear why and how all the afore-mentioned *E. coli* K1 determinants contribute to crossing of the blood-brain barrier *in vivo*. As described

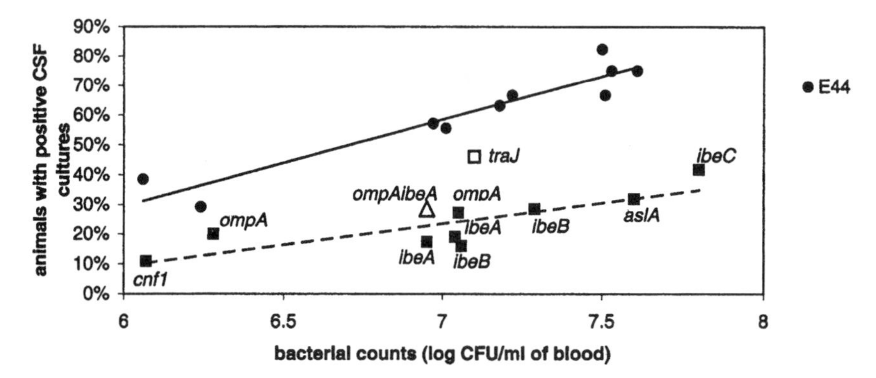

FIGURE 1 Relationship between the magnitude of bacteremia and the percentage of positive CSF cultures among 5-day-old rats receiving *E. coli* K1 (strain E44) or its isogenic single-gene deletion mutants (*cnf1, ompA, ibeA, ibeB, aslA, ibeC, traJ*) or double knockouts (*ompA, ibeA*). The results are expressed as the mean of 17–51 animals for each circle (E44), each square (single-gene deletion mutants), and each triangle (double knockouts).

previously, a high degree of bacteremia is a primary determinant for *E. coli* K1 crossing of the blood-brain barrier *in vivo* [14], and *E. coli* K1's ability to cross the blood-brain barrier is enhanced as the magnitude of bacteremia increases (Fig. 1). As also shown previously, isogenic single-gene deletion mutants of *E. coli* K1 have been shown to be significantly less efficient in their penetration into the CNS in infant rats despite having similar degrees of bacteremia compared with their parent K1 strains (see Table I). Of interest, although these isogenic deletion mutants were significantly less able to cross the blood-brain barrier, their ability to traverse the blood-brain barrier seemed to increase with higher magnitudes of bacteremia (see Fig. 1). As shown in Fig. 1, the *traJ* mutant's ability to penetrate into the CNS appears to be somewhat greater than that of other single-gene deletion mutants.

We next examined whether or not double knockouts were significantly less invasive in their penetration into the CNS compared with their individual deletion mutants. This was done by comparing the invasion phenotypes of *ompA* and *ibeB* deletion mutants, alone and in combination, in BMECs *in vitro* and in the experimental hematogenous meningitis animal model *in vivo* [36]. As expected, the *ompA* and *ibeB* deletion mutants were significantly less invasive in BMECs and less able to cross the blood-brain barrier *in vivo* compared with their parent *E. coli* K1 strain. Our unexpected finding, however, was that the invasion phenotypes of the *ompA* and *ibeB* double knockouts were similar to those of the single-gene deletion mutants both *in vitro* and *in vivo* (see Fig. 1). The reasons for no additive/synergetic effect observed with the Δ*ompA*, Δ*ibeB* mutant are not clear. As described earlier, we anticipate finding that OmpA and IbeB interact with different receptors on BMECs, and thus the effects of OmpA and IbeB on

BMEC invasion will be at least additive. Identification and characterization of specific BMEC receptors for OmpA and IbeB should clarify why the contributions of OmpA and IbeB to *E. coli* K1 crossing of the blood-brain barrier are not additive. For example, OmpA and IbeB interactions with their receptors involve common signaling pathways contributing to host cell actin cytoskeletal rearrangements and invasion of BMECs; thus no additive effects are observed with Δ*ompA* and Δ*ibeB*.

E. COLI K1 GENOME

Comparative macrorestriction mapping and subtractive hybridization of the chromosomes of meningitis-causing *E. coli* K1 (e.g., strains RS 218 and C5) with serotype of O18:K1:H7 compared with nonpathogenic *E. coli* K12 (MG 1655) have identified a 500-kb spread over at least 12 chromosomal loci specific to *E. coli* K1, one 20-kb deletion (so-called black hole) from *E. coli* K1, and a large plasmid (approximately 100 kb) specific to E. *coli* K1 [37–39] (Fig. 2). As shown in Table I and Fig. 1, several structures contributing to *E. coli* K1 invasion of BMECs are unique to *E. coli* K1, such as *ibeA*, *traJ*, and *cnfl*, whereas other structures critical to *E. coli* K1 crossing of the blood-brain barrier are shown to have K12 homologues such as *ompA*, *ibeB*, *ibeC*, and *aslA*. Thus *E. coli* K1 determinants contributing to invasion of BMECs include K1-specific genes as well as K12 homologues. Mapping studies reveal that those *E. coli* loci involved in BMEC invasion are located at different regions of *E. coli* K1 chromosome (see Fig. 2).

BACTERIAL TRAFFICKING OF BMEC

Transcytosis of BMECs by *E. coli* K1 occurs without any change in the integrity of monolayers. For example, *E. coli* K1 strains E44 and C5 (O18:K1:H7) were able to cross BMEC monolayers without affecting transendothelial electrical resistance (TEER) [11], suggesting that traversal of *E. coli* K1 across BMECs occurs using a transcellular mechanism. Transmission electron microscopy revealed that *E. coli* K1 invades BMECs via a zipper-like mechanism and transmigrates through BMECs in an enclosed vacuole without intracellular multiplication (Fig. 3). *E. coli* K1 invasion of BMEC requires actin cytoskeletal rearrangements and induces tyrosine phophorylation of focal adhesion kinase (FAK) and paxillin [40], a cytoskeletal protein known to associate with FAK. FAK is a 125-kDa nonreceptor tyrosine kinase that contains a central kinase domain flanked by an N-terminal domain and a C-terminal domain. The N-terminal domain contains an autophosphorylation site (Tyr 397), whereas the C-terminal domain contains a region required for its localization to focal adhesions

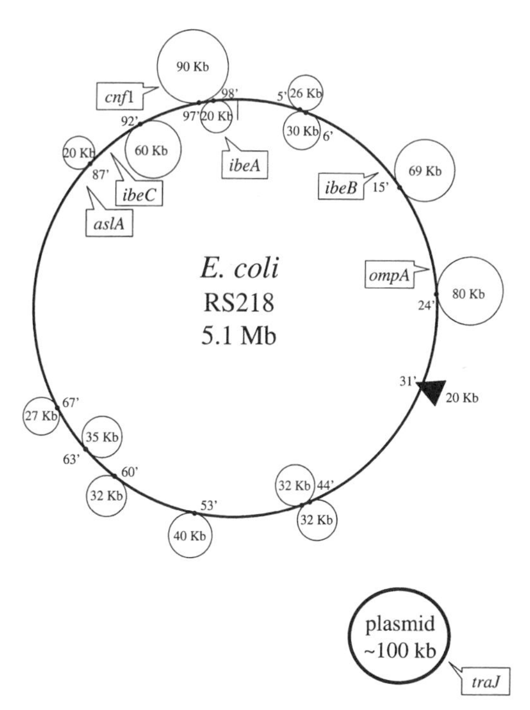

FIGURE 2 Sizes and chromosomal locations of *E.coli* K1 strain RS 218 (018:K1:H7). Specific segments identified by comparative macrorestriction mapping [37] are shown by 10 small circles. Numbers in the circles represent sizes of K1-specific DNA segments. The sizes of two additional segments, shown by arrows labeled 1 and 4, identified by subtractive hybridization [38] are currently unknown. A20-kb deletion (so-called black hole) and a large plasmid (~100 kb) [39] are also depicted. Also shown are seven *E. coli* K1 genes contributing to invasion of BMECs, indicated by squares that include K1-specific genes (e.g., *ibeA*, *cnf1*, *traJ*) as well as genes that have K12 homologues (e.g., *ompA*, *ibeB*, *ibeC*, *aslA*). (Modified with permission from *Infection and Immunity* [47].)

and binding sites for the cytoskeletal proteins such as paxillin. Using FAK dominant-negative mutants, FAK activity and its autophosphorylation site tyrosine 397 (Tyr 397) were shown to be critical for *E. coli* K1 invasion of BMECs [40]. The autophosphorylation site (Tyr 397) of FAK has been shown to bind Src kinases and phosphatidylinositol 3-kinase, and binding of one or both is required for FAK-mediated functions. Src kinases, however, were not critical in *E. coli* K1 invasion of BMECs. This was shown by the demonstration that pretreatment of BMECs with the Src kinases–specific inhibitor PP1 did not affect *E. coli* K1 invasion of BMECs and also that overexpression of Src kinase dominant-negative mutants did not block *E. coli* K1 invasion of BMECs [40].

In contrast, phosphatidylinositol 3-kinase activation and its association with FAK are required for *E. coli* K1 invasion of BMECs [41]. This was shown by blockade of both phosphatidylinositol 3-kinase activation and *E. coli* K1 invasion of BMECs with specific phosphatidylinositol 3-kinase inhibitor (LY294002) as

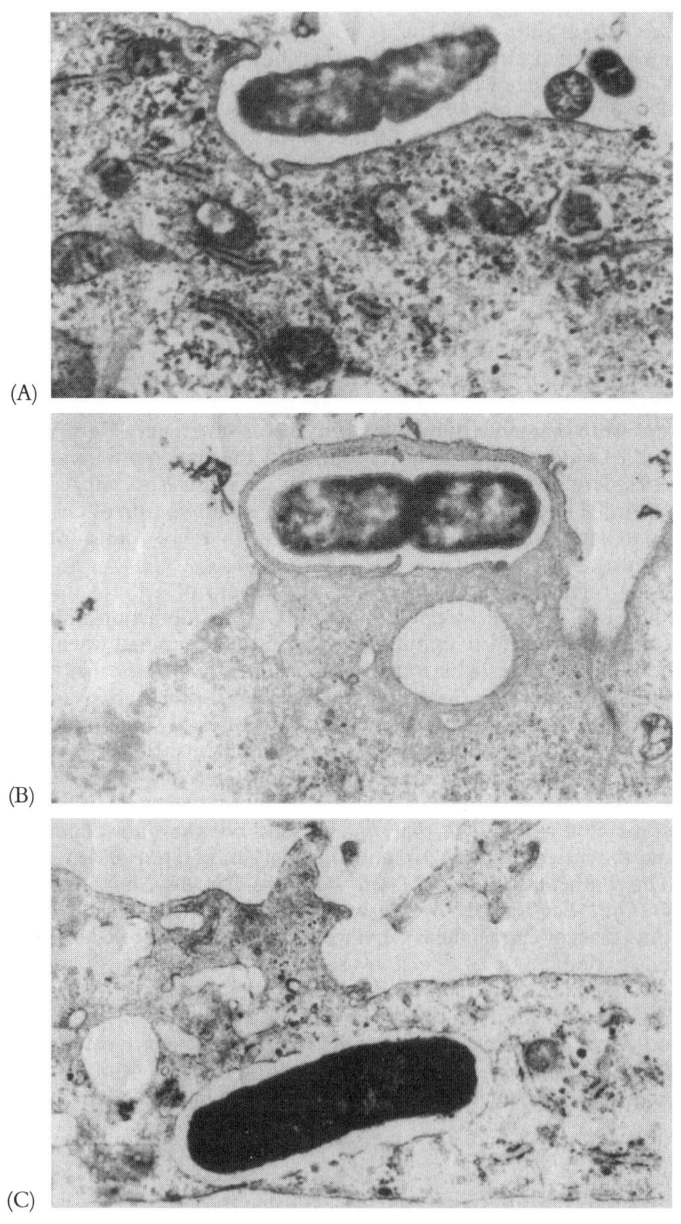

FIGURE 3 Transmission electron micrographs of BMEC illustrating *E. coli* K1 invasion and transmigration across BMEC. The bacterium appeared to closely contact with the BMEC membrane (*A*), elicit its own uptake (*B*), reside in vacuoles (*C*), and transmigrate across BMEC. BMEC vacuoles containing *E. coli* K1 were found to have markers for early and late endosomes but were devoid of lysosomal markers. (Modified with permission from *Infection and Immunity* [35].)

well as with dominant-negative mutants of phosphatidylinositol 3-kinase and FAK (Fig. 4). Phosphatidylinositol 3-kinase activation was abolished by FAK dominant-negative mutants [41] (see Fig. 4*D*), indicating that FAK is upstream of phosphatidylinositol 3-kinase in *E. coli* K1 invasion of BMECs. Phosphatidylinositol 3-kinase has been shown to participate in actin reorganization, recruitment of early endosome proteins, and movement of the endosomes along the microtubules. It remains to be determined how FAK and phosphatidylinositol 3-kinase activation contribute to *E. coli* K1 invasion of BMECs.

Phospholipase A_2, particularly cytosolic phospholipase A_2 ($cPLA_2$), has been shown to contribute to *E. coli* K1 invasion of BMECs [42]. This was shown by the demonstration that $AACOCF_3$, a selective $cPLA_2$ inhibitor, blocked *E. coli* K1 invasion of BMECs, and *E. coli* K1 invasion was decreased significantly in BMECs derived from $cPLA_2$ knockout mice compared with BMECs from control mice (Fig. 5). Phospholipase A_2 hydrolyzes phospholipids at their sn-2 position, resulting in the release of fatty acids, e.g., arachidonic acid. Actin cytoskeletal rearrangements in mammalian cells have been linked to intracellular signaling via metabolites of arachidonic acid. These findings indicate that FAK, phosphatidylinositol 3-kinase, and $cPLA_2$ activation contribute to *E. coli* K1 invasion of BMECs, presumably via affecting the signaling mechanisms associated with BMEC actin cytoskeletal arrangements.

It should be noted that bacterial trafficking mechanisms in BMECs differ between *E. coli* K1 and other meningitis-causing bacteria such as *Listeria monocytogenes* and group B *Streptococcus* (Table II). For example, BMEC actin cytoskeletal rearrangements are a prerequisite for BMEC invasion by *E. coli* K1, *L. monocytogenes*, and group B *Streptococcus* (see Table II) [35,43,44]. However,

TABLE II Comparison of Host Cell Cytoskeleton and Signaling Mechanisms in Bacterial Invasion of BMECs[a]

Mechanisms	*E. coli* K1	Group B streptococci	L. monocytogenes
Actin cytoskeletal rearrangements	+[b]	+	+
RhoA activation	+	ND[d]	ND
FAK activation	+	ND	–
Src activation	–[c]	ND	+
PI3K activation	+	–	+
cPLA2 activation	+	–	–
References	9, 40, 41	9, 37, 40	9, 15, 40, 41

[a]Modified with permission from *Infection and Immunity* [47].
[b]Denotes active participation in BMEC invasion.
[c]Denotes no role in BMEC invasion.
[d]ND denotes not examined.

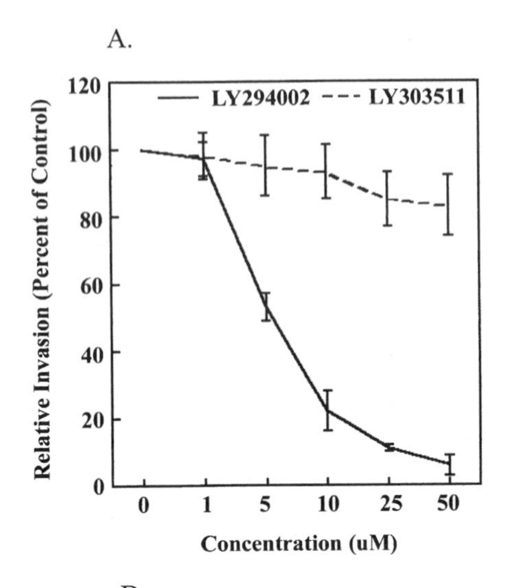

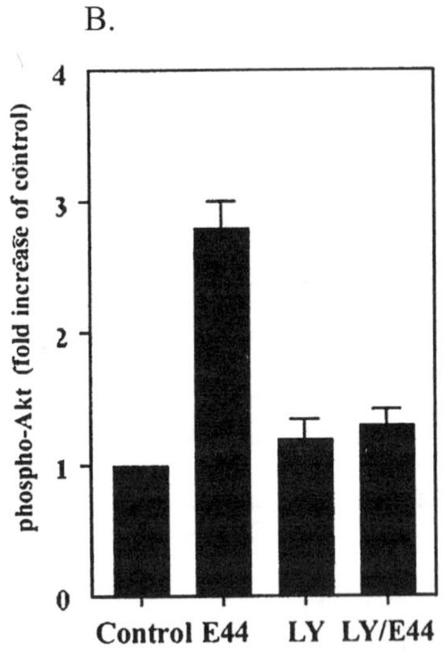

FIGURE 4 Blockade of (*A*) *E. coli* K1 invasion of BMECs and (*B*) phosphatidylinositol 3-kinase activation with specific phosphatidylinositol 3-kinase inhibitor (LY 294002) as well as using (*C*) dominant-negative mutants of phosphatidylinositol 3-kinase and (*D*) FAK dominant negative mutants. (*A*) LY294002-inhibited *E. coli* K1 invasion of BMEC in a dose-dependent manner, whereas LY303511, an inactive analogue of LY294002, had very little effect on *E. coli* K1 invasion of BMECs. (*B*) LY 294002 also inhibited phosphatidylinositol 3-kinase activation expressed as Akt activation. (*C*) Overexpression of dominant-negative mutants of either p85 ($\Delta p85$) or p110 ($p110\Delta K$) subunits of phosphatidylinositol 3-kinase significantly inhibited *E. coli* K1 invasion of BMECs. (*D*) Phosphatidylinositol 3-kinase activation is blocked by FAK dominant-negative mutants, FRNK (FAK-related kinase), and autophosphorylation mutant Phe397 FAK. (Modified with permission from the *Journal of Biological Chemistry* [41].)

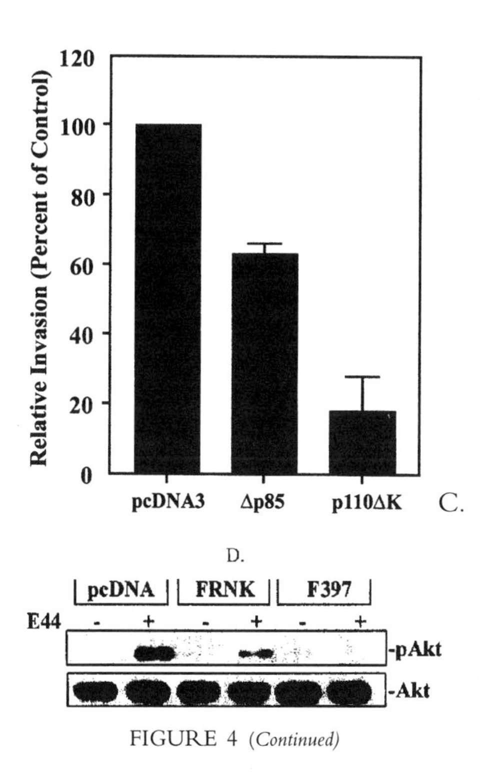

FIGURE 4 (*Continued*)

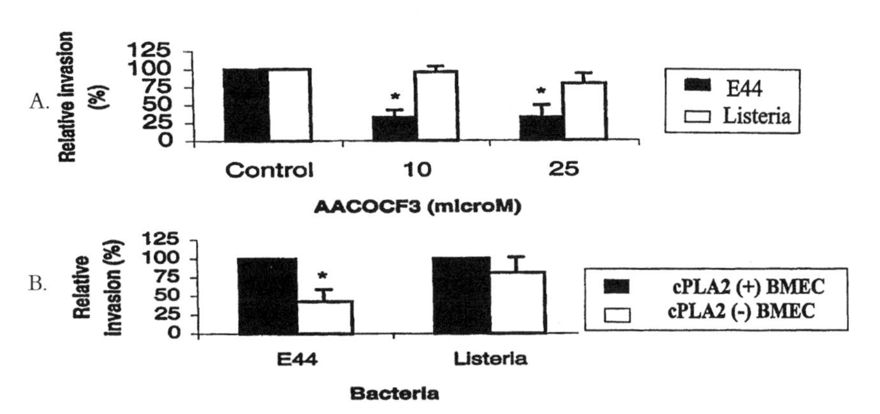

FIGURE 5 The role of $cPLA_2$ in *E. coli* K1 invasion of BMECs. (*A*) $AACOCF_3$, a selective $cPLA_2$ inhibitor, blocked *E. coli* K1 invasion of BMECs but exhibited no effect on *L. monocytogenes* invasion of BMECs. (*B*) *E. coli* K1 invasion was significantly less in BMECs derived from $cPLA_2$ knockout mice compared with BMECs derived from control mice. In contrast, *L. monocytogenes* invasion of BMECs did not differ between knockout mice and control mice. (Modified with permission from the *Journal of Infectious Diseases* [42].)

L. monocytogenes invasion of BMECs depends on Src kinases, not on FAK and $cPLA_2$ [35,42,45]. In contrast, group B streptococcal invasion of BMECs is independent of phosphatidylinositol 3-kinase and $cPLA_2$ [41,42]. BMEC vacuoles containing *E. coli* K1 were found to have markers for early and late endosomes but are devoid of lysosomal markers (K. J. Kim and K. S. Kim, unpublished observations), suggesting that there is an escape from transport to and/or a blockade of fusion to the lysosome. Additional studies are needed to understand the trafficking mechanisms involved in bacterial transcytosis of BMECs.

TRAVERSAL OF THE BLOOD-BRAIN BARRIER AS LIVE BACTERIA

Previous studies of *E. coli* K1 meningitis have shown that the K1 capsule is a critical determinant in the development of meningitis, defined as a positive CSF culture [14]. This was shown by the demonstration of sterile CSF cultures from animals infected with $K1^-$ strains, which was interpreted to indicate that the K1 capsule was necessary for the bacterial crossing of the blood-brain barrier. A recent study, however, has shown that both *E. coli* $K1^+$ and $K1^-$ strains are able to traverse BMECs *in vitro* and enter the CNS *in vivo*, but infections caused by $K1^+$ strains resulted in positive CSF cultures [46]. Thus the K1 capsule has, in addition to its well-recognized serum resistance and antiphagocytic properties, a role in the traversal of *E. coli* K1 across the blood-brain barrier as live bacteria. The nature of this BMEC activity that is bactericidal to *E. coli* strains without a capsule is currently unknown. This has been shown not to be related to nitric oxide (NO), peroxynitrites, superoxides, and other oxygen radicals [46].

CONCLUSION

A major limitation to advances in prevention and therapy of bacterial meningitis is our incomplete understanding of the pathogenesis of this disease, such as how circulating bacteria cross the blood-brain barrier. Successful isolation and cultivation of BMECs, which constitute the blood-brain barrier, and the development of an experimental hematogenous animal model that mimics closely the pathogenesis of human meningitis enabled dissection of the mechanisms of bacterial translocation across the blood-brain barrier. As summarized in Table III, the studies so far have indicated that *E. coli* K1 crossing of the blood-brain barrier requires a high degree of bacteremia. However, a high degree of bacteremia alone is not sufficient for the development of meningitis. Recent studies with *E. coli* K1 have shown that several microbial determinants such as the K1 capsule, OmpA, Ibe proteins, AslA, TraJ, and CNF1 contribute to invasion of BMECs that

TABLE III Pathogenetic Mechanisms Involved in *E. coli* K1 Traversal of the Blood-Brain Barrier

1. Magnitude of bacteremia (i.e., a threshold level)
2. *E. coli* invasion of BMECs a. Requirement of certain microbial structures (i.e., OmpA, Ibe, AslA, TraJ, CNF1) b. Presence of BMEC receptors (i.e., receptors for OmpA and IbeA)
3. Host cell cytoskeletal rearrangements and related signal transduction pathways (i.e., Rho GTPases, FAK, P13K, cPLA2) for *E. coli* K1 invasion of BMECs
4. Traversal of the blood-brain barrier as live bacteria

is required for successful penetration into the CNS in experimental hematogenous meningitis *in vivo*. Some of these *E. coli* determinants interact with specific receptors present in BMECs, suggesting that their contributions to invasion of BMECs most likely occur via ligand-receptor interactions (e.g., OmpA, IbeA). CNF1 activates RhoA GTPase, resulting in actin cytoskeletal rearrangements and invasion of BMECs. In contrast, it is unclear how other *E. coli* determinants contribute to invasion of BMECs (e.g., AslA, TraJ). In addition, bacterial trafficking of BMECs by *E. coli* K1 requires BMEC actin cytoskeletal reorganizations and activations of FAK, phosphatidylinositol 3-kinase, and $cPLA_2$ (see Table III). Of interest, these *E. coli* trafficking mechanisms differ from those of other meningitis-causing bacteria such as *L. monocytogenes* and group B *Streptococcus*. Structural genomic studies have identified DNA segments specific to the prototypes of meningitis-causing *E. coli* K1 (e.g., strains RS 218 and C5), and their sequencing is in progress. It is unclear, however, whether the sequence information specific to *E. coli* K1 will identify all the microbial determinants relevant to the pathogensis of *E. coli* K1 meningitis. This was exemplified by the identification of *E. coli* K1 structures contributing to crossing of the blood-brain barrier but having highly homologous structures in the *E. coli* K12 genome (e.g., *ompA*, *ibeB*, *ibeC*, and *aslA*). Thus it is likely that *E. coli* K1 determinants contributing to crossing of the blood-brain barrier are not clustered within K1-specific segments and include K1-specific genes as well as K12 homologues. In the meantime, it seems prudent to suggest that the prevention of bacterial multiplication in the blood necessary for bacterial entry into the CNS would be one potential approach to the prevention of serious *E. coli* meningitis.

ACKNOWLEDGMENTS

I would like to acknowledge that the information contained in this review was derived from studies carried out by my former and current laboratory members. This work was supported by NIH Grants RO1 NS 26310, AI 47225, and HL 61951.

REFERENCES

1. Unhanand, M., Musatafa, M. M., McCracken, G. H., and Nelson, J. D. (1993). Gram-negative enteric bacillarymeningitis: A twenty-one year experience. *J. Pediatr.* 122:15–21.
2. Dawson, K. G., Emerson, J. C., and Burns, J. L. (1999). Fifteen years of experience with bacterial meningitis. *Pediatr. Infect. Dis. J.* 18:816–822.
3. Robbins, J. B., McCracken, G. H., Gotschlich, E. C., *et al.* (1974). *Escherichia coli* K1 capsular polysaccharide associated with neonatal meningitis. *New Engl. J. Med.* 290:1216–1220.
4. Gross, R. J., Ward, L. R., Threlfall, E. J., *et al.* (1982). Drug resistance among *Escherichia coli* strains isolated from cerebrospinal fluid. *J. Hyg.* 90:195–198.
5. Korhonen, T. K., Valtonen, M. V., Parkinen, J., *et al.* (1985). Serotypes hemolysin production and receptor recognition of *Escherichia coli* strains associated with neonatal sepsis and meningitis. *Infect. Immun.* 488:486–491.
6. Sarff, L. D., McCracken, G. H., Jr., Schiffer, M. S., *et al.* (1975). Epidemiology of *Escherichia coli* in healthy and diseased newborns. *Lancet* 1:1099–1104.
7. Feigin, R. D., McCracken, G. H., Jr., and Klein, J. O. (1992). Diagnosis and management of meningitis. *Pediatr. Infect. Dis. J.* 11:785–814.
8. Quagliarello, V. Q., and Scheld, W. M. (1992). Bacterial meningitis: Pathogenesis, pathophysiology and progress. *New Engl. J. Med.* 327:864–871.
9. Kim, K. S. (2000). *E. coli* invasion of brain microvascular endothelial cells as a pathogenetic basis of meningitis. *Subcell Biochem.* 33:47–59.
10. Kim, K. S., Wass, C. A., and Cross, A. S. (1997). Blood-brain barrier permeability during the development of experimental bacterial meningitis in the rat. *Exp. Neurol.* 45:253–257.
11. Stins, M. F., Badger, J. L., and Kim, K. S. (2001). Bacterial invasion and transcytosis in transfected human brain microvascular endothelial cells. *Microb. Pathogenesis.* 30:19–28.
12. Stins, M. F., Gilles, F., and Kim, K. S. (1997). Selective expression of adhesion molecules on human brain microvascular endothelial cells. *J. Neuroimmunol.* 76:81–90.
13. Stins, M. F., Nemani, P., Wass, C., and Kim, K. S. (1999). *E. coli* binding to and invasion of brain microvascular endothelial cells derived from humans and rats of different ages. *Infect. Immun.* 67:5522–5525.
14. Kim, K. S., Itabashi, H., Gemski, P., *et al.* (1992). The K1 capsule is the critical determinant in the development of *Escherichia coli* meningitis in the rat. *J. Clin. Invest.* 90:897–905.
15. Huang, S. H., Wass, C. A., Fu, Q., *et al.* (1995). *E. coli* invasion of brain microvascular endothelia cell *in vitro*: Molecular cloning and characterization of *E. coli* invasion gene Ibe10. *Infect. Immun.* 63:4470–4475.
16. Huang, S. H., Chen, Y. H., Fu, Q., *et al.* (1999). Identification and characterization of an *E. coli* invasion gene locus *ibeB* required for penetration of brain microvascular endothelia cells. *Infect. Immun.* 67:2103–2109.
17. Wang, Y., Huang, S. H., Wass, C., and Kim, K. S. (1999). The gene locus *yijP* contributes to *E. coli* invasion of brain microvascular endothelial cells. *Infect. Immun.* 67:4751–4756.
18. Hoffman, J. A., Badger, J. L., Zhang, Y., *et al.* (2000). *E. coli* K1 *aslA* contributed to invasion of brain microvascular endothelial cells *in vitro* and *in vivo*. *Infect. Immun.* 68:5062–5067.
19. Badger, J. L., Wass, C., Weissman, S., and Kim, K. S. (2000). Application of signature-tagged mutagenesis for the identification of *E. coli* K1 genes that contribute to invasion of the blood-brain barrier. *Infect. Immun.* 68:5056–5061.
20. Dietzman, D. E., Fischer, G. W., and Schoenknecht, F. D. (1974). Neonatal *Escherichia coli* septicemia-bacterial counts in blood. *J. Pediatr.* 85:128–130.
21. Parkkinen, J., Korhonen, T. K., Pere, A., *et al.* (1988). Binding sites of the rat brain for *Escherichia coli* S-fimbriae associated with neonatal meningitis. *J. Clin. Invest.* 81:860–865.
22. Garner, A. M., and Kim, K. S. (1996). The effects of *E. coli* S-fimbriae and outer membrane protein A on rat pial arterioles. *Pediatr. Res.* 39:604–608.

23. Hacker, J. (1990). Genetic determinants coding for fimbriae and adhesins of extraintestinal *Escherichia coli*. *Curr. Top. Microbiol. Immunol.* 151:1–27.
24. Stins, M. F., Prasadarao, N. V., Ibric, L., *et al.* (1994). Binding characteristics of S-fimbriated *E. coli* to isolated brain microvascular endothelial cells. *Am. J. Pathol.* 145:1228–1236.
25. Prasadarao, N. V., Wass, C. A., Hacker, J., *et al.* (1993). Adhesion of S-fimbriated *Escherichia coli* to brain glycolipids mediated by *sfaA* gene-encoded protein of S-fimbriae. *J. Biol. Chem.* 268:10356–10363.
26. Moch, T., Hoschutzky, H., Hacker, J., *et al.* (1987). Isolation and characterization of the α-sialyl-β-galactosyl-specific adhesin from fimbriated *Escherichia coli*. *Proc. Natl. Acad. Sci. USA* 84:3462–3466.
27. Nemani, P. V., Wass, C., Stins, M. F., *et al.* (1996). Outer membrane protein A of *E. coli* contributes to invasion of brain microvascular endothelia cells. *Infect. Immun.* 64:146–153.
28. Nemani, P. V., Wass, C. A., and Kim, K. S. (1996). Endothelia cell GlcNAc1–4 GlcNAc epitopes for outer membrane protein A enhance traversal of *E. coli* across the blood-brain barrier. *Infect. Immun.* 64:154–160.
29. Nemani, P. V., Huang, S. H., Wass, C. A., and Kim, K. S. (1999). Identification and characterization of a novel Ibe10 binding protein contribution to *E. coli* invasion of brain microvascular endothelial cells. *Infect. Immun.* 67:1131–1138.
30. Badger, J. L., and Kim, K. S. (1998). Environmental growth conditions influence the ability of K1 *Escherichia coli* to invade brain microvascular endothelia cells (BMECs) and confer serum resistance. *Infect. Immun.* 66:5692–5697.
31. Badger, J. L., Wass, C. A., and Kim, K. S. (2000). Identification of *E. coli* K1 genes contributing to human brain microvascular endothelia cell invasion by differential fluorescence induction. *Mol. Microbiol.* 36:174–182.
32. Khau, N. A., Wang, Y., Kim, K. I., Chung, J. W., Wass, C. A., and Kim, K. S. (2002). Cytoxic necrotizing factor 1 contributes to *Escherichia coli* K1 invasion of the central nervous system. *J. Biol. Chem.* 277:15607–15612.
33. Flatau, G., Lemichez, E., Gauthier, M., *et al.* (1997). Toxin-induced activation of the G protein p21 Rho by deamidation of Glutamine. *Nature* 387:729–733.
34. Schmidt, G., Sehr, P., Wilm, M., *et al.* (1997). Gln 63 of Rho is deamidated by *Escherichia coli* cytotoxic necrotizing factor 1. *Nature* 387:725–729.
35. Nemani, P. V., Stins, M., Wass, C. A., *et al.* (1999). Outer membrane protein A promoted cytoskeletal rearrangement of brain microvascular endothelial cells is required for *E. coli* invasion. *Infect. Immun.* 67:5775–5783.
36. Wang, Y., and Kim, K. S. (2002). Role of OmpA and IbeB in *Escherichia coli* K1 invasion of brain microvascular endothelial cells *in vitro* and *in vivo*. *Ped. Res.* 51:559–563.
37. Rode, C. K., Welkerson-Watson, L. J., Johnson, A. T., and Bloch, C. A. (1999). Type-specific contributions to chromosome size differences in *Escherichia coli*. *Infect. Immun.* 19:230–236.
38. Bonacorse, S. P. P., Clermont, O., Tinsley, C., *et al.* (2000). Identification of regions of the *Escherichia coli* chromosome specific for neonatal meningitis-associated strains. *Infect. Immun.* 68:2096–2101.
39. Mercer, A. A., Morelli, G., Heuzenroeder, M., *et al.* (1984). Conservation of plasmids among *Escherichia coli* K1 isolates of diverse origins. *Infect. Immun.* 46:649–657.
40. Reddy, M. A., Wass, C. A., Kim, K. S., *et al.* (2000). Involvement of focal adhesion kinases in *E. coli* invasion of human brain microvascular endothelial cells. *Infect. Immun.* 68:6419–6422.
41. Reddy, M. A., Nemani, P. V., Wass, C. A., and Kim, K. S. (2000). Phosphatidylinositol 3-kinase activation and interaction with focal adhesion kinase in *E. coli* K1 invasion of human brain microvascular endothelial cells. *J. Biol. Chem.* 275:36769–36774.
42. Das, A., Asatryan, L., Reddy, M. A., *et al.* (2000). Differential role of cytosolic phospholipase A_2 in the invasion of brain microvascular endothelial cells by *Escherichia coli* and *Listeria monocytogenes*. *J. Infect. Dis.* 184:732–737.

43. Nizet, V., Kim, K. S., Stins, M., *et al.* (1997). Invasion of brain microvascular endothelial cells by group B streptococci. *Infect. Immun.* 65:5074–5081.
44. Greiffenberg, L., Goebel, W., Kim, K. S., *et al.* (1998). Interaction of *Listeria monocytogenes* with human brain microvascular endothelial cells: In1B-dependent invasion, long-term intracellular growth, and spread from macrophages to endothelial cells. *Infect. Immun.* 66:5260–5267.
45. Ireton, K., Payrastre, B., and Cossart, P. (1999). The *Listeria monocytogenes* protein InlB is an agonist of mammalian phosphoinositide 3-kinase. *J. Biol. Chem.* 274:17025–17032.
46. Hoffman, J., Wass, C., Stins, M. F., *et al.* (1999). The capsule supports survival but not traversal of K1 *E. coli* across the blood-brain barrier. *Infect. Immun.* 67:3566–3570.
47. Kim, K. S. (2001). *E. coli* translocation at the blood-brain barrier. *Infect. Immun.* 69:5217–5222.

PART III

Escherichia coli Virulence Factors

CHAPTER 10

Adhesive Pili of the Chaperone-Usher Family

James G. Bann
Department of Biochemistry and Molecular Biophysics, Washington University School of Medicine, St. Louis, Missouri

Karen W. Dodson
Department of Molecular Microbiology and Immunology, Washington University School of Medicine, St. Louis, Missouri

Carl Frieden
Department of Biochemistry and Molecular Biophysics, Washington University School of Medicine, St. Louis, Missouri

Scott J. Hultgren
Department of Molecular Microbiology and Immunology, Washington University School of Medicine, St. Louis, Missouri

INTRODUCTION

The mechanisms by which pathogens are able to recognize and enter host tissues and the subsequent responses from the host tissue to combat the invading microbe are subjects of intense investigation from researchers worldwide (Donnenberg, 2000). For a variety of gram-negative bacteria, the initial establishment of infection requires the presence of long, hairlike structures called *pili* that extend from the surface of the bacterium (Fig. 1) (Hultgren *et al.*, 1996). These pili, which include the chaperone-usher family of pili and type IV pili, have at the distal tip of the pilus an adhesin protein that binds to a specific receptor on the host cell membrane (Lund *et al.*, 1987; Hanson *et al.*, 1988). The adhesin is the key determinant in the ability of the bacterium to recognize and bind host cells and tissues and is the primary target of researchers whose aim is to develop vaccines that will prevent this binding event (Langermann *et al.*, 1997). The recent crystal structures of the receptor binding domains of the adhesins PapGII for P-pili and FimH for type 1 pili, in complex with their natural carbohydrate ligands, will greatly aid in the design of novel therapeutics directed

Escherichia coli: Virulence Mechanisms of a Versatile Pathogen
ISBN 0-12-220751-3

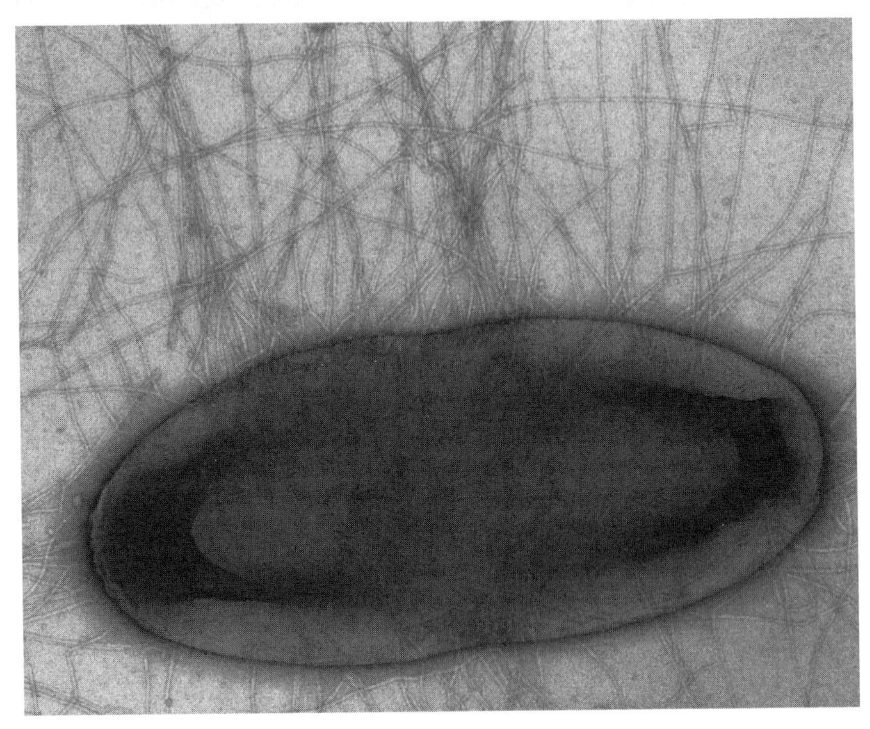

FIGURE 1 Negative-stain electron micrograph of *Escherichia coli* expressing P pili [MC4100 (Casadaban, 1976)/pPap5 (Lindberg *et al.*, 1986)].

at blocking host cell—adhesin interactions (Dodson *et al.*, 2001; Choudhury *et al.*, 1999; Hung *et al.*, 2002).

The pili that are assembled via the chaperone-usher family are composite structures made up of individual subunits that are coordinated to assemble into a thin, flexible-tip fibrillum connected to a thicker pilus rod (Kuehn, Heuser, Normark and Hultgren, 1992). The crystallographic structures of the P-pilus tip fibrillum subunit PapK bound to the chaperone PapD and the type 1 pilus adhesin FimH bound to the chaperone FimC show that each pilin domain structure exhibits an immunoglobulin (Ig) fold but with an unexpected difference (Sauer *et al.*, 1999; Choudhury *et al.*, 1999). The seventh G β-strand, which is required to complete the Ig fold, is absent from the subunit itself. Remarkably, this seventh β-strand is provided by the G_1 β-strand of the chaperone. This mechanism has been termed *donor-strand complementation* to indicate that the chaperone provides the necessary missing steric information required to complete the Ig fold. Since the chaperone is not incorporated into the pilus, the steric information provided by the chaperone must be filled in by some other protein in the final structure. This happens by a process termed *donor-strand exchange*, in which during polymerization of the subunits the G_1 β-strand of the chaperone, bound to the subunit, is replaced by the N-terminal strand of another subunit, again completing the subunit Ig fold. This N-terminal strand has a very similar pattern of alternating hydrophobic residues as is found in the chaperone G_1 β-strand and is well conserved among the different subunits (Soto *et al.*, 1998). Furthermore, biochemical and genetic data have shown an important role of this strand in subunit-subunit interactions (Soto *et al.*, 1998; Barnhart *et al.*, 2000). Based on the donor-strand complementation mechanism, a model of the pilus has been constructed that is in excellent agreement with the data obtained using both high-resolution electron microscopy and x-ray diffraction (Sauer *et al.*, 1999; Choudhury *et al.*, 1999). Thus donor-strand complementation is a common theme encountered in both the biogenesis and the final structure of pili.

The process of assembling the subunits begins when the immature subunits are first secreted into the periplasmic space through the Sec pathway (Dood *et al.*, 1984; Roosendal *et al.*, 1987). The subunit encounters the chaperone, which facilitates folding by binding and completing the Ig fold of the subunit. Without the chaperone present, the subunits are highly susceptible to proteolytic degradation. The chaperone-subunit complex is then brought to the usher, an integral outer membrane protein that forms a channel to the outside of the cell. Through a mechanism not currently understood, binding to the usher facilitates the donor-strand exchange mechanism mentioned previously, allowing the dissociation and the subsequent polymerization of the subunits to occur. The subunits then pass through the channel of the usher into the outside environment.

In this chapter we will discuss what is known about the adhesive pili of the chaperone-usher family, beginning with the structure of the pilus itself and in particular the structure of the adhesin at the tip of the pilus. We will discuss the structure of the chaperone and the usher, focusing on the two best-characterized systems, the Pap and type 1 systems. Through this we hope to provide a succinct and up-to-date overview of what are known about the adhesive pili of the chaperone-usher pathway.

CHARACTERISTICS OF PILI

Molecular Biology

A number of gram-negative pathogens have been found to produce adhesive pili from their surfaces; however, the best-characterized systems of the chaperone-usher family of pili are the P (pylenophritis) pilus and the type 1 pilus systems found in uropathogenic strains of *Escherichia coli* (Sauer, *et al.*, 2000). These systems are encoded on gene clusters that contain all the specific information necessary for the assembly and structure of the pilus (Fig. 2). Comparing the P-pilus (*pap*) and type 1 (*fim*) gene clusters, the genes that are required to form the structure of the pilus include a major rod subunit (*papA* or *fimA*), an adhesin

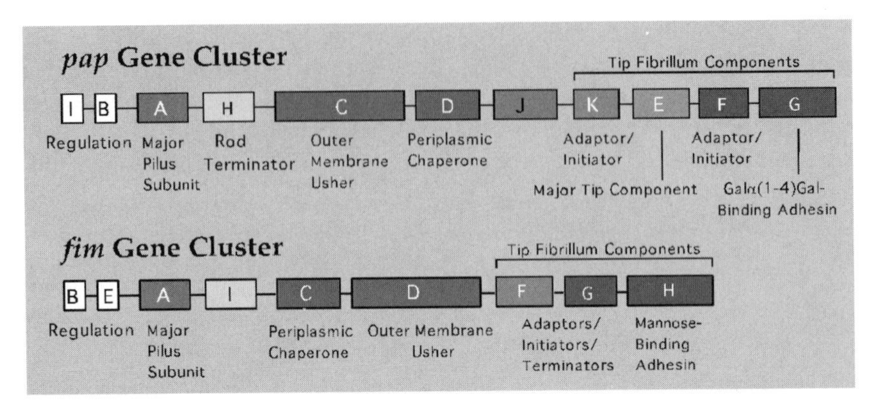

FIGURE 2 Gene clusters encoding P and type 1 pili.

(*papG* or *fimH*), adapter and tip fibrillar subunits (*papEFK* or *fimFG*), and an anchoring gene (*papH*).

Two proteins also are necessary for assembly of the pili, an 80- to 90-kDa outer membrane protein usher (*papC* or *fimD*) and a smaller, 24- to 28-kDa chaperone protein (*papD* or *fimC*) (Norgren *et al.*, 1987; Lindberg *et al.*, 1989; Jones *et al.*, 1993). The genes that encode the chaperone and usher are highly conserved among different organisms expressing different types of pili. PapD is the prototype of a large superfamily of chaperones involved in the production of adhesive organelles (Hung *et al.*, 1996). The superfamily is composed of two distinct classes depending on whether the chaperone assembles pili or non-fimbrial-like structures and the primary structure of the chaperone (Hung *et al.*, 1996). Specifically, the presence of a short F_1–G_1 loop (FGS) in the N-terminal domain of the chaperone correlates with the formation of rodlike pili, whereas the presence of a long F_1–G_1 loop (FGL) correlates with the formation of non-fimbrial-like structures (Hung *et al.*, 1996). PapD and FimC are both FGS family members. A similar kind of difference is observed in the subunits that are assembled from these two subfamilies. The subunits in the FGS subfamily contain a highly conserved patch of alternating hydrophobic residues at positions 4, 6, and 8, residues from the C-terminal end that includes a penultimate tyrosine as well as a conserved glycine at position 14 (Kuehn *et al.*, 1993; Hung *et al.*, 1996). From crystallographic studies of chaperone-subunit complexes, this patch is now known to be the sixth F-strand of the subunit Ig fold, forming a β-zipper with the G_1 strand of the chaperone (Sauer *et al.*, 1999; Choudhury *et al.*, 1999). In the FGL subfamily this alternating hydrophobic patch is less extensive but includes a tyrosine at position 3 as well as the conserved glycine at position 14. Since most of these residues from both subfamilies are conserved, the β-zipper motif also may be conserved.

The usher is also part of a large conserved family (Van Rosmalen and Saier Jr., 1993). Phylogenetic data suggest that the chaperone and usher proteins from each subfamily may have coevolved with each other (Van Rosmalen and Saier Jr., 1993). Structurally, the usher is known to have membrane-spanning, periplasmic, and pore-forming domains (Valent *et al.*, 1995). Through interactions with the pore-forming domains, the usher can oligomerize to form ring-shaped complexes that have a central pore of approximately 2 nm (Thanassi *et al.*, 1998). It is through this pore that the subunits pass, giving rise eventually to the long oligomeric complex of the pilus.

Structure of Pili

In the P and type 1 systems, the major rod, adhesin, and adapter subunits are assembled into a composite fiber consisting of two distinct subassemblies differing in diameter and length. The rod portion of P-pili or type 1 pili is rigid and is composed of repeating subunits of PapA or FimA. Both are approximately 68 Å in diameter and approximately 1000 nm in length. The tip fibrillae, on the other hand, exhibit variations in length. While both are 20 Å in diameter, the length of the tip fibrillum of P-pili is 40–60 nm, whereas for type 1 pili it is approximately 16 nm (Jones *et al.*, 1995). Although the rod is very rigid, early electron micrographs of P or type 1 pili showed that mechanical shearing can cause sharp kinks in the structure where the rod has been partially broken (Fig. 3) (Gong and Makowski, 1992; Bullitt and Makowski, 1995). In some instances, the points in between the break are large, and a small, thin filament is observed spanning the break. This filament has the same dimensions as the tip fibrillae, suggesting that the thin filament is analogous to the tip fibrillae formed by the repeating subunits of PapE (Bullitt and Makowski, 1995). The observation of local unraveling of the pilus rod suggested a mechanism for how the pilus was constructed. Based on these early electron microgaphs, as well as x-ray studies of purified pili (Mitsui *et al.*, 1973; Gong and Makowski, 1992; Bullitt and Makowski, 1995), it was proposed that the rod is generated by the winding up of a thin, tip fibrillae like structure that forms a right-handed helical cylinder with 3.28 subunits per turn (Bullitt and Makowski, 1995). This helical cylinder has a central cavity that winds through the center of the pilus and is approximately 25 by 15 Å. The tip fibrillum, however, is an unwound, or open, form of this helical structure (Bullitt and Makowski, 1995). Recently, a higher-resolution model of the pilus has been constructed based on the crystallographic structures of the chaperone-subunit complexes PapD-PapK or FimC-FimH (Fig. 4) (Sauer *et al.*, 1999; Choudhury *et al.*, 1999). In these structures, an N-terminal β-strand from a neighboring subunit occupies the groove that is also occupied during biogenesis by the G_1 strand of the chaperone. The N-terminal strand, present in each subunit, has been shown to have a conserved sequence of

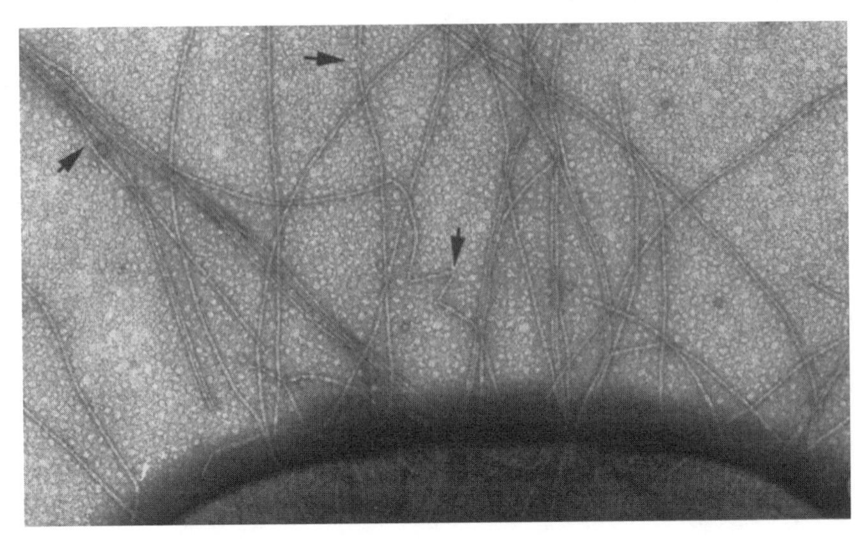

FIGURE 3 Negative-stain electron micrograph of *Eschrichia coli* expressing P pili (MC4100/pPap5) (see Fig. 1). The arrows point to sharp bends in the structure or to thin filaments spanning a break.

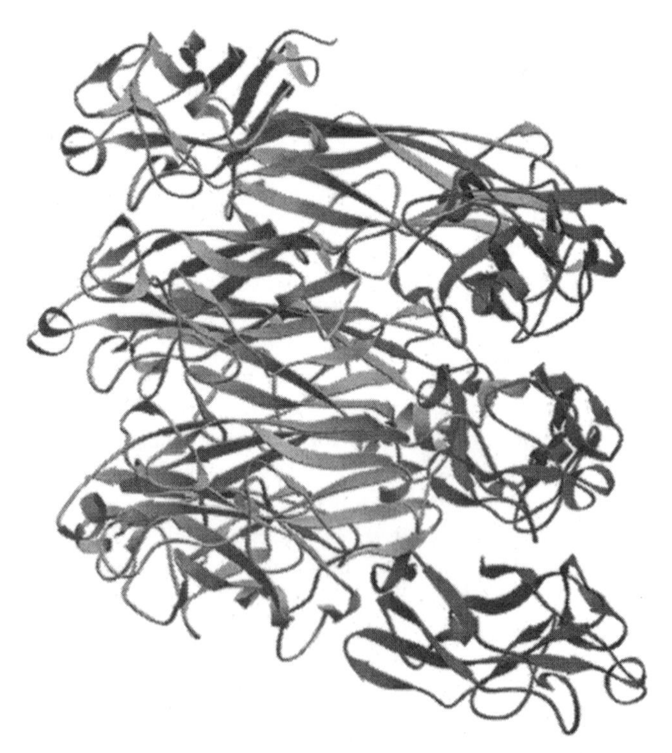

FIGURE 4 Ribbon model of PapK subunits assembled into a pilus rod. Note that the N-terminal β-strand of the preceding subunit is inserted into the G groove that in the PapD-PapK crystal structure is occupied by PapD. (Reprinted with permission from Sauer *et al.*, 1999.)

alternating hydrophobic residues similar to that found for the G_1 strand of the chaperone. Mutational analysis of these alternating hydrophobic residues in the rod subunit PapA has shown that these residues are critical for subunit-subunit interactions (Soto *et al.*, 1998). The interactions that occur between a chaperone-subunit complex and the usher must then facilitate the ordered unzipping of the chaperone from the subunit and the subsequent rezipping of the next subunit's N-terminal strand (Bullitt *et al.*, 1996). The removal of the G_1 strand of the chaperone from the subunit exposes the hydrophobic core of the subunit. The subsequent insertion of the N-terminal strand of the next subunit covers the hydrophobic core and stabilizes the fold. This process of exchanging the G_1 strand of the chaperone with the N-terminal strand of the next subunit has been termed *donor-strand exchange* (Sauer *et al.*, 1999 Choudhury *et al.*, 1999; Barnhart *et al.*, 2000). The final pilus structure is thus an assembly of approximately 1000 interlocking Ig folds, in which each domain is made up of parts from two different subunits.

Donor-Strand Complementation and Exchange

In an effort to show that subunit-subunit interactions in the pilus depend on a donor-strand exchange mechanism, a donor-strand-complemented form of the adhesin FimH (dscFimH) was engineered genetically (Barnhart *et al.*, 2000). FimH is a two-domain protein with an N-terminal receptor-binding domain that binds to D-mannose and a C-terminal pilin domain that binds to the chaperone and presumably to FimG, the penultimate component of the type 1 tip fibrillum (Jones *et al.*, 1995). The donor-strand complementation of FimH was achieved by genetically engineering onto the C-terminal domain of FimH the 13-residue N-terminal extension of FimG, including a 4-amino-acid linker. The new polypeptide, when expressed in the absence of the chaperone, was stable. Previous attempts at synthesizing subunits in the absence of the chaperone yielded either no or very little of the subunit, presumably due to proteolytic degradation (Hultgren *et al.*, 1989; Slonim *et al.*, 1992). The dscFimH did not bind to the chaperone, nor did it interact with the subunits to form pili, since the dscFimH was not able to complement a *fimH*–type 1 gene cluster to give type 1 pili. Presumably, the occupation of the groove where the G strand is normally located prevents interactions with the chaperone or other subunits. Further, urea-denatured dscFimH did not aggregate on dilution in pH 6.5 buffer, and it regained much of its native structure, whereas urea-denatured wild-type FimH, when purified away from FimC, aggregated on dilution. These data indicate that the subunit requires a seventh β-strand to fold into a stable structure and that this G β-strand is provided only by the chaperone or another subunit. Although several proteins require accessory proteins or additional domains to fold (Shinde *et al.*, 1997), to our knowledge, this is the first observation of a protein that is synthesized in an incomplete form, only to have that form made complete by another protein.

Structure of Adhesins

For the invading bacterium, the ability to sense the host tissue depends on an adhesin protein at the tip of the pilus (Lund *et al.*, 1987; Hultgren *et al.*, 1989). This protein functions to bind to a receptor, frequently a carbohydrate or a carbohydrate complex located on the host cell surface, and initiate the colonization and/or invasion of the host tissue. As mentioned, for type 1 pili the adhesin is FimH, whereas for P-pili the adhesin is PapG. Recently, the crystal structures of FimC-FimH bound to D-mannose and the receptor binding domain of PapGII bound to globoside (GbO4) (Kihlberg *et al.*, 1989) have been solved by x-ray crystallography (Hung *et al.*, 2002; Dodson *et al.*, 2001). PapGII is one of three different alleles of PapG (I, II, and III), each exhibiting different binding specificities to different receptor isotypes (Dodson *et al.*, 2001). These structures have revealed for the first time the structural basis for carbohydrate recognition by two different types of pili and, interestingly, show two different modes of binding.

While the lower halves of the molecules are similar, being comprised of a β-barrel, there are very distinct differences in the size and shape of the region where the carbohydrate binds (Fig. 5). The FimH binding pocket resides at the tip of the molecule, where a negative anion hole is formed that may direct interactions with positively charged molecules on the surface of the host. This binding pocket is made up of residues Phe1, Asn46, Asp47, Asp54, Gln133, Asn135, Asp140, and Phe142. Mutational analysis of these residues indicates that the pocket is optimized for binding to the small D-mannose rather than a larger, more complex sugar because mutations that specifically block interactions with D-mannose also block interactions with the host cell membrane (Hung *et al.*, 2002). The crystal structure of PapGII bound to GbO4 shows that a different mechanism of recognizing and binding carbohydrate is used. Rather than a binding pocket residing at the tip of the molecule, the pocket of PapGII lies on the side but like FimH involves aromatic as well as charged residues that interact with the carbohydrate. The residues in PapGII that are critical for binding include Trp107, which forms contacts with the hydrophobic surface of GbO4 (Striker *et al.*, 1995), and Arg170 and Lys172, which form several direct and water-mediated contacts with the hydroxyls of the sugar. Mutations at these positions to alanine were shown to abolish the ability to cause hemagglutination of human erythrocytes. Based on the predicted orientation of the GbO4 sugar in the eukaryotic membrane, the location of the binding site on the side of the protein suggests that PapGII would bind parallel to the cell surface. This may explain in part why

FIGURE 5 Ribbon diagram of the receptor-binding domains of FimH (*left*) and PapGII (*right*) in complex with carbohydrate (*sticks*). Note the similarity in the bottom half of the structures.

the tip fibrillum for P-pili is longer and more flexible than for type 1 pili in order to accommodate this side-on mode of binding.

BIOGENESIS OF PILI

The ability to assemble the subunits into the pilus is critically dependent on the formation of a complex between the chaperone and a subunit. Early studies on the role of the chaperone PapD showed that it was required for the stability of the subunits (Norgren *et al.*, 1987). When the subunits of the pilus are synthesized independently without coexpression of PapD, very little protein is observed in periplasmic extracts, but when both PapD and subunit genes are present and induced concomitantly, a stable chaperone-subunit complex is obtained. The sensitivity of the subunits to proteolysis indicates that the subunit is unable to achieve a correctly folded, stable conformation in the absence of the chaperone. The structural details of how the chaperone is able to confer this resistance have been elucidated recently with the crystal structures of the PapD-PapK and FimC-FimH (see above).

Crystal Structure of PapD

The first of the superfamily of chaperones to have its structure determined by x-ray crystallography was PapD (Holmgren and Branden, 1989). Because it was the first structure to be determined, it has become the prototype for all chaperones that belong to the chaperone-usher family. Since then, the structure of FimC, which is responsible for assembly of type 1 pili, has been determined by both nuclear magnetic resonance (NMR) and crystallography, as well as the chaperone SfaE (Pellecchia *et al.*, 1998; Choudhury *et al.*, 1999; Stefan Knight, unpublished data). These structures show that the chaperone is a two-domain protein with the two domains oriented together in the shape of a boomerang. The N-terminal domain exhibits the s-type Ig found in such proteins as CD4 and the fibronectin type III repeat and closely resembles the major sperm protein (MSP) (Bullock *et al.*, 1996). The fold contains seven antiparallel β-strands but with a distinctive kink in the D_1 strand that is imparted by the presence of *cis*-proline residues. The C-terminal domain exhibits low structural homology with the Ig set (Bork *et al.*, 1994). The C-terminal domain, however, is structurally similar to the small α-amylase inhibitor tendamistat, a protein whose folding pathway has been studied extensively (Wiegand *et al.*, 1995; Pappenberger *et al.*, 2001). At the end of the C-terminal domain is a disulfide bond that functions to link the seventh G_2 strand to a short additional strand, H_2. Although a C-terminal disulfide is not a common feature among the chaperones, it has been shown to be critical for the *in vivo* stability of PapD (Jacob-Dubuisson *et al.*, 1994). This disulfide is oriented toward the hydrophobic core of the C-terminal domain, and mutation of both of these cysteine residues to serines (C207S + C212S) resulted in very little production of PapD. Also, these mutants were unable to bind subunits or to complement a *papD*–P-pilus gene cluster. Finally, periplasmic disulfide-bond isomerase DsbA is required for formation of this disulfide in PapD because a DsbA null mutant resulted in very little PapD production *in vivo* (Jacob-Dubuisson *et al.*, 1994). These results indicate that while the disulfide is not strictly conserved among the chaperone family, for those chaperones with the disulfide, it provides a necessary stabilizing effect. Possible mechanisms for the role of the disulfide include increasing the conformational stability or decreasing the susceptibility to endopeptidases, mechanisms that have been proposed to explain the enhanced stability of a cytotoxic ribonuclease (Leland, 2000). Further work on the folding and stability of PapD both *in vivo* and *in vitro* will shed light on the role of the disulfide on chaperone stability.

The domains of PapD are oriented in the shape of a boomerang with the help of three conserved residues that form a unique interdomain salt bridge: Glu83, Arg116, and Asp196. The role of these residues in the stability of the chaperone has been explored using site-directed mutagenesis, in which each of these has been changed to Ala or to the more conserved amides (Glu83Gln and Asp196Asn) or Arg116Lys (Hung *et al.*, 1999). The Glu83 and Arg116 mutations resulted in very little protein being found in the periplasm, suggesting that the integrity of the salt bridge is critical for the stability of the molecule. Interestingly, while mutations at Asp196 did not significantly alter the stability of the chaperone, they completely diminished the ability of the chaperone to assemble pili. The Asp196 mutants are still able to bind subunits, however, and it was suggested that Asp196 may play a role in a step occurring after binding to the subunit. The COOH of Asp196 potentially could interact with Arg8, a critical residue for subunit binding, by competing for the C-terminal COOH of a subunit and facilitating dissociation of the subunit (Hung *et al.*, 1999).

Structural Basis of Chaperone-Subunit Interactions

It was not until the crystal structure of PapD in complex with a C-terminal peptide of PapG (1′–19′, corresponding to residues 296–314) that the interactions with the subunit began to be elucidated (Kuehn *et al.*, 1993). In this structure, the peptide formed an extended β-zipper along the G_1 strand of the chaperone, stabilized by hydrophobic, van der Waals contacts, and hydrogen bonds. This β-zipper is locked into place by hydrogen bonds between the C-terminal carboxylate group of the peptide and Arg8 and Lys112, present in the cleft of the chaperone (Slonim *et al.*, 1992). Arg8 and Lys112 are invariant between the FGS and FGL subfamilies (Hung *et al.*, 1996; Holmgren *et al.*, 1992). Along the G_1 strand is a pattern of alternating hydrophobic residues, conserved throughout all 30 members of the chaperone superfamily. Later mutational analysis of two of these residues, Ile105 and Leu107, to alanine or glutamate resulted in decreased or no subunit binding and the ability to assemble pili, respectively (Hung, 1999). Since the mutant PapDs were produced at levels similar to wild type, the mutations reflected solely an inability to stabilize the chaperone-subunit complex. The same panoply of interactions was observed again in the crystal structure of PapD in complex with the peptide PapK (1′–19′, corresponding to residues 139–157) (Soto *et al.* 1998). In this structure, the hydrogen bonds between the G_1 strand of the chaperone and that of the peptide were the same as those observed in the crystal structure with the PapG peptide, forming a β-zipper. Additionally, in comparing the apo-PapD and the PapD–G-peptide and PapD–K-peptide complexes, it was observed that there is a substantial amount of ordering and elongation of the F_1–G_1 loop on binding. Recent studies using ^{19}F-NMR also have revealed that the structure is induced to a more ordered conformation on ligand binding, and this stabilization extends not just to the F_1–G_1 loop but also to the entire protein (J. Bann and C. Frieden, unpublished data).

The structural basis for chaperone-subunit interaction has been revealed by the recent crystal structures of PapD-PapK and FimC-FimH (Sauer *et al.*, 1999; Choudhury *et al.*, 1999). These structures show that the binding surface is similar to what is observed with the peptides, forming contacts with the F strand of the subunit and the G_1 strand of the chaperone, as mentioned earlier (Fig. 6). The G_1 strand, however, is aligned parallel to, rather than antiparallel as in other Ig folds, the F strand of the subunit. More extensive chaperone-subunit contacts are observed and include a substantial contribution to the hydrophobic core of the subunit. In the crystal structure of PapD-PapK the subunit also makes several contacts with residues from the C-terminal domain of the chaperone, in contrast to transverse relaxation-optimized spectroscopy (TROSY) NMR experiments on FimC-FimH (Pellecchia *et al.*, 1999). Interestingly, the N-terminal strand of the PapK subunit, which is postulated to insert into the groove of PapE, was disordered and not able to be observed in structure. Many proteins, including DNA binding proteins, undergo transitions that couple local folding with ligand binding (Spolar and Record, 1994). A transition from a disordered peptide to an ordered β-sheet may contribute substantially to the energetics involved in stabilizing the pilus structure.

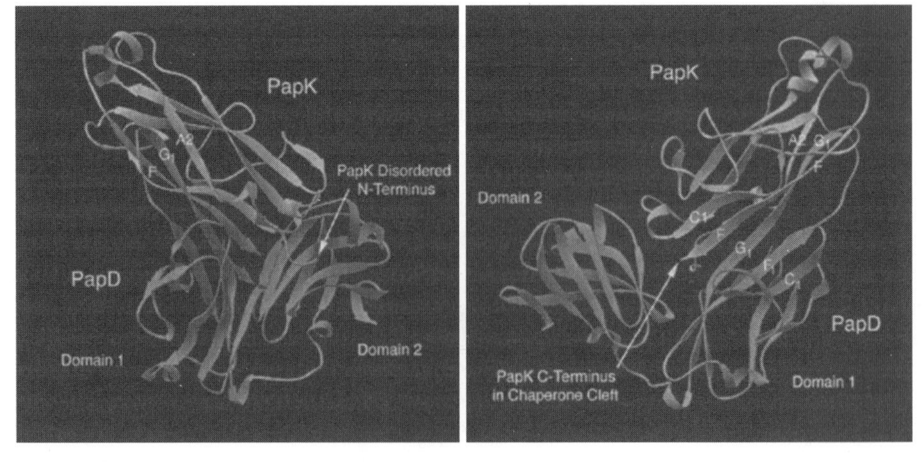

FIGURE 6 Ribbon diagrams of the PapD-PapK complex from two different views. Note the parallel orientation of the G_1 strand of the chaperone and the F strand of the subunit.

The Usher

As mentioned previously, the usher is an integral outer membrane protein. The structure is likely similar to that of other outer membrane porin proteins, comprised of a β-barrel (Koebnik *et al.*, 2000). PapC, the outer membrane usher for P-pili, is approximately 9 nm in diameter and forms a pore that is approximately 2 nm in diameter (Fig. 7) (Thanassi *et al.*, 1998). The pore is large enough that the adhesin and the rest of the tip fibrillum can pass through in a fully folded state. In order for the rod to pass through, it has been proposed that it does so in a fibrillar structure and that the side-to-side PapA interactions that form the helical structure would occur only after passage through the pore. Experiments on whether a cellular energy source is required to drive the movement of the subunits through the pore have indicated that the export is independent of ATP, a pH gradient, or an electrochemical gradient (Jacob-Dubuisson *et al.*, 1994). A possible mechanism for driving the export of the subunits is the actual process of winding the major-rod helix outside the bacterium. This would facilitate movement of the subunits through the pore by the favorable and possibly cooperative transition from a linear fiber to a helical rod (Thanassi *et al.*, 1998).

The name *usher* was obtained through experiments showing that the assembly of the subunits of the pilus occurs in an ordered fashion (Dodson *et al.*, 1993; Saulino *et al.*, 1998). Using an *in vitro* ELISA-based assay, the binding affinities of various chaperone-subunit complexes to PapC could be determined (Dodson *et al.*, 1993). Specifically, chaperone-subunit complexes bound to PapC with affinities in the order PapD-PapG > PapD-PapF > PapD-PapE, whereas PapD-PapK, PapD-PapA, or PapD alone did not bind to the usher at any measurable level. Furthermore, purified tip fibrillae composed of PapG, PapF, PapE, and PapK could bind to the usher independent of PapD, suggesting that the interactive surfaces required for binding to the usher are located primarily on the subunit.

In PapD, two surface-exposed residues that are not part of the subunit-binding interface have been studied, Thr53 and Arg68, that may influence in some way interactions with the usher (Hung *et al.*, 1999). These residues are conserved

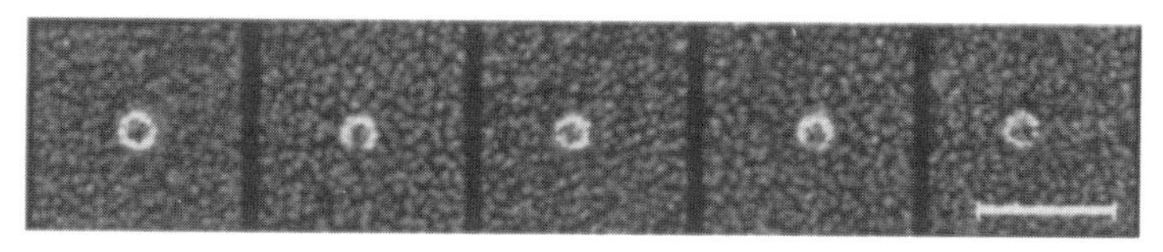

FIGURE 7 Electron micrographs of ring-shaped complexes of PapC. (Reprinted with permission from Thanassi *et al.*, 1998.)

throughout the chaperone superfamily, with Thr53 located along the D_1 strand of the N-terminal domain and Arg68 along the E_1 strand, both residues being within 5 Å of each other. Mutation of Thr53 to alanine, serine, or aspartate or Arg68 to alanine resulted in a reduction in the ability to assemble pili, which was abrogated completely in the Thr53Asp mutant. Recent *in vitro* studies of this mutant have shown that it has identical secondary structure and exhibits identical folding properties to the wild-type protein (J. Bann, unpublished results). These mutations did not inhibit the ability to form chaperone-subunit complexes (PapD-PapG or PapD-PapA), nor did they affect the *in vivo* stability of the chaperone. In addition, these mutations did not inhibit the ability of a PapD-PapG complex to bind to the outer membrane usher PapC. One hypothesis to explain these data is that once PapD has bound to the usher, there are secondary interactions that involve Thr53 that facilitate the release of the subunit (Hung *et al.*, 1999). Whether or not these secondary interactions involve the usher or an incoming subunit remains to be determined.

Based on these results, the ordered assembly of P-pili begins with an initial interaction between the chaperone-adhesin complex. Subsequently, binding of the PapD-PapF complex to the usher facilitates chaperone uncapping of PapG and allows the N-terminal strand of PapF to insert into the groove of PapG. Similarly, PapD-PapE binding facilitates uncapping of PapF, and this is followed by multiple rounds of binding, uncapping, and incorporation of PapE, resulting in the tip fibrillum. Once the tip fibrillum is complete, the PapD-PapK complex binds. Formation of the tip fibrillum is a critical point for further pilus assembly because deletion of the tip-fibrillum genes prevented the assembly of the major rod (Jacob-Dubuisson *et al.*, 1993). After this, the PapD-PapA complex binds, uncapping PapK and initiating growth and polymerization of the major rod. Thus the usher functions not only to provide a pore for the subunits to pass through but also, like an usher at a concert or play, to allow entry only if one has the right ticket (or affinity).

CONCLUSION

The ordered assembly of the subunits of pili managed by the chaperone-usher pathway is becoming better understood as more structures of key components are resolved and integrated into the framework provided by genetic and biochemical data. Of particular interest currently is understanding how the subunit-subunit interactions differ energetically from the chaperone-subunit interactions. Perhaps the subunit in the chaperone-subunit complex is in an intermediate conformation, and the subunit assumes a more stable complex once the N-terminal strand of the subunit has bound. Thermodynamic studies of the differences in binding between chaperone-subunit complexes and subunit-subunit complexes will help in determining whether subunit-subunit complexes are indeed more energetically favored.

In this chapter we have focused on the pili themselves and what is currently understood about the structure of the pilus and the mechanism of pilus assembly. The recent crystal structures of the receptor-binding domains of PapGII and FimH bound to carbohydrate complexes has provided a structural basis through which interactions between the host and the pilus can be better understood. These interactions are critical and in some cases (e.g., FimH) are sufficient for entry of the bacterium into the host. Further FimH binding induces a plethora of host responses, including the exfoliation of epithelial cells and cytoskeletal rearrangements resulting in the engulfment of the bacterium (Martinez *et al.*, 2000; Mulvey *et al.*, 2000). These responses are further augmented by secondary interactions between lipopolysaccharide from the bacterium and the host (Schilling *et al.*, 2000). A firm understanding of the structural basis of these and other interactions, e.g., between FimH and collagen (Pouttu, 1999), will aid in the therapeutic approach to treating diseases caused by bacteria expressing adhesive pili of the chaperone-usher pathway.

ACKNOWLEDGMENTS

We thank Wandy Beatty of the Molecular Microbiology Imaging Facility for producing the electron micrographs. We are also grateful to Fred Sauer and Julie Bouckaert for their help with figures. This work supported by a Keck Foundation Postdoctoral Fellowship (JGB), NIH grants AI29549, DK51406, AI48689 (SJH) and DK13332 (CF).

REFERENCES

Barnhardt, M. M., Pinkner, J., Soto, G., *et al.* (2000). PapD-like chaperones provide the missing information for folding of pilin proteins. *Proc. Natl. Acad. Sci. USA* 97:7709–7714.

Bork, P., Holm, L., and Sander, C. (1994). The immunoglobulin fold: Structural classification, sequence patterns and common core. *J. Mol. Biol.* 242:309–320.

Bullitt, E., and Makowski, L. (1995). Structural polymorphism of bacterial adhesion pili. *Nature* 373:164–167.

Bullitt, E., Jones, C. H., Striker, R., *et al.* (1996). Development of pilus organelle subassemblies *in vitro* depends on chaperone uncapping of a beta zipper. *Proc. Natl. Acad. Sci. USA* 93:12890–12895.

Bullock, T. L., Roberts, T. M., and Stewart, M. (1996). 2.5-Å-resolution crystal structure of the motile sperm protein (MSP) of *Ascaris suum*. *J. Mol. Biol.* 263:284–296.

Casadaban, M. J. (1976). Transposition and fusiion of the *lac* genes to selected promoters in *Escherichia coli* using bacteriophage λ and μ. *J. Mol. Biol.* 104:541–555.

Choudhury, D., Thompson, A., Stojanoff, V., *et al.* (1999). X-ray structure of the FimC-FimH chaperone-adhesin complex from uropathogenic *Escherichia coli*. *Science* 285:1061–1066.

Dodson, K. W., Jacob-Dubuisson, F., Striker, R. T., and Hultgren, S. J. (1993). Outer-membrane PapC molecular usher discriminately recognizes periplasmic chaperone-pilus subunit complexes. *Proc. Natl. Acad. Sci. USA* 90:3670–3674.

Dodson, K. W., Pinkner, J. S., Rose, T. G., *et al.* (2001). Structural basis of the interactions of the pylenophritic *E. coli* adhesin to its human kidney receptor. *Cell* 105:733–743.

Donnenberg, M. S. (2000). Pathogenic strategies of enteric bacteria. *Nature* 406:768–774.

Dood, D. C., Bassford, P. J. J., and Eisenstein, B. I. (1984) Dependence of secretion and assembly of type 1 fimbrial subunits of *Escherichia coli* on normal protein export. *J. Bacteriol.* 159:1077–1079.

Gong, M., and Makowski, L. (1992). Helical structure of P pili from *Escherichia coli:* Evidence from x-ray fiber diffraction and scanning transmission electron microscopy. *J. Mol. Biol.* 228:735–742.

Hanson, M. S., Hempel, J., and Brinton, C. C., Jr. (1988). Identification and characterization of *E. coli* type 1 pilus adhesion protein. *Nature* 332:265–268.

Holmgren, A., and Branden, C. (1989). Crystal structure of chaperone protein PapD reveals an immunoglobulin fold. *Nature* 342:248–251.

Hultgren, S. J., Jones, C. H., and Normark, S. N. (1996). In F. C. Neidhardt (ed.), *Escherichia coli and Salmonella Cellular and Molecular Biology*, 2d ed., pp. 2730–2756. Washington: ASM Press.

Hultgren, S. J., Lindberg, F., Magnusson, G., *et al.* (1989). The PapG adhesin of uropathogenic *E. coli* contains separate regions for receptor binding and for the incorporation into the pilus. *Proc. Natl. Acad. Sci. USA* 86:4357–4361.

Hung, D. L., Knight, S. D., Woods, R. M., *et al.* (1996). Molecular basis of two subfamilies of immunoglobulin-like chaperones. *EMBO J.* 15:3792–3805.

Hung, D. L., Knight, S. D., and Hultgren, S. J. (1999). Probing conserved surfaces on PapD. *Mol. Microbiol.* 31:773–783.

Hung, C., Bouckaert, J., Hung, D., *et al.* (2002). Structural basis of tropism of *Escherichia coli* to the bladder during urinary tract infection. *Mol. Microbiol.* 44:903–915.

Jacob-Dubuisson, F., Striker, R., and Hultgren, S. J. (1994). Chaperone-assisted self-assembly of pili independent of cellular energy. *J. Biol. Chem.* 29:12447–12455.

Jacob-Dubuisson, F., Pinkner, J., Xu, Z., *et al.* (1994). PapD chaperone function in pilus biogenesis depends on oxidant and chaperone-like activities of DsbA. *Proc. Natl. Acad. Sci. USA* 91:11552–11556.

Jones, C. H., Pinkner, J. S., Roth, R., *et al.* (1995). FimH adhesin of type 1 pili is assembled into a fibrillar tip structure in the Enterobacteriaceae. *Proc. Natl. Acad. Sci. USA* 92:2081–2085.

Jones, C. H., Pinkner, J. S., Nicholes, A. V., *et al.* (1993). FimC is a periplasmic PapD-like chaperone that directs assembly of type 1 pili in bacteria. *Proc. Natl. Acad. Sci. USA* 90:8397–8401.

Jones, C. H., Danese, P. N., Pinkner, J. S., *et al.* (1997). The chaperone-assisted membrane release and folding pathway is sensed by two signal transduction systems. *EMBO J.* 16:6394–6406.

Kihlberg, J., Hultgren, S. J., Normark, S., and Magnusson, G. (1989). Probing of the combining site of the PapG adhesion of uropathogenic *Escherichia coli* bacteria by synthetic analogues of gal-abiose. *J. Am. Chem. Soc.* 111:6364–6368.

Koebnik, R., Locher, K. P., and Van Gelder, P. (2000). Structure and function of bacterial outer membrane proteins: Barrels in a nutshell. *Mol. Microbiol.* 37:239–253.

Kuehn, M. J., Ogg, D. J., Kihlberg, J., *et al.* (1993). Structural basis of pilus subunit recognition by the PapD chaperone. *Science* 262:1234–1241.

Kuehn, M. J., Heuser, J., Normark, S., and Hultgren, S. J. (1992). P pili in uropathogenic E. coli are composite fibres with distinct fibrillar adhesive tips. *Nature* 356(6366):252–255.

Langermann, S., *et al.* (2000). Vaccination with FimH adhesin protects cynomolgus monkeys from colonization and infection by uropathogenic *Escherichia coli. J. Infect. Dis.* 181:774–778.

Leland, P. A., Staniszewski, K. E., Kim, B., and Raines, R. T. (2000). A synapomorphic disulfide bond is critical for the conformational stability and cytotoxicity of an amphibian ribonuclease. *FEBS Lett.* 477:203–207.

Lindberg, F., Tennent, J. M., Hultgren, S. J., *et al.* (1989). PapD, a periplasmic transport protein in P-pilus biogenesis. *J. Bacteriol.* 171:6052–6058.

Lindberg, F., Lund, B., and Normark, S. (1986). Gene products specifying adhesion of uropathogenic *Escherichia coli* are minor components of pili. *Proc. Natl. Acad. Sci. USA* 83:1891–1895.

Lund, B., Lindberg, F., Marklund, B. I., and Normark, S. (1987). The PapG protein is the α-D-galactopyranosyl-(1–4)-β-D-galactopyranose-binding adhesin of uropathogenic *Escherichia coli. Proc. Natl. Acad. Sci. USA* 84:5898–5902.

Martinez, J. J., Mulvey, M. A., Schilling, J. D., *et al.* (2000). Type 1 pilus—mediated bacterial invasion of bladder epithelial cells. *EMBO J.* 19:2803–2812.

Mitsui, Y., Dyer, F. P., and Langridge, R. (1973). X-ray diffraction studies of bacterial pili. *J. Mol. Biol.* 79:57–64.

Mulvey, M. A., Lopez-Boado, Y. S., Wilson, C. L., *et al.* (1998). Induction and evasion of host defenses by type 1 piliated uropathogenic *Escherichia coli. Science* 282:1494–1497.

Norgren, M., Băga, M., Tennent, J. M., and Normark, S. (1987). Nucleotide sequence, regulation and functional analysis of the *papC* gene required for cell surface localization of Pap pili of uropathogenic *Escherichia coli. Mol. Microbiol.* 1:169–178.

Pappenberger, G., Aygun, H., Engels, J. W., *et al.* (2001). Nonprolyl *cis* peptide bonds in unfolded proteins cause complex folding kinetics. *Nature Struct. Biol.* 8:452–458.

Pellecchia, M., Guntert, P., Glockshuber, R., and Wuthrich, K. (1998). NMR solution structure of the periplasmic chaperone FimC. *Nature Struct. Biol.* 5:885–890.

Pellecchia, M., Sebbel, P., Hermanns, U., *et al.* (1999). Pilus chaperone FimC-adhesin FimH interactions mapped by TROSY-NMR. *Nature Struct. Biol.* 6:336–339.

Pouttu, R., Puustinen, T., Virkola, R., *et al.* (1999). Amino acid residue Ala-62 in the FimH fimbrial adhesin is critical for the adhesiveness of meningitis-associated *Escherichia coli* to collagens. *Mol. Microbiol.* 31:1747–1757.

Roosendal, E., Jacobs, A. A. C., Rathman, P., *et al.* (1987). Primary structure and subcellular localization of two fimbrial subunits involved in the biogenesis of K99 fimbrillae. *Mol. Microbiol.* 1:211–217.

Sauer, F. G., Barnhart, M., Choudhury, D., Knight, S. D., Waksman, G., and Hultgren, S. J. (2000). Chaperone-assisted pilus assembly and bacterial attachment. *Curr. Opin. Struct. Biol.* 5:548–556.

Sauer, F. G., Futterer, K., Pinkner, J. S., *et al.* (1999). Structural basis of chaperone function and pilus biogenesis. *Science* 285:1058–1061.

Saulino, E. T., Thanassi, D. G., Pinkner, J., and Hultgren, S. J. (1998). Ramifications of kinetic partitioning on usher-mediated pilus biogenesis. *EMBO J.* 17:2177–2185.

Schilling, J. D., Mulvey, M. A., Vincent, C. D., *et al.* (2001). Bacterial invasion augments epithelial cytokine responses to *Escherichia coli* through a lipopolysaccharide-dependent mechanism. *J. Immunol.* 166:1148–1155.

Shinde, U. P., Liu, J. J., and Inouye, M. (1997). Protein memory through altered folding mediated by intramolecular chaperones. *Nature* 389:520–522.

Slonim, L., Pinkner, J. S., Branden, C., and Hultgren, S. J. (1992). Interactive surface of PapD chaperone cleft is conserved in pilus chaperone superfamily and essential in subunit recognition and assembly. *EMBO J.* 11:4747–4756.

Soto, G. E., Dodson, K. W., Ogg, D., *et al.* (1998). Periplasmic chaperone recognition motif of subunits mediates quaternary interactions in the pilus. *EMBO J.* 17:6155–6167.

Spolar, R. S., and Record, M. T., Jr. (1994). Coupling of local folding to site-specific binding of proteins. *Science* 11:769–770.

Striker, R., Nilsson, U., Stonecipher, A., *et al.* (1995). Structural requirements for the glycolipid receptor of human uropathogenic *Escherichia coli. Mol. Microbiol.* 16:1021–1029.

Thanassi, D. G., Saulino, E. T., Lombardo, M., *et al.* (1998). The PapC usher forms an oligomeric channel: implications for pilus biogenesis across the outer membrane. *Proc. Natl. Acad. Sci. USA* 95:3146–3151.

Valent, Q. A., Zaal. J., de Graaf, F. K., and Oudega, B. (1995). Subcellular localization and topology of the K88 usher FaeD in *Escherichia coli*. *Mol. Microbiol.* 16:1243–1257.

Van Rosmalen, M., and Saier, M. H., Jr. (1993). Structural and evolutionary relationships between two families of bacterial extracytoplasmic chaperone proteins which function cooperatively in fimbrial assembly. *Res. Microbiol.* 144:507–527.

Wiegand, G., Epp, O., and Huber, R. (1995). The crystal structure of porcine pancreatic alpha-amylase in complex with the microbial inhibitor tendamistat. *J. Mol. Biol.* 247:99–110.

CHAPTER 11

Type IV Pili

W. Schreiber
Division of Infectious Diseases, Department of Medicine, University of Maryland School of Medicine, Baltimore, Maryland

Michael S. Donnenberg
Division of Infectious Diseases, Department of Medicine, University of Maryland School of Medicine, Baltimore, Maryland

INTRODUCTION

Type IV pili (Tfps) or fimbriae are nonflagellar, filamentous surface appendages that are distributed widely among bacterial species. By mediating interactions to other bacterial cells of the same species, to eukaryotic cells, or to other surfaces, they function in diverse processes such as motility [1–4], virulence [5–7], biofilm formation [8], and horizontal gene transfer [9–12]. Originally, Tfps were classified by their related morphology and polar location on the cell [13] and their association with a special kind of bacterial locomotion designated as *twitching motility* (see below). Interestingly, this rather nonspecific classification scheme was confirmed by later studies showing that these pili have conserved sequence signatures and biogenesis machineries. Tfps are mainly homopolymeric structures consisting of a structural subunit that is synthesized as a preprotein (prepilin) with a short, positively charged leader peptide. The mature protein is characterized by a conserved hydrophobic N-terminus including an *N*-methylated residue. Tfp assembly seems to be mediated by similar systems because replacement of the pilin gene by one of another species often results in the formation of functional heterologous fimbriae [14–16]. Identification of a number of accessory genes

Escherichia coli: Virulence Mechanisms of a Versatile Pathogen
ISBN 0-12-220751-3

required for pilus biogenesis and/or function confirmed the existence of conserved elements among different Tfp assembly machineries. These include prepilin peptidases, polytopic inner membrane proteins, prepilin-like proteins, putative nucleotide-binding proteins, and outer membrane proteins of the secretin family. This set of proteins is not unique for Tfp biogenesis but has counterparts in the main terminal branch of the general secretion pathway, in DNA uptake systems, in the assembly of filamentous phages, and perhaps even in the biogenesis of flagellae within the archea. Thus it appears that there was a common ancestor for these different systems involved in macromolecular transport [17,18]. Nevertheless, the role of most of the proteins participating in pilus biogenesis remains to be elucidated.

The ability to promote adherence to eukaryotic host cells, a prerequisite for successful colonization, explains the wide distribution of Tfps among gram-negative pathogens. The significance of pili-mediated adhesion for several species has been demonstrated by the drastically reduced virulence of mutants without pili or with nonfunctional pili [5,19,20] and by the fact that purified pili can trigger immunoprotective responses by the host [21,22]. Whereas in some species the structural subunit itself can act in adhesion, other organisms such as *Neisseria gonorrhoeae* possess a distinct tip adhesin [23] to mediate host cell contact.

The main focus of this chapter are Tfps of pathogenic *Escherichia coli*. So far the best-studied system is the bundle-forming pilus (BFP) of enteropathogenic *E. coli* (EPEC), which mediates reversible bacterial aggregation and the localized adherence phenotype (Fig. 1*A*). The latter describes the formation of microcolonies on the surface of epithelial cells [24]. A cluster of 14 genes together with regulatory genes or an exogenous promoter is sufficient to express BFP in an *E. coli* K-12 host strain [25–27]. Both loci are located on the 50- to 70-MDa EPEC adherence factor (EAF) plasmid [28,29], which is now completely sequenced in EPEC strain B171 and therefore designated pB171 [30]. Two other Tfps have been described for human enterotoxigenic *E. coli* (ETEC) [31–33]. The *colonization factor antigen III* (CFA/III) is composed of semiflexible, 5- to 10-μm-long pili that have a peritrichious distribution on the bacterial cell (see Fig. 1*B*). Recently, a cluster of 14 genes required for CFA/III formation has been identified [34]. *Longus* stands for "long pilus," describing its striking length of over 20 μm. Longus shows the Tfp characteristic polar distribution and intertwines like BFP to form a network of large bundles (see Fig. 1*C*). Both ETEC Tfps are encoded on large virulence plasmids. Although closely related, as confirmed by the primary sequences [33,35], they most likely represent different pilus types because their expression is induced under different conditions. Furthermore, they lack common native epitopes, as confirmed by immunologic studies. Recently, Tfps also have been identified in the closely related *Salmonella enterica* serovar *typhi* [36], where they seem to be involved in cell adherence and invasion of human intestinal epithelial cells [36]. Beside their role in pathogenicity, Tfps also have been described in nonpathogenic *E. coli* strains, in which

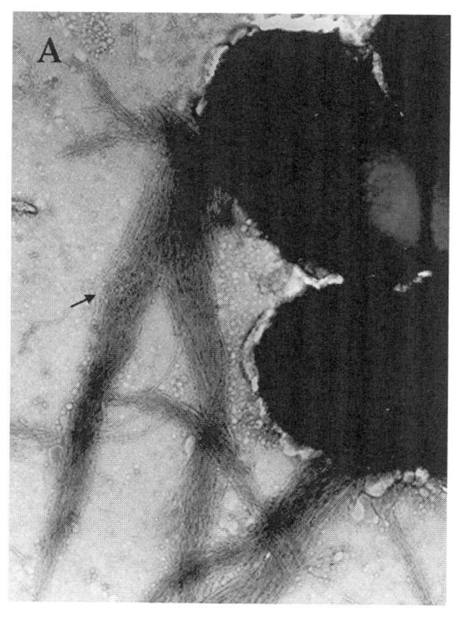

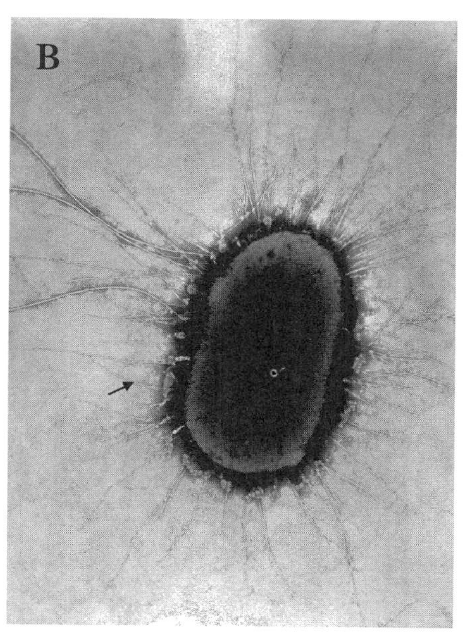

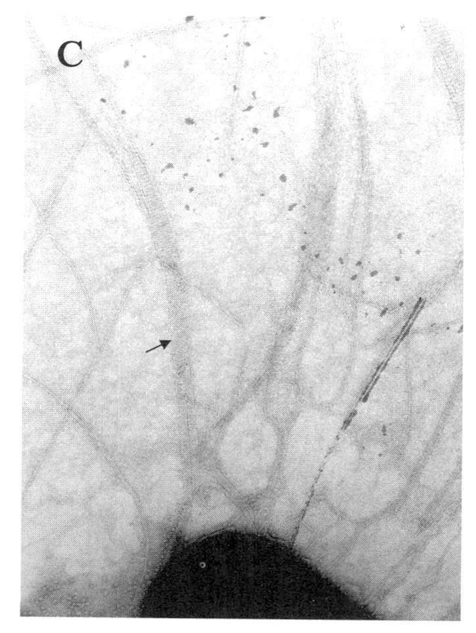

FIGURE 1 Transmission electron micrographs of (*A*) BFP, (*B*) CFA/III, and (*C*) longus.

they are encoded on large conjugative plasmids and function in conjugation in liquid media [37].

This chapter will discuss the distribution, function, and biogenesis of Tfps, with a focus on the bundle-forming pilus (BFP) of EPEC. The chapter will summarize the current knowledge of these organelles with reference to other well-known Tfps and related systems.

DISTRIBUTION

Type IV fimbriae are produced by a variety of gram-negative bacteria, including important pathogens of plants, animals, and humans. In addition to conserved structural, biochemical, and functional features, the classification is based primarily on amino acid sequence homologies among the major structural pilus subunits, pilins, that are especially striking within the highly hydrophobic N-terminal domain of the mature protein. Strictly conserved amino acid residues are a glycine at position −1 and a glutamate at position 5 relative to the leader sequence cleavage site. Tfps can be subdivided into two groups, A and B, based on differences in the signal peptide, cleavage site, and genomic organization.

Class A pilins are characterized by unusual short basic N-terminal leader sequences (6–7 amino acids) that are cleaved endoproteolytically between an invariant glycine and phenylalanine. The latter residue is modified subsequently by *N*-methylation [38]. Class A Tfps are produced by *Pseudomonas aeruginosa*, *N. gonorrhoeae*, *N. meningitis*, *Moraxella bovis*, *M. lacunata*, *M. nonliquefaciens*, *Dichelobacter nodosus*, *Eikanella corrodens*, *Kingella kingae*, *K. denitrificans*, *Branhamella catarrhalis*, *Aeromonas hydrophila*, and *Myxococcus xanthus* [39]. Beside the toxin-coregulated pilus (TCP) (see below), *V. cholerae* is also able to express class A Tfps: the maltose-sensitive hemagglutinin pilus of El Tor biotypes [40] and a putative second one that is encoded by the *pilA* gene cluster [41]. The arrangement of genes involved in fimbrial biogenesis in these species is quite varied, with the genes often scattered throughout the genome [20].

Features of TfpB pilins include longer prepeptides (13–30 amino acids) that undergo cleavage between the invariant glycine and a variable amino acid at position +1. *N*-Methylation of the first residue is so far only proven for the TCP of *V. cholerae* [42] and BFP (Donnenberg, unpublished data). An N-terminal modification is also suggested for the ColIb-P9 pilin, although methylation can be excluded in this case [37]. TfpB pilins are represented by the BFP of EPEC strains [43–45], longus and CFA/III from ETEC [31,33], thin pili of *S. enterica* serovar *typhi* [36], and thin pili of the conjugative IncI plasmid R64 [46]. Genes encoding the assembly machinery for class B Tfps generally are organized in clusters and are located in large pathogenicity islands in the case of TCP [47] and the thin pili of *S. enterica* [48] or on large plasmids for the others.

GENETIC ORGANIZATION

All Tfps described so far and expressed by *E. coli* belong to class B and are encoded by genes that are organized on large plasmids. Limited information is available on the biogenesis machinery of longus, but one would assume that a similar set of gene products might be required for pilus assembly and function as reported for BFP, CFA/III, and R64 thin pili. The biogenesis of BFP and R64 thin pili depends on 12 gene products encoded by gene clusters of about 10 kb (Fig. 2). Which of 14 Cof proteins are essential for CFA/III formation is not yet known. Curiously, it has been reported that a 5-kb fragment containing *lngA* enables *E. coli* K-12 to produce longus [49]. This observation implies that a smaller set of proteins might be sufficient for the biogenesis of this pilus.

The relative location of the cistrons encoding the outer membrane protein of the secretin family, the putative NTP-binding protein involved in assembly and the polytopic inner membrane protein in that order, seems to be conserved among class B Tfps (see Fig. 2) including TCP [46]. Genes involved in the regulation of TfpB expression are mainly located outside the region encoding the biogenesis machinery.

Sequencing of the genomes of *E. coli* K-12 and EHEC O157:H7 [50,51] revealed the existence of silent homologues of other Tfp genes that are distrib-

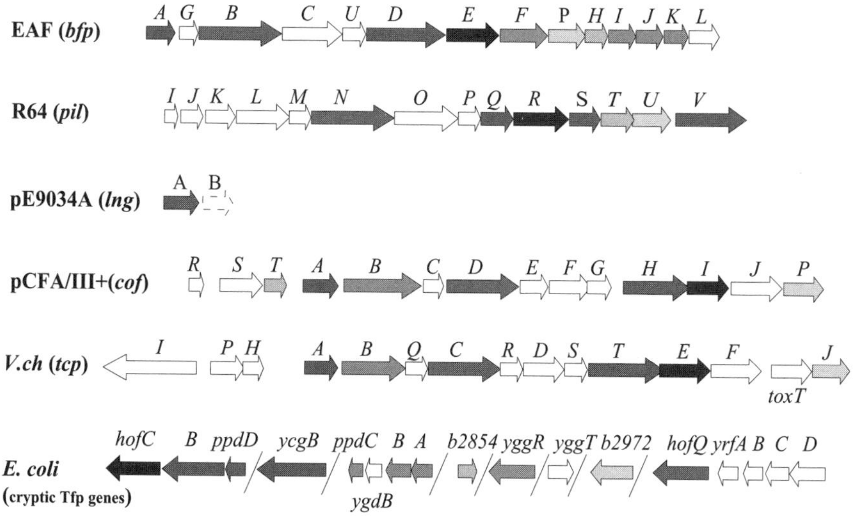

FIGURE 2 Scheme of the BFP gene cluster of EPEC in comparison with the gene organization of other type IVB biogenesis clusters. Lower part shows genes within the *E. coli* K-12 genome with homologies to other Tfp genes. The putative major pilin gene *ppdD* belongs into the class A of Tfp. Arrows: red (prepilin); magenta (secretin); dark green (putative NTP-binding protein); light green (putative NTP-binding protein/retraction), dark blue (polytopic inner membrane protein); yellow (prepilin peptidase); light blue (lytic transglycosylase); orange (prepilin-like proteins). See Color Plate 8.

uted over noncontiguous sections of the genome, as is typical for TfpA gene organization (shown in Fig. 2 for *E. coli* K-12). These genes are expressed at very low levels under standard laboratory conditions and do not seem to be affected by global regulators such as IHF, H-NS, or Fis [52]. Although no conditions have been found to express the respective pilus in K-12, the hypothetical major pilin (PpdD) could be assembled into fimbriae by *P. aeruginosa*. In addition, overexpression of both the pullulanase secreton of *Klebsiella oxytoca* and *ppdD* in *E. coli* led to a secreton-dependent incorporation of PpdD into pilus-like structures [53].

ROLE IN VIRULENCE

Tfps of pathogenic *E. coli* biotypes seem to play an important role in the pathogenesis of diarrheal diseases, perhaps by promoting colonization of mucosal surfaces in the intestinal tract (for details, see Chaps. 3 and 5 and ref. 54). CFA/III, for example, have been shown to mediate adherence of ETEC to intestinal epithelial cells of infant rabbits or human enterocytes [55]. The observation that antibodies directed against the respective Tfps reduce adherence of EPEC and other pathogens to epithelial cells underscores their potential role in pathogenesis [20,43]. Moreover, *bfpA* mutants, which are not able to produce BFP, show reduced adherence to epithelial cells [56] and are less virulent in volunteers [5]. Various models for EPEC infection suggested different roles for BFP. In the traditional three-step model (i.e., localized adherence, cell signaling, and intimate adherence), BFP act as a nonspecific adhesin in the first stage, mediating nonintimate attachment of preformed three-dimensional bacterial aggregates to the host cells [57,58]. *In vitro* organ culture studies using intestinal tissue of children that presumably better mimicks *in vivo* conditions favor a four-stage model (i.e., nonintimate attachment, signal transduction, intimate adherence, and microcolony formation), in which BFP is responsible for the formation of three-dimensional colonies in the fourth step of EPEC pathogenesis [59]. The fimbriae form the interbacterial connections within microcolonies [43,60]. Whether recruitment of further single bacteria leads to the formation of growing microcolonies or cell division is primarily responsible is not yet clear [24,59]. The existence of these aggregates is a transient phenomenon in infection ending with dispersal of the bacteria from the microcolonies to the apical surfaces of the host cells. Interestingly, *in vitro*, the BFP morphology changes concomitantly with these events, as demonstrated in standard adherence assays. Typical pilus bundles in a microcolony have a diameter of about 40 nm. With dispersal, bundles become thicker (100 nm in diameter) and longer and tend to intertwine to even more complex structures. It is possible that this drastic alteration in the quaternary structure is essential to free the bacteria and enable diffuse adherence [60]. A fiber network remains at the locus of initial attachment, capturing some bacteria. A mutation in the *bfpF* gene interferes with dispersal, arresting bacteria in microcolonies. In contrast to the wild type, the interbacterial linkages in the *bfpF* mutant aggregates appear to be static [5]. Nevertheless, progression of the infection *in vitro* is not influenced by this mutation [60]. Intimate attachment to the epithelial cell surfaces by intimin still occurs. This allows bacteria-to-cell signal transduction, resulting in the appearance of typical attachment and effacement (AE) lesions, with the effacement of microvilli, pedestal formation caused by actin polymerization and other cytoskeletal rearrangements [54,61–63]. Surprisingly, the BfpF paralogue in *N. meningitidis*, PilT, is required for intimate attachment and the formation of AE lesions [6]. In this context, it is important to mention that BfpF is essential for virulence in EPEC because mutants are attenuated in their ability to cause diarrhea [5]. A detailed characterization of BfpF and paralogous proteins in other systems, as well as their contribution to Tfp function, is given later in this chapter.

PILUS STRUCTURE

Tfps are polymeric, flexible fibers with lengths of 1 to 4 μm described for pili of group A [7] and 5 to over 20 μm for group B [49]. They are formed by the ordered association of thousands of identical subunits. In some species, minor components are integrated in the pilus structure, such as the tip adhesin of the pilus on *N. gonorrhoeae* [23] or a second pilin protein in the case of the thin pili encoded by the conjugative plasmid R64 of *E. coli* [46]. So far there are no reports of proteins that copurify with type IV fimbriae of pathogenic *E. coli* strains.

The primary sequence of a pilin gene can be subdivided into three regions. Region 1 can be further divided into three parts: the species-specific sequence encoding the leader peptide [1A: 13 amino acids for bundlin (the pilin protein of BFP), 20 amino acids for LngA and CofA of ETEC]; the highly conserved hydrophobic domain (1B), which is proposed to anchor pilin monomers into the inner membrane prior to assembly and is buried in the assembled pilus fiber; and another species-specific region (1C). The central and C-terminal regions of the pilin are highly variable. The C-terminal third contains two conserved cysteines [20]. There exists considerable sequence variation in regions 2 and 3 of bundlin from different EPEC strains, leading to the definition of α and β subgroups [64]. Assembly into fibers requires cleavage of the signal peptide by a prepilin peptidase [38], which will be described in detail later. Stability of bundlin requires disulfide linkage of the C-terminal cysteines catalyzed by the periplasmic DsbA protein, which implies that the C-terminal part is exposed to this cell compartment [65]. The simultaneous but independent processing of prebundlin by the prepilin peptidase at the cytoplasmic face of the inner membrane and by

DsbA at the periplasmic face of the inner membrane indicates that the prepilin protein exists as a bitopic inner membrane protein.

Despite such obvious differences as the molecular weight (A: 15–18 kDa; B: 18.5–22 kDa) or such structural features of the assembled fibers as lateral aggregation (TfpB), all type IV pilins are believed to share the same basic structure that allows their polymerization into the typical Tfp. Therefore, it is justifiable, despite the absence of crystallographic data on pili of pathogenic *E. coli* or generally on class B pili, to describe at this point the current knowledge about class A pili based on x-ray crystal structures combined with biophysical, immunologic, and genetic information. According to studies of the Tfps of *N. gonorrhoeae*, the pilin monomers, which resemble a ladle with a long N-terminal handle and a globular head, are arranged into a three-layer spiraling fiber such that one helical turn around the longitudinal axis is built by five pilin monomers [66,67]. This arrangement implies an interaction of the N-terminal region of monomers with C-terminal regions of other subunits in the helix. The outermost layer is fully exposed to the environment and corresponds to a hypervariable region in the C-terminal third of the primary sequence. Although a few conserved interactions to the central layer are proposed, this part of the "head" does not seem to be an integral element, which means variations in this region would not be expected to interfere with the main pilus architecture. Antiparallel β-strands characterize the central layer and are very likely responsible for the impressive mechanical stability of the pilus. The structural elements of the innermost layer include the highly conserved N-terminal α-helices of the pilin monomers, which are arranged as parallel coiled coils and are essential for helix packing.

Since the N-terminal hydrophobic region of the mature pilins is highly conserved among Tfps, the gonorrheal x-ray crystallization data of this region were combined with the available genetic and immunologic information to develop a theoretical model of the polymeric structure of TCP [68] (R. Chattopadhyaya and A.C. Ghose, unpublished results; Protein Data Bank Accession Number 1QQZ). This model, in accordance with the results of detailed mutation analysis [69], suggests that the C-terminal disulfide loop, which can be subdivided into two domains, contributes to the structure and function of TCP. The first, or structural, domain is very likely involved in intramolecular fiber stabilization, whereas the second, or functional, domain seems to be surface-exposed and contributes to intermolecular interactions affecting colonization or bacteriophage sensitivity. The same authors also submitted a bundlin (BFP) model (Protein Data Bank Accession Number 1QT2). According to the predicted structure, highly variable amino acids of BfpA in different EPEC strains are, as expected, predominantly located on the surface of the theoretical BFP structure [64] (Fig. 3). The role of the C-terminal disulfide loop in stability of the bundlin monomer has already been demonstrated [65], but whether it also contributes to the structure and function of the fiber in analogy to TCP remains to be elucidated. Nevertheless, replacement of this region with that of TCP resulted in unstable hybrids [70].

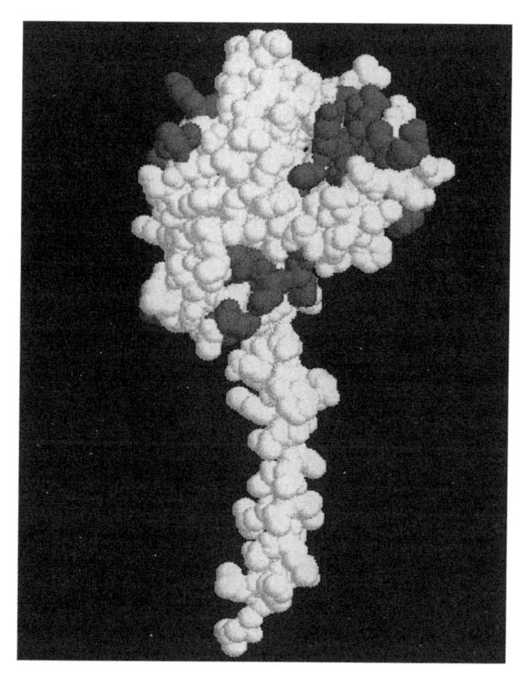

FIGURE 3 Model of bundlin (Chattopadhyaya and Ghose, unpublished data). Gray molecules represent variable amino acids, which are clustered on the surface of the theoretical 3D structure.

PROTEINS INVOLVED IN PILUS BIOGENESIS AND FUNCTION

The biogenesis of Tfps appears to be a quite complicated process requiring a variety of proteins that have to be assembled coordinately to provide an infrastructure that allows pilin polymerization, fiber stabilization, and surface translocation. Other than the fact that mutations in the respective genes resulted in nonfimbriated phenotypes, the precise functions of most of these proteins remain unknown. Recent studies using conditional double-knockout mutants have allowed dissection of the process into three steps: pilin processing, pilus formation, and pilus extrusion [71,72]. The following subsections summarize the current knowledge about components involved in BFP biogenesis and function, including information about homologues in other secretion and DNA transfer systems (overview in Table I).

The Prepilin Peptidase

The prepilin peptidase, which is encoded in the BFP operon by the ninth gene, designated *bfpP* [26], is the best-characterized accessory protein in Tfp

TABLE I Overview of Proteins Involved in Tfp Biogenesis and Their Putative Function in Comparison with Paralogues in Type II Secretion and DNA Transfer Systems

	Tfp biogenesis						Type II secretion			DNA transfer
	Class B				Class A					
	EPEC	ETEC	*V. cholerae*	*E. coli*/R64	*N. gonorrhoe*	*P. aeruginosa*	*P. aeruginosa*	*K. oxytoca*	*Erwinia* sp.	*B. subtilis*
Pilin	BfpA	CofA	TcpA	PilS PilVA'	PilE	PilA				
Prepilin like proteins	BfpI BfpK BfpJ	CofB	TcpB		PilA	FimT, FimU PilE PilV PilW PilX	XcpT XcpU XcpV XcpW XcpX	PulG PulH PulI PulJ PulK	OutG OutH OutI OutJ OutK	ComGC ComGD ComGE ComGG
Prepilin-peptidase	BfpP	CofP	TcpJ	PilU	PilD	PilD	XcpA(=PilD)	PulO	OutO	ComC
Secretin	BfpB*	CofD*	TcpC*	PilN*	PilQ	PilQ	XcpQ	PulD	OutD	—
Secretin—spec. lipoprotein					PilP		XcpP	PulS	OutS	
NTP-binding proteins	BfpD	CofH	TcpT	PilQ	PilF	PilB	XcpR	PulE	OutE	ComGA
NTP-binding proteins/ retraction	BfpF				PilT	PilT PilU				
Polytopic inner membrane protein	BfpE	CofI	TcpE	PilR	PilG	PilC	XcpS	PulF	OutF	ComGB
Lytic *trans*-glycosylase	BfpH	CofT		PilT						

*Lipoprotein.

biogenesis. In contrast to other proteins involved in biogenesis and function of Tfps, homologies within the family of prepilin peptidases are in general considerably high and extend over the entire amino acid sequence. Table I demonstrates that the role of these enzymes is not restricted to pilus formation but extends to other systems, such as type II secretion by gram-negative bacteria (GspO) or competence development in the gram-positive species *B. subtilis* and *Streptococcus pneumoniae* (not in the table) [73]. The close relationship of Tfp biogenesis and type II secretion is especially striking in *P. aeruginosa* and *Vibrio cholerae*, where the prepilin peptidase is shared between the two systems [74,75].

The prepilin peptidase was first identified and characterized in *P. aeruginosa* [76,77]. It probably functions in the very first steps in pilus biogenesis by hydrolyzing the peptide bond between the leader peptide and the characteristic hydrophobic domain of the prepilins and prepilin-like proteins. This reaction is carried out 50 to 100 times faster than cleavage of the classic signal peptides by the *E. coli* leader peptidase I [78]. A prerequisite for efficient cleavage is the highly conserved glycine in position −1 of all prepilins [79], whereas a number of amino acids are tolerated in position +1, as demonstrated by analysis of pilin mutants in *P. aeruginosa* [77] and its variability in class B pilins. Nevertheless, mutant analysis of *pilS*, the prepilin of the thin pili encoded by plasmid R64, revealed that other residues, which are distributed over the entire sequence, also are essential for processing [80]. Surprisingly, the long leader sequence characteristic of group B pilins does not seem to influence the cleavage efficiency.

Following endoproteolytic cleavage, the prepilin peptidase catalyzes the methylation of the mature N-terminus, a process that requires the highly conserved glutamate at position +5 [7]. The two reactions are independent, suggesting the presence of two active centers [38]. The significance of this modification remains dubious because the assembly of functional pili or secretion of proteins is unaffected in the absence of the methylase activity, although a role in the natural habitat cannot be excluded [81]. Parge and colleagues [66] suggested that this amino acid might be involved in registration or recognition of subunits during the assembly process.

Prepilin peptidases are polytopic, cytoplasmic membrane proteins (see Fig. 4 for BfpE). For example, eight transmembrane domains have been suggested for the OutO protein from *Erwinia carotovora* [82]. Since the signal sequence of prepilins is directed toward the cytoplasm, in accord with the availability of the CH_3-group donor *S*-adenosylmethionine, the processing has to be carried out on the cytoplasmic side of the inner membrane [38]. This notion is supported by the topologic model of OutO showing considerable transmembrane loops only toward the cytoplasm.

A recent study [69] based on mutational analysis of TcpJ from *V. cholerae* identified two aspartate residues that are critical for protease activity of TcpJ. Changing the corresponding sites in VcpD, which processes mainly TfpA prepilins in *V. cholerae*, also abolishes processing. These amino acids are located in the puta-

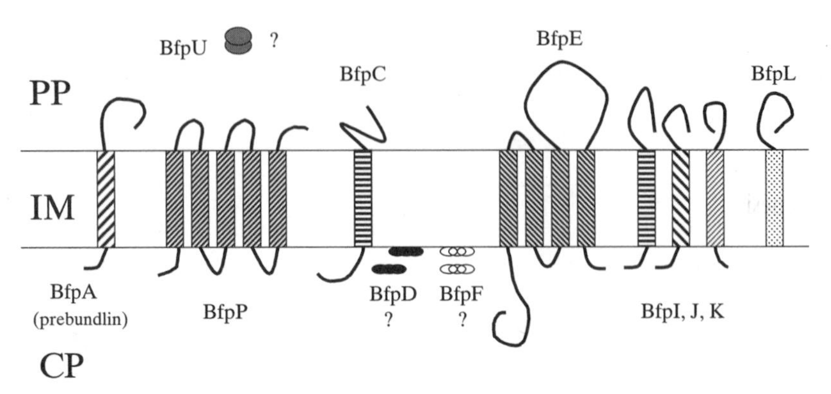

FIGURE 4 Subcellular localization of the BFP proteins and topology of the inner membrane components (PP, periplasm; IM, inner membrane; CP, cytoplasm).

tive second and third cytoplasmic loops of the protein and are conserved throughout the whole family (i.e., BfpP: amino acids 129 + 190). In addition, inhibitor studies support the involvement of the aspartates in activity. Nevertheless, typical aspartic proteases are characterized by a D(T/S)G motif, acidic pH optima, and inhibition by pepstatin, attributes that do not apply to prepilin peptidases. Therefore, it was suggested that these enzymes represent a novel family of bilobed aspartate proteases and may be attractive targets for pathogen-directed therapeutic approaches.

Since the overall structure of all prepilins is very similar and cleavage occurs in a highly conserved region, it is not surprising that prepilin peptidases are interchangeable between many systems. For example, BfpP can replace PilD in *P. aeruginosa*, and PilD can process prebundlin [83]. The prepilin peptidases PulO and ComC (see Table I) involved in type II secretion and genetic transfer, respectively, are able to process the type IV prepilin of *N. gonorrhoeae* in an *E. coli* host [84]. However, not all prepilin peptidases are interchangeable: TcpJ cannot process PilA of *P. aeruginosa* in *E. coli*, although the sequences of TcpJ and PilD are more similar than those of BfpP and PilD.

Prepilin-Like Proteins

The biogenesis of BFP in EPEC requires the presence of three proteins, designated as BfpI, BfpJ, and BfpK (Schreiber *et al.*, unpublished results), that are highly related to pilins [25,26]. They are members of a large family of proteins referred as *prepilin-like proteins* or *pseudopilins* and participate not only in Tfp assembly but also in secretion and DNA transfer. As implicated by the presence of the invariant glycine at −1 and glutamate at +5, they can be processed by the respective prepilin peptidase [85,86]. Cleavage of the leader peptide is essential for the function of these proteins [87]. Sequence similarities to pilins in general are restricted to the characteristic highly hydrophobic N-terminal domain of about 20 amino acids that is required for pilus assembly. Topologic studies with fusion proteins indicate that the prepilin-like proteins are type II bitopic inner membrane proteins [88] with the extreme N-terminus exposed to the cytoplasm (see Fig. 4). It was suggested that the N-terminal signal sequence might function in anchoring of prepilins prior to putative assembly [18]. Fractionation studies, nevertheless, revealed a partitioning of the *P. aeruginosa* pseudopilin XcpT to both membrane fractions on overexpression. The mature protein was found to be enriched in the outer membrane [89,90], supporting the idea of relocalization of the pseudopilins after processing. PulG, the major pseudopilin involved in pullulanase secretion of *K. oxytoca*, also was detectable in both membranes but without a preferred localization of the mature form [91].

Although most pseudopilins are absolutely required for functioning of their respective systems, the precise role they play remains mysterious. It is tempting to think that they might be minor components of Tfps, but experimental evidence argues against that hypothesis.

The simultaneous existence in Tfp biogenesis, secretion, and DNA transfer suggests a more general role of these molecules, i.e., as part of a oligomeric structure far more complicated than fimbriae that transports macromolecules across the cellular borders. In such a complex, pseudopilins were proposed to form a pilus-like structure or pseudopilus involved in interconnecting the inner and outer membrane to direct exoproteins to the outer membrane pore. They also could provide a scaffold on which the pilus or secreton components can be assembled [85,87,92]. This scenario proceeds on the assumption that pseudopilins interact with themselves and other components of the respective system. In favor of *in vivo* interactions are the results of cross-linking experiments using whole *P. aeruginosa* cells that demonstrate the formation of homodimers and mixed dimers of the most abundant secretion pseudopilin XcpT with itself and the Tfp pilin protein PilA. Mixed dimer formation with different combinations of Xcp pseudopilins occured when those were present in higher amounts [93]. However, these interactions and dimerization of PulG occur in the absence of processing, which is required for proper functioning and still occurs when the C-terminus has been modified [93,94]. On the other hand, the facts that XcpT-W is produced in different and strict stoichiometric amounts [95] and that exchange of the conserved glutamate +5 in XcpT, which is probably critical for pilus assembly, results in secretion defects support an involvement in a higher-ordered structure. Suppressor mutant analysis suggests a direct or indirect interaction of XcpT and the putative NTPase XcpR, although the proteins reside in different cell compartments [96]. The potential ability of pseudopilins to polymerize to a pilus structure was demonstrated recently by overexpression of the *pul* system in *E. coli*, resulting in the formation of Tfp-like cell appendages consisting of PulG monomers [53]. Whereas pili formation required only two further pseudopilins, four were involved in secretion. Nevertheless, these structures cannot be detected under physiologic conditions but may exist in a rudimentary form to support secretion. Based on cross-linking studies, PulG also interacts with the PulELM complex (involving two inner membrane proteins and the cytoplasmic NTP-binding protein) in the presence of other proteins required for pullulanase secretion [97].

Altogether these results support the existence of interactions between pseudopilins as homo- or heteropolymeric molecules and other proteins conserved in the respective system in order to provide an infrastructure allowing pili formation, secretion, or DNA-uptake.

Additional Inner Membrane Proteins

BfpC, encoded by the fourth gene in the *bfp* gene cluster, is a bitopic inner membrane protein with no obvious paralogues in other Tfp systems [25,26] (Anantha and Donnenberg, unpublished results; see Fig. 4). Nevertheless, PilO, a component of the closely related R64 thin pili biogenesis system, is similar in size and predicted localization (PilO, 48 kDa; BfpC, 45 kDa) and might serve the same purpose [46].

The center of interactions within a putative inner membrane subassembly might be represented by BfpE, the product of the seventh gene in the *bfp* operon. It is a member of the GspF family, which is common to type II secretion systems and Tfp biogenesis machineries and also is involved in DNA transfer via conjugation or transformation (see also Table I). It is interesting to note that *gspE*, encoding the putative NTP-binding protein, and *gspF* genes are linked in many systems, as are *bfpD* and *bfpE* in the *bfp* operon. Topology studies with alkaline phosphatase and β-galactosidase fusion proteins revealed that BfpE contains four membrane-spanning (TM) domains [98] (see Fig. 4) differing substantially from the three membrane-transversing elements of the OutF paralogue [99]. Both proteins show a large N-terminal cytoplasmic domain. The first three TM domains in BfpE are connected by short loops. The fourth TM segment is located near the C-terminus, resulting in a large periplasmic fragment between the third and fourth TM domains. In contrast, OutF is predominantly cytoplasmic because the third TM domain is close to the C-terminus. Two potential roles have been assigned for GspF proteins. In one, GspF serves as anchor for the putative NTP-binding proteins [97,100]. Alternatively, GspF might be required for efficient localization of the prepilin peptidase to the inner membrane [101]. However, no defect in prepilin processing was noted in a *bfpE* mutant strain [98]. A large periplasmic and cytoplasmic domain, like that postulated for BfpE, might allow interactions with many other components of the assembly complex to form an interbridging structure between the inner and outer membrane.

Assembly Proteins with Nucleotide-Binding Motifs

It seems obvious that assembly of the complicated infrastructure necessary for Tfp biogenesis, type II secretion, or DNA transfer, as well as pilus polymerization or the transmembrane transport reaction itself, would require a source of energy. This energy can be provided principally by ATP hydrolysis, the proton motive force (pmf) generated over the cytoplasmic membrane or by the favorable entropy generated by the interactions among pilin subunits in the mature structure, such as for usher-chaperone fimbriae (see Chap. 10).

One apparent candidate to provide energy from ATP hydrolysis in BFP biogenesis is BfpD, a protein of 60.5 kDa that is predicted to be localized in the cytoplasm because it exhibits a mainly hydrophilic character and no signal sequence or putative membrane-spanning elements [25,26]. BfpD is a member of the so-called PulE superfamily, which has paralogues in all the above-mentioned systems. These proteins share several conserved features, namely, a Walker box A or P loop implicated in nucleotide binding and a second less conserved Walker box (B), which is characteristic for traffic ATPases. The region between contains two aspartate boxes and a tetracysteine motif that defines typical PulE paralogues [92,102,103]. To date, there is no unequivocal proof of an ATP-hydrolyzing activity of PulE paralogues in type II secretion and Tfp biogenesis and function. Only the purified EpsE protein exhibits autokinase activity [104]. Mutational analysis has revealed that the exchange of key residues in the Walker A box abolishes *in vivo* function of many PulE homologues, e.g., PulE itself [92,103], EpsE [104], OutE [105], PilB [106], and BfpD [5]. In addition, ATPase activity has been demonstrated for more distant PulE-related proteins in type IV secretion [107]. It is possible that ATPase activity of PulE and similar proteins requires interaction with additional components of the machinery. Support for this hypothesis has been given by numerous interaction studies on type II secretion systems. These studies suggest the existence of a subcomplex in the inner membrane consisting of the general secretion components (GPS) GspE (PulE homologue), GspL, and GspM. GspE is primarily a cytoplasmic protein, but fractionation data also revealed an association with the inner membrane mediated by the bitopic transmembrane protein GspL, which also stabilizes GspE [97,103,104,108]. GspL interacts through its periplasmic and/or transmembrane domain with GspM, another bitopic cytoplasmic membrane protein that is mainly oriented toward the periplasm [109]. In this way, a conformational change, i.e., induced by the binding and/or hydrolysis of ATP, could be transformed to the periplasm [105]. It is tempting to assume a similar arrangement in pilus biogenesis, although no GspL and M homologues are obvious in the *bfp* gene cluster.

Purification and structural analysis of PulE relatives in type IV secretion revealed homohexameric ring-shaped complexes in the presence of nucleotides and divalent cations. Interestingly, TrbB involved in conjugative transfer of the plasmid RP4 has a more elliptical shape and was suggested based on its cross-linking pattern to consist of dimers as the initial building unit [107]. This is reminiscent of PulE, which forms dimers in the absence of reducing agents [103], and studies of XcpR, which suggest at least the formation of dimers via the N-termini [110]. Homomultimerization also was proposed for OutL of *E. chrysanthemi* based on results of yeast-two-hybrid studies [105]. Recently, the crystal structure of another NTP-binding protein (TrwB of *E. coli* R300 conjugative system) has been resolved: Six equivalent monomers form a ringlike structure with a central channel [111]. A hexameric ringlike structure seems to be common to many NTP-binding proteins and, based on the findings of Krause and colleagues [107], can be suggested for the whole PulE superfamily. Furthermore, this structure also implies that these enzymes catalyze a repetitive process that could be linked to DNA transfer, secretion, or pilus formation.

Besides BfpD, the *bfp* operon encodes another putative nucleotide-binding protein designated BfpF. Like members of the already described PulE family, BfpF is characterized by a Walker box A and the two aspartate motifs, although the second motif seems to be less conserved [112]. BfpD and BfpF can be distinguished by their size and the absence of the typical cysteine motif and the second Walker box in the latter. Whereas gene interruption or exchange of a conserved residue in the Walker box A resulted in a nonfimbriated phenotype in the case of *bfpD*, the opposite, hyperfimbriation, is true for *bfpF* [5,112,113]. The impact of the *bfpF* mutation on EPEC pathogenesis was described earlier (i.e., its role in virulence).

BfpF paralogues can be found in many other Tfp systems, especially in organisms that exhibit twitching motility or social gliding. Twitching motility describes a flagella-independent type of movement on solid surfaces that requires about 100 cells and therefore is very similar to social gliding, a cooperative smooth kind of motion exhibited by fruiting bodies of myxobacteria [2–4]. In suspension, twitching motility is perceptible as small sporadic twitches of single cells. This is reminiscent of the jerky movements of EPEC cells within a microcolony [5].

Mutations affecting homologues of *bfpF* such as *pilT*, the prototype of the twitching motility gene in *P. aeruginosa*, result in the formation of pili that are unable to mediate twitching motility, transformation, or other pilus related functions. Interestingly, the pattern of bacteriophage attachment, which occurs primarily at the proximal ends of pili in the wild type, changes in hyperfimbriated phage-resistant *P. aeruginosa* mutants: They are distributed over the whole length of the filaments. These observations led to the assumption that the pili are retracted to bring the phages toward the cell surface and that retraction is also essential for twitching motility [4,12,114]. Since hyperfimbriation is the result of a mutation in *pilT* [115], this protein might be a key element in the retraction process. Furthermore, the presence of a putative NTP-binding site implies that PilT might power retraction from the pilus base antagonizing PilB-dependent assembly. Recent biophysical studies [1] confirm the pilus-retraction hypothesis in *N. gonorrhoeae* and suggest that the process may be described as repeated cycles of assembly and disassembly of the pilin fibers. Retraction and therefore motility were dependent on *de novo* protein synthesis and the presence of functional PilT. Considering that pilus retraction generated mechanical force and that mechanical forces can trigger specific responses in eukaryotic cells [116] as well as in bacteria (*V. haemolyticus*, [117]), pilus retraction may be a method of internal communication between pathogenic bacteria and their host cells. Thus the arrest of *N. meningitis* in localized adherence on PilT mutation could be caused by the failure to induce an appropriate response in the host cell because of a defect in signaling [6]. It is also conceivable that microcolony formation in EPEC is a way to instruct the cells for their further role in infection and that the mechanical forces between them trigger the necessary gene expression [5].

Recent studies [71,72] have supported the idea that PilT is also responsible for quality control in Tfp biogenesis. If essential components of the assembly machinery are missing, PilT prevents fiber formation in order to minimize growth defects. Therefore, all mutations of genes involved in Tfp biogenesis resulted in one phenotype: nonpiliated bacteria. In support of this role, PilT, like other NTP-binding proteins involved in the assembly and disassembly of macromolecular complexes, shares structural motifs with ATP-dependent Clp chaperones and eukaryotic proteases involved in posttranslational quality control [118]. The absence of PilT and the secretin PilQ resulted in the formation of membrane protrusions containing pili-like structures, demonstrating that fiber formation occurs inside the cell, probably initiated and anchored from a base in the inner membrane.

According to current knowledge, the mechanism underlying retraction is best described with a molecular ratchet model [119] in which force is generated by polymerization [72] and depolymerization of pilin subunits [1]. The balance between PilB (BfpD) and PilT (BfpF) probably dictates whether competent pilin subunits are assembled into a fiber or retranslocated from an existing fimbria to the inner membrane in a protease-accessible form. Whether contact with other surfaces alone is enough to induce retraction and what events occur concomitantly inside the cells are questions that remain to be addressed.

The Outer Membrane Protein of the Secretin Family

In gram-negative bacteria, the outer membrane represents the final barrier to the export of macromolecules, such as exoproteins, toxins, S-layers, phages, or Tfp fibers. Type I secretion systems realize outer membrane translocation via a homotrimeric gated channel tunnel spanning the outer membrane and the periplasm [120] (see Chap. 13). A common feature of type II/III/IV secretion systems as well as Tfp and filamentous phage assembly machineries is an outer membrane protein that is very likely involved in the translocation of macromolecules to the surface and therefore is designated the *secretin* [121,122].

The fact that secretins are often the only integral outer membrane component of these systems makes them the obvious candidates to form the translocation pore. Indeed, pili polymerization in the absence of PilT (see Chap. XX) and PilQ, the secretin in *N. gonorrhoeae*, results in a failure to extrude the growing fiber to the surface and causes membrane deformations [72]. The putative secretin of the BFP machinery is encoded by the third gene of the operon, *bfpB*. Most secretins form SDS-resistant multimeric complexes that differ from each other in their ability to be dissociated on heat treatment [122–130]. Electron microscopic analysis of purified complexes reveals large, ring-shaped oligomeric structures with a central pore big enough to accommodate the transport of the respective substrate or with a central structured mass that probably represents a closed conformation of the pore [123,127,128,131,132]. Some preparations contain only "open" complexes, indicating that the purification conditions may have led to partial degradation or loss of an additional factor. Side views of the type II secretin PulD and the phage secretin pIV resemble each other in that a substantial part of the ring extends laterally in a cuplike or horseshoe-like fashion. It was postulated that this part is not integrated into the membrane, presenting potential interaction sites to inner membrane proteins. In contrast to pIV, PulD seem to be composed of two stacked rings.

Purified secretins form ion-conductive channels in planar-lipid bilayers that are characterized by a complex electrophysiologic behavior [132–134]. This complexity may be due to variable pore sizes or incomplete pore opening in the experiments. Since the expression of BfpB confers sensitivity to vancomycin, it was postulated that BfpB might form an incomplete gated channel [130].

On sequence analysis, a modular organization of secretins was suggested, consisting of a variable N-terminal region that can be conserved within related transport systems, an optional highly variable linker region rich in serine and glycine residues, and a conserved C-terminal domain predicted to contain strongly amphipathic β-sheets [121]. This organization implies different functions for the N- and C-terminal domains. Based on the high conservation and β-sheet prediction in the C-terminus, this part was postulated to build the pore, whereas the N-terminal region was suggested to be involved in interactions with other components of the system and/or in gating of the channel. The membrane localization and integration of the pIV secretin is indeed mediated by its C-terminal part [135], and exchanging distinct amino acids in this region influences multimerization or function of the pore [134,136]. Insertion and deletion analysis, especially in regions of predicted β-strands, underlined the role of the C-terminal domain in the complex formation of type II secretins such as PulD, XcpQ, and XpsD (*X. campestris*) and for outer membrane localization of XcpQ and XpsD [123,137,138]. Recent protease protection studies strongly support the concept that the C-terminal domain determines the main three-dimensional structure of the translocation channel [133,137,139]. Proteolysis of purified complexes resulted in a stable C-terminal product that lacks the very C-terminus (which extends over the S-domain in the case of PulD) but retains the multimeric organization. *In vivo*, this domain is embedded in the membrane, which is consistent with the predicted structure and circular dichroism analysis [133]. Proteolysis did not affect the overall structure of the complex, as visualized by electron microscopy. Nevertheless, the hypothetical plug, represented by the structured mass inside the ring, disappeared in the protease-treated PulD complexes [139]. Similarly, the electrophysiologic behavior of the XcpQ pore, integrated in planar lipid bilayers, changed on protease treatment. Based on these observations, it is tempting to speculate that the N-terminal region represents a kind of plug that could gate the pore by folding back into the channel. This also would explain the relatively minor changes observed in side views of digested complexes. A similar mechanism has been suggested for TonB-dependent siderophore receptors: An N-terminal extension inside the closed barrel obstructs the pore but is able to mediate conformational changes on substrate binding to TonB with subsequent ferrisiderophore uptake [140].

Whereas the N-terminus of the siderophore receptors is highly conserved, the equivalent region in secretins reveals a certain degree of conservation only in closely related systems and therefore possibly is responsible for recognition specificity [121]. Experimental data from the out system of *E. chrysanthemi* and *E. carotovora* support a model in which the translocated macromolecule confers this specificity by recognizing exclusively the N-terminal GspD domain of its own secretion machinery [126,141], thus explaining the inability of both systems to secrete the almost identical exoenzymes of each other. On the other hand, the N-terminal domain of PulD can be replaced with the equivalent region of OutD without affecting secretion [137]. In filamentous phage export, specificity seems to be conferred by the interaction of pIV, the secretin, with pI, a bitopic inner membrane protein with an ATP-binding motif [17,142]. Similarly, GspD (secretin) and GspC (bitopic inner membrane protein) interact in type II secretion systems and/or cannot be replaced by other paralogues [137,141,143]. GspC as well as pI might function in a similar fashion as TonB in transducing conformational changes to open the channel [143]. It remains to be shown whether Tfp biogenesis requires a gating process because the mechanical force of the growing fiber may be sufficient to remove the hypothetical lid and open the translocation channel [72].

Some secretins depend on chaperone-like lipoproteins (GspS) for stabilization and/or outer membrane localization. These lipoproteins often are encoded on an adjacent gene within one operon. A typical feature of lipoprotein-dependent secretins is a C-terminal extension that serves as an interaction domain. PilQ, a Tfp secretin, requires only its C-terminal domain for proper localization and multimerization [125], but formation and stabilization of the complex also depend on the presence of the lipoprotein PilP [144]. The lipoprotein paralogue in *K. oxytoca*, PulS, not only protects PulD against proteolysis but is also indispensable for piloting. These functions are separable because PulS has to be membrane-associated to localize PulD into the outer membrane but not to protect the secretin. Interestingly, tagging a protein with the interaction or S-domain renders their stability dependent on PulS without affecting the localization [17,145]. Copurification and electron microscopic analysis of purified complexes clearly demonstrated a stable interaction of the two proteins, which are present in about equal amounts in the purified complex. PulS is visible as radial spikes around an end view of the channel [132,139,145]. BfpB is able to form multimeric porelike complexes in the absence of other Bfp proteins. Since BfpB and other TfpB secretins are lipoproteins themselves, their localization and stability may be independent of GspS homologues [46,146,147]. Nevertheless, other studies [130] suggest that multimerization of BfpB depends on BfpG, a small protein that is partially located in the outer membrane but is not a lipoprotein.

Additional Ancillary Proteins

All Tfp systems contain a number of proteins that lack significant homologies to proteins in the current databases but which are involved in pilus biogenesis because mutations in their genes result in non-fimbriated phenotypes. One group of these proteins is characterized by a signal-peptidase I cleavage site indicating a putative localization in the periplasm. BfpU, a small protein encoded by the fifth gene in the *bfp* operon, fits into this category. BfpU is detectable in the soluble cell fraction and can be released partly into the pool of periplasmic proteins after suitable extraction procedures (Schreiber *et al.*, unpublished observations). Furthermore, the preprotein is indeed cleaved at the predicted signal-peptidase I cleavage site. Besides the fact that successful pilus assembly depends on strict stoichiometric amounts of BfpU, nothing is known about its precise function. Nevertheless, the requirement for precise amounts of BfpU indicates that it may be involved in interactions with the other assembly factors.

Because of its localization immediately upstream of *bfpB*, the secretin-encoding gene, putative signal-peptidase II cleavage sites, and its predicted size, the *bfpG* gene product resembled very much the small lipoproteins influencing function and localization of secretins [26]. However, N-terminal sequence and mass analyses indicate that BfpG is processed by signal-peptidase I. Mutation of the conserved cysteines in the putative signal-peptidase II motifs did not affect the function of BfpG, confirming that it is not a lipoprotein. Interestingly, the localization of BfpG depends on the presence of other Bfp proteins. Whereas in an *E. coli* K-12 background, it is mostly soluble, the protein also colocalizes with the outer membrane fraction in EPEC. Furthermore, immunoprecipitation and cross-linking studies indicate an interaction of BfpG and the secretin BfpB [130].

CofC, encoded by the sixth gene of the CFA/III gene cluster, is not only the closest homologue to BfpG but is also located upstream of the putative secretin CofD and therefore may serve the same function [34].

The product of *bfpH*, the tenth gene within the Bfp cluster, is apparently not involved in pilus biogenesis. The only phenotype that can be observed in a non-polar mutant is a modest reduction in autoaggregation [113]. Nevertheless, the absence of the paralogue (PilT) in the R64 thin pili system led to a reduction in the amount of pilin in the shear fraction, indicating that fewer pili were produced that were furthermore unable to support DNA transfer [11]. Based on sequence analysis, *bfpH*, *pilT*, and *cofT* encode lytic transglycosylases. Similar enzymes are not found in other pilus assembly machineries but apparently are involved in transfer of IncF plasmids, invasion of eukaryotic host cells by *Shigella flexneri* or *S. thyphi*, and more generally in maintaining the integrity of the bacterial cell wall during growth [148]. Whereas PilT activity may support plasmid passage by localized peptidoglycan hydrolysis [46], the function of BfpH remains enigmatic. Accumulation of mutations may have rendered BfpH unstable and/or nonfunctional [113,148]. This would be in a line with a theory that Bfp may have been acquired by horizontal gene transfer mediated by a bacteriophage that required a lytic transglycosylase to hydrolyze the bacterial cell wall from the outside.

REGULATION OF BFP EXPRESSION

Tfps represent a potential target for the host immune system. Therefore, it is important in the natural habitat that the expression of BFP is confined to the stages of infection that require functional pili. Coordinated expression of bundlin and other proteins involved in assembly and function is achieved by their organization in an operon. Nevertheless, it is obvious that the main structural subunit of the pilus is needed in much greater amounts than the remaining components. Whereas a *bfpA*-specific transcript can be detected easily on Northern blots, more sensitive methods are required to visualize the much less abundant *bfpB* mRNA. This differential expression probably is achieved by a putative secondary structure downstream of *bfpA* that might function as a transcriptional attenuator [147]. Putative stem-loop structures with similar structural free-energy values also have been identified in other *tfpB* gene clusters (*tcp*, *cof*) directly downstream of the respective prepilin-encoding gene [34,146].

The importance of expression of various Bfp proteins in precise relative proportions is also obvious in complementation experiments of *bfp* mutants, which were successful only when the expression did not exceed a certain threshold level [113]. This phenomenon also was described in other systems as *trans*- or dominant-negative inhibition [94,108].

The transcription of the *bfp* gene cluster depends on logarithmic growth, temperature (35–37°C), the presence of calcium, and the absence of ammonium. These are conditions the bacteria would be expected to find in the small intestine of their host but not in the colon or external environments [149]. The transcriptional activator of the *bfp* operon is PerA, which binds to a region downstream (−55 to −94) of the σ^{70}-dependent promoter. PerA belongs to the XylS/AraC family of transcriptional regulators with higher homology to a subfamily that is involved in the activation of virulence-associated genes [27]. *perA* is part of the *perABC* operon [150], which is located 6 kb downstream of *bfpL*. The gene products of *perB* and *perC* represent cofactors that are required for full transcriptional activation of the *bfp* gene cluster, although their exact function with respect to their role in enhancing BFP expression is not yet known. The transcription of the *perABC* operon itself is not regulated by ammonium, indicating that other factors are involved in BFP expression.

Coordinated expression and regulation of virulence factors are important for successful colonization of the host organism and pathogenesis. Thus it is not surprising that the regulatory function of the *per* operon is not limited to activate BFP expression. The regulator was first identified by its ability to activate transcription of *eae*, which encodes the outer membrane protein intimin within the locus of enterocyte effacement (LEE) pathogenicity island [150] (see Chap. 3). Per is also required to upregulate other LEE promoters mediated by a LEE-encoded regulator Ler [151]. The LEE is of central importance for EPEC pathogenesis because it contains genes required for the attaching and effacing phenotype [152].

It remains to be elucidated how EPEC is able to sense the appropriate environmental conditions that are required to induce regulatory cascades for coordinated expression of major virulence factors. Whereas pilus formation and function in *P. aeruginosa* are controlled by a network of signal-transduction systems similar to chemotaxis proteins [153,154], no classic sensor-regulator pairs have yet been identified in EPEC. Therefore, it is tempting to speculate about a potential role of the N-terminal domain of PerA in the integration of environmental information that also could include mechanical force, as suggested earlier.

ACKNOWLEDGMENTS

We are grateful to Jorge Girón for providing the electron microscopic images of CFA/III and longus and to Ravi Anantha for providing the image of BFP.

REFERENCES

1. Merz, A. J., So, M., and Sheetz, M. P. (2000). Pilus retraction powers bacterial twitching motility. *Nature* 407:98–102.

2. Wall, D., and Kaiser, D. (1999). Type IV pili and cell motility. *Mol. Microbiol.* 32:1–10.
3. Henrichsen, J. (1983). Twitching motility. *Ann. Rev. Microbiol.* 37:81–93.
4. Bradley, D. E. (1980). A function of *Pseudomonas aeruginosa* PAO polar pili: Twitching motility. *Can. J. Microbiol.* 26:146–154.
5. Bieber, D., Ramer, S. W., Wu, C. Y., *et al.* (1998). Type IV pili, transient bacterial aggregates, and virulence of enteropathogenic *Escherichia coli*. *Science* 280:2114–2118.
6. Pujol, C., Eugene, E., Marceau, M., and Nassif, X. (1999). The meningococcal PilT protein is required for induction of intimate attachment to epithelial cells following pilus-mediated adhesion. *Proc. Natl. Acad. Sci. USA* 96:4017–4022.
7. Strom, M. S., and Lory, S. (1993). Structure-function and biogenesis of the type IV pili. *Annu. Rev. Microbiol.* 47:565–596.
8. O'Toole, G. A., and Kolter, R. (1998). Flagellar and twitching motility are necessary for *Pseudomonas aeruginosa* biofilm development. *Mol. Microbiol.* 30:295–304.
9. Seifert, H. S., Ajioka, R. S., Marchal, C., *et al.* (1988). DNA transformation leads to pilin antigenic variation in *Neisseria gonorrhoeae*. *Nature* 336:392–395.
10. Dubnau, D. (1999). DNA uptake in bacteria. *Annu. Rev. Microbiol.* 53:217–244.
11. Yoshida, T., Kim, S. R., and Komano, T. (1999). Twelve *pil* genes are required for biogenesis of the R64 thin pilus. *J. Bacteriol.* 181:2038–2043.
12. Bradley, D. E. (1972). Stimulation of pilus formation in *Pseudomonas aeruginosa* by RNA bacteriophage adsorption. *Biochem. Biophys. Res. Commun.* 47:1080–1087.
13. Ottow, J. C. (1975). Ecology, physiology, and genetics of fimbriae and pili. *Annu. Rev. Microbiol.* 29:79–108.
14. Mattick, J. S., Bills, M. M., Anderson, B. J., *et al.* (1987). Morphogenetic expression of *Bacteroides nodosus* fimbriae in *Pseudomonas aeruginosa*. *J. Bacteriol.* 169:33–41.
15. Beard, M. K., Mattick, J. S., Moore, L. J., *et al.* (1990). Morphogenetic expression of *Moraxella bovis* fimbriae (pili) in *Pseudomonas aeruginosa*. *J. Bacteriol.* 172:2601–2607.
16. Hoyne, P. A., Haas, R., Meyer, T. F., *et al.* (1992). Production of *Neisseria gonorrhoeae* pili (fimbriae) in *Pseudomonas aeruginosa*. *J. Bacteriol.* 174:7321–7327.
17. Russel, M. (1998). Macromolecular assembly and secretion across the bacterial cell envelope: Type II protein secretion systems. *J. Mol. Biol.* 279:485–499.
18. Nunn, D. (1999). Bacterial type II protein export and pilus biogenesis: more than just homologies? *Trends Cell Biol.* 9:402–408.
19. Farinha, M. A., Conway, B. D., Glasier, L. M. G., *et al.* (1994). Alteration of the pilin adhesin of *Pseudomonas aeruginosa* PAO results in normal pilus biogenesis but a loss of adherence to human pneumocyte cells and decreased virulence in mice. *Infect. Immun.* 62:4118–4123.
20. Tennent, J. M., and Mattick, J. S. (1994). Type 4 fimbriae. In P. Klemm (ed.), *Fimbriae: Adhesion, Genetics, Biogenesis, and Vaccines*, pp. 127–146. Boca Raton, FL: CRC Press.
21. Cachia, P. J., Glasier, L. M., Hodgins, R. R., *et al.* (1998). The use of synthetic peptides in the design of a consensus sequence vaccine for *Pseudomonas aeruginosa*. *J. Pept. Res.* 52:289–299.
22. Hahn, H. P. (1997). The type-4 pilus is the major virulence-associated adhesin of *Pseudomonas aeruginosa*: A review. *Gene* 192:99–108.
23. Rudel, T., Scheuerpflug, I., and Meyer, T. F. (1995). *Neisseria* PilC protein identified as type-4 pilus tip-located adhesin. *Nature* 373:357–359.
24. Scaletsky, I. C. A., Silva, M. L. M., and Trabulsi, L. R. (1984). Distinctive patterns of adherence of enteropathogenic *Escherichia coli* to HeLa cells. *Infect. Immun.* 45:534–536.
25. Sohel, I., Puente, J. L., Ramer, S. W., *et al.* (1996). Enteropathogenic *Escherichia coli*: Identification of a gene cluster coding for bundle-forming pilus morphogenesis. *J. Bacteriol.* 178:2613–2628.
26. Stone, K. D., Zhang, H.-Z., Carlson, L. K., and Donnenberg, M. S. (1996). A cluster of fourteen genes from enteropathogenic *Escherichia coli* is sufficient for biogenesis of a type IV pilus. *Mol. Microbiol.* 20:325–337.

27. Tobe, T., Schoolnik, G. K., Sohel, I., *et al.* (1996). Cloning and characterization of *bfpTVW* genes required for the transcriptional activation of *bfpA* in enteropathogenic *Escherichia coli. Mol. Microbiol.* 21:963–975.
28. Baldini, M. M., Kaper, J. B., Levine, M. M., *et al.* (1983). Plasmid-mediated adhesion in enteropathogenic *Escherichia coli. J. Pediatr. Gastroenterol. Nutr.* 2:534–538.
29. Nataro, J. P., Maher, K. O., Mackie, P., and Kaper, J. B. (1987). Characterization of plasmids encoding the adherence factor of enteropathogenic *Escherichia coli. Infect. Immun.* 55:2370–2377.
30. Tobe, T., Hayashi, T., Han, C. G., *et al.* (1999). Complete DNA sequence and structural analysis of the enteropathogenic *Escherichia coli* adherence factor plasmid. *Infect. Immun.* 67:5455–5462.
31. Girón, J. A., Levine, M. M., and Kaper, J. B. (1994). Longus: A long pilus ultrastructure produced by human enterotoxigenic *Escherichia coli. Mol. Microbiol.* 12:71–82.
32. Honda, T., Arita, M., and Miwatani, T. (1984). Characterization of new hydrophobic pili of human enterotoxigenic *Escherichia coli*: A possible new colonization factor. *Infect. Immun.* 43:959–965.
33. Taniguchi, T., Fujino, Y., Yamamoto, K., *et al.* (1995). Sequencing of the gene encoding the major pilin of pilus colonization factor antigen III (CFA/III) of human enterotoxigenic *Escherichia coli* and evidence that CFA/III is related to type IV pili. *Infect. Immun.* 63:724–728.
34. Taniguchi, T., Akeda, Y., Haba, A., *et al.* (2001). Gene cluster for assembly of pilus colonization factor antigen III of enterotoxigenic *Escherichia coli. Infect. Immun.* 69:5864–5873.
35. Gómez-Duarte, O. G., Ruiz-Tagle, A., Gómez, D. C., *et al.* (1999). Identification of *lngA*, the structural gene of longus type IV pilus of enterotoxigenic *Escherichia coli. Microbiology* 145:1809–1816.
36. Zhang, X. L., Tsui, I. S., Yip, C. M., *et al.* (2000). *Salmonella enterica* serovar *typhi* uses type IVB pili to enter human intestinal epithelial cells. *Infect. Immun.* 68:3067–3073.
37. Yoshida, T., Furuya, N., Ishikura, M., *et al.* (1998). Purification and characterization of thin pili of IncI1 plasmids ColIb-P9 and R64: Formation of PilV-specific cell aggregates by type IV pili. *J. Bacteriol.* 180:2842–2848.
38. Strom, M. S., Nunn, D. N., and Lory, S. (1993). A single bifunctional enzyme, PilD, catalyzes cleavage and N-methylation of proteins belonging to the type IV pilin family. *Proc. Natl. Acad. Sci. USA* 90:2404–2408.
39. Alm, R. A., and Mattick, J. S. (1997). Genes involved in the biogenesis and function of type-4 fimbriae in *Pseudomonas aeruginosa. Gene* 192:89–98.
40. Jonson, G., Lebens, M., and Holmgren, J. (1994). Cloning and sequencing of *Vibrio cholerae* mannose-sensitive haemagglutinin pilin gene: Localization of *mshA* within a cluster of type 4 pilin genes. *Mol. Microbiol.* 13:109–118.
41. Fullner, K. J., and Mekalanos, J. J. (1999). Genetic characterization of a new type IV-A pilus gene cluster found in both classical and El Tor biotypes of *Vibrio cholerae. Infect. Immun.* 67:1393–1404.
42. Shaw, C. E., and Taylor, R. K. (1990). *Vibrio cholerae* O395 *tcpA* pilin gene sequence and comparison of predicted protein structural features to those of type 4 pilins. *Infect. Immun.* 58:3042–3049.
43. Girón, J. A., Ho, A. S. Y., and Schoolnik, G. K. (1991). An inducible bundle-forming pilus of enteropathogenic *Escherichia coli. Science* 254:710–713.
44. Donnenberg, M. S., Girón, J. A., Nataro, J. P., and Kaper, J. B. (1992). A plasmid-encoded type IV fimbrial gene of enteropathogenic *Escherichia coli* associated with localized adherence. *Mol. Microbiol.* 6:3427–3437.
45. Sohel, I., Puente, J. L., Murray, W. J., *et al.* (1993). Cloning and characterization of the bundle-forming pilin gene of enteropathogenic *Escherichia coli* and its distribution in *Salmonella* serotypes. *Mol. Microbiol.* 7:563–575.
46. Kim, S. R., and Komano, T. (1997). The plasmid R64 thin pilus identified as a type IV pilus. *J. Bacteriol.* 179:3594–3603.

47. Brown, R. C., and Taylor, R. K. (1995). Organization of *tcp*, *acf*, and *toxT* genes within a ToxT-dependent operon. *Mol. Microbiol.* 16:425–439.
48. Zhang, X. L., Morris, C., and Hackett, J. (1997). Molecular cloning, nucleotide sequence, and function of a site-specific recombinase encoded in the major "pathogenicity island" of *Salmonella typhi*. *Gene* 202:139–146.
49. Girón, J. A., Gómez-Duarte, O. G., Jarvis, K. G., and Kaper, J. B. (1997). Longus pilus of enterotoxigenic *Escherichia coli* and its relatedness to other type-4 pili: A minireview. *Gene* 192:39–43.
50. Blattner, F. R., Plunkett, G., III, Bloch, C. A., *et al.* (1997). The complete genome sequence of *Escherichia coli* K-12. *Science* 277:1453–1462.
51. Perna, N. T., Plunkett, G., III, Burland, V., *et al.* (2001). Genome sequence of enterohaemorrhagic *Escherichia coli* O157:H7. *Nature* 409:529–533.
52. Sauvonnet, N., Gounon, P., and Pugsley, A. P. (2000). PpdD type IV pilin of *Escherichia coli* K-12 can be assembled into pili in *Pseudomonas aeruginosa*. *J. Bacteriol.* 182:848–854.
53. Sauvonnet, N., Vignon, G., Pugsley, A. P., and Gounon, P. (2000). Pilus formation and protein secretion by the same machinery in *Escherichia coli*. *EMBO J.* 19:2221–2228.
54. Nataro, J. P., and Kaper, J. B. (1998). Diarrheagenic *Escherichia coli*. *Clin. Microbiol. Rev.* 11:142–201.
55. Knutton, S., McConnell, M. M., Rowe, B., and McNeish, A. S. (1989). Adhesion and ultrastructural properties of human enterotoxigenic *Escherichia coli* producing colonization factor antigens III and IV. *Infect. Immun.* 57:3364–3371.
56. Donnenberg, M. S., Girón, J. A., Schoolnik, G. K., and Kaper, J. B. (1992) A plasmid-encoded fimbrial gene of enteropathogenic *Escherichia coli* (EPEC) associated with localized adherence (abstract D-280). In *Abstracts of the 92nd General Meeting of the American Society for Microbiology*, p. 142. Washington: American Society for Microbiology Press.
57. Donnenberg, M. S., and Kaper, J. B. (1992). Minireview: Enteropathogenic *Escherichia coli*. *Infect. Immun.* 60:3953–3961.
58. Vuopio-Varkila, J., and Schoolnik, G. K. (1991). Localized adherence by enteropathogenic *Escherichia coli* is an inducible phenotype associated with the expression of new outer membrane proteins. *J. Exp. Med.* 174:1167–1177.
59. Hicks, S., Frankel, G., Kaper, J. B., *et al.* (1998). Role of intimin and bundle-forming pili in enteropathogenic *Escherichia coli* adhesion to pediatric intestinal tissue in vitro. *Infect. Immun.* 66:1570–1578.
60. Knutton, S., Shaw, R. K., Anantha, R. P., *et al.* (1999). The type IV bundle-forming pilus of enteropathogenic *Escherichia coli* undergoes dramatic alterations in structure associated with bacterial adherence, aggregation and dispersal. *Mol. Microbiol.* 33:499–509.
61. Moon, H. W., Whipp, S. C., Argenzio, R. A., *et al.* (1983). Attaching and effacing activities of rabbit and human enteropathogenic *Escherichia coli* in pig and rabbit intestines. *Infect. Immun.* 41:1340–1351.
62. Donnenberg, M. S., Kaper, J. B., and Finlay, B. B. (1997). Interactions between enteropathogenic *Escherichia coli* and host epithelial cells. *Trends Microbiol.* 5:109–114.
63. Frankel, G., Phillips, A. D., Rosenshine, I., *et al.* (1998). Enteropathogenic and enterohaemorrhagic *Escherichia coli*: more subversive elements. *Mol. Microbiol.* 30:911–921.
64. Blank, T. E., Zhong, H., Bell, A. L., *et al.* (2000). Molecular variation among type IV pilin (*bfpA*) genes from diverse enteropathogenic *Escherichia coli* strains. *Infect. Immun.* 68:7028–7038.
65. Zhang, H.-Z., and Donnenberg, M. S. (1996). DsbA is required for stability of the type IV pilin of enteropathogenic *Escherichia coli*. *Mol. Microbiol.* 21:787–797.
66. Parge, H. E., Forest, K. T., Hickey, M. J., *et al.* (1995). Structure of the fibre-forming protein pilin at 2.6 Å resolution. *Nature* 378:32–38.
67. Forest, K. T., and Tainer, J. A. (1997). Type-4 pilus-structure: Outside to inside and top to bottom—A minireview. *Gene* 192:165–169.
68. Sun, D., Lafferty, M. J., Peek, J. A., and Taylor, R. K. (1997). Domains within the *Vibrio cholerae* toxin coregulated pilin subunit that mediate bacterial colonization. *Gene* 192:79–85.

69. Kirn, T. J., Lafferty, M. J., Sandoe, C. M., and Taylor, R. K. (2000). Delineation of pilin domains required for bacterial association into microcolonies and intestinal colonization by *Vibrio cholerae. Mol. Microbiol.* 35:896–910.
70. McNamara, B. P., and Donnenberg, M. S. (2000). Evidence for specificity in type 4 pilus biogenesis by enteropathogenic *Escherichia coli. Microbiology* 146:719–729.
71. Wolfgang, M., Lauer, P., Park, H. S., *et al.* (1998). PilT mutations lead to simultaneous defects in competence for natural transformation and twitching motility in piliated *Neisseria gonorrhoeae. Mol. Microbiol.* 29:321–330.
72. Wolfgang, M., van Putten, J. P., Hayes, S. F., *et al.* (2000). Components and dynamics of fiber formation define a ubiquitous biogenesis pathway for bacterial pili. *EMBO J.* 19:6408–6418.
73. Pestova, E. V., and Morrison, D. A. (1998). Isolation and characterization of three *Streptococcus pneumoniae* transformation-specific loci by use of a *lacZ* reporter insertion vector. *J. Bacteriol.* 180:2701–2710.
74. Nunn, D. N., and Lory, S. (1992). Components of the protein-excretion apparatus of *Pseudomonas aeruginosa* are processed by the type IV prepilin peptidase. *Proc. Natl. Acad. Sci. USA* 89:47–51.
75. Marsh, J. W., and Taylor, R. K. (1998). Identification of the *Vibrio cholerae* type 4 prepilin peptidase required for cholera toxin secretion and pilus formation. *Mol. Microbiol.* 29:1481–1492.
76. Nunn, D., Bergman, S., and Lory, S. (1990). Products of three accessory genes, *pilB*, *pilC*, and *pilD*, are required for biogenesis of *Pseudomonas aeruginosa* pili. *J. Bacteriol.* 172:2911–2919.
77. Nunn, D. N., and Lory, S. (1991). Product of the *Pseudomonas aeruginosa* gene *pilD* is a prepilin leader peptidase. *Proc. Natl. Acad. Sci. USA* 88:3281–3285.
78. Strom, M. S., and Lory, S. (1992). Kinetics and sequence specificity of processing of prepilin by PilD, the type IV leader peptidase of *Pseudomonas aeruginosa. J. Bacteriol.* 174:7345–7351.
79. Strom, M. S., and Lory, S. (1991). Amino acid substitutions in pilin of *Pseudomona aeruginosa*: Effect on leader peptide cleavage, amino-terminal methylation, and pilus assembly. *J. Biol. Chem.* 266:1656–1664.
80. Horiuchi, T., and Komano, T. (1998). Mutational analysis of plasmid R64 thin pilus prepilin: the entire prepilin sequence is required for processing by type IV prepilin peptidase. *J. Bacteriol.* 180:4613–4620.
81. Pepe, J. C., and Lory, S. (1998). Amino acid substitutions in PilD, a bifunctional enzyme of *Pseudomonas aeruginosa:* Effect on leader peptidase and N-methyltransferase activities *in vitro* and *in vivo*. *J. Biol. Chem.* 273:19120–19129.
82. Reeves, P. J., Douglas, P., and Salmond, G. P. C. (1994). Beta-lactamase topology probe analysis of the OutO NMePhe peptidase, and six other Out protein components of the *Erwinia carotovora* general secretion pathway apparatus. *Mol. Microbiol.* 12:445–457.
83. Zhang, H.-Z., Lory, S., and Donnenberg, M. S. (1994). A plasmid-encoded prepilin peptidase gene from enteropathogenic *Escherichia coli. J. Bacteriol.* 176:6885–6891.
84. Dupuy, B., Taha, M.-K., Possot, O., *et al.* (1992). PulO, a component of the pullulanase secretion pathway of *Klebsiella oxytoca*, correctly and efficiently processes gonococcal type IV prepilin in *Escherichia coli. Mol. Microbiol.* 6:1887–1894.
85. Hobbs, M., and Mattick, J. S. (1993). Common components in the assembly of type 4 fimbriae, DNA transfer systems, filamentous phage and protein-secretion apparatus: A general system for the formation of surface-associated protein complexes. *Mol. Microbiol.* 10:233–243.
86. Pugsley, A. P. (1993). Processing and methylation of PulG, a pilin-like component of the general secretory pathway of *Klebsiella oxytoca. Mol. Microbiol.* 9:295–308.
87. Alm, R. A., and Mattick, J. S. (1995). Identification of a gene, *pilV*, required for type 4 fimbrial biogenesis in *Pseudomonas aeruginosa*, whose product possesses a pre-pilin-like leader sequence. *Mol. Microbiol.* 16:485–496.
88. Bleves, S., Lazdunski, A., and Filloux, A. (1996). Membrane topology of three Xcp proteins involved in exoprotein transport by *Pseudomonas aeruginosa. J. Bacteriol.* 178:4297–4300.

89. Bally, M., Filloux, A., Akrim, M., *et al.* (1992). Protein secretion in *Pseudomonas aeruginosa*: Characterization of seven *xcp* genes and processing of secretory apparatus components by prepilin peptidase. *Mol. Microbiol.* 6:1121–1131.
90. De Groot, A., Heijnen, I., De Cock, H., *et al.* (1994). Characterization of type IV pilus genes in plant growth-promoting *Pseudomonas putida* WCS358. *J. Bacteriol.* 176:642–650.
91. Pugsley, A. P., and Possot, O. (1993). The general secretory pathway of *Klebsiella oxytoca*: No evidence for relocalization or assembly of pilin-like PulG protein into a multiprotein complex. *Mol. Microbiol.* 10:665–674.
92. Pugsley, A. P. (1993). The complete general secretory pathway in gram-negative bacteria. *Microbiol. Rev.* 57:50–108.
93. Lu, H. M., Motley, S. T., and Lory, S. (1997). Interactions of the components of the general secretion pathway: Role of *Pseudomonas aeruginosa* type IV pilin subunits in complex formation and extracellular protein secretion. *Mol. Microbiol.* 25:247–259.
94. Pugsley, A. P. (1996). Multimers of the precursor of a type IV pilin-like component of the general secretory pathway are unrelated to pili. *Mol. Microbiol.* 20:1235–1245.
95. Nunn, D. N., and Lory, S. (1993). Cleavage, methylation, and localization of the *Pseudomonas aeruginosa* export proteins XcpT, -U, -V, and -W. *J. Bacteriol.* 175:4375–4382.
96. Kagami, Y., Ratliff, M., Surber, M., *et al.* (1998). Type II protein secretion by *Pseudomonas aeruginosa*: Genetic suppression of a conditional mutation in the pilin-like component XcpT by the cytoplasmic component XcpR. *Mol. Microbiol.* 27:221–233.
97. Possot, O. M., Vignon, G., Bomchil, N., *et al.* (2000). Multiple interactions between pullulanase secreton components involved in stabilization and cytoplasmic membrane association of PulE. *J. Bacteriol.* 182:2142–2152.
98. Blank, T. E., and Donnenberg, M. S. (2001). Novel topology of BfpE, a cytoplasmic membrane protein required for type IV fimbrial biogenesis in enteropathogenic *Escherichia coli*. *J. Bacteriol.* 183:4435–4450.
99. Thomas, J. D., Reeves, P. J., and Salmond, G. P. C. (1997). The general secretion pathway of *Erwinia carotovora* subsp. *carotovora*: Analysis of the membrane topology of OutC and OutF. *Microbiology* 143:713–720.
100. Iredell, J. R., and Manning, P. A. (1997). Outer membrane translocation arrest of the TcpA pilin subunit in *rfb* mutants of *Vibrio cholerae* O1 strain 569B. *J. Bacteriol.* 179:2038–2046.
101. Koga, T., Ishimoto, K., and Lory, S. (1993). Genetic and functional characterization of the gene cluster specifying expression of *Pseudomonas aeruginosa* pili. *Infect. Immun.* 61:1371–1377.
102. Possot, O., d'Enfert, C., Reyss, I., and Pugsley, A. P. (1992). Pullulanase secretion in *Escherichia coli* K-12 requires a cytoplasmic protein and a putative polytopic cytoplasmic membrane protein. *Mol. Microbiol.* 6:95–105.
103. Possot, O., and Pugsley, A. P. (1994). Molecular characterization of PulE, a protein required for pullulanase secretion. *Mol. Microbiol.* 12:287–299.
104. Sandkvist, M., Bagdasarian, M., Howard, S. P., and DiRita, V. J. (1995). Interaction between the autokinase EpsE and EpsL in the cytoplasmic membrane is required for extracellular secretion in Vibrio cholerae. *EMBO J.* 14:1664–1673.
105. Py, B., Loiseau, L., and Barras, F. (1999). Assembly of the type II secretion machinery of *Erwinia chrysanthemi*: Direct interaction and associated conformational change between OutE, the putative ATP-binding component and the membrane protein OutL. *J. Mol. Biol.* 289:659–670.
106. Turner, L. R., Lara, J. C., Nunn, D. N., and Lory, S. (1993). Mutations in the consensus ATP-binding sites of XcpR and PilB eliminate extracellular protein secretion and pilus biogenesis in *Pseudomonas aeruginosa*. *J. Bacteriol.* 175:4962–4969.
107. Krause, S., Barcena, M., Pansegrau, W., *et al.* (2000). Sequence-related protein export NTPases encoded by the conjugative transfer region of RP4 and by the *cag* pathogenicity island of *Helicobacter pylori* share similar hexameric ring structures. *Proc. Natl. Acad. Sci. USA* 97: 3067–3072.

108. Ball, G., Chapon-Hervé, V., Bleves, S., *et al.* (1999). Assembly of XcpR in the cytoplasmic membrane is required for extracellular protein secretion in *Pseudomonas aeruginosa. J. Bacteriol.* 181:382–388.
109. Sandkvist, M., Hough, L. P., Bagdasarian, M. M., and Bagdasarian, M. (1999). Direct interaction of the EpsL and EpsM proteins of the general secretion apparatus in *Vibrio cholerae. J. Bacteriol.* 181:3129–3135.
110. Turner, L. R., Olson, J. W., and Lory, S. (1997). The XcpR protein of *Pseudomonas aeruginosa* dimerizes via its N-terminus. *Mol. Microbiol.* 26:877–887.
111. Gomis-Ruth, F. X., Moncalian, G., Perez-Luque, R., *et al.* (2001). The bacterial conjugation protein TrwB resembles ring helicases and F1- ATPase. *Nature* 409:637–641.
112. Anantha, R. P., Stone, K. D., and Donnenberg, M. S. (1998). The role of BfpF, a member of the PilT family of putative nucleotide-binding proteins, in type IV pilus biogenesis and in interactions between enteropathogenic *Escherichia coli* and host cells. *Infect. Immun.* 66:122–131.
113. Anantha, R. P., Stone, K. D., and Donnenberg, M. S. (2000). Effects of *bfp* mutations on biogenesis of functional enteropathogenic *Escherichia coli* type IV pili. *J. Bacteriol.* 182:2498–2506.
114. Bradley, D. E. (1974). The adsorption of *Pseudomonas aeruginosa* pilus-dependent bacteriophages to a host mutant with nonretractile pili. *Virology* 58:149–163.
115. Whitchurch, C. B., Hobbs, M., Livingston, S. P., *et al.* (1991). Characterisation of a *Pseudomonas aeruginosa* twitching motility gene and evidence for a specialised protein export system widespread in eubacteria. *Gene* 101:33–44.
116. Felsenfeld, D. P., Schwartzberg, P. L., Venegas, A., *et al.* (1999). Selective regulation of integrin-cytoskeleton interactions by the tyrosine kinase Src. *Nature Cell Biol.* 1:200–206.
117. Kawagishi, I., Imagawa, M., Imae, Y., *et al.* (1996). The sodium-driven polar flagellar motor of marine *Vibrio* as the mechanosensor that regulates lateral flagellar expression. *Mol. Microbiol.* 20:693–699.
118. Wickner, S., Maurizi, M. R., and Gottesman, S. (1999). Posttranslational quality control: folding, refolding, and degrading proteins. *Science* 286:1888–1893.
119. Mahadevan, L., and Matsudaira, P. (2000). Motility powered by supramolecular springs and ratchets. *Science* 288:95–100.
120. Koronakis, V., Sharff, A., Koronakis, E., *et al.* (2000). Crystal structure of the bacterial membrane protein TolC central to multidrug efflux and protein export. *Nature* 405:914–919.
121. Genin, S., and Boucher, C. A. (1994). A superfamily of proteins involved in different secretion pathways in gram-negative bacteria: modular structure and specificity of the N-terminal domain. *Mol. Gen. Genet.* 243:112–118.
122. Hardie, K. R., Lory, S., and Pugsley, A. P. (1996). Insertion of an outer membrane protein in *Escherichia coli* requires a chaperone-like protein. *EMBO J.* 15:978–988.
123. Bitter, W., Koster, M., Latijnhouwers, M., *et al.* (1998). Formation of oligomeric rings by XcpQ and PilQ, which are involved in protein transport across the outer membrane of *Pseudomonas aeruginosa. Mol. Microbiol.* 27:209–219.
124. Newhall, W. J., Wilde, C. E., III, Sawyer, W. D., and Haak, R. A. (1980). High-molecular-weight antigenic protein complex in the outer membrane of *Neisseria gonorrhoeae. Infect. Immun.* 27: 475–482.
125. Drake, S. L., and Koomey, M. (1995). The product of the pilQ gene is essential for the biogenesis of type IV pili in *Neisseria gonorrhoeae. Mol. Microbiol.* 18:975–986.
126. Shevchik, V. E., Robert-Baudouy, J., and Condemine, G. (1997). Specific interaction between OutD, an *Erwinia chrysanthemi* outer membrane protein of the general secretory pathway, and secreted proteins. *EMBO J.* 16:3007–3016.
127. Koster, M., Bitter, W., De Cock, H., *et al.* (1997). The outer membrane component, YscC, of the Yop secretion machinery of *Yersinia enterocolitica* forms a ring-shaped multimeric complex. *Mol. Microbiol.* 26:789–797.
128. Crago, A. M., and Koronakis, V. (1998). *Salmonella* InvG forms a ring-like multimer that requires the InvH lipoprotein for outer membrane localization. *Mol. Microbiol.* 30:47–56.

129. Linderoth, N. A., Model, P., and Russel, M. (1996). Essential role of a sodium dodecyl sulfate-resistant protein IV multimer in assembly-export of filamentous phage. *J. Bacteriol.* 178: 1962–1970.
130. Schmidt, S. A., Bieber, D., Ramer, S. W., *et al.* (2001). Structure-function analysis of BfpB, a secretin-like protein encoded by the bundle-forming-pilus operon of enteropathogenic *Escherichia coli. J. Bacteriol.* 183:4848–4859.
131. Linderoth, N. A., Simon, M. N., and Russel, M. (1997). The filamentous phage pIV multimer visualized by scanning transmission electron microscopy. *Science* 278:1635–1638.
132. Nouwen, N., Ranson, N., Saibil, H., *et al.* (1999). Secretin PulD: association with pilot PulS, structure, and ion-conducting channel formation. *Proc. Natl. Acad. Sci. USA* 96:8173–8177.
133. Brok, R., Van Gelder, P., Winterhalter, M., *et al.* (1999). The C-terminal domain of the *Pseudomonas* secretin XcpQ forms oligomeric rings with pore activity. *J. Mol. Biol.* 294: 1169–1179.
134. Marciano, D. K., Russel, M., and Simon, S. M. (1999). An aqueous channel for filamentous phage export. *Science* 284:1516–1519.
135. Russel, M., and Kazmierczak, B. (1993). Analysis of the structure and subcellular location of filamentous phage pIV. *J. Bacteriol.* 175:3998–4007.
136. Russel, M. (1994). Phage assembly: A paradigm for bacterial virulence factor export. *Science* 265:612–614.
137. Guilvout, I., Hardie, K. R., Sauvonnet, N., and Pugsley, A. P. (1999). Genetic dissection of the outer membrane secretin PulD: Are there distinct domains for multimerization and secretion specificity? *J. Bacteriol.* 181:7212–7220.
138. Chen, L. Y., Chen, D. Y., Miaw, J., and Hu, N. T. (1996). XpsD, an outer membrane protein required for protein secretion by *Xanthomonas campestris* pv. *campestris*, forms a multimer. *J. Biol. Chem.* 271:2703–2708.
139. Nouwen, N., Stahlberg, H., Pugsley, A. P., and Engel, A. (2000). Domain structure of secretin PulD revealed by limited proteolysis and electron microscopy. *EMBO J.* 19:2229–2236.
140. Koebnik, R., Locher, K. P., and Van Gelder, P. (2000). Structure and function of bacterial outer membrane proteins: Barrels in a nutshell. *Mol. Microbiol.* 37:239–253.
141. Lindeberg, M., Salmond, G. P. C., and Collmer, A. (1996). Complementation of deletion mutations in a cloned functional cluster of *Erwinia chrysanthemi* out genes with *Erwinia carotovora* out homologues reveals OutC and OutD as candidate gatekeepers of species-specific secretion of proteins via the type II pathway. *Mol. Microbiol.* 20:175–190.
142. Feng, J. N., Model, P., and Russel, M. (1999). A *trans*-envelope protein complex needed for filamentous phage assembly and export. *Mol. Microbiol.* 34:745–755.
143. Bleves, S., Gérard-Vincent, M., Lazdunski, A., and Filloux, A. (1999). Structure-function analysis of XcpP, a component involved in general secretory pathway-dependent protein secretion in *Pseudomonas aeruginosa. J. Bacteriol.* 181:4012–4019.
144. Drake, S. L., Sandstedt, S. A., and Koomey, M. (1997). PilP, a pilus biogenesis lipoprotein in *Neisseria gonorrhoeae*, affects expression of PilQ as a high-molecular-mass multimer. *Mol. Microbiol.* 23:657–668.
145. Hardie, K. R., Seydel, A., Guilvout, I., and Pugsley, A. P. (1996). The secretin-specific, chaperone-like protein of the general secretory pathway: Separation of proteolytic protection and piloting functions. *Mol. Microbiol.* 22:967–976.
146. Ogierman, M. A., Zabihi, S., Mourtzios, L., and Manning, P. A. (1993). Genetic organization and sequence of the promoter-distal region of the *tcp* gene cluster of *Vibrio cholerae. Gene* 126:51–60.
147. Ramer, S. W., Bieber, D., and Schoolnik, G. K. (1996). BfpB, an outer membrane lipoprotein required for the biogenesis of bundle-forming pili in enteropathogenic *Escherichia coli. J. Bacteriol.* 178:6555–6563.
148. Lehnherr, H., Hansen, A. M., and Ilyina, T. (1998). Penetration of the bacterial cell wall: a family of lytic transglycosylases in bacteriophages and conjugative plasmids. *Mol. Microbiol.* 30: 454–457.

149. Puente, J. L., Bieber, D., Ramer, S. W., *et al.* (1996). The bundle-forming pili of enteropathogenic *Escherichia coli*: Transcriptional regulation by environmental signals. *Mol. Microbiol.* 20: 87–100.
150. Gómez-Duarte, O. G., and Kaper, J. B. (1995). A plasmid-encoded regulatory region activates chromosomal *eaeA* expression in enteropathogenic *Escherichia coli*. *Infect. Immun.* 63:1767–1776.
151. Mellies, J. L., Elliott, S. J., Sperandio, V., *et al.* (1999). The Per regulon of enteropathogenic *Escherichia coli*: Identification of a regulatory cascade and a novel transcriptional activator, the locus of enterocyte effacement (LEE)–encoded regulator (Ler). *Mol. Microbiol.* 33:296–306.
152. McDaniel, T. K., and Kaper, J. B. (1997). A cloned pathogenicity island from enteropathogenic *Escherichia coli* confers the attaching and effacing phenotype on K-12 *E. coli*. *Mol. Microbiol.* 23:399–407.
153. Hobbs, M., Collie, E. S. R., Free, P. D., *et al.* (1993). PilS and PilR, a two-component transcriptional regulatory system controlling expression of type 4 fimbriae in *Pseudomonas aeruginosa*. *Mol. Microbiol.* 7:669–682.
154. Darzins, A., and Russell, M. A. (1997). Molecular genetic analysis of type-4 pilus biogenesis and twitching motility using *Pseudomonas aeruginosa* as a model system: A review. *Gene* 192:109–115.

CHAPTER 12

The LEE-Encoded Type III Secretion System in EPEC and EHEC: Assembly, Function, and Regulation

J. Adam Crawford

Center for Vaccine Development, Department of Microbiology and Immunology, University of Maryland School of Medicine, Baltimore, Maryland

T. Eric Blank

Division of Infectious Diseases, Department of Medicine, University of Maryland School of Medicine, Baltimore, Maryland

James B. Kaper

Center for Vaccine Development and Department of Microbiology and Immunology, Department of Medicine, University of Maryland School of Medicine, Baltimore, Maryland

INTRODUCTION

Many gram-negative pathogens use type III secretion (TTS) systems to inject virulence factors into the cytosol of eukaryotic host cells to subvert cellular processes to the advantage of the bacterium. During the process of TTS, bacteria adhering to the surface of a eukaryotic cell transport virulence factors across three cell membranes: the inner and outer membranes of the bacterium and the eukaryotic cell membrane. In fact, contact with the host cell can activate TTS systems (Pettersson *et al.*, 1996; Rosqvist *et al.*, 1994; Zierler and Galan, 1995), and the delivery of virulence factors into the host cell occurs at the site of contact between the bacterium and the host cell (Rosqvist *et al.*, 1994; Persson *et al.*, 1995).

TTS systems are found in a variety of human pathogens, including enteropathogenic *Escherichia coli* (EPEC) and enterohemorrhagic *E. coli* (EHEC), *Yersinia*, *Shigella*, *Salmonella*, *Pseudomonas*, and *Chlamydia*. Plant pathogens, such as *Erwinia* spp., *Psuedomonas syringae*, *Xanthomonas campestris*, and *Ralstonia solanacearum*, also use TTS as a virulence mechanism. Interestingly, components of the flagellum are exported by a TTS system, suggesting that the virulence-related systems and

Escherichia coli: Virulence Mechanisms of a Versatile Pathogen
ISBN 0-12-220751-3

the flagellar system evolved from a common ancestral secretion system. Many genes involved in the secretion process are strongly conserved among all type III systems described to date, and thus the mechanism of secretion seems to be conserved. However, many types of virulence factors (effectors) are secreted by TTS systems, which reflect the broad number of different pathogenic strategies used by bacteria to cause disease. Several in-depth reviews of TTS have been written, and for a comprehensive discussion, the reader is referred to these (Cheng and Schneewind, 2000; Cornelis and Van Gijsegem, 2000; Cornelis, 2000; Hueck, 1998; Plano *et al.*, 2001).

SECRETION STRATEGIES IN GRAM-NEGATIVE BACTERIA

Gram-negative bacteria have evolved several other strategies in addition to TTS for transporting proteins across the inner and outer membranes. The major secretion systems described to date have been numbered I to V. Type I secretion systems involve only three proteins, an inner membrane ABC transporter, an outer membrane protein, and an additional inner membrane protein that spans the periplasm (reviewed in Young and Holland, 1999). The best known type I secretion substrate is the *E. coli* hemolysin (see Chap. 14). Type II secretion systems are an extension of the general secretory pathway and function to transport enzymes or toxins from the periplasm and through the outer membrane into the extracellular space (reviewed in Sandkvist, 2001). Thus type II secretion is a two-step process: transport of the protein across the inner membrane by the general secretory pathway, followed by transport across the outer membrane by the type II system. EHEC contains a type II secretion system that is encoded on a virulence plasmid, pO157, although proteins secreted by it have not been identified to date (Schmidt *et al.*, 1997). *E. coli* K-12 also contains a type II secretion system that is capable of secreting an endochitinase, ChiA (Francetic *et al.*, 2000). Additionally, type IV pilus biogenesis systems (see Chap. 12) contain several components that are homologous to type II components (Nunn, 1999). Type IV secretion systems, which several pathogens use to deliver virulence factors into host eukaryotic cells, are related to the conjugation systems responsible for DNA transfer between two bacterial cells (reviewed in Christie, 2001). Type V secretion consists of a group of secreted proteins called *autotransporters* that are delivered to the periplasm via the general secretory pathway (reviewed in Henderson and Nataro, 2001). Once in the periplasm, autotransporters are thought to form a pore in the outer membrane through which they pass to gain access to the extracellular space. The autotransporters EspC and EspP have been described in EPEC and EHEC, respectively (Mellies *et al.*, 2001; Stein *et al.*, 1996; Brunder *et al.*, 1997).

GENERAL FEATURES OF TYPE III SECRETION

Components of the Secretion Apparatus

Virulence-associated TTS systems contain approximately 10 highly conserved genes that encode an apparatus that functions to transport proteins through the inner and outer membranes. The genes encoding the secretion apparatus are highly conserved among bacterial pathogens, although unique components are found in each (Hueck, 1998). Virulence-associated TTS systems are also highly homologous with the flagellar TTS system. In fact, visualization of the *Salmonella typhimurium* and *Shigella flexneri* secretion apparatuses bears a striking resemblance to the flagellar hook–basal body structure (Blocker *et al.*, 2001; Kubori *et al.*, 1998). The conserved components of type III systems have been classified into several protein families and will be referred to using the *Yersinia* protein name because this is the best-characterized type III system to date. The YscV, -R, -S, -T, -U, and -D protein families have been demonstrated or are predicted to be inner membrane proteins that may form a channel in the inner membrane (Allaoui *et al.*, 1995; Plano *et al.*, 1991). Interestingly, YscV and YscU contain large cytoplasmic domains that could interact with cytoplasmic components of the TTS system (Allaoui *et al.*, 1995; Plano *et al.*, 1991). Members of the YscC family are outer membrane proteins that form a ring-shaped multimeric complex with an apparent central core through which proteins are secreted (Koster *et al.*, 1997; Daefler and Russel, 1998; Crago and Koronakis, 1998). YscC is homologous to outer membrane translocator proteins (secretins) from type II secretion systems and from filamentous phage assembly and extrusion pathways (Linderoth *et al.*, 1997; Nouwen *et al.*, 1999; Russel, 1998). The flagellar TTS system does not contain a YscC homologue. YscJ family members are predicted to predominantly reside in the periplasm, with the C-terminus anchored in the inner membrane and the N-terminus anchored to the outer membrane by a lipid-attachment-site motif (Allaoui *et al.*, 1992). Thus YscJ may function as a periplasmic bridge that connects the inner and outer membrane secretion channels, thereby allowing secretion to occur without a periplasmic intermediate. The YscN family consists of cytoplasmic ATPases that may provide energy for the secretion apparatus (Woestyn *et al.*, 1994; Eichelberg *et al.*, 1994).

Structure of the Secretion Apparatus

As mentioned earlier, the macromolecular structure of virulence-associated TTS systems and the flagellar TTS system is strikingly similar. Visualization of the secretion apparatus from *S. typhimurium* and *S. flexneri* has revealed that it consists of a cylindrical base that spans the inner and outer membrane, which resembles the flagellar basal body, and a hollow needle-like structure that projects outward from the base and extends away from the surface of the bacterium (Blocker *et al.*, 2001; Kubori *et al.*, 1998; Kubori *et al.*, 2000; Tamano *et al.*, 2000). Such structures have been termed *needle complexes.* Recently, a needle complex from EPEC has been described, and it is discussed later. Biochemical analysis of the needle complex from *S. flexneri* and *S. typhimurium* has revealed that it is composed of a homologous protein (PrgI and MxiH, respectively) and requires a functional TTS system for assembly (Sukhan *et al.*, 2001; Kubori *et al.*, 2000). The needle may serve as a hollow conduit that allows the passage of type III secreted proteins through the inner and outer membranes. In support of this model, it has been shown that secretion of type III substrates depends on an assembled needle structure (Tamano *et al.*, 2000; Kubori *et al.*, 2000; Blocker *et al.*, 2001). Components of the cylindrical base (which includes homologues of the outer membrane YscC secretin and two membrane-associated lipoproteins, PrgK and PrgH) contain *sec*-dependent secretion signals, and therefore, assembly of the base does not require a functional TTS system (Tamano *et al.*, 2000; Kubori *et al.*, 1998; Kubori *et al.*, 2000; Blocker *et al.*, 2001). The YscN ATPase and the inner membrane proteins YscV, -R, -S, -T, -U, and -D described earlier may interact with the base of the needle structure as part of the type III secretion mechanism (Blocker *et al.*, 1999; Macnab, 1999).

Translocation Apparatus: Transport Across the Eukaryotic Cell Membrane

A subset of proteins secreted by TTS systems forms the translocation apparatus, which is an extracellular extension of the TTS system that functions to translocate effector proteins across the eukaryotic cell membrane. In the absence of a functional translocation apparatus, effector molecules are secreted into the culture supernatant but are not delivered into host cells. It is thought that the translocation apparatus forms a pore in the eukaryotic cell membrane through which effector proteins move to gain access to the host cell cytosol (Hakansson *et al.*, 1996; Neyt and Cornelis, 1999a). In support of this model, it has been shown that YopB and YopD of *Yersinia* exhibit contact-dependent lytic activity on sheep erythrocytes (Hakansson *et al.*, 1996; Neyt and Cornelis, 1999a) and are required for the formation of channels in liposomes by *Yersinia* (Tardy *et al.*, 1999). YopB and YopD contain hydrophobic domains, which further supports the idea that these proteins insert into the membrane of the host cell (Hakansson *et al.*, 1993). Similar to YopB and YopD, IpaB and IpaC of *Shigella* are required for contact-dependent hemolysis of red blood cells and therefore may function as part of the translocation apparatus of *Shigella* (Blocker *et al.*, 1999). In *Salmonella*, SipB, -C, and -D appear to function as part of the translocation apparatus (Collazo and Galan, 1997; Fu and Galan, 1998).

The relationship between the needle structure described earlier and the process of translocation is not clear. One group has claimed that the *Yersinia* needle, which is composed of YscF, actually punctures the eukaryotic cell membrane to form a conduit between the bacterium and the host cell through which Yops move to gain access to the host cell cytosol (Hoiczyk and Blobel, 2001). However, as discussed later, in EPEC the needle may not interact directly with the host cell. Instead, a large filamentous structure on the cell surface composed of EspA, which is distinct from the EPEC needle homologue EscF, may function as the conduit for delivery of effectors into host cells. Large surface appendages involved in type III secretion also have been reported for *S. typhimurium* (Ginocchio *et al.*, 1994) and the plant pathogens *P. syringae* (Roine *et al.*, 1997) and *R. solanacearum* (Van Gijsegem *et al.*, 2000).

Mechanism of Secretion

Type III secreted proteins are not subjected to N-terminal processing during secretion, and it appears that proteins are secreted directly from the cytoplasm and through the inner and outer membranes without a distinct periplasmic intermediate step. Most of the details regarding secretion and translocation signals come from studies on Yop proteins of *Yersinia*. It has been shown that the first 15–20 N-terminal codons of various Yop proteins contain a signal that marks the protein for secretion into the culture supernatant (Rosqvist *et al.*, 1994; Sory *et al.*, 1995). However, no amino acid homology at the N-terminus is shared between these various Yop proteins. Anderson and colleagues have reported that the secretion signal is not protein in nature but instead resides in the mRNA structure encoding the secreted protein (Anderson and Schneewind, 1997, 1999). However, another group has claimed that mRNA has no role in secretion, but instead, an amphipathic N-terminal structure functions as the secretion signal (Lloyd *et al.*, 2001). Type III secreted proteins also contain a binding site, typically within the first 100 N-terminal residues, for a small acidic protein that exhibits functional features of a chaperone (Sory *et al.*, 1995; Wattiau and Cornelis, 1993; Wattiau *et al.*, 1994, 1996; Neyt and Cornelis, 1999b). However, the exact function of these chaperone proteins is in dispute. Several groups have shown that type III chaperones are important for the stability of the cognate secreted protein (Elliott *et al.*, 1999a; Frithz-Lindsten *et al.*, 1995), but this is not true for all type III secreted proteins (Wattiau *et al.*, 1994). Also, it has been reported that chaperones can function as secretion "pilots" by directing the secreted protein to the secretion apparatus (Cheng *et al.*, 1997; Cheng and Schneewind, 1999), although other groups dispute this (Frithz-Lindsten *et al.*, 1995). In the flagellar TTS system, the chaperone FlgN is required for translation of the type III secreted protein FlgM (Karlinsey *et al.*, 2000). In *Yersinia*, it appears that chaperones may control the timing and order in which various Yops are delivered into host cells (Boyd *et al.*, 2000).

TYPE III SECRETION IN EPEC AND EHEC

The LEE Pathogenicity Island and the TTS System

The locus of enterocyte effacement (LEE) is a 35.6-kb pathogenicity island found in EPEC and EHEC that is required for the formation of attaching and effacing (AE) lesions (see Chap. 4). The LEE contains 41 open reading frames (ORFs) (Elliott *et al.*, 1998) (Fig. 1). Most of these ORFs are organized into five major operons (see below) (Elliott *et al.*, 1999a; Mellies *et al.*, 1999; Sanchez-SanMartin *et al.*, 2001). LEE genes designated *esc* (*E. coli* secretion) encode components of a type III protein secretion apparatus that are highly homologous with the *ysc* genes described earlier and are highly conserved between EPEC and EHEC (Elliott *et al.*, 1998; Jarvis *et al.*, 1995). Genes designated *esp* encode *E. coli* secreted proteins that are the substrates of the LEE-encoded TTS system. Interestingly, EPEC and EHEC contain significant sequence variation in the *esp* genes, which are predicted to interact with host tissue. The *ces* genes encode TTS chaperones. CesT has been shown to function as a chaperone for Tir (translocated intimin receptor; see Chap. 4), CesD functions as a chaperone for EspD and perhaps EspB, and CesF functions as a chaperone for EspF (Elliot *et al.*, submitted, 1999a; Abe *et al.*, 1999; Wainwright and Kaper, 1998). Mutations in genes of the EPEC TTS system cause deficiencies in AE, invasion, host cell signal transduction, and host cell cytoskeletal rearrangements, in addition to the expected deficiencies in protein secretion (De Rycke *et al.*, 1997; Donnenberg *et al.*, 1990, 1997; Foubister *et al.*, 1994; Jarvis *et al.*, 1995; Kenny and Finlay, 1995; Rosenshine *et al.*, 1992).

The EspABD Translocation Apparatus

To date, seven proteins have been identified that are encoded by genes of the LEE and are secreted or translocated via the TTS system: EspA, EspB (previously EaeB), EspD, Tir (EspE), EspF, EspG, and Map (Elliott *et al.*, 2001; Haigh *et al.*, 1995; Kenny and Finlay, 1995; Kenny *et al.*, 1996, 1997a; Kenny and Jepson, 2000; Lai *et al.*, 1997; McNamara and Donnenberg, 1998; Taylor *et al.*, 1998; Wolff *et al.*, 1998). These proteins can be separated into two functional groups: (1) translocated effectors (Tir, EspF, EspG, Map, and possibly EspB; see Chap. 4) and (2) components of a translocation apparatus (EspA, EspB, and EspD).

As described earlier, a subset of type III secreted proteins inserts into the host cell membrane and forms a pore through which effector proteins gain access to

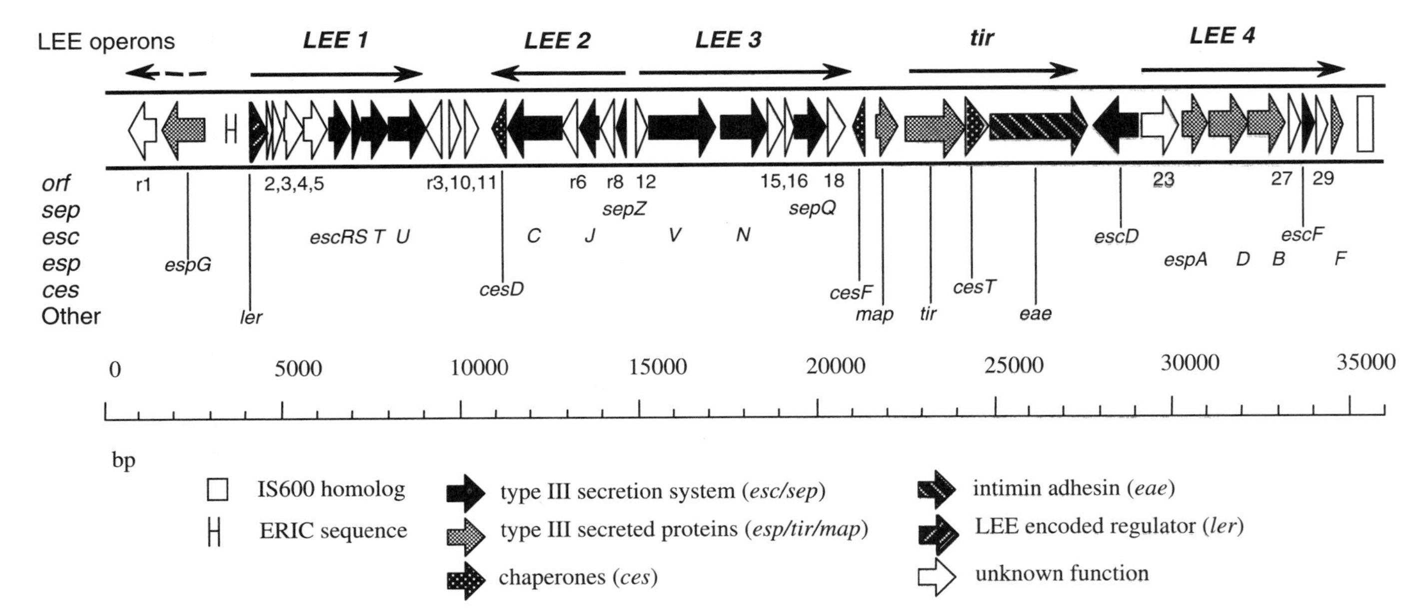

FIGURE 1 The locus of enterocyte effacement (LEE) pathogenicity island. The major operons (*LEE1–4*, *tir*) are shown. *esc* genes encode the secretion apparatus, *esp* genes encode secreted proteins, *ces* genes encode TTS chaperones, and unnamed ORFs are numbered. Reproduced with permission from *Current Topics in Microbiology & Immunology*, 264/1. Copyright Springer-Verlag.

the cytosol of the host cell. In EPEC and EHEC, EspA, EspB, and EspD are believed to be components of the translocation apparatus (Knutton *et al.*, 1998). *espA*, *espB* (*eaeB*), and *espD* mutants of EPEC have similar phenotypes. Each of these mutants has been shown to be deficient in the translocation of one or more of the effector proteins into host cells (Kenny *et al.*, 1997a; Kenny and Jepson, 2000; Knutton *et al.*, 1998; Taylor *et al.*, 1998; Wolff *et al.*, 1998). Each is also deficient in AE, invasion, as well as host cell signaling and cytoskeletal reorganization (Abe *et al.*, 1997, 1998; Donnenberg *et al.*, 1990, 1993; Foubister *et al.*, 1994; Kenny *et al.*, 1996, 1997a; Lai *et al.*, 1997; Manjarrez-Hernandez *et al.*, 1996; Eichelberg *et al.*, 1994; Rosenshine *et al.*, 1992). The latter phenotypes are presumably a result of being unable to translocate specific effector proteins into the host cell. EspA, EspB, and EspD each have been shown to be required for virulence in the rabbit model (Abe *et al.*, 1998; Nougayrède, Ph.D. thesis).

The putative translocation apparatus has been visualized as a large filamentous appendage of EPEC that appears to make direct contact between EPEC and host cell membranes. Most TTS systems lack a corresponding structure. The EspA protein is a required component of this appendage and probably represents the major structural subunit (Knutton *et al.*, 1998; Neves *et al.*, 1998). EspA has a moderate degree of sequence similarity to flagellins in a region predicted to form a coiled-coil domain that could be involved in protein-protein interaction (Knutton *et al.*, 1998; Pallen *et al.*, 1997). Mutational analysis has demonstrated that amino acid substitutions in this region interfere with both EspA multimerization and filament formation (Delahay *et al.*, 1999; Shaw *et al.*, 2001). The EspA filament is attached to the tip of the TTS needle complex. Both the EspA filament and the EPEC needle complex have been purified as a single macromolecular structure. In this structure, the EPEC needle complex is visible and is similar to the *Shigella* needle complex. Extending from the EPEC needle complex is the EspA filament (Sekiya *et al.*, 2001). The LEE contains a gene, *escF*, whose product is presumed to be the predominant subunit of the EPEC needle structure (Sekiya *et al.*, 2001). EscF is similar in sequence to PrgI and MxiH from *S. typhimurium* and *S. flexneri.* Interestingly, an *espA* mutant EPEC strain was defective in needle complex formation, suggesting that an interaction between EspA and EscF is required for stability of the EPEC needle complex (Sekiya *et al.*, 2001). An *escF* mutant EPEC strain was deficient in TTS and did not form AE lesions on eukaryotic cells (Sekiya *et al.*, 2001).

The functions of EspB and EspD in translocation are less clear. These proteins are somewhat similar to each other in sequence (Elliott *et al.*, 1998) and in other characteristics. Both EspB and EspD fractionate with the host cell membrane after EPEC infection (Knutton *et al.*, 1998; Wachter *et al.*, 1999; Wolff *et al.*, 1998). Both proteins contain hydrophobic domains that may integrate into the host cell membrane, as well as coiled-coil domains that may mediate functional protein-protein interactions (Daniell *et al.*, 2001; Pallen *et al.*, 1997;

Wachter *et al.*, 1999; Wolff *et al.*, 1998). EspB and EspD exhibit divergent properties as well, suggesting that they each play unique roles in EPEC infection. EspB has been shown to partially fractionate to the host cell cytoplasm as well as to the membrane (Knutton *et al.*, 1998; Taylor *et al.*, 1998; Wolff *et al.*, 1998), is inaccessible to antibody without membrane permeabilization (Wolff *et al.*, 1998), and acquires a protease-resistant form (Kenny and Finlay, 1995). EspD localizes to the membrane and to extracellular aggregates and is both accessible to antibodies and sensitive to protease (Daniell *et al.*, 2001; Shaw *et al.*, 2001; Wachter *et al.*, 1999). The EspA filament is shortened in *espD* but not in *espB* mutants (Daniell *et al.*, 2001; Hartland *et al.*, 2000; Knutton *et al.*, 1998). An *espD* mutant also exhibits reduced expression and secretion of EspA and EspB (Abe *et al.*, 1997; Lai *et al.*, 1997). EspB interacts with EspA and appears to colocalize with EspA filaments during infection. EspB is not required, however, for the attachment of EspA filaments to the host cell, suggesting that it is involved in translocation but not adhesion (Hartland *et al.*, 2000). EspD monomers are capable of interacting with each other (Daniell *et al.*, 2001). EspD does not localize with EspA filaments but does appear to be involved in their adhesion to the host cell (Daniell *et al.*, 2001). It has been hypothesized that EspB and EspD form a transmembrane channel or pore through the host cell membrane at the distal end of the EspA filament, allowing effector proteins to enter the host cell cytoplasm (Knutton *et al.*, 1998; Wolff *et al.*, 1998, Hartland *et al.*, 2000). Support for the pore concept is provided by the findings that the TTS system, EspA filament, EspD, and to some extent EspB are required for the hemolysis of red blood cells (RBCs) by EPEC (Shaw *et al.*, 2001; Warawa *et al.*, 1999). RBC hemolysis by EPEC, unlike that promoted by other bacterial pathogens having a TTS system, does not require close contact between bacteria and host cells presumably because of the EspA filament (Shaw *et al.*, 2001). TTS systems of other bacterial pathogens contain translocator proteins with weak sequence similarity to EspD [PopB of *Pseudomonas*, SipB (SspB) of *Salmonella*, IpaB of *Shigella*, and YopB of *Yersinia*] and EspB (PopD of *Pseudomonas* and YopD of *Yersinia*) (Daniell *et al.*, 2001; Elliott *et al.*, 1998; Hartland *et al.*, 2000; Wachter *et al.*, 1999). Various studies are consistent with the notion that these proteins are components of translocation pore complexes (reviewed by Cornelis and Van Gijsegem, 2000; Hueck, 1998). A model of the TTS system encoded by the LEE is shown in Fig. 2.

An SPI-I Homologous Type III System in EHEC

Sequencing of the EHEC genome revealed the presence of an additional type III system that is homologous to the *Salmonella* pathogenicity island I (SPI-I) TTS system (Perna *et al.*, 2001), which is required for invasion of host cells by *Salmonella*. However, genes encoding secreted effector proteins and components

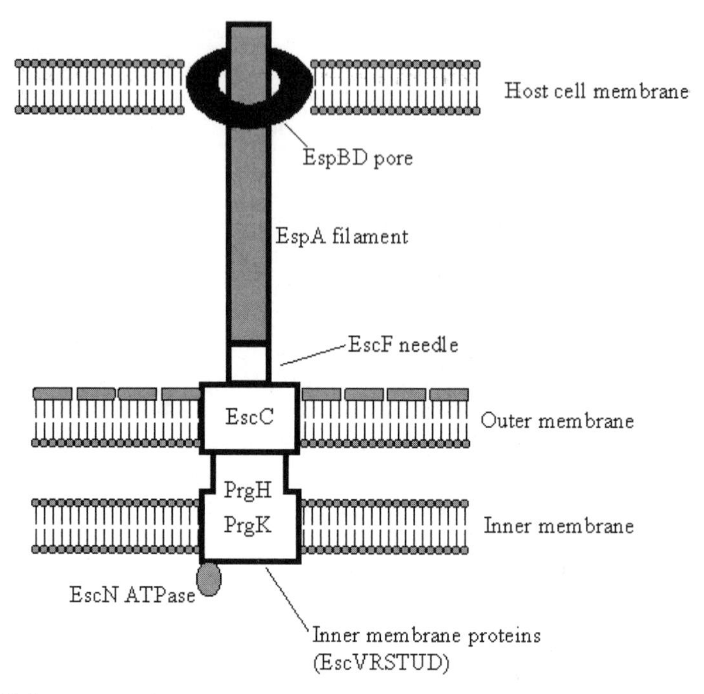

FIGURE 2 Drawing of a TTS system associated with a host cell membrane. Protein names correspond to EPEC and EHEC, except for PrgH and PrgK, of which homologues have not yet been described in EPEC or EHEC. The translocation apparatus, consisting of EspB, the EspD pore, and the EspA filament, is shown. The putative needle consisting of EscF is labeled. The base of the needle structure consists of the outer membrane secretin EscC and two membrane-associated lipoproteins, PrgH and PrgK, from *Salmonella*. Associated with the base of the needle structure are the ATPase EscN and several inner membrane proteins.

of the translocation apparatus are not present in the EHEC SPI-I homologue. Additionally, an important regulator of SPI-I expression, HilA, is also absent. It appears that an intact secretion apparatus could be encoded by the EHEC SPI-I homologue, but secretion substrates have not been identified to date.

REGULATION OF LEE GENE EXPRESSION

Gene Arrangement in the LEE

Investigation of LEE gene regulation has revealed that a majority of the LEE genes are organized into five major operons: *LEE1* (*ler*, *orf2–5*, and *escRSTU*), *LEE2* (*sepZ*, *rorf8*, *escJ*, *rorf6*, *escC*, and *cesD*), *LEE3* (*orf12*, *escVN*, *orf15*, *orf16*, *sepQ*, and *orf18*), *LEE4* [*orf23* (*sepL*), *espADB*, *orf27*, *escF*, *orf29*, and *espF*], and *tir* (*tir*, *cesT*, and *eae*; the *tir* operon also has been called *LEE5*) (Elliott *et al.*, 1999a; Mellies *et al.*, 1999; Sperandio *et al.*, 1999; Sanchez-SanMartin *et al.*, 2001). A few genes in the LEE lie outside of these major operons. The first gene of *LEE1*, *ler*, encodes a critical regulator of the other LEE operons, which will be discussed below. Promoters for each of the major operons just described have been defined experimentally, as well as for *orf19* and *escD* (known as *pas* in EHEC), which are transcribed as monocistronic messages (Beltrametti *et al.*, 2000; Sanchez-SanMartin *et al.*, 2001). Additionally, based on transcription-initiation sites, putative promoters have been assigned to *rorf3* and *orf10* (Mellies *et al.*, 1999). One interesting feature of the gene arrangement in the LEE is the presence of three sets of overlapping promoters that control divergently transcribed operons (*rorf3* and *orf10*, *LEE2* and *LEE3*, *escD* and *LEE4*) (Mellies *et al.*, 1999). Additionally, a *cesT* promoter and a second *LEE4* promoter located in between *orf23* and *espA* have been reported (Beltrametti *et al.*, 1999; Sanchez-SanMartin *et al.*, 2001). Also, an *eae* promoter controlling intimin expression has been reported, although another group has failed to confirm this (Gomez-Duarte and Kaper, 1995; Sanchez-SanMartin *et al.*, 2001).

Environmental Regulation of LEE Expression

In the laboratory, cell culture medium such as DMEM is used to activate the LEE TTS system, with maximal secretion occurring at mid-exponential phase, whereas secretion is not supported by growth in a rich medium such as LB (Kenny *et al.*, 1997b; Rosenshine *et al.*, 1996). In addition to growth in cell culture medium, LEE expression is optimal at 37°C, physiologic osmolarity, and pH 7.0, conditions similar to those in the gastrointestinal tract (Kenny *et al.*, 1997b; Rosenshine *et al.*, 1996; Kresse *et al.*, 2000). Maximal secretion also requires sodium bicarbonate, calcium, and $Fe(NO_3)_3$ (Kenny *et al.*, 1997b). In the presence of eukaryotic cells, synthesis and transport of LEE-encoded type III secreted proteins are enhanced (Beltrametti *et al.*, 1999; Wachter *et al.*, 1999; Wolff *et al.*, 1998), features that have been observed with other type III systems (Pettersson *et al.*, 1996; Rosqvist *et al.*, 1994; Zierler and Galan, 1995).

Regulators of the LEE

Ler: Master Regulator of the LEE

ler (LEE-encoded regulator), the first gene in the *LEE1* operon, encodes a critical regulator of LEE expression in both EPEC and EHEC (Elliott *et al.*, 2000; Mellies *et al.*, 1999; Friedberg *et al.*, 1999; Bustamante *et al.*, 2001). Ler is homologous to the H-NS family of DNA-binding proteins (Elliott *et al.*, 2000). The strongest region of homology is between the C-terminus of Ler and the DNA-binding domain of H-NS family members, whereas the N-terminus of Ler varies considerably relative to the H-NS sequence (Elliott *et al.*, 2000). Except for one amino acid difference in the N-terminus, the sequence of Ler is identical between EPEC and EHEC (Elliott *et al.*, 2000). The essential role Ler plays in LEE gene expression was demonstrated by mutating *ler* in both EHEC and EPEC. The *ler* mutation in both strains abolished synthesis and secretion of LEE-encoded proteins such as Tir, EspA, EspB, and EspD, and consequently, both mutant strains were negative in the fluorescent actin staining (FAS) assay, which measures actin accumulation beneath adherent EPEC or EHEC (Elliott *et al.*, 2000). Ler has been shown to activate the *LEE2*, *LEE3*, *tir*, and *orf19* promoters in a K-12 background but not the *LEE4* promoter (Elliott *et al.*, 1999a; Mellies *et al.*, 2001; Sanchez-SanMartin *et al.*, 2001). However, *LEE4* mRNA is drastically reduced in *ler* mutants of EPEC and EHEC, showing that Ler is required for *LEE4* expression (Elliott *et al.*, 2000). Ler does not activate the *LEE1* promoter, suggesting that Ler expression is not autoregulated (Mellies *et al.*, 1999). However, as discussed later, many factors control the *LEE1* promoter and therefore expression of Ler. Additionally, Ler controls genes located outside the LEE, such as *espC*, which encodes a type V secreted enterotoxin in EPEC, and genes encoding several morphologically distinct types of fimbriae (Elliott *et al.*, 2000; Mellies *et al.*, 2001).

Studies examining the mechanism by which Ler controls transcription have revealed that Ler antagonizes H-NS–mediated repression of the *LEE2*, *LEE3*, and *orf19* promoters (Bustamante *et al.*, 2001; Sanchez-SanMartin *et al.*, 2001). Thus one function of Ler is to counteract the repressing effects of H-NS at several LEE promoters, which may occur through Ler-mediated interference of DNA binding by H-NS. However, at the *tir* promoter, Ler antagonizes an unidentified repressor that is distinct from H-NS, and Ler also seems to counteract a repressor protein at the *LEE4* promoter (Elliott *et al.*, 2000; Sanchez-SanMartin *et al.*, 2001). Thus it appears that Ler functions as a master regulator of LEE expression by acting as an antirepressor. Ler has been purified and characterized biochemically. Purified Ler binds specifically to the *LEE2* promoter region, and interestingly, this *LEE2* binding site controls transcription of the divergently transcribed *LEE3* operon (Sperandio *et al.*, 2000). The N-terminus of Ler contains a predicted coiled-coil domain that may facilitate homodimerization. Disruption of this coiled-coil domain abolished *in vitro* DNA binding and promoter activation, suggesting that dimerization is critical in the function of Ler (Sperandio *et al.*, 2000).

Regulation of LEE Gene Expression by Quorum Sensing

In both EPEC and EHEC, quorum sensing controls expression of the LEE. Quorum sensing is a mechanism in which production of a small molecule known as an *autoinducer* accumulates in the extracellular environment and, on reaching a critical concentration, activates transcription of target genes (for reviews, see Bassler, 1999; de Kievit and Iglewski, 2000; Withers *et al.*, 2001; Schauder and Bassler, 2001). One quorum-sensing system found in both EPEC and EHEC involves an autoinducer known as *autoinducer-2* (AI-2) (Sperandio *et al.*, 1999) that was first described in *Vibrio harveyi* (Bassler *et al.*, 1994). The structure of AI-2 is unknown, but it appears to be a furanone (Schauder *et al.*, 2001). An essential gene required for its production, *luxS*, has been identified (Surette *et al.*, 1999). *luxS* is found in over 30 species of gram-positive and gram-negative bacteria (Bassler, 1999). The AI-2 quorum sensing system directly stimulates transcription of the *LEE1* and *LEE2* operons, whereas the other LEE operons are controlled indirectly by AI-2 via the *ler* gene product (Sperandio *et al.*, 1999). In this regulatory cascade model, the AI-2 quorum-sensing system activates *ler* transcription, and Ler subsequently turns on the other operons in the LEE, as discussed earlier. The factors that sense AI-2 and activate transcription of the *LEE1* and *LEE2* promoters are currently unknown.

Since the density of normal flora in the large intestine is relatively high, it is possible that EHEC responds to AI-2 produced by the resident coliforms or other bacterial species. This could account in part for the low infectious dose of EHEC required to cause disease (Sperandio *et al.*, 1999; Tilden *et al.*, 1996). EPEC that colonize the small intestine, where the density of normal flora is much lower, require a much larger infectious dose to cause disease relative to EHEC (Simon and Gorbach, 1995). The larger infectious dose may be necessary to produce sufficient concentrations of AI-2 to induce LEE expression.

Additional regulatory targets of AI-2 in EHEC include genes involved in cell division, motility, and the SOS response (Sperandio *et al.*, 2001). Upregulation of the SOS response genes leads to increased production of Shiga toxin, which is contained within a lambda-like phage genome (Neely and Friedman, 1998) and is an important virulence factor in EHEC pathogenesis (see Chap. 5).

In addition to the AI-2 quorum-sensing system just described, a second quorum-sensing system may be involved in downregulating expression of LEE genes in EHEC (Kanamaru *et al.*, 2000a, 2000b). This second system involves the *sdiA* gene, which is homologous to *luxR* of *V. fischeri* (Kanamaru *et al.*, 2000a). LuxR is a DNA-binding transcription activator that senses production of a homoserine lactone autoinducer (AI-1) and activates transcription of target genes when AI-1 reaches a threshold concentration (Stevens *et al.*, 1999; Trott and Stevens, 2001). Unlike AI-2 (discussed earlier), which activates LEE expression, quorum sensing involving SdiA may repress transcription of genes in the LEE (Kanamaru *et al.*, 2000a, 2000b), although the mechanism by which SdiA represses expression of the LEE has not been described. The relationship between these two systems has not been investigated thoroughly.

Plasmid-Encoded Regulator (Per)

The *per* (plasmid-encoded regulator) operon consists of three genes: *perA*, *-B*, and *-C* (Gomez-Duarte and Kaper, 1995) (these genes also have been called *bfpTVW* by Tobe *et al.*, 1996) located on the EAF plasmid (Gomez-Duarte and Kaper, 1995). The EAF plasmid, which is found in EPEC but not in EHEC strains, also contains genes that encode the bundle-forming pilus (BFP; see Chap. 12). PerA is a member of the AraC transcription factor family, PerB shows limited homology with several eukaryotic DNA-binding proteins, and PerC exhibits no homology with any known sequence (Gomez-Duarte and Kaper, 1995). The *per* operon is essential for BFP production (Tobe *et al.*, 1996) (regulation of BFP pili is discussed in Chap. 12). In addition, PerA, -B, and -C have been implicated in controlling expression of the LEE by directly activating the *LEE1* operon, which contains *ler* (Mellies *et al.*, 1999). Thus a cascade model of LEE activation can be proposed in which PerA, -B, and -C induce *ler* transcription, and then Ler activates the additional LEE operons. PerC appears to be the most critical factor involved in activation of the *LEE1* promoter (Bustamante *et al.*, 2001), whereas the role of the additional *per* genes is not clear. One group reported that PerC activity was enhanced by the additional Per proteins (Gomez-Duarte and Kaper, 1995), whereas another group did not observe this (Bustamante *et al.*, 2001). Interestingly, EPEC strains that are cured of the EAF plasmid and therefore are unable to express the *per* operon are capable of forming AE lesions and also express LEE genes to high levels (Bustamante *et al.*, 2001; Knutton *et al.*, 1997). However, *per* mutants that contain an otherwise intact EAF plasmid exhibit low levels of *LEE1* promoter activity (Bustamante *et al.*, 2001), suggesting that the Per proteins, in addition to directly activating the *LEE1* promoter, counteract the effect of an EAF-encoded repressor of *LEE1* expression. PerA, -B, and -C expression is also controlled by the *luxS*/AI-2 quorum-sensing system described earlier (Sperandio *et al.*, 1999).

Integration Host Factor

In EPEC, integration host factor (IHF) is required for expression of the LEE (Friedberg *et al.*, 1999). LEE-encoded proteins, such as Tir, EspB, and EspF, are greatly reduced in an IHF mutant, and as a result, this strain is negative in the FAS assay (Friedberg *et al.*, 1999). IHF exerts its control over the LEE by binding to and directly activating the *LEE1* promoter, thereby upregulating Ler expression (Friedberg *et al.*, 1999).

BipA

BipA is a tyrosine-phosphorylated GTPase in EPEC, whereas BipA homologues found in *E. coli* K-12 and *Salmonella* are not tyrosine-phosphorylated (Farris *et al.*, 1998). A *bipA* mutant EPEC strain does not induce the cytoskeletal rearrangements in host cells seen with wild-type EPEC, suggesting that BipA controls the AE phenotype (Farris *et al.*, 1998); however, the mechanism by which BipA accomplishes this is unclear. BipA shares sequence homology with ribosome-binding GTPases, suggesting that it acts at the level of the ribosome (Farris *et al.*, 1998).

Fis

In EPEC, Fis (factor for inversion stimulation) is involved in regulating LEE expression (Goldberg *et al.*, 2001). Fis is a member of the histone-like protein family because it has the capability of altering DNA topology (Gille *et al.*, 1991). As measured by the FAS assay, a Δ*fis* EPEC strain did not alter the host cell cytoskeleton, demonstrating that Fis is a critical regulator of the AE phenotype. *LEE4* and *ler* mRNA was severely reduced in the Δ*fis* strain, although the exact target of Fis regulation in the LEE has not been described. Interestingly, *tir* mRNA was not reduced in the Δ*fis* strain, even though *ler* is not expressed in the mutant. Reasons for this are not clear.

Additional Regulators

GadX, which is a member of the AraC transcription factor family, controls transcription of *gadA* and *gadB*, which encode glutamate decarboxylase (GAD) isoforms involved in acid resistance (Shin *et al.*, 2001). In EPEC, GadX also has a regulatory influence on LEE gene expression by repressing transcription of the *per* operon in acidic conditions while activating *per* in alkaline conditions (Shin *et al.*, 2001). This may serve to coordinate GAD and LEE gene expression in response to environmental conditions.

In EHEC, transcription of the *LEE3* and *LEE4* operons is enhanced by the stationary-phase sigma factor σ^s, but these observations have not been investigated in detail (Beltrametti *et al.*, 1999; Sperandio *et al.*, 1999). A model of LEE gene regulation is presented in Fig. 3.

EPEC and EHEC May Regulate LEE Gene Expression Differently

An EPEC LEE cosmid clone confers on *E. coli* K-12 the ability to form AE lesions on host cells, thereby demonstrating that all the genes required to cause AE lesions are contained in the LEE (McDaniel and Kaper, 1997). However, although the EPEC and EHEC LEE sequences are highly homologous and share identical operon arrangements, the EHEC LEE does not confer the AE phenotype on *E. coli* K-12 (Elliott *et al.*, 1999b), suggesting that additional EHEC factors are necessary to express the AE phenotype. Additional evidence supporting the hypothesis that differences in LEE regulation exist between EPEC and

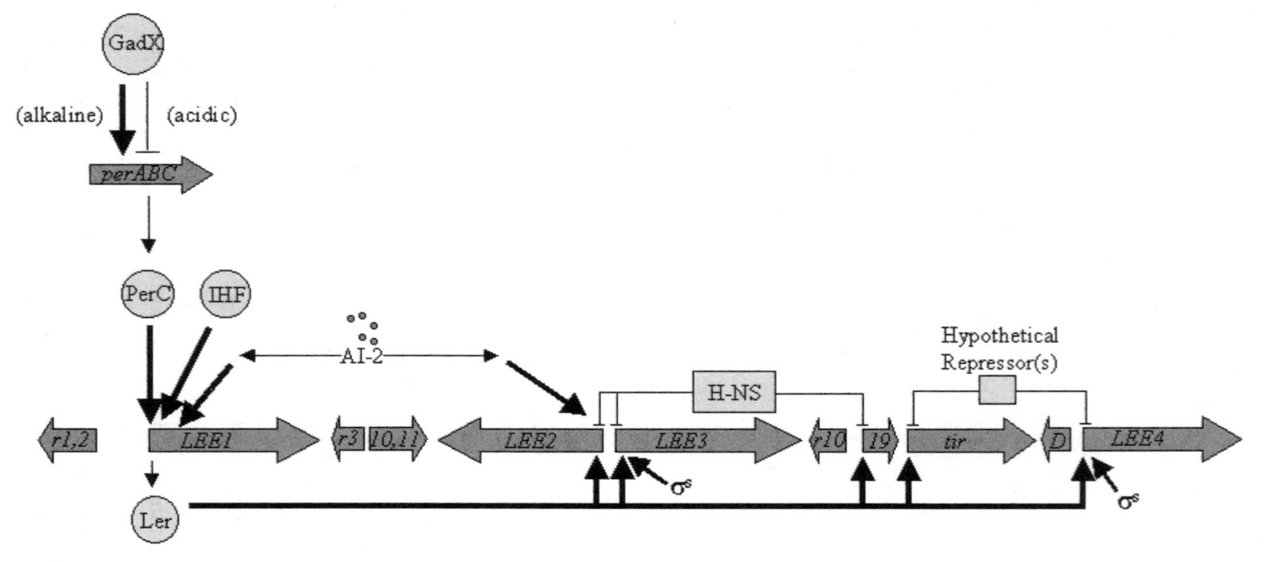

FIGURE 3 Regulatory model of LEE expression. The *LEE1–4* and *tir* operons are labeled. The gene names r*orf1*, *-2*, *-3*, and *-10*; *orf10*, *-11*, and *-19*; and *escD* have been abbreviated in the model. Thick arrows indicate activation of transcription at a given promoter, whereas solid lines represent repression of transcription. Not shown are the regulatory factors Fis, SdiA, and BipA. See text for details.

EHEC is the observation that in M9 medium EHEC but not EPEC is capable of secreting Tir into the medium (DeVinney *et al.*, 1999).

CONCLUSIONS

In the past decade, TTS has been one of the most extensively studied fields in bacterial pathogenesis and has greatly increased our understanding of how a wide range of pathogens interact with host tissues to cause disease. In particular, investigation into how TTS systems work has uncovered several fascinating features, including contact-dependent gene expression, a unique mechanism of protein secretion and translocation into host cells, and molecular characterization of the secretion machine as a macromolecular complex. Investigation of EPEC and EHEC has made several important contributions to the TTS field. Among these has been identification of the EspA filament, characterization of which undoubtedly will reveal further mechanistic features of protein delivery into host cells. Also, regulation of TTS in EPEC and EHEC continues to be investigated vigorously by several laboratories, which, in addition to showing that TTS is controlled by quorum sensing, undoubtedly will reveal additional interesting features of TTS regulation. Collectively, research into the structure, function, and regulation of TTS systems potentially could lead to the development of new vaccines and antigen-delivery systems.

REFERENCES

Abe, A., de Grado, M., Pfuetzner, R. A., *et al.* (1999). Enteropathogenic *Escherichia coli* translocated intimin receptor, Tir, requires a specific chaperone for stable secretion. *Mol. Microbiol.* 33:1162–1175.

Abe, A., Heczko, U., Hegele, R. G., and Brett, F. B. (1998). Two enteropathogenic *Escherichia coli* type III secreted proteins, EspA and EspB, are virulence factors. *J. Exp. Med.* 188:1907–1916.

Abe, A., Kenny, B., Stein, M., and Finlay, B. B. (1997). Characterization of two virulence proteins secreted by rabbit enteropathogenic *Escherichia coli*, EspA and EspB, whose maximal expression is sensitive to host body temperature. *Infect. Immun.* 65:3547–3555.

Allaoui, A., Sansonetti, P. J., and Parsot, C. (1992). MxiJ, a lipoprotein involved in secretion of *Shigella* Ipa invasins, is homologous to YscJ, a secretion factor of the *Yersinia* Yop proteins. *J. Bacteriol.* 174:7661–7669.

Allaoui, A., Schulte, R., and Cornelis, G. R. (1995). Mutational analysis of the *Yersinia enterocolitica virC* operon: characterization of *yscE, F, G, I, J, K* required for Yop secretion and *yscH* encoding YopR. *Mol. Microbiol.* 18:343–355.

Anderson, D. M., and Schneewind, O. (1997). A mRNA signal for the type III secretion of Yop proteins by *Yersinia enterocolitica*. *Science* 278:1140–1143.

Anderson, D. M., and Schneewind, O. (1999). *Yersinia enterocolitica* type III secretion: An mRNA signal that couples translation and secretion of YopQ. *Mol. Microbiol.* 31:1139–1148.

Bassler, B. L. (1999). How bacteria talk to each other: regulation of gene expression by quorum sensing. *Curr. Opin. Microbiol.* 2:582–587.

Bassler, B. L., Wright, M., and Silverman, M. R. (1994). Multiple signaling systems controlling expression of luminescence in *Vibrio harveyi*: Sequence and function of genes encoding a second sensory pathway. *Mol. Microbiol.* 13:273–286.

Beltrametti, F., Kresse, A. U., and Guzman, C. A. (1999). Transcriptional regulation of the *esp* genes of enterohemorrhagic *Escherichia coli*. *J. Bacteriol.* 181:3409–3418.

Beltrametti, F., Kresse, A. U., and Guzman, C. A. (2000). Transcriptional regulation of the *pas* gene of enterohemorrhagic *Escherichia coli*. *FEMS Microbiol. Lett.* 184:119–125.

Blocker, A., Gounon, P., Larquet, E., *et al.* (1999). The tripartite type III secreton of *Shigella flexneri* inserts IpaB and IpaC into host membranes. *J. Cell Biol.* 147:683–693.

Blocker, A., Jouihri, N., Larquet, E., *et al.* (2001). Structure and composition of the *Shigella flexneri* "needle complex," a part of its type III secreton. *Mol. Microbiol.* 39:652–663.

Boyd, A. P., Lambermont, I., and Cornelis, G. R. (2000). Competition between the Yops of *Yersinia enterocolitica* for delivery into eukaryotic cells: Role of the SyeE chaperone binding domain of YopE. *J. Bacteriol.* 182:4811–4821.

Brunder, W., Schmidt, H., and Karch, H. (1997). EspP, a novel extracellular serine protease of enterohaemorrhagic *Escherichia coli* O157:H7 cleaves human coagulation factor V. *Mol. Microbiol.* 24:767–778.

Bustamante, V., Santana, F., Calva, E., and Puente, J. (2001). Transcriptional regulation of type III secretion genes in enteropathogenic *Escherichia coli*: Ler antagonizes H-NS–dependent repression. *Mol. Microbiol.* 39:664–678.

Cheng, L. W., Anderson, D. M., and Schneewind, O. (1997). Two independent type III secretion mechanisms for YopE in *Yersinia enterocolitica*. *Mol. Microbiol.* 24:757–765.

Cheng, L. W., and Schneewind, O. (1999). *Yersinia enterocolitica* type III secretion: On the role of SycE in targeting YopE into HeLa cells. *J. Biol. Chem.* 274:22102–22108.

Cheng, L. W., and Schneewind, O. (2000). Type III machines of gram-negative bacteria: Delivering the goods. *Trends Microbiol.* 8:214–220.

Christie, P. J. (2001). Type IV secretion: Intercellular transfer of macromolecules by systems ancestrally related to conjugation machines. *Mol. Microbiol.* 40:294–305.

Collazo, C. M., and Galan, J. E. (1997). The invasion-associated type III system of *Salmonella typhimurium* directs the translocation of Sip proteins into the host cell. *Mol. Microbiol.* 24:747–756.

Cornelis, G. R. (2000). Type III secretion: A bacterial device for close combat with cells of their eukaryotic host. *Philos. Trans. R. Soc. Lond. [B]* 355:681–693.

Cornelis, G. R., and Van Gijsegem, F. (2000). Assembly and function of type III secretory systems. *Annu. Rev. Microbiol.* 54:735–774.

Crago, A. M., and Koronakis, V. (1998). *Salmonella* InvG forms a ring-like multimer that requires the InvH lipoprotein for outer membrane localization. *Mol. Microbiol.* 30:47–56.

Daefler, S., and Russel, M. (1998). The *Salmonella typhimurium* InvH protein is an outer membrane lipoprotein required for the proper localization of InvG. *Mol. Microbiol.* 28:1367–1380.

Daniell, S. J., Delahay, R. M., Shaw, R. K., *et al.* (2001). Coiled-coil domain of enteropathogenic *Escherichia coli* type III secreted protein EspD is involved in EspA filament-mediated cell attachment and hemolysis. *Infect. Immun.* 69:4055–4064.

de Kievit, T. R., and Iglewski, B. H. (2000). Bacterial quorum sensing in pathogenic relationships. *Infect. Immun.* 68:4839–4849.

De Rycke, J., Comtet, E., Chalareng, C., *et al.* (1997). Enteropathogenic *Escherichia coli* O103 from rabbit elicits actin stress fibers and focal adhesions in HeLa epithelial cells, cytopathic effects that are linked to an analog of the locus of enterocyte effacement. *Infect. Immun.* 65:2555–2563.

Delahay, R. M., Knutton, S., Shaw, R. K., *et al.* (1999). The coiled-coil domain of EspA is essential for the assembly of the type III secretion translocon on the surface of enteropathogenic *Escherichia coli*. *J. Biol. Chem.* 274:35969–35974.

DeVinney, R., Stein, M., Reinscheid, D., *et al.* (1999). Enterohemorrhagic *Escherichia coli* O157:H7 produces Tir, which is translocated to the host cell membrane but is not tyrosine phosphorylated. *Infect. Immun.* 67:2389–2398.

Donnenberg, M. S., Calderwood, S. B., Donohue-Rolfe, A., *et al.* (1990). Construction and analysis of Tn*phoA* mutants of enteropathogenic *Escherichia coli* unable to invade HEp-2 cells. *Infect. Immun.* 58:1565–1571.

Donnenberg, M. S., Lai, L. C., and Taylor, K. A. (1997). The locus of enterocyte effacement pathogenicity island of enteropathogenic *Escherichia coli* encodes secretion functions and remnants of transposons at its extreme right end. *Gene* 184:107–114.

Donnenberg, M. S., Yu, J., and Kaper, J. B. (1993). A second chromosomal gene necessary for intimate attachment of enteropathogenic *Escherichia coli* to epithelial cells. *J. Bacteriol.* 175:4670–4680.

Eichelberg, K., Ginocchio, C. C., and Galan, J. E. (1994). Molecular and functional characterization of the *Salmonella typhimurium* invasion genes *invB* and *invC*: Homology of InvC to the F0F1 ATPase family of proteins. *J. Bacteriol.* 176:4501–4510.

Elliott, S. J., Hutcheson, S. W., Dubois, M. S., *et al.* (1999a). Identification of CesT, a chaperone for the type III secretion of Tir in enteropathogenic *Escherichia coli*. *Mol. Microbiol.* 33:1176–1189.

Elliott, S. J., Krejany, E. O., Mellies, J. L., *et al.* (2001). EspG, a novel type III system-secreted protein from enteropathogenic *Escherichia coli* with similarities to VirA of *Shigella flexneri*. *Infect. Immun.* 69:4027–4033.

Elliott, S. J., Sperandio, V., Giron, J. A., *et al.* (2000). The locus of enterocyte effacement (LEE)–encoded regulator controls expression of both LEE- and non-LEE-encoded virulence factors in enteropathogenic and enterohemorrhagic *Escherichia coli*. *Infect. Immun.* 68:6115–6126.

Elliott, S. J., Wainwright, L. A., McDaniel, T. K., *et al.* (1998). The complete sequence of the locus of enterocyte effacement (LEE) From enteropathogenic *Escherichia coli* E2348/69. *Mol. Microbiol.* 28:1–4.

Elliott, S. J., Yu, J., and Kaper, J. B. (1999b). The cloned locus of enterocyte effacement from enterohemorrhagic *Escherichia coli* O157:H7 is unable to confer the attaching and effacing phenotype upon *E. coli* K-12. *Infect. Immun.* 67:4260–4263.

Farris, M., Grant, A., Richardson, T. B., and O'Connor, C. D. (1998). BipA: A tyrosine-phosphorylated GTPase that mediates interactions between enteropathogenic *Escherichia coli* (EPEC) and epithelial cells. *Mol. Microbiol.* 28:265–279.

Foubister, V., Rosenshine, I., Donnenberg, M. S., and Finlay, B. B. (1994). The *eaeB* gene of enteropathogenic *Escherichia coli* is necessary for signal transduction in epithelial cells. *Infect. Immun.* 62:3038–3040.

Francetic, O., Belin, D., Badaut, C., and Pugsley, A. P. (2000). Expression of the endogenous type II secretion pathway in *Escherichia coli* leads to chitinase secretion. *EMBO J.* 19:6697–6703.

Friedberg, D., Umanski, T., Fang, Y., and Rosenshine, I. (1999). Hierarchy in the expression of the locus of enterocyte effacement genes of enteropathogenic *Escherichia coli*. *Mol. Microbiol.* 34:941–952.

Frithz-Lindsten, E., Rosqvist, R., Johansson, L., and Forsberg, A. (1995). The chaperone-like protein YerA of *Yersinia pseudotuberculosis* stabilizes YopE in the cytoplasm but is dispensible for targeting to the secretion loci. *Mol. Microbiol.* 16:635–647.

Fu, Y., and Galan, J. E. (1998). The *Salmonella typhimurium* tyrosine phosphatase SptP is translocated into host cells and disrupts the actin cytoskeleton. *Mol. Microbiol.* 27:359–368.

Gille, H., Egan, J. B., Roth, A., and Messer, W. (1991). The FIS protein binds and bends the origin of chromosomal DNA replication, *oriC*, of *Escherichia coli*. *Nucleic Acids Res.* 19:4167–4172.

Ginocchio, C. C., Olmsted, S. B., Wells, C. L., and Galan, J. E. (1994). Contact with epithelial cells induces the formation of surface appendages on *Salmonella typhimurium*. *Cell* 76:717–724.

Goldberg, M. D., Johnson, M., Hinton, J. C. D., and Williams, P. H. (2001). Role of the nucleoid-associated protein Fis in the regulation of virulence properties of enteropathogenic *Escherichia coli*. *Mol. Microbiol.* 41:549–559.

Gomez-Duarte, O. G., and Kaper, J. B. (1995). A plasmid-encoded regulatory region activates chromosomal *eaeA* expression in enteropathogenic *Escherichia coli*. *Infect. Immun.* 63:1767–1776.

Haigh, R., Baldwin, T., Knutton, S., and Williams, P. H. (1995). Carbon dioxide regulated secretion of the EaeB protein of enteropathogenic *Escherichia coli*. *FEMS Microbiol. Lett.* 129:63–67.

Hakansson, S., Bergman, T., Vanooteghem, J. C., *et al.* (1993). YopB and YopD constitute a novel class of *Yersinia* Yop proteins. *Infect. Immun.* 61:71–80.

Hakansson, S., Schesser, K., Persson, C., *et al.* (1996). The YopB protein of *Yersinia pseudotuberculosis* is essential for the translocation of Yop effector proteins across the target cell plasma membrane and displays a contact-dependent membrane disrupting activity. *EMBO J.* 15:5812–5823.

Hartland, E. L., Daniell, S. J., Delahay, R. M., *et al.* (2000). The type III protein translocation system of enteropathogenic *Escherichia coli* involves EspA-EspB protein interactions. *Mol. Microbiol.* 35:1483–1492.

Henderson, I. R., and Nataro, J. P. (2001). Virulence functions of autotransporter proteins. *Infect. Immun.* 69:1231–1243.

Hoiczyk, E., and Blobel, G. (2001). Polymerization of a single protein of the pathogen *Yersinia enterocolitica* into needles punctures eukaryotic cells. *Proc. Natl. Acad. Sci. USA* 98:4669–4674.

Hueck, C. J. (1998). Type III protein secretion systems in bacterial pathogens of animals and plants. *Microbiol. Mol. Biol. Rev.* 62:379–433.

Jarvis, K. G., Giron, J. A., Jerse, A. E., *et al.* (1995). Enteropathogenic *Escherichia coli* contains a putative type III secretion system necessary for the export of proteins involved in attaching and effacing lesion formation. *Proc. Natl. Acad. Sci. USA* 92:7996–8000.

Kanamaru, K., Kanamaru, K., Tatsuno, I., *et al.* (2000b). Regulation of virulence factors of enterohemorrhagic *Escherichia coli* O157:H7 by self-produced extracellular factors. *Biosci. Biotechnol. Biochem.* 64:2508–2511.

Kanamaru, K., Kanamaru, K., Tatsuno, I., *et al.* (2000a). SdiA, an *Escherichia coli* homologue of quorum-sensing regulators, controls the expression of virulence factors in enterohaemorrhagic *Escherichia coli* O157:H7. *Mol. Microbiol.* 38:805–816.

Karlinsey, J. E., Lonner, J., Brown, K. L., and Hughes, K. T. (2000). Translation/secretion coupling by type III secretion systems. *Cell* 102:487–497.

Kenny, B., Abe, A., Stein, M., and Finlay, B. B. (1997b). Enteropathogenic *Escherichia coli* protein secretion is induced in response to conditions similar to those in the gastrointestinal tract. *Infect. Immun.* 65:2606–2612.

Kenny, B., DeVinney, R., Stein, M., *et al.* (1997a). Enteropathogenic *E. coli* (EPEC) transfers its receptor for intimate adherence into mammalian cells. *Cell* 91:511–520.

Kenny, B., and Finlay, B. B. (1995). Protein secretion by enteropathogenic *Escherichia coli* is essential for transducing signals to epithelial cells. *Proc. Natl. Acad. Sci. USA* 92:7991–7995.

Kenny, B., and Jepson, M. (2000). Targeting of an enteropathogenic *Escherichia coli* (EPEC) effector protein to host mitochondria. *Cell Microbiol.* 2:579–590.

Kenny, B., Lai, L. C., Finlay, B. B., and Donnenberg, M. S. (1996). EspA, a protein secreted by enteropathogenic *Escherichia coli*, is required to induce signals in epithelial cells. *Mol. Microbiol.* 20:313–323.

Knutton, S., Adu-Bobie, J., Bain, C., *et al.* (1997). Down regulation of intimin expression during attaching and effacing enteropathogenic *Escherichia coli* adhesion. *Infect. Immun.* 65:1644–1652.

Knutton, S., Rosenshine, I., Pallen, M. J., *et al.* (1998). A novel EspA-associated surface organelle of enteropathogenic *Escherichia coli* involved in protein translocation into epithelial cells. *EMBO J.* 17:2166–2176.

Koster, M., Bitter, W., de Cock, H., *et al.* (1997). The outer membrane component, YscC, of the Yop secretion machinery of *Yersinia enterocolitica* forms a ring-shaped multimeric complex. *Mol. Microbiol.* 26:789–797.

Kresse, A. U., Beltrametti, F., Muller, A., *et al.* (2000). Characterization of SepL of enterohemorrhagic *Escherichia coli*. *J. Bacteriol.* 182:6490–6498.

Kubori, T., Matsushima, Y., Nakamura, D., *et al.* (1998). Supramolecular structure of the *Salmonella typhimurium* type III protein secretion system. *Science* 280:602–605.

Kubori, T., Sukhan, A., Aizawa, S. I., and Galan, J. E. (2000). Molecular characterization and assembly of the needle complex of the *Salmonella typhimurium* type III protein secretion system. *Proc. Natl. Acad. Sci. USA* 97:10225–10230.

Lai, L. C., Wainwright, L. A., Stone, K. D., and Donnenberg, M. S. (1997). A third secreted protein that is encoded by the enteropathogenic *Escherichia coli* pathogenicity island is required for transduction of signals and for attaching and effacing activities in host cells. *Infect. Immun.* 65:2211–2217.

Linderoth, N. A., Simon, M. N., and Russel, M. (1997). The filamentous phage pIV multimer visualized by scanning transmission electron microscopy. *Science* 278:1635–1638.

Lloyd, S. A., Norman, M., Rosqvist, R., and Wolf-Watz, H. (2001). *Yersinia* YopE is targeted for type III secretion by N-terminal, not mRNA, signals. *Mol. Microbiol.* 39:520–531.

Macnab, R. M. (1999). The bacterial flagellum: reversible rotary propellor and type III export apparatus. *J. Bacteriol.* 181:7149–7153.

Manjarrez-Hernandez, H. A., Baldwin, T. J., Williams, P. H., *et al.* (1996). Phosphorylation of myosin light chain at distinct sites and its association with the cytoskeleton during enteropathogenic *Escherichia coli* infection. *Infect. Immun.* 64:2368–2370.

McDaniel, T. K., and Kaper, J. B. (1997). A cloned pathogenicity island from enteropathogenic *Escherichia coli* confers the attaching and effacing phenotype on *E. coli* K-12. *Mol. Microbiol.* 23:399–407.

McNamara, B. P., and Donnenberg, M. S. (1998). A novel proline-rich protein, EspF, is secreted from enteropathogenic *Escherichia coli* via the type III export pathway. *FEMS Microbiol. Lett.* 166:71–78.

Mellies, J. L., Elliott, S. J., Sperandio, V., *et al.* (1999). The Per regulon of enteropathogenic *Escherichia coli*: Identification of a regulatory cascade and a novel transcriptional activator, the locus of enterocyte effacement (LEE)–encoded regulator (Ler). *Mol. Microbiol.* 33:296–306.

Mellies, J. L., Navarro-Garcia, F., Okeke, I., *et al.* (2001). *espC* pathogenicity island of enteropathogenic *Escherichia coli* encodes an enterotoxin. *Infect. Immun.* 69:315–324.

Neely, M. N., and Friedman, D. I. (1998). Functional and genetic analysis of regulatory regions of coliphage H-19B: Location of Shiga-like toxin and lysis genes suggest a role for phage functions in toxin release. *Mol. Microbiol.* 28:1255–1267.

Neves, B. C., Knutton, S., Trabulsi, L. R., *et al.* (1998). Molecular and ultrastructural characterisation of EspA from different enteropathogenic *Escherichia coli* serotypes. *FEMS Microbiol. Lett.* 169:73–80.

Neyt, C., and Cornelis, G. R. (1999a). Insertion of a Yop translocation pore into the macrophage plasma membrane by *Yersinia enterocolitica*: Requirement for translocators YopB and YopD, but not LcrG. *Mol. Microbiol.* 33:971–981.

Neyt, C., and Cornelis, G. R. (1999b). Role of SycD, the chaperone of the *Yersinia* Yop translocators YopB and YopD. *Mol. Microbiol.* 31:143–156.

Nouwen, N., Ranson, N., Saibil, H., *et al.* (1999). Secretin PulD: Association with pilot PulS, structure, and ion-conducting channel formation. *Proc. Natl. Acad. Sci. USA* 96:8173–8177.

Nunn, D. (1999). Bacterial type II protein export and pilus biogenesis: More than just homologies? *Trends Cell Biol.* 9:402–408.

Pallen, M. J., Dougan, G., and Frankel, G. (1997). Coiled-coil domains in proteins secreted by type III secretion systems. *Mol. Microbiol.* 25:423–425.

Perna, N. T., Plunkett, G., III, Burland, V., *et al.* (2001). Genome sequence of enterohaemorrhagic *Escherichia coli* O157:H7. *Nature* 409:529–533.

Persson, C., Nordfelth, R., Holmstrom, A., *et al.* (1995). Cell-surface-bound *Yersinia* translocate the protein tyrosine phosphatase YopH by a polarized mechanism into the target cell. *Mol. Microbiol.* 18:135–150.

Pettersson, J., Nordfelth, R., Dubinina, E., *et al.* (1996). Modulation of virulence factor expression by pathogen target cell contact. *Science* 273:1231–1233.

Plano, G. V., Barve, S. S., and Straley, S. C. (1991). LcrD, a membrane-bound regulator of the *Yersinia* pestis low-calcium response. *J. Bacteriol.* 173:7293–7303.

Plano, G. V., Day, J. B., and Ferracci, F. (2001). Type III export: New uses for an old pathway. *Mol. Microbiol.* 40:284–293.

Roine, E., Wei, W., Yuan, J., *et al.* (1997). Hrp pilus: An *hrp*-dependent bacterial surface appendage produced by *Pseudomonas syringae* pv. *tomato* DC3000. *Proc. Natl. Acad. Sci. USA* 94:3459–3464.

Rosenshine, I., Donnenberg, M. S., Kaper, J. B., and Finlay, B. B. (1992). Signal transduction between enteropathogenic *Escherichia coli* (EPEC) and epithelial cells: EPEC induces tyrosine phosphorylation of host cell proteins to initiate cytoskeletal rearrangement and bacterial uptake. *EMBO J.* 11:3551–3560.

Rosenshine, I., Ruschkowski, S., and Finlay, B. B. (1996). Expression of attaching/effacing activity by enteropathogenic *Escherichia coli* depends on growth phase, temperature, and protein synthesis upon contact with epithelial cells. *Infect. Immun.* 64:966–973.

Rosqvist, R., Magnusson, K. E., and Wolf-Watz, H. (1994). Target cell contact triggers expression and polarized transfer of *Yersinia* YopE cytotoxin into mammalian cells. *EMBO J.* 13:964–972.

Russel, M. (1998). Macromolecular assembly and secretion across the bacterial cell envelope: Type II protein secretion systems. *J. Mol. Biol.* 279:485–499.

Sanchez-SanMartin, C., Bustamante, V. H., Calva, E., and Puente, J. L. (2001). Transcriptional regulation of the *orf19* gene and the *tir-cesT-eae* operon of enteropathogenic *Escherichia coli*. *J. Bacteriol.* 183:2823–2833.

Sandkvist, M. (2001). Biology of type II secretion. *Mol. Microbiol.* 40:271–283.

Schauder, S., and Bassler, B. L. (2001). The languages of bacteria. *Genes Dev.* 15:1468–1480.

Schauder, S., Shokat, K., Surette, M. G., and Bassler, B. L. (2001). The LuxS family of bacterial autoinducers: Biosynthesis of a novel quorum-sensing signal molecule. *Mol. Microbiol.* 41:463–476.

Schmidt, H., Henkel, B., and Karch, H. (1997). A gene cluster closely related to type II secretion pathway operons of gram-negative bacteria is located on the large plasmid of enterohemorrhagic *Escherichia coli* O157 strains. *FEMS Microbiol. Lett.* 148:265–272.

Sekiya, K., Ohishi, M., Ogino, T., *et al.* (2001). Supermolecular structure of the enteropathogenic *Escherichia coli* type III secretion system and its direct interaction with the EspA-sheath-like structure. *Proc. Natl. Acad. Sci. USA* 98:11638–11643.

Shaw, R. K., Daniell, S., Ebel, F., *et al.* (2001). EspA filament-mediated protein translocation into red blood cells. *Cell Microbiol.* 3:213–222.

Shin, S., Castanie-Cornet, M., Foster, J. W., *et al.* (2001). An activator of glutamate decarboxylase genes regulates the expression of enteropathogenic *Escherichia coli* virulence genes thorugh control of the plasmid-encoded regulator, Per. *Mol. Microbiol.* 41:1–20.

Simon, G. L., and Gorbach, S. (1995). In M. J. Blaster, P. D. Smith, J. I. Ravdin, H. B. Greenberg, and R. L. Guerrant (eds.), *Infections of the Gastrointestinal Tract*, pp. 53–69. New York: Raven.

Sory, M. P., Boland, A., Lambermont, I., and Cornelis, G. R. (1995). Identification of the YopE and YopH domains required for secretion and internalization into the cytosol of macrophages, using the *cyaA* gene fusion approach. *Proc. Natl. Acad. Sci. USA* 92:11998–12002.

Sperandio, V., Mellies, J. L., Delahay, R. M., *et al.* (2000). Activation of enteropathogenic *Escherichia coli* (EPEC) LEE2 and LEE3 operons by Ler. *Mol. Microbiol.* 38:781–793.

Sperandio, V., Mellies, J. L., Nguyen, W., *et al.* (1999). Quorum sensing controls expression of the type III secretion gene transcription and protein secretion in enterohemorrhagic and enteropathogenic *Escherichia coli*. *Proc. Natl. Acad. Sci. USA* 96:15196–15201.

Sperandio, V., Torres, A. G., Giron, J. A., and Kaper, J. B. (2001). Quorum sensing is a global regulatory mechanism in enterohemorrhagic *Escherichia coli* O157:H7. *J. Bacteriol.* 183:5187–5197.

Stein, M., Kenny, B., Stein, M. A., and Finlay, B. B. (1996). Characterization of EspC, a 110-kilodalton protein secreted by enteropathogenic *Escherichia coli* which is homologous to members of the immunoglobulin A protease-like family of secreted proteins. *J. Bacteriol.* 178:6546–6554.

Stevens, A. M., Fujita, N., Ishihama, A., and Greenberg, E. P. (1999). Involvement of the RNA polymerase alpha-subunit C-terminal domain in LuxR-dependent activation of the *Vibrio fischeri* luminescence genes. *J. Bacteriol.* 181:4704–4707.

Sukhan, A., Kubori, T., Wilson, J., and Galan, J. E. (2001). Genetic analysis of assembly of the *Salmonella enterica* serovar *typhimurium* type III secretion-associated needle complex. *J. Bacteriol.* 183:1159–1167.

Surette, M. G., Miller, M. B., and Bassler, B. L. (1999). Quorum sensing in *Escherichia coli, Salmonella typhimurium*, and *Vibrio harveyi*: A new family of genes responsible for autoinducer production. *Proc. Natl. Acad. Sci. USA* 96:1639–1644.

Tamano, K., Aizawa, S., Katayama, E., *et al.* (2000). Supramolecular structure of the *Shigella* type III secretion machinery: The needle part is changeable in length and essential for delivery of effectors. *EMBO J.* 19:3876–3887.

Tardy, F., Homble, F., Neyt, C., *et al.* (1999). *Yersinia enterocolitica* type III secretion-translocation system: Channel formation by secreted Yops. *EMBO J.* 18:6793–6799.

Taylor, K. A., O'Connell, C. B., Luther, P. W., and Donnenberg, M. S. (1998). The EspB protein of enteropathogenic *Escherichia coli* is targeted to the cytoplasm of infected HeLa cells. *Infect. Immun.* 66:5501–5507.

Tilden, J., Jr., Young, W., McNamara, A. M., *et al.* (1996). A new route of transmission for *Escherichia coli*: Infection from dry fermented salami. *Am. J. Public Health* 86:1142–1145.

Tobe, T., Schoolnik, G. K., Sohel, I., *et al.* (1996). Cloning and characterization of *bfpTVW*, genes required for the transcriptional activation of *bfpA* in enteropathogenic *Escherichia coli*. *Mol. Microbiol.* 21:963–975.

Trott, A. E., and Stevens, A. M. (2001). Amino acid residues in LuxR critical for its mechanism of transcriptional activation during quorum sensing in *Vibrio fischeri*. *J. Bacteriol.* 183:387–392.

Van Gijsegem, F., Vasse, J., Camus, J. C., *et al.* (2000). *Ralstonia solanacearum* produces *hrp*-dependent pili that are required for PopA secretion but not for attachment of bacteria to plant cells. *Mol. Microbiol.* 36:249–260.

Wachter, C., Beinke, C., Mattes, M., and Schmidt, M. A. (1999). Insertion of EspD into epithelial target cell membranes by infecting enteropathogenic *Escherichia coli*. *Mol. Microbiol.* 31:1695–1707.

Wainwright, L. A., and Kaper, J. B. (1998). EspB and EspD require a specific chaperone for proper secretion from enteropathogenic *Escherichia coli*. *Mol. Microbiol.* 27:1247–1260.

Warawa, J., Finlay, B. B., and Kenny, B. (1999). Type III secretion-dependent hemolytic activity of enteropathogenic *Escherichia coli*. *Infect. Immun.* 67:5538–5540.

Wattiau, P., Bernier, B., Deslee, P., *et al.* (1994). Individual chaperones required for Yop secretion by *Yersinia*. *Proc. Natl. Acad. Sci. USA* 91:10493–10497.

Wattiau, P., and Cornelis, G. R. (1993). SycE, a chaperone-like protein of *Yersinia enterocolitica* involved in Ohe secretion of YopE. *Mol. Microbiol.* 8:123–131.

Wattiau, P., Woestyn, S., and Cornelis, G. R. (1996). Customized secretion chaperones in pathogenic bacteria. *Mol. Microbiol.* 20:255–262.

Withers, H., Swift, S., and Williams, P. (2001). Quorum sensing as an integral component of gene regulatory networks in gram-negative bacteria. *Curr. Opin. Microbiol.* 4:186–193.

Woestyn, S., Allaoui, A., Wattiau, P., and Cornelis, G. R. (1994). YscN, the putative energizer of the *Yersinia* Yop secretion machinery. *J. Bacteriol.* 176:1561–1569.

Wolff, C., Nisan, I., Hanski, E., *et al.* (1998). Protein translocation into host epithelial cells by infecting enteropathogenic *Escherichia coli*. *Mol. Microbiol.* 28:143–155.

Young, J., and Holland, I. B. (1999). ABC transporters: bacterial exporters-revisited five years on. *Biochim. Biophys. Acta* 1461:177–200.

Zierler, M. K., and Galan, J. E. (1995). Contact with cultured epithelial cells stimulates secretion of *Salmonella typhimurium* invasion protein InvJ. *Infect. Immun.* 63:4024–4028.

CHAPTER 13

Hemolysin

Vassilis Koronakis

Department of Pathology, Cambridge University, Cambridge, United Kingdom

Colin Hughes

Department of Pathology, Cambridge University, Cambridge, United Kingdom

INTRODUCTION

Escherichia coli Hemolysin: An Old Story with Novel Twists

Hemolytic *E. coli* cultures and filtrates were described as long ago as the early 1900s, but despite a substantial literature from medical microbiology and virulence studies on clinical isolates, there was in the mid-1970s little understanding of the nature and role of the responsible toxin. Analysis of hemolysin production indicated that a cellular pool of active hemolysin protein was specifically exported across the gram-negative cell envelope, and one of the earliest molecular genetics studies of a bacterial pathogenicity factor was initiated using transposon mutagenesis and complementation with recombinant DNA [1,2]. This established that the hemolysin determinant comprises four genes, *hlyC* and *hlyA*, encoding synthesis of active intracellular toxin, and *hlyB* and *hlyD*, required for its secretion out of the cell [2,3]. Defined hemolysin mutants and plasmids were used to confirm the importance of the toxin in virulence [4,5]. Research up to this time was reviewed in 1984–1985 [6,7]. DNA sequencing defined the four-gene *hlyCABD* operon [8,9] and indicated that export of the large (110-kDa)

Escherichia coli: Virulence Mechanisms of a Versatile Pathogen
ISBN 0-12-220751-3

HlyA toxin did not follow the N-terminal signal-dependent secretion pathway across the inner membrane [10]. The 39% (G+C) DNA sequence of *hlyCABD* suggested that it had been acquired recently by *E. coli*, and the synthesis and export of HlyA were soon found to be models for the production of many related toxins forming pores in mammalian cell membranes [e.g., 11–13; reviewed in 1991–1992 in 14–16]. Chromosome walking with cosmid clones revealed that the *hly*CABD operon was linked closely to other virulence genes in what are now widely recognised as *pathogenicity islands* [17].

Subsequently, it emerged that the successive mechanisms underlying synthesis and export of active hemolysin are remarkable [variously reviewed in 16,18–20]. The export process was shown to be a previously unknown means of traversing both cytoplasmic and outer membranes, and this direct type I mechanism was found to direct export of other virulence proteins by pathogens such as *Erwinia*, *Serratia*, and *Pseudomonas* [e.g., 21]. HlyB was noted to be a member of a large superfamily of ATP-binding proteins [22], giving a first view of an energy-requiring transport process. In some of the related operons, a *tolC* homologue was incorporated as the fifth gene [11], and mutagenesis of the *E. coli* cellular *tolC* gene showed TolC to be the third hemolysin export component [23]. In the 1990s, two other cellular genes were found to be required for the synthesis of active hemolysin, the transcriptional regulator *rfaH* [24] and the acyl carrier protein gene *acpP* [25], as substantial progress was made in establishing the biochemistry underlying toxin biogenesis and action. The stages in biogenesis are represented in the summary Fig. 1 and are detailed below.

E. coli Hemolysin and a Family of Bacterial Pore-Forming Toxins

Hemolysin is an important virulence factor in the extremely common *E. coli* infections of the urinary tract (UTI) and other extraintestinal sites [4,5,14] and is also produced by enterohemorrhagic *E. coli* 0157 [reviewed in 26]. It is a member of a family of membrane-targeted toxins influential in juvenile periodontitis, pneumoniae, whooping cough, and wound infections of humans and animals, i.e., the leukotoxin of *Pasteurella*, the hemolysins and leukotoxins of *Actinobacillus*, the bifunctional adenylate cyclase hemolysin of *Bordetella pertussis*, and the hemolysins of *Proteus*, *Morganella*, and *Moraxella* [11–14,26]. These toxins have 30–75% sequence identity with *E. coli* HlyA, and each requires Ca^{2+} binding by a conserved C-terminal domain of acidic glycine-rich nonapeptide repeats (in HlyA, there are between 11 and 17 repeats [26]), which has led to the RTX (repeat toxin) family nomenclature. When the repeats bind Ca^{2+} (one ion per repeat), they form a parallel β-barrel or β-superhelix [27]. All the RTX toxins are made as an inactive precursor and must undergo the common posttranslational maturation step before export by the type I pathway.

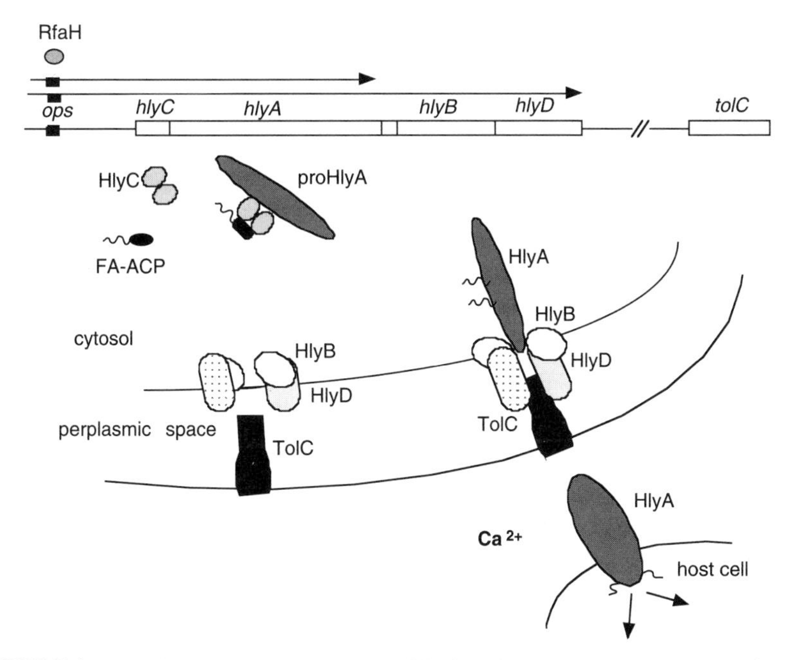

FIGURE 1 Hemolysin biogenesis. Expression of the hemolysin operon is governed by RfaH and the *ops* element. Gene *hlyA* encodes inactive prohemolysin (proHlyA) that is activated by HlyC (FA-ACP, acyl-acyl carrier protein). Mature HlyA is secreted by the type I mechanism. Ca^{2+} binds to the glycine-rich repeats of the toxin, which then interacts with the mammalian cell membrane.

Hemolysin Action: Cell Subversion and Pore Formation

E. coli HlyA has a wide spectrum of cytocidal activity, attacking erythrocytes, granulocytes, monocytes, endothelial cells, and renal epithelial cells, whereas related leukotoxins have a narrower range of target cell specificities, with little erythrolytic activity but a potent cytotoxic activity toward phagocytic cells [reviewed in 26]. How such reported host cell specificities arose is not clear, but it is possible that receptor-ligand interactions play a role. The involvement of target cell components such as β-integrins [28] would be consistent with a lack of receptor involvement in erythrocytes; action against erythrocytes could be an intrinsic property, but glycophorin has been suggested recently to act as the HlyA erythrocyte receptor [29]. At very low, sublytic concentrations, HlyA is a potent trigger of G-protein-dependent generation of inositol triphosphate and diacylglycerol in granulocytes and endothelial cells, stimulating the respiratory burst and the secretion of vesicular constituents [30,31]. It stimulates the release of interleukin 1β (IL-1β) and tumor necrosis factor (TNF) from human monocytes, the lipoxygenase products leukotriene B_4 and 5-hydroxyeicosatetraenoic acid and nitric oxide from endothelial cells, and IL-1β (but not TNF-α) from cultured monocytes, whereas it inhibits the release of IL-1β, IL-6, and TNF-α from human leukocytes [32]. HlyA produces an *in vivo* cytokine response similar to that seen *in vitro*, elevating serum levels of IL-1 and TNF [33]. These effects may occur without necrotic cell death because HlyA causes apoptosis of human lymphocytes [34]. Recently, it was reported that hemolysin induces in primary renal epithelial cells low-frequency Ca^{2+} oscillations [35] that could increase the efficacy of signaling and determine activation of inflammatory processes. Inflammation may disrupt epithelial cell junctions, favor translocation of bacteria through the intestinal barrier, and even induce presentation of a receptor used by the pathogen for binding. It is nevertheless possible that some effects assigned to HlyA reflect cooperative responses to both toxin and LPS, which are commonly coisolated, perhaps significantly because host cells attacked by HlyA show rapid shedding of the LPS receptor CD14 [26].

E. coli HlyA alteration of membrane permeability also causes lysis and death, which may provide iron and prevent phagocytosis [26]. Once inserted, HlyA behaves as an integral membrane protein and causes target cell lysis by forming transmembrane pores that are cation-selective, pH-dependent, and apparently asymmetric [36,37]. The pore in artificial lipid bilayers has a diameter of about 1.0 nm, a conductance of about 500 pS in 0.15 *M* KCl, and a mean lifetime of 2 s at small transmembrane voltages [38]. HlyA pores in erythrocytes similarly have a predicted cutoff of approximately 2 kDa [37]. HlyA recovered from detergent-solubilized erythrocyte membranes is monomeric, indicating either that oligomerization is not required for pore formation or that oligomers dissociate [37]. Membrane insertion of HlyA is believed to occur through a monomolecular mechanism, with oligomerization, if any, occurring by the subsequent addition of monomers within the membrane. It has been estimated that only 1–3 HlyA molecules form the pore [38]. It should be noted nevertheless that under conditions leading to contents leakage from unilamellar vescicles, HlyA has been reported to occupy only one of the phospholipid monolayers [39], suggesting that HlyA may disrupt membranes without discrete pore formation.

A conserved region toward the N-terminus of HlyA is essential for lysis and has been supposed to be involved in pore formation because it spans the only pronounced hydrophobic sequences in the otherwise hydrophilic HlyA protein [26]. It is predicted to include four membrane-spanning α-helices, and mutations altering its hydrophobicity attenuate pore-forming activity [40]. Recently, direct evidence for the key role of HlyA sequence 177–411 in binding and insertion has been achieved by the use of stably active toxin and derived peptides in both liposome floatation and insertion-dependent labeling by a photoactivatable probe incorporated into the target lipid bilayer [41]. These results are complementary with parallel spectroscopic study of HlyA derivatives coupled to flourescent probes via single cysteine residues introduced throughout the toxin. Emission of each derivative HlyA in solution and following toxin insertion revealed insertion only by the cysteines located in the same N-terminal hydrophobic region [42]. The C-terminal glycine-rich repeat domain that binds Ca^{2+} is required for hemolysis but not for pore formation in lipid bilayers [43,44], and direct assay of binding and insertion reveals no clear role for Ca^{2+} in lipid interaction [41,42]. It therefore seems possible that Ca^{2+}-induced conformational change may promote a subsequent irreversible insertion of the toxin into cell membranes [26,45].

SYNTHESIS OF *E. COLI* PROHEMOLYSIN

RfaH-Dependent Expression of Virulence and Fertility Operons

Transposon mutagenesis identified a locus required for the transcription of the *hlyCABD* operon [24]. Mapping and complementation showed that this was allelic to *rfaH* and *sfrB*, positive regulators of lipolysaccharide (LPS) and F (fertility) pilus biogenesis [reviewed in 46]. RfaH loss showed a distinctive effect on *hlyCABD* transcription, modestly decreasing transcription of the first two genes, *hlyC* and *hlyA*, but virtually abolishing transcription of the distal *hlyB* and *hlyD* [24,47,48], similar to earlier observations on the *tra* and *rfa* operons [reviewed in 46]. RfaH was found to be a basic 18-kDa protein not obviously related to any other transcriptional regulator [46].

Comparing the RfaH-dependent *hly*, *rfa*, and *tra* operons of *E. coli* identified a single common 8-bp motif, 5′ GGCGGTAG 3′, and its deletion from its position 2 kbp upstream of the *hlyC, -A, -B,* and *-D* genes gave the same transcriptional phenotype as that of an *rfaH* null mutation [46,47,50], i.e., increased transcriptional polarity within the operon. The motif was termed the *ops* (operon polarity suppressor) element. To function, it must be downstream of an active promoter and be transcribed in its native orientation; however, it is effective downstream of a nonnative promoter such as the p*tac* promoter and can increase distal gene transcription over at least 10 kbp [46–48]. The *ops* element is dispersed among gram-negative bacterial operons that direct synthesis and secretion/assembly of hemolysin, the F pilus, the LPS core (*rfa*) and O-antigen (*rfb*), and colanic acid (*cps*) and type II (*kps*) polysaccharide capsules, and where transcript-initiation sites are known, the *ops* element lies in a nontranslated leader sequence (Fig. 2).

RfaH and the Nucleic Acid Element *ops* Direct Transcript Elongation

The mechanistic link between the two activation components, *ops* and RfaH, was substantiated *in vitro* by using cytosolic extracts from wild-type and *rfaH*

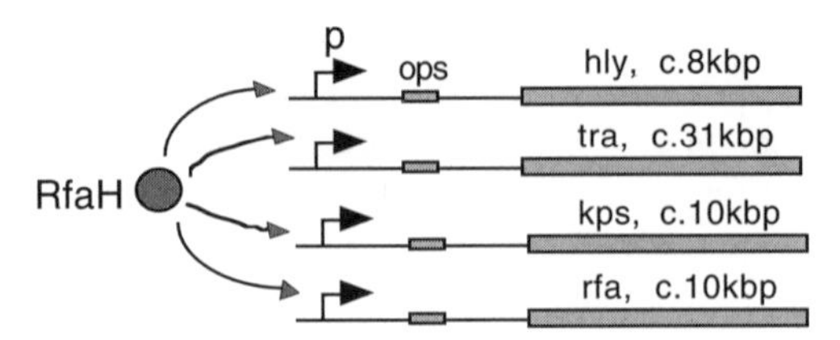

FIGURE 2 Expression of virulence and fertility operons (hly, hemolysin; tra, DNA transfer; kps, capsular polysaccharide; rfa, lipopolysaccharide) governed by RfaH and *ops*.

mutant *E. coli* strains to transcribe the *hlyCABD* operon [48]. The *rfaH*$^-$ extract determined a similar transcriptional polarity to that seen in *rfaH*$^-$ cells *in vivo*, and this polarity was abolished by purified RfaH protein. Furthermore, an *ops*$^-$ *hlyCABD* template required much higher concentrations of RfaH for suppression than did the *ops*$^+$ template [48]. These data suggested that *ops* lowers the concentration of RfaH required to abolish transcriptional polarity.

RfaH (162 amino acids) is weakly homologous to the 181-residue NusG [48], which is suggested to act as a bridge between RNA polymerase and Rho and thus recruits Rho into the termination complex, i.e., to modulate the switch from elongation to termination [46]. Strongest identity is between amino acids 109–135 of RfaH and 129–155 of NusG, containing a postulated RNA-binding motif [46]. Another view of RfaH function would have it acting analogous to NusA, the general elongation factor believed to prevent termination by maintaining transcription uncoupling to translation. However, this is not supported by the observation that RfaH stimulates distal gene transcription *in vitro* when translation is inhibited [48]. The atypical base composition of *hlyCABD* may cause RNA polymerase pausing, and RfaH may act to suppress this. This could be mediated by a direct interaction between RfaH and the RNA polymerase core, or alternatively, RfaH could interact with Rho to inhibit termination of the paused complex [46].

These possibilities depict RfaH directing a specialized elongation-control mechanism in which the RNA-polymerase complex is modified to resist premature termination. The key role of the *ops* element would be to recruit RfaH and potentially other factors to the complex, and because it is effective only 3′ of the promoter [47,48], *ops* mayt be recognized as either DNA or as part of the untranslated leader. *In vivo* footprinting upstream of the *hly* operon has shown that during transcription, the DNA immediately downstream of *ops* is melted to single strands (50), consistent with an RNA-polymerase complex initiating at the upstream promoter and pausing at this sequence to be modified by RfaH. This modification may be novel or resemble that of either of the λ antitermination protein Q, which binds to its recognition sequence as a DNA but only when RNA polymerase is paused at the site, or protein N, which recognizes the *nut*

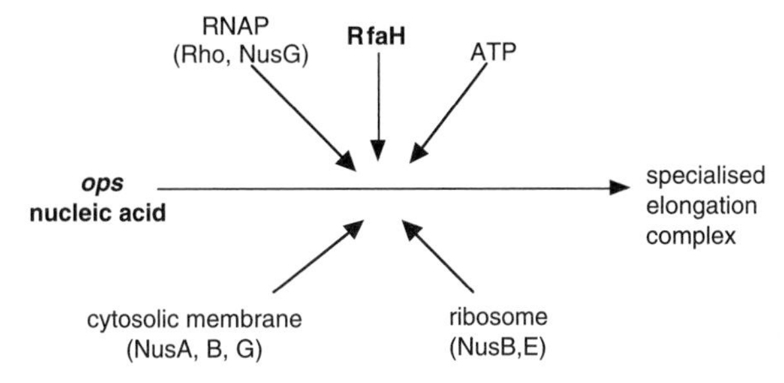

FIGURE 3 *ops*-dependent assembly of the RfaH transcript elongation complex.

site as an RNA after it has been transcribed, looping out the intervening RNA to contact the RNA-polymerase complex. The effect in either case would be to increase the processivity of the complex [46].

Indirect evidence for recruitment of RfaH into a multicomponent RNA-polymerase complex is provided by second-site suppressors of *rfaH* mutation in *rho*, in loci at or near *nusG* and *rpoBC* (encoding the β and β′ subunits of RNA polymerase), and in other loci of unknown function [51,52]. Direct experimental support was provided by showing that RfaH-dependent elongation occurs concomitant with recruitment of RfaH into a high-molecular-weight transcription complex specifically directed by *ops* [53]. Neither recruitment nor elongation was observed *in vitro* reactions containing only *ops,* RfaH, and purified core RNA polymerase; both processes required subcellular fractions containing the RNA-polymerase complex, the cytoplasmic membrane, and the ribosome (Fig. 3).

Whatever the details of this mechanism, RfaH and *ops* cooperate to control transcription elongation in a specialized subset of operons (see Fig. 2). They control the synthesis, export, and assembly of cell surface and extracellular components. The coregulation of the *hly* and *tra* operons with those encoding surface polysaccharides possibly may have arisen from the dependence of both hemolysin and the F pilus on LPS for full activity [26]; i.e., these operons may have evolved to coutilize the LPS regulator RfaH.

TOXIN MATURATION

Fatty Acylation Directed by HlyC

Hemolysin is inactive until matured intracellularly by the cotranslated HlyC. In an *in vitro* system using recombinant HlyC and protoxin (pro-HlyA), it was shown that maturation occurs by fatty acylation and that HlyC is a homodimeric putative acyltransferase that uses acyl-acyl carrier protein (ACP) as the fatty acid donor [16,25,54]. In reactions containing only acyl-ACP, HlyC, and pro-HlyA, acquisition of hemolytic activity is directly related to the binding of fatty acid by proHlyA. Mass spectrometry and Edman degradation of proteolytic products from mature HlyA toxin activated *in vitro* by HlyC using radiolabeled acyl-ACP revealed two fatty-acylated internal lysine residues, K564 (KI) and K690 (KII), and resistance of the acylation to hydroxylamine suggested an amide linkage [55]. Although the two lysines are acylated independently, both are required for *in vivo* toxin activity [55]. The modified K564 and K690 residues subsequently were confirmed in *in vivo*—secreted HlyA. The other HlyA-related toxins all require HlyC-type protoxin activation, and the CyaA secreted by *B. pertussis* is modified by amide-linked palmitoylation of K983 (analogous to HlyA KII, K690) [26,56].

HlyC, a Novel Virulence Lysyl-Acyltransferase

The ability to transfer an acyl group to an internal lysine of a target protein distinguishes HlyC from all other bacterial acyltransferases. HlyC has no significant homology with known acyltransferases and therefore may be structurally and functionally distinct [26]. Bacterial and eukaryotic acyltransferases generally accept either acyl-CoA or acyl-ACP as an acyl donor, but ACP is a strict requirement for HlyC. Myristoyl-ACP gave the highest hemolytic activity of HlyA acylated *in vitro* with a range of fatty acids [25], and heterogeneous acylation is evident *in vivo* [57,58]. The specificity for acyl-ACP is shared in *E. coli* by the acyltransferases involved in lipid A biosynthesis [26], but the use of a protein as a substrate is unique to HlyC.

Pro-HlyA has two independent HlyC recognition domains (FAI and FAII) spanning target lysines KI and KII [59] (Fig. 4). Each domain requires 15–30 amino acids for basal recognition and 50–80 amino acids for full wild-type acy-

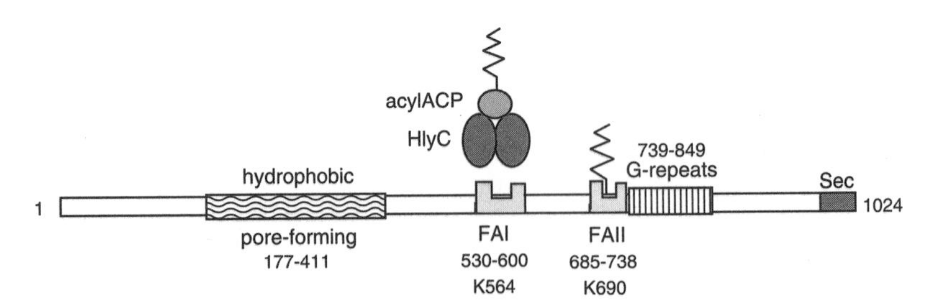

FIGURE 4 Maturation of pro-HlyA by HlyC. The HlyC homodimer associates with the acylated acyl carrier protein. Following binding to the HlyC recognition domains FAI and FAII, the acyl chain is transferred to the corresponding acyl modification sites. Also shown are the hydrophobic pore-forming domain, glycine-rich Ca^{2+}-binding repeats, and the C-terminal secretion signal.

lation. No other HlyA sequences are required for toxin maturation, and Ca^{2+} ions prevent *in vitro* acylation [59], compatible with the view that Ca^{2+} does not bind until toxin is secreted. Sequences of the two acylation sites have little in common apart from the central GK motif, suggesting that similarity between FAI and FAII may be of a higher structural order.

In vitro kinetic data show that HlyC acyltransferase activity is fully described by Michaelis-Menten analysis [60]. The V_{max} at saturating levels of both substrates is approximately 115 nmol acyl group per minute per milligram, with a $K_m^{\text{acyl-ACP}}$ of 260 n*M* and $K_m^{\text{pro-HlyA}}$ of 27 n*M*, parameters sufficient to explain why *in vivo* HlyC is required at a concentration equimolar to its substrate pro-HlyA. HlyC binds the fatty acyl group from acyl-ACP to generate an acylated HlyC intermediate that is depleted in the presence of pro-HlyA but enriched in the presence of pro-HlyA derivatives lacking acylation sites [26,60]. HlyC is able to bind tightly but it seems noncovalently both the acyl chain from acyl-ACP and phosphopantetheine [26]. Substitution of conserved amino acids that could act as covalent attachment sites does not prevent binding of the fatty acyl or 4′-phosphopantetheine groups. These data and reaction analyses suggest that the acylation proceeds via a sequential ordered Bi Bi reaction mechanism requiring the formation of a noncovalent ternary acyl-ACP—HlyC—pro-HlyA complex [26,60]. This view of the mechanism (Fig. 5) would explain why the host bacterial cell (with its own ACP) appears to influence cross-maturation of different protoxins and also why native CyaA from *B. pertussis* is exclusively palmitoylated at K983 (KII), whereas recombinant pro-CyaA modified by CyaC in *E. coli* is acylated at K983 and at K860 (analogous to KI) [61].

Role of Lipidation in Hemolysin Action?

Lipidation is involved in the maturation of many prokaryotic and eukaryotic proteins but is achieved by diverse mechanisms. The functions of the HlyA

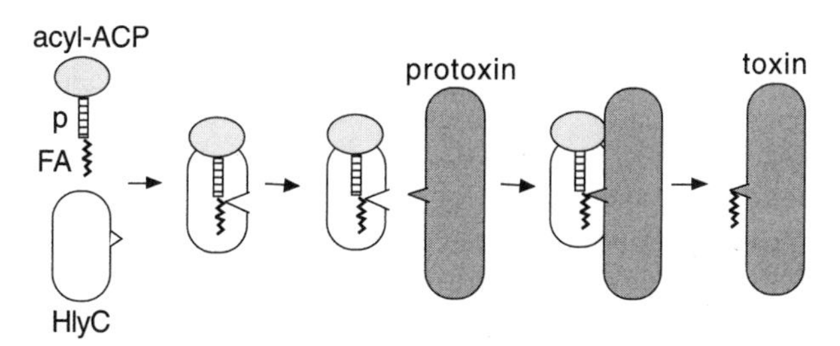

FIGURE 5 The toxin fatty acylation mechanism (FA, fatty acid; P, panthothenate; ACP, acyl carrier protein).

toxin suggest that its acylation could be central to each of the functions typically influenced by such posttranslational modification, i.e., anchoring to membranes, signal transduction, and oligomerization [reviewed in 26].

The most obvious effect of HlyA fatty acylation appears to be at the pore-forming step. Both acyl modifications must occur for HlyA to form channels. Loss of either acyl group results in the virtual abolition (>99.5%) of cytolytic activity [55]. If monomers of the toxin must combine within the membrane in order to form pores, the acyl groups may be involved in protein-protein interactions to bring about oligomerization. The influence of acylation on the primary interaction and binding of HlyA to target membranes has been investigated, but the results are contradictory. Most recently, direct assay of HlyA and pro-HlyA revealed no influence of acylation on liposome binding or insertion, and this was confirmed following *in vitro* acylation [41]. We have suggested that in a Ca^{2+}-myristoyl switch, the distal binding of Ca^{2+} could dictate a conformational change in mature HlyA that could expose the acyl groups and make them available to the lipid bilayer [26]. The highly charged nature of the residues either side of KI and KII, 126 amino acids apart, increases the instability of potential membrane association, and single acyl groups are unlikely to provide sufficient hydrophobic interaction to stabilize a transmembrane region, so they may act more likely as anchorage points onto the surface of the lipid bilayer. In this way, acyl groups may prevent essential domains looping away from the membrane surface and may enhance lateral diffusion to accelerate contact. HlyA also acts as a pseudochemokine. The major component of HlyA binding to granulocytes triggers the respiratory burst apparently through a short-circuiting of the classic signal-transduction pathway, although where it interacts with the host signaling pathway is unknown. It may be significant that membrane immunoglobulin, nicotinic acetylcholine, and insulin receptors appear to possess comparable amide-linked internal fatty acylation as found on HlyA, whereas three inflammatory cytokines, tumor necrosis factor (TNF-α) and interleukin 1α (IL-1α) and IL-1β possess myristoyl groups attached through amide bonds to internal lysines [26]. These eukaryotic proteins with internal amide-linked acyl groups form active membrane-associated complexes. Could this be linked to HlyA toxin function in eukaryotic membranes?

SECRETION OF HEMOLYSIN

Export by the Type I Mechanism

Proteins destined for the cell surface or the surrounding medium must cross the cytoplasmic (inner) and outer membranes and the periplasm between them. Several pathways achieve this by two-step mechanisms that employ periplasmic intermediates (e.g., adhesion pilus assembly) or by using a large number of proteins in machineries spanning the envelope (e.g., the type III assembly of flagella). The type I export mechanism used to secrete the 110-kDa *E. coli* hemolysin, related toxins, and other virulence proteins does not generate periplasmic intermediates [19,62,63] and requires only three conserved exporter components. These are integral membrane proteins, a traffic ATPase (in the hemolysin system HlyB), an accessory or adaptor protein (HlyD) in the inner membrane (IM), and the outer membrane (OM) protein TolC. HlyB has a large, enzymatically active cytosolic domain typical of traffic ATPases [64], whereas HlyD has a large periplasmic domain connected to a small cytosolic domain by a single transmembrane domain [65]. Together HlyB and HlyD form an energized IM substrate-specific complex, but this complex does not function alone. Without TolC, there is no export to the periplasm or from spheroplasts [66]; i.e., TolC is an integral part of the machinery.

HlyA does not use N-terminal signal peptide—dependent translocation across the IM, and the HlyC-determined maturation does not influence export. HlyA translocation depends on an uncleaved C-terminal signal within the final 50 residues and is believed to be multicomponent and contain an amphipathic helix [62,67,68]. Despite a lack of identity between signal primary sequences, interchangeability of type I exporter proteins suggests that higher-order structures in the C-terminal signals are shared among the export substrates. Mutations in the secretion signal are partially compensated for by suppressor mutations in HlyB [69] compatible with the export signal interacting with HlyB. Direct analysis by *in vivo* cross-linking has shown that HlyA binds to both HlyB and HlyD [70]. Hemolysin export requires hydrolysis of ATP bound by HlyB [71]. Mutations in highly conserved residues in the ATP-binding fold and glycine-rich linker peptide cause a complete loss of export, and ATPase activity in HlyB is still able to bind ATP effectively and undergo ATP-induced conformational change. Hemolysin export also shares with the Sec secretion across the cytoplasmic membrane an early requirement for the total proton motive force (PMF) but also has a late stage that does not require PMF. A translocation intermediate identified in this late stage is closely associated with the IM, putatively in the assembled translocation complex spanning both membranes [63].

Cross-linking has defined the sequence of protein-protein interactions underlying export. TolC is recruited by the IM complex (translocase) in response to engagement of the HlyA substrate [70]. Contact between the IM complex and TolC is mediated principally by HlyD, and the active type I export complex contains the substrate and all three export proteins, each of which undergoes conformational change. This complex is transient; once the substrate passes out of the cell, the IM and OM components disengage. It nevertheless remains unclear how the IM complexes could physically connect with TolC to bridge the periplasm, especially if TolC had a simple OM porin-like structure.

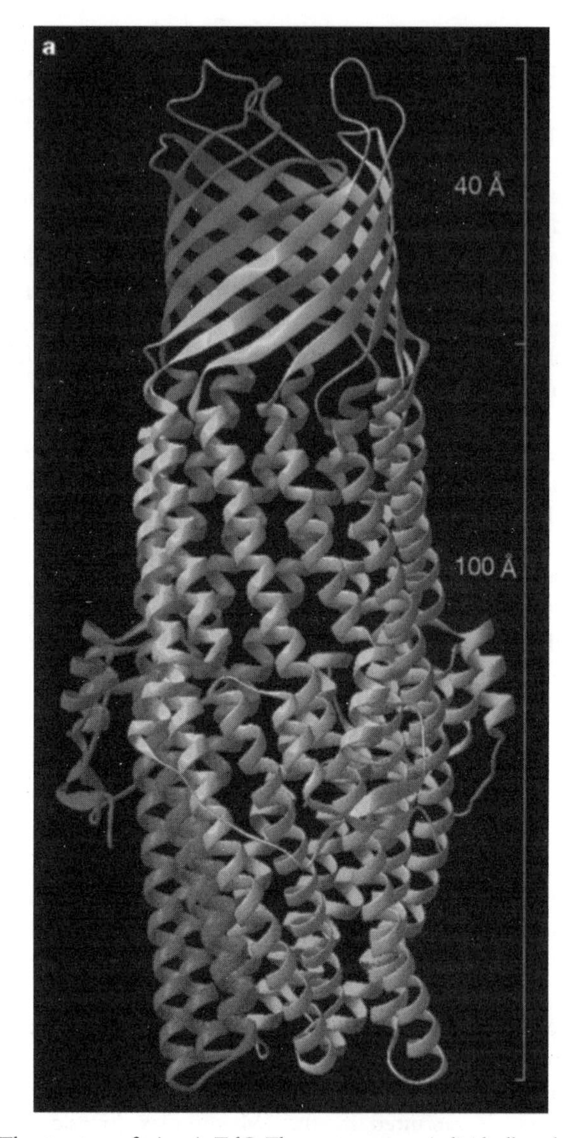

FIGURE 6 The structure of trimeric TolC. The protomers are individually colored. The β-barrel channel (outer membrane) domain is at the top, and the α-helical tunnel (periplasmic) domain is at the bottom. Reprinted with permission from Nature 405:914–919, Fig 1a. See Color Plate 9.

The TolC Channel-Tunnel Underlying Toxin Export and Drug Efflux

When TolC was purified from the *E. coli* OM and crystallized in two-dimensional lattices in phospholipid bilayers, electron microscopy generated a projection at 12 Å of resolution [65]. This showed that TolC was a trimeric structure with a 58-Å outer diameter reminiscent of porins but with a single central pool of stain, hinting at a single pore rather than three. Side views parallel to the membrane plane revealed an additional domain outside the membrane, suggesting the possible existence of a periplasmic domain [66] that together with the periplasmic domain of the IM HlyD could form the periplasmic bypass. Subsequent x-ray crystallography [72] revealed that TolC is a homotrimer forming a 140-Å-long hollow tapered cylinder that comprises a 100-Å-long α-helical barrel (the tunnel domain) anchored in the OM by a 40-Å-long β-barrel (the channel domain) (Fig. 6). This conduit has a 35-Å internal diameter, forming a water-filled 43,000-$Å^3$ duct open to the cell exterior. The TolC channel-tunnel decreases to a virtual close at its periplasmic entrance, compatible with electrophysiologic measurements that show that membrane TolC has a relatively low conductance [73,74].

The TolC β-barrel architecture is quite distinct from other OM proteins, which form one barrel per monomer. Three TolC molecules each contribute four β-strands to form a single 12-strand β-barrel. The TolC interior is also different. The cross-sectional area of the channel is 960 $Å^2$, 15-fold larger than that of the general diffusion pore OmpF, and TolC lacks both the inward-folded loop that typically constricts the internal diameter of β-barrels and the plug domain that closes the larger β-barrels of the FhuA and FepA iron transporters [75,76]. The most striking difference from other known OM proteins is the 100-Å-long periplasmic tunnel composed of a previously unknown fold, a 12-strand α-helical barrel. The tunnel is also assembled from four antiparallel strands per monomer (two continuous long helices and two pairs of shorter helices) so that TolC forms a continuous proteinaceous exit duct.

The periplasmic entrance of TolC must undergo conformational change to allow passage of substrate. This is envisaged to occur by an allosteric mechanism [72,75,76] in which realignment of the inner pair of helices of each TolC monomer with respect to the outer pair could open the entrance, like an iris, to a diameter of 30 Å. The cross-linking experiments indicate that this tunnel opening is triggered and stabilized by the recruiting IM complex in response to engagement of the substrate allowing gated exit from the cell interior to the external environment. Cross-linking has shown that the principal contact between the substrate-bound IM complex and TolC is made by the HlyD accessory protein, which, like TolC, is trimeric [70]. Since this protein connects the substrate-binding energy-providing traffic ATPase to TolC, it can be regarded as the type I adaptor protein. A potential site for the key contact of HlyD is the

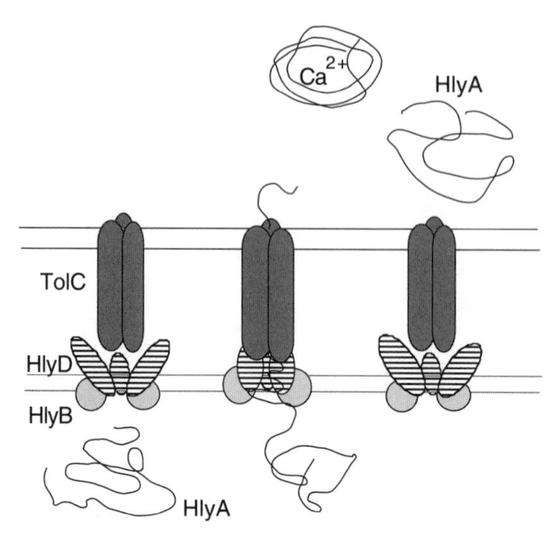

FIGURE 7 Export of hemolysin involving reversible recruitment of the TolC channel-tunnel by the toxin-engaged IM complex of the HlyB traffic ATPase and the adaptor protein HlyD. An animation can be viewed at *http://archive.bmn.com/supp/ceb/ani1.html* [75].

TolC tunnel entrance, but HlyD is also highly likely to form coiled-coil structures that potentially could repack against the coiled-coils of the tunnel β-barrel and/or reach to the TolC equatorial domain to effect opening of the entrance (Fig. 7). The most recent analysis show that this recruiment trigger specifically involves binding of the HlyA substrate to the small HlyD cytosolic domain [77].

A closely related TolC-dependent mechanism effects efflux of small noxious compounds such as detergents, organic solvents, and antibacterial drugs. Analogous to the type I protein export machineries, these efflux pumps are composed of an OM TolC protein cooperating with a substrate-specific IM complex containing a HlyD-like adaptor protein and an energy-providing protein, in this case a proton transporter [75]. TolC homologues seem to be ubiquitous among gram-negative bacteria [75], and these retain the α-helices and β-strands of the channel-tunnel structure, suggesting that the essentials of the structure are conserved [76]. One can conclude that the core functions of the channel-tunnel are common to export and efflux throughout gram-negative bacteria and that current analyses of the *E. coli* TolC opening mechanism will highlight important features underlying both types of transport machinery.

ACKNOWLEDGMENTS

Our work is supported by a Medical Research Council Program Grant.

REFERENCES

1. Springer, W., and Goebel, W. (1980). Synthesis and secretion of hemolysin by *E. coli*. *J. Bacteriol.* 144:53–59.
2. Noegel, A., Rdest, U., Springer, W., and Goebel, W. (1979). Plasmid cistrons controlling synthesis and excretion of the exotoxin α-hemolysin of *Escherichia coli*. *Mol. Gen. Genet.* 175:343–350.
3. Wagner, W., Vogel, M., and Goebel, W. (1983). Transport of hemolysin across the outer membrane of *Escherichia coli* requires two functions. *J. Bacteriol.* 154:200–210.
4. Welch, R. A., Dellinger, E. P., Minshew, B., and Falkow, S. (1981). Hemolysin contributes to virulence of extraintestinal *Escherichia coli* infections. *Nature* 294:665–667.
5. Hacker, J., Hughes, C., Hof, H., and Goebel, W. (1983). Cloned hemolysin genes from *Escherichia coli* that cause urinary tract infection determine different levels of toxicity in mice. *Infect. Immun.* 42:57–63.
6. Cavalieri, S. J., Bohach, G. A., and Snyder, I. S. (1984). *Escherichia coli* α-hemolysin: Characteristics and probable role in pathogenicity. *Microbiol. Rev.* 48:326–343.
7. Hacker, J., and Hughes, C. (1985) Genetics of *E. coli* hemolysin. *Curr. Top. Microbiol. Immunol.* 118:139–162.
8. Felmlee, T., Pellett, S., and Welch, R. A. (1985). Nucleotide sequence of an *Escherichia coli* chromosomal hemolysin. *J. Bacteriol.* 163:94–105.
9. Hess, J., Wels, W., Vogel, M., and Goebel, W. (1986). Nucleotide sequence of a plasmid encoded hemolysin determinant and its comparison with a corresponding chromosomal hemolysin sequence. *FEMS Microbiol. Lett.* 34:1–11.
10. Felmlee, T., Pellett, S., Lee, Y.-R., and Welch, R. (1985). *Escherichia coli* hemolysin is released extracellularly without cleavage of a signal peptide. *J. Bacteriol.* 163:88–93.
11. Glaser, P., Sakamoto, H., Bellalou, J., *et al.* (1988). Secretion of cyclolysin, the calmodulin-sensitive adenylate cyclase-haemolysin bifunctional protein of *Bordetella pertussis*. *EMBO J.* 7:3997–4004.
12. Strathdee, C. A., and Lo, R. Y. C. (1987). Extensive homology between the leukotoxin of *Pasteurella haemolytica* A1 and the α-hemolysin of *Escherichia coli*. *Infect. Immun.* 55:3233–3236.
13. Koronakis, V., Cross, M., Senior, B., *et al.* (1987). The secreted hemolysins of *Proteus mirabilis*, *Proteus vulgaris*, and *Morganella morganii* are genetically related to each other and to the α-hemolysin of *Escherichia coli*. *J. Bacteriol.* 169:1509–1515.
14. Welch, R. A. (1991). Pore-forming cytolysins of gram-negative bacteria. *Mol. Microbiol.* 5:521–528.
15. Coote, J. G. (1992). Structural and functional relationships among the RTX toxin determinants of gram-negative bacteria. *FEMS Microbiol. Rev.* 88:137–162.
16. Hughes, C., Stanley, P., and Koronakis, V. (1992). Bacterial hemolysin interactions with prokaryotic and eukaryotic membranes. *Bioessays* 14:519–526.
17. Knapp, S., Hacker, J., Then, I., *et al.* (1984). Multiple copies of hemolysin genes and associated sequences in the chromosome of uropathogenic *E. coli* strains. *J. Bacteriol.* 159:1027–1033.
18. Holland, I. B., Blight, M. A., and Kenny, B. (1990). The mechanism of secretion of hemolysin and other polypeptides from gram-negative bacteria. *J. Bioenerget. Biomemb.* 22:473–491.
19. Koronakis, V., and Hughes, C. (1993). Bacterial signal peptide-independent protein export: HlyB-directed secretion of hemolysin. *Semin. Cell Biol.* 4:7–15.
20. Hughes, C., and Koronakis, V. (1993). Export of hemolysin and other proteins from the gram-negative cell. In J. M. Ghuysen (ed.), *New Comprehensive Biochemistry: Bacterial Cell Wall,* pp. 425–446. New York: Elsevier.
21. Letoffe, S., Delepelaire, P., and Wandersman, C. (1990). Protease secretion by *Erwinia chrysanthemi*: The specific secretion functions are analogous to those of *E. coli* alpha-haemolysin. *EMBO J.* 9:1375–1382.
22. Hyde, S. C., Emsley, P., Hartshon, M. J., *et al.* (1990). Structural model of ATP-binding proteins associated with cystic fibrosis, multidrug resistance and bacterial transport. *Nature* 346:362–365.

23. Wandersman, C., and Delepelaire, P. (1990). TolC, an *Escherichia coli* outer membrane protein required for hemolysin secretion. *Proc. Natl. Acad. Sci. USA* 87:4776–4780.
24. Bailey, M. J. A., Koronakis, V., Schmoll, T., and Hughes, C. (1992). *Escherichia coli* HlyT protein, a transcriptional activator of haemolysin synthesis and secretion, is encoded by the rfaH(sfrB) locus required for expression of sex factor and lipopolysaccharide genes. *Mol. Microbiol.* 6:1003–1012.
25. Issartel, J. P., Koronakis, V., and Hughes, C. (1991). Activation of *Escherichia coli* prohaemolysin to the mature toxin by acyl carrier protein-dependent fatty acylation. *Nature* 351:759–761.
26. Stanley, P., Koronakis, V., and Hughes C. (1998). Acylation of *Escherichia coli* hemolysin: A unique protein lipidation mechanism underlying toxin function. *Microbiol. Mol. Biol. Rev.* 62:309–333.
27. Baumann, U., Wu, S., Flaherty, K. M., and McKay, D. B. (1993). Three-dimensional structure of the alkaline protease of *Pseudomonas aeruginosa:* A two-domain protein with a calcium binding parallel beta roll motif. *EMBO J.* 12:3357–3364.
28. Lally, E. T., Kieba, I. R., Sato, A., *et al.* (1997). RTX toxins recognize a β_2 integrin on the surface of human target cells. *J. Biol. Chem.* 272:30463–30469.
29. Cortajarena, A. L., Goni, F. M., and Ostolaza, H. (2001). Glycophorin as a receptor for *Escherichia coli* alpha-hemolysin in erythrocytes. *J. Biol. Chem.* 276:12513–12519.
30. Bhakdi, S., and Martin, E. (1991). Superoxide generation by human neutrophils induced by low doses of *Escherichia coli* hemolysin. *Infect. Immun.* 59:2955–2962.
31. Konig, B., Ludwig, A., Goebel, W., and Konig, W. (1994). Pore formation by the *Escherichia coli* alpha-hemolysin: Role for mediator release from human inflammmatory cells. *Infect. Immun.* 62:4611–4617.
32. Suttorp, N., Fuhrmann, M., Tannertotto, S., *et al.* (1993). Pore-forming bacterial toxins potently induce release of nitric oxide in porcine endothelial cells. *J. Exp. Med.* 178:337–341.
33. May, A. K., Sawyer, R. G., Gleason, T., *et al.* (1996). *In vivo* cytokine response to *Escherichia coli* α-hemolysin determined with genetically engineered hemolytic and nonhemolytic *Escherichia coli* variants. *Infect. Immun.* 64:2167–2171.
34. Jonas, D., Schultheis, B., Klas, C., *et al.* (1993). Cytocidal effects of *Escherichia coli* hemolysin on human T lymphocytes. *Infect. Immun.* 61:1715–1721.
35. Uhlen, P. *et al.* (2000). α-Hemolysin of uropathogenic *E. coli:* An inducer of calcium oscillations in renal epithelial cells. *Nature* 405:694–697.
36. Menestrina, G., Ropele, M., Dalla Serra, M., *et al.* (1995). Binding of antibodies to functional epitopes on the pore formed by *Escherichia coli* hemolysin in cells and model membranes. *Biochim. Biophys. Acta. Biomemb.* 1238:72–80.
37. Bhakdi, S., Mackman, N., Nicaud, J. M., and Holland, I. B. (1986). *Escherichia coli* hemolysin may damage target cell membranes by generating transmembrane pores. *Infect. Immun.* 52:63–69.
38. Benz, R., Schmid, A., Wagner, W., and Goebel, W. (1989). Pore formation by the *Escherichia coli* hemolysin: Evidence for an association-dissociation equilibrium of the pore-forming aggregates. *Infect. Immun.* 57:887–895.
39. Soloaga, A., Veiga, M. P., Garcia-Segura, L. M., *et al.* (1999). Insertion of *Escherichia coli* alpha-haemolysin in lipid bilayers as a nontransmembrane integral protein: Prediction and experiment. *Mol. Microbiol.* 31:1013–1024.
40. Ludwig, A., Schmid, A., Benz, R., and Goebel, W. (1991). Mutations affecting pore formation by hemolysin from *Escherichia coli*. *Mol. Gen. Genet.* 226:198–208.
41. Hyland, C., Vuillard, L., Hughes, C., and Koronakis, V. (2001). Membrane interaction of *Escherichia coli* hemolysin: Flotation and insertion-dependent labeling by phospholipid vesicles. *J. Bacteriol.* 183:5364–5370.
42. Schindel, C., Zitzer, A., Schulte, B., *et al.* (2001). Interaction of *E. coli* hemolysin with biological membranes: A study using cysteine scanning mutagenesis. *Eur. J. Biochem.* 268:800–808.
43. Boehm, D. F., Welch, R. A., and Snyder, I. S. (1990). Domains of *Escherichia coli* hemolysin (HlyA) involved in binding of calcium and erythrocyte membranes. *Infect. Immun.* 58:1959–1964.

44. Ludwig, A., Jarchau, T., Benz, R., and Goebel, W. (1988). The repeat domain of *Escherichia coli* hemolysin (HlyA) is responsible for its Ca^{2+}-dependent binding to erythrocytes. *Mol. Gen. Genet.* 214:553–561.
45. Bakas, L., Veiga, M. P., Soloaga, A., *et al.* (1998). Calcium-dependent conformation of *E. coli* alpha-hemolysin: Implications for the mechanism of membrane insertion and lysis. *Biochim. Biophys. Acta.* 19:225–234.
46. Bailey, M. A., Hughes, C., and Koronakis, V. (1997). RfaH and the *ops* element, components of a novel system controlling bacterial transcription elongation. *Mol. Microbiol.* 26:845–852.
47. Nieto, J. M., Bailey, M. J. A., Hughes, C., and Koronakis, V. (1996). Suppression of transcription polarity in the *Escherichia coli* haemolysin operon by a short upstream element shared by polysaccharide and DNA transfer determinants. *Mol. Microbiol.* 19:705–713.
48. Bailey, M. J. A., Hughes, C., and Koronakis, V. (1996). Increased distal gene transcription by the elongation factor RfaH, a specialised homologue of NusG. *Mol. Microbiol.* 22:729–737.
49. Koronakis, V., Cross, M., and Hughes, C. (1989). Transcription antitermination in an *E. coli* hemolysin operon is directed progressively by *cis*-acting DNA sequences upstream of the promoter region. *Mol. Microbiol.* 3:1397–1404.
50. Leeds, J. A., and Welch, R. A. (1997). Enhancing transcription through the *Escherichia coli* hemolysin operon, hlyCABD: RfaH and upstream JUMPstart DNA sequences function together via a postinitiation mechanism. *J. Bacteriol.* 179:3519–3527.
51. Farewell, A., Brazas, R., Davie, E., *et al.* (1991). Suppression of the abnormal phenotype of *Salmonella typhimurium* rfaH mutants by mutations in the gene for transcription termination factor Rho. *J. Bacteriol.* 173:5188–5193.
52. Wong, K. R., Hughes, C., and Koronakis, V. (1998). A gene, *yaeQ*, suppressing mutations in *rfaH*, the transcription elongation factor gene. *Mol. Gen. Genet.* 257:693–696.
53. Bailey, M. J. A., Hughes, C., and Koronakis, V. (2000). *In vitro* recruitment of the RfaH regulatory protein into a specialised transcription complex, directed by the nucleic acid *ops* element. *Mol. Gen. Gen.* 262:1052–1059.
54. Hardie, K. R., Issartel, J. P., Koronakis, E., *et al.* (1991). *In vitro* activation of *Escherichia coli* prohaemolysin to the mature membrane-targeted toxin requires HlyC and a low-molecular-weight cytosolic polypeptide. *Mol. Microbiol.* 5:1669–1679.
55. Stanley, P., Packman, L. C., Koronakis, V., and Hughes, C. (1994). Fatty acylation of two internal lysine residues required for the toxic activity of *Escherichia coli* hemolysin. *Science* 266:1992–1996.
56. Hackett, M., Guo, L., Shabanowitz, J., *et al.* (1994). Internal lysine palmitoylation in adenylate cyclase toxin from *Bordetella pertussis*. *Science* 266:433–435.
57. Ludwig, A., Garcia, F., Bauer, S., *et al.* (1996). Analysis of the *in vivo* activation of hemolysin (HlyA) from *Escherichia coli*. *J. Bacteriol.* 178:5422–5430.
58. Lim, K. B., Walker, C. R., Guo, L., *et al.* (2000). *Escherichia coli* alpha-hemolysin (HlyA) is heterogeneously acylated *in vivo* with 14-, 15-, and 17-carbon fatty acids. *J. Biol. Chem.* 275:36698–36702.
59. Stanley, P., Koronakis, V., Hardie, K., and Hughes, C. (1996). Independent interaction of the acyltransferase HlyC with two maturation domains of the *Escherichia coli* toxin HlyA. *Mol. Microbiol.* 20:813–822.
60. Stanley, P., Hyland, C., Koronakis, V., and Hughes, C. (1999). An ordered reaction mechanism for bacterial toxin acylation by the specialized acyltransferase HlyC: Formation of a ternary complex with acylACP and protoxin substrates. *Mol. Microbiol.* 34:887–901.
61. Hackett, M., Walker, C. B., Guo, L., *et al.* (1995). Hemolytic, but not cell-invasive activity, of adenylate-cyclase toxin is selectively affected by differential fatty acylation in *Escherichia coli*. *J. Biol. Chem.* 270:20250–20253.

62. Koronakis, V., Koronakis, E., and Hughes, C. (1989). Isolation and analysis of the C-terminal signal directing export of *Escherichia coli* hemolysin protein across both bacterial membranes. *EMBO J.* 8:595–605.
63. Koronakis, V., Hughes, C., and Koronakis, E. (1991). Energetically distinct early and late stages of HlyB/HlyD-dependent secretion across both *Escherichia coli* membranes. *EMBO J.* 10:3263–3272.
64. Wang, R. C., Seror, S. J., Blight, M., *et al.* (1991). Analysis of the membrane organization of an *Escherichia coli* protein translocator, HlyB, a member of a large family of prokaryote and eukaryote surface transport proteins. *J. Mol. Biol.* 217:441–454.
65. Schulein, R., Gentschev, I., Mollenkopf, H. J., and Goebel, W. (1992). A topological model for the hemolysin translocator protein HlyD. *Mol. Gen. Genet.* 234:155–163.
66. Koronakis, V., Li, J., Koronakis, E., and Stauffer, K. (1997). Structure of TolC, the outer membrane component of the bacterial type I efflux system, derived from two-dimensional crystals. *Mol. Microbiol.* 23:617–626.
67. Stanley, P., Koronakis, V., and Hughes, C. (1991). Mutational analysis supports a role for multiple structural features in the C-terminal secretion signal of *Escherichia coli* hemolysin. *Mol. Microbiol.* 5:2391–2403.
68. Kenny, B., Taylor, S., and Holland, I. B. (1992). Identification of individual amino acids required for secretion within the hemolysin (HlyA) C-terminal targeting region. *Mol. Microbiol.* 6:1477–1489.
69. Sheps, J. A., Cheung, I., and Ling, V. (1995). Hemolysin transport in *Escherichia coli:* Point mutants in HlyB compensate for a deletion in the predicted amphiphilic helix region of the HlyA signal. *J. Biol. Chem.* 270:14829–14834.
70. Thanabalu, T., Koronakis, E., Hughes, C., and Koronakis, V. (1998). Substrate-induced assembly of a contiguous channel for protein export from *E. coli*: Reversible bridging of an inner-membrane translocase to an outer membrane exit pore. *EMBO J.* 17:6487–6496.
71. Koronakis, E., Hughes, C., Milisav, I., and Koronakis, V. (1995). Protein exporter function and *in vitro* ATPase activity are correlated in ABC-domain mutants of HlyB. *Mol. Microbiol.* 16:87–96.
72. Koronakis, V., Sharff, A., Koronakis, E., *et al.* (2000). Crystal structure of the bacterial membrane protein TolC central to multidrug efflux and protein export. *Nature* 405:914–919.
73. Benz, R., Maier, E., and Gentschev, I. (1993). TolC of *Escherichia coli* functions as an outer membrane channel. *Zentralbl. Bakteriol.* 278:187–196.
74. Andersen, C., Hughes, C., and Koronakis, V. (2002). Electrophysiological behavior of the TolC channel-tunnel in planar lipid bilayers. *J. Membr. Biol.* 185:83–92.
75. Andersen C., Hughes C., and Koronakis, V. (2001) Protein export and drug efflux through bacterial channel-tunnels. *Curr. Opin. Cell Biol.* 13:412–416.
76. Koronakis V., Andersen C., and Hughes, C. (2001). Channel-tunnels. *Curr. Opin. Struct. Biol.* 11:403–407.
77. Balakrishnan, L., Hughes, C., and Koronakis, V. (2001). Substrate-triggered recruitment of the TolC channel-tunnel during type I export of hemolysin by *E. coli*. *J. Mol. Biol.* 313:501–510.

CHAPTER 14

Capsule and Lipopolysaccharide

Thomas A. Russo

Department of Medicine and Microbiology, The Center for Microbial Pathogenesis, Veterau's Administration Medical Center, SUNY Buffalo, Buffalo, New York

INTRODUCTION

As we learn more about virulence genes, it is becoming clear that hosts and their pathogens have been coadapting throughout evolutionary history. Over time, microbial pathogens have acquired a variety of virulence factors (VFs) that are requisite to be an effective pathogen. In fact, the possession of specialized virulence genes not only defines a pathogen but also dictates host range and host sites at risk for infection. In turn, host factors also have evolved to counteract these bacterial traits. This host-pathogen "chess match" is never-ending.

The ability of different strains of *Escherichia coli* to express specific types of capsular polysaccharides and/or a complete lipopolysaccharide (LPS) is an important trait that in part dictates their pathogenic potential. A new classification for *E. coli* capsules (groups 1–4) has been proposed [1]. Nearly all strains of *E. coli* possess genes that code for group 1 capsule or the group 1–like capsule colanic acid. Only selected strains possess genes for and express group 2 or group 3 (primarily extraintestinal pathogenic strains of *E. coli*) or group 4 capsules. Further, group 1 capsules or colanic acid can be coexpressed with capsule groups 2–4 [1,2]. LPS is a unique glycolipid present in the outer membrane of gram-negative bacteria. It consists of lipid A, a core oligosaccharide region, and a

Escherichia coli: Virulence Mechanisms of a Versatile Pathogen
ISBN 0-12-220751-3

serotype-specific O-antigen. All strains of *E. coli* possess lipid A and at least two 3-deoxy-D-manno-octulosonic acid (KDO) residues from the core region because these moieties appear to be needed for structural integrity. However, most naturally occurring strains of *E. coli* possess a complete LPS.

LPS and group 2 and 3 capsules play major roles in the pathogenesis of infection, contributing at several levels. Classically, capsule and LPS are considered to be VFs that protect *E. coli* against various host bactericidal defense factors. However, LPS and likely capsule are also pathogen-associated molecular patterns (PAMPs). Host recognition of these polysaccharides results in stimulation of the innate immune response. Further, recent data from my laboratory and others demonstrate that these surface polysaccharides not only induce the innate inflammatory response but also are capable of modulating this response in an advantageous manner. Lastly, capsule and LPS may mask underlying or other surface antigens, thereby subverting the development of an optimal acquired immune response.

CLINICAL CLASSIFICATION OF *E. COLI*

From a genomic and clinical perspective, *E. coli* can be divided into three separate categories: (1) commensal strains, (2) intestinal pathogenic (enteric or diarrheagenic) strains, and (3) extraintestinal pathogenic strains (ExPEC) [3]. Commensal strains of *E. coli* constitute the bulk of the facultative fecal flora but usually do not cause disease outside the intestinal tract except in the presence of precipitating factors such as an indwelling foreign body or some impairment of host defenses. In contrast to commensal *E. coli*, intestinal pathogenic strains of *E. coli* are rarely encountered in the fecal flora of healthy hosts and instead appear to be essentially obligate pathogens, causing gastroenteritis or colitis whenever ingested in sufficient quantities by a naive host. Despite their ability to cause enteric disease, these strains are largely incapable of causing disease outside the intestinal tract [4]. From both pathogenic and clinical viewpoints, ExPEC are distinct from commensal and intestinal pathogenic strains of *E. coli*. ExPEC also make up part of the normal human fecal flora, but in contrast to commensal strains, ExPEC possess specialized genes that encode VFs that enable them to cause extraintestinal infections. Entry into an extraintestinal site (e.g., lungs, urinary tract, or peritoneum), not acquisition, is the limiting factor for infection. Considering the differences in local environment and host defenses between intestinal and extraintestinal sites, it is not surprising that apparently distinct evolutionary pathways have emerged between intestinal pathogenic strains and extraintestinal pathogenic strains of *E. coli* (ExPEC).

E. coli is also an important veterinary pathogen causing considerable morbidity and mortality in other mammals and birds [5–10]. These veterinary pathogens, at least in part, possess common virulence properties with human isolates, including capsule and complete LPS. Further, certain veterinary strains may be capable of causing human infection [10]. However, the role of these surface polysaccharides in the pathogenesis of veterinary infections will not be addressed in this chapter.

PHYSICAL AND GENETIC CHARACTERISTICS OF CAPSULE AND LPS

Capsule

Over 80 serologically and chemically unique capsular polysaccharides can be produced by *E. coli* [11,12]. The structures of many these have been determined. Initially, these polysaccharides were divided into group 1 and group 2 based on chemical, physical, genetic, and microbiologic distinctions [13,14]. Subsequently, division of the group 2 capsules into groups 2 and 3 (formerly I and II) was proposed and has been accepted [15]. More recently, a fourth capsule group (formerly group 1B) has been proposed (Table I) [1].

Capsule Groups 1 and 4

At least 16 structurally distinct group 1 capsules have been described and usually are coexpressed with a limited range of O-antigens (O8, O9, O20, O101). In *E. coli* they exist as high- (K-antigen, >100,000 kDa) and low-molecular-weight (K_{LPS}) forms. The K-antigen is a polymer that forms the surface-exposed capsule, whereas the K_{LPS} possesses the same chemical composition as the K-antigen but is an oligosaccharide with one or a few repeat units. The cellular attachment of K-antigen is unknown, whereas the K_{LPS} is linked to lipid A-core [16]. Group 1 capsules are acidic (usually due to hexuronic acid or pyruvate), possess a low charge density and electrophoretic mobility, and are stable at pH 5–6 at 100°C. The group 1 capsule locus (*cps*) is adjacent to the *rfb* and *his* loci and possesses a conserved genetic organization [17]. The first region of the gene cluster is highly conserved and consists of genes (*orfX-wza-wzb-wzc*), required for translocation and surface assembly. The second region contains polymer-synthesis genes and is predictably variable for each group 1 capsule. Group 1 capsules are positively regulated by the Rcs system, which has been described in detail [18–20]. A characteristic JUMPstart sequence precedes group 1 capsule gene clusters [17], suggesting that RfaH-mediated antitermination also may play a regulatory role [21–23].

Colanic acid possesses many physical, genetic, and regulatory similarities with group 1 capsules. Although its inclusion as a member of the group 1 capsules has been debated [1], it should be thought of, at least, as group 1–like.

TABLE I Features of *E. coli* Capsules

Characteristic	Group 1	Group 2	Group 3	Group 4
Former capsule group	IA	II	I/II	IB
Genetic locus	*cps* near *his, rfb*	*kps* near *serA*	*kps* near *serA*	*rfb* near *his*
O-antigen coexpression	Limited (O8, O9, O20, O101)	Many	Many	Limited (O8, O9) or none
Potential for coexpression with colanic acid	No	Yes	Yes	Yes
Thermostability at 100°C	Yes	No	No	Yes
Molecular mass	>100,00 kDa (K-antigen)	<50,000 kDa	<50,000 kDa	>100,000 kDa (K-antigen)
Cellular attachment	K-antigen unknown K_{LPS} via lipid A-core	α-Glycerophosphate	?α-Glycerophosphate	K-antigen unknown K_{LPS} via lipid A-core
Polymerization	Wzy-dependent	Processive	?Processive	Wzy-dependent
Translocation and surface assembly	Wzx-Wza, c, ?b	KpsM, T-KpsD, E, ?F	?KpsM, T-KpsD, E	Unknown
Regulation	Rcs system (+) ?RfaH-antitermination	Temperature (at <20°C) RfaH-antitermination Integration host factor (+)	RfaH-antitermination ?Rcs system (–)	Unknown
Model systems	K30, colanic acid	K1, K5	K54, K10	K40

Source: Modified from Whitfield and Roberts [1].

A separate category for group 4 capsules (formerly group 1B) has been proposed [1]. Group 1 and 4 capsules share many similarities [1]. In *E. coli*, like group 1 capsules, group 4 capsules exist as both LPS-independent capsules (originally described as O-antigen capsules [1]) and as lipid A-core-linked capsules. The basis for separating group 4 capsules from group 1 is differences in transport and assembly genes and in regulation. Homologues for the group 1 genes *wza-wzb-wzc* have not been identified in group 4 gene clusters. Further, group 4 capsules are not positively regulated by the Rcs system [1].

Capsule Groups 2 and 3

Group 2 and 3 capsules, which are physically, genetically, and biologically distinct from group 1 and 4 capsules, share more similarities than differences. At least 50 unique group 2 and 3 capsules have been reported that are coexpressed with many O-specific antigens. Group 2 capsules and likely group 3 capsules are linked to the cell via α-glycerophosphate [1,24]. Typically, these capsules are characterized physically by a strong negative charge; a high hydrophilicity; a water content of greater than 50%; molecular weights of less than 50,000 kDa; hexuronic acids, *N*-acetyl neuraminic acid, phosphate, glucuronic acid, or 2-keto-3-deoxy mannosoctonic acid (KDO) as acidic components; a higher charge density and electrophoretic mobility; and a general lack of stability at pH 5–6 at 100°C [11,12]. The group 2 and 3 capsule gene clusters share the same locus adjacent to *serA* [25–27].

The group 2 capsule gene clusters evaluated to date (K1, K4, K5, K7, K12, and K92) possess a conserved genetic organization that consists of three functional regions [24,28–30]. Region 2, which is unique for a given capsular antigen, encodes for genes whose products are responsible for the synthesis of the K-specific serotype. This region is flanked by regions 1 and 3, which are highly conserved among the group 2 capsule gene clusters evaluated to date (Fig. 1). Region 1 contains 6 genes (*kpsF, -E, -D, -U, -C,* and *-S*), and region 3 contains two genes (*kpsM* and *kpsT*). The gene products from these regions are needed for transport of the capsular polysaccharide across the cytoplasmic membrane (ABC-2 transporter, *kpsM* and *kpsT*) and assembly onto the cell's surface (*kpsD* and *kpsE*). Models for this complex process have been proposed and reviewed recently [1,31]. Regions 1 and 3 are transcribed as single transcriptional units and are temperature-regulated with no detectable transcription at less than 20°C [23,32]. Integration host factor enhances region 1 transcription [23], and region 3 is regulated by RfaH antitermination [23].

Group 3 capsules are distinguished from group 2 capsules primarily by differences in the homology of transport and assembly genes, genetic organization, and regulation. At the DNA level, group 2 capsule genes from regions 1 and 3 are not highly homologous with their group 3 counterparts [25,33]; however, protein homologues have been identified for *kpsMTDECS*. The genetic organ-

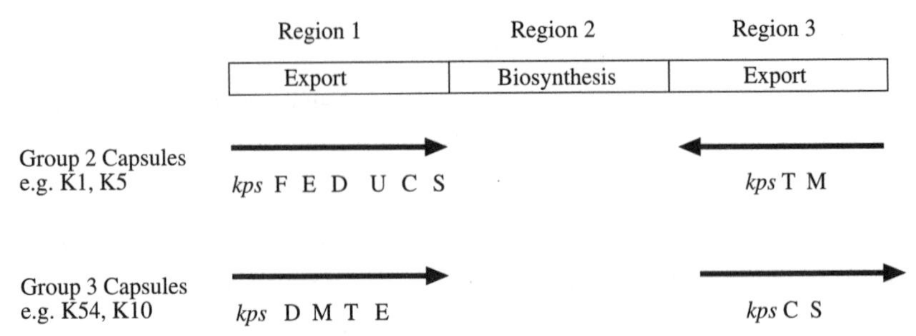

FIGURE 1 Genetic organization of group 2 and group 3 capsule gene clusters. A modular organization exists for these capsule gene clusters. Regions 1 and 3 contain genes that encode products necessary for assembly and transport of the capsular polysaccharide. Regions 1 and 3 appear to consist of single transcriptional units defined by the arrows. These regions are conserved within a given capsular group (e.g., K1, K4, K5, K92 for group 2 and K10 and K54 for group 3) but are divergent between groups 2 and 3. KpsM (membrane component) and KpsT (ATP-binding component) are ABC transporters. KpsD and KpsE are required for translocation. KpsC and KpsS appear to contribute to the modification of the capsule polymer. KpsU is a CMP-Kdo synthetase, and the role of KpsF is not yet defined. In contrast, region 2 contains a variable number (six for K1 and four for K5) of unique serotype-specific biosynthetic genes. It is logical to envision that recombination within region 2 or between homologous regions flanking region 2 account for its diversity. Analysis of the K10, K54 group 3 region 1 capsule genes suggests that this cluster was acquired by horizontal transfer, and the genes replaced a preexisting group 2 region 1 capsule gene cluster.

ization for these group 3 genes also differs from their group 2 counterparts in the two serotypes analyzed to date (K54, K10) [25,33]. The serotype-specific region remains flaked by regions 1 and 3; however, region 1 consists of *kpsDMTE,* and region 3 codes for *kpsCS* homologues (see Fig. 1). Lastly, differences in regulation also exist. Serotypes K3, K10, K11, K19, and K54 do not show temperature regulation of capsule expression, which correlates with constitutive levels of CMP-KDO activity [34], whereas group 2 capsules have increased capsule expression and CMP-KDO activity at 37°C. However, in at least the group 3 K54 and K10 capsule gene clusters, region 3 appears to be regulated in part by RfaH antitermination [25,33].

The distinctions between group 2 and 3 capsules may prove to be artificial as more capsule gene clusters are studied in detail. For example, the K2 capsule is not temperature-regulated and possesses constitutive CMP-KDO activity, similar to group 3 capsules. Via hybridization, however, $kpsMT_{K2}$ is more homologous to its group 2 than group 3 counterparts. The full sequence of the K2 capsule gene cluster from CFT073 will be available in the near future. Hopefully this information will further clarify our understanding of group 2 and 3 capsule genes.

LPS

Unlike capsules, the various LPS serotypes are considered as a single group. The three components of LPS are lipid A, which is covalently attached to a core oligosaccharide region, which in turn is usually capped with the O-specific antigen (Fig. 2). The lipid A moiety is a hydrophobic molecule consisting of a backbone of glucosaminyl-β(1 → 6)glucosamine substituted with 6–7 saturated fatty acid residues. Lipid A anchors the LPS via its insertion into the outer membrane phospholipid bilayer. As described earlier, it also anchors groups 1 and 4 K_{LPS} as well as a quantitatively minor form of enterobacterial common antigen (ECA_{LPS}). Lipid A is chemically and antigenically conserved between different *E. coli* strains. The core region of LPS can be considered as an inner (2 KDOs and heptose) and outer core. The inner core is phosphorylated and is essentially conserved not only between *E. coli* strains but also among the Enterobacteriaceae. The outer core displays some variability in *E. coli* (6–7 variably linked sugar

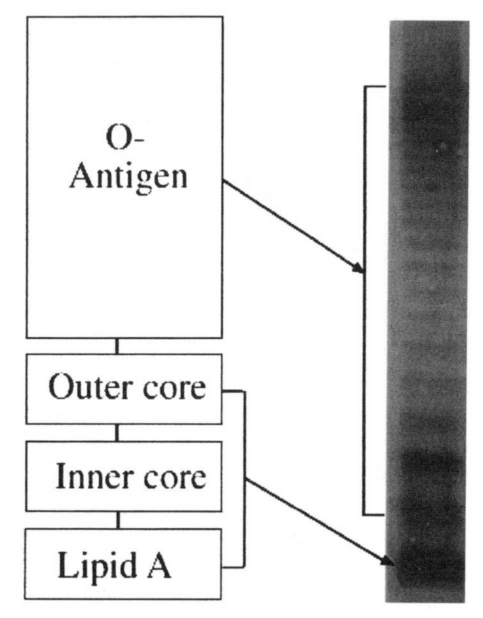

FIGURE 2 Structural organization of lipopolysaccharide. This figure depicts lipid A, the core region (inner and outer), and the O-antigen, which are the three components of LPS. These structures are represented in cartoon fashion on the left, and the corresponding regions are seen when separated on a SDS-14% PAGE gel and silver stained. On this gel, the lipid A and core regions are not resolved. The variable number of repeating units in the O-antigen region results in the generation of the LPS ladder. The variation in the specific distribution of O-antigen chain length also can be appreciated, which has been termed *modality*.

residues), with five types (R1-4, K-12) described to date [35]. An increased proportion of intestinal pathogenic *E. coli* strains possess a type R3 outer core. However, the significance of this from a pathogenic point of view is unclear [36,37]. In natural isolates, the LPS core usually is capped by repeating tri-pentasaccharide units termed the *O-specific antigen.* Over 150 antigenically distinct O-antigen moieties of LPS have been described in *E. coli* [12,38]. The number of O-antigen repeating units for a given strain is variable (0–40), thereby giving rise to the LPS "ladder" seen on LPS gels. Further, a number of strain-specific distributions of O-antigen chain length exist, which has been termed *modality* [39]. The biologic significance of these variations in chain length distribution, if any, remains unclear. The O-antigen is hydrophilic, neutral or acidic, and typically negatively charged. Divalent cations enable cross-bridging of LPS molecules, which contributes to LPS excluding potentially deleterious molecules (e.g. certain antibiotics, detergents). The O-antigen is extremely antigenic and confers serospecificity for a given LPS serotype.

At least 50 genes that are dispersed throughout the *E. coli* chromosome are required for the synthesis of LPS. In the laboratory strain of *E.coli* (K-12), three gene clusters involved with LPS synthesis have been identified. The *lpx* gene cluster near minute 4 contains lipid A biosynthetic genes, the *waa* (formerly *rfa*) operon near minute 81 encodes genes for core synthesis and assembly, and the *rfb* operon near minute 45 includes most of the genes needed for O-antigen synthesis and assembly [40]. Despite the large number of different O-antigens, LPS biosyntheis, assembly, and transport appear to be fairly conserved in the strains evaluated to date [35,39,41,42]. The O-antigen is synthesized and exported to the periplasm, where it is ligated to the lipid A-core oligosaccharide moieties. In *E. coli,* at least two pathways exist for O-antigen polymerization and export (Wzy-dependent and -independent). For O-antigens synthesized via the Wzy-dependent pathway, modality is regulated in part by Wzz [39].

EVOLUTIONARY ROOTS OF CAPSULE AND O-SPECIFIC ANTIGEN GENE CLUSTERS

The increasing availability of DNA sequence data has resulted in some insights as to the origins of capsule and O-antigen gene clusters. Many virulence genes appear to have been acquired either alone or en bloc by horizontal transfer [43]. Mechanisms that mediate these transfers include transposition events, plasmids, and lysogenic phages. Capsule and O-antigen gene clusters are additional examples of virulence traits whose acquisition appears to have occurred by horizontal transfer.

Group 2 and 3 capsule genes possess a lower (37–43%) G + C content than is typically observed for *E. coli* K-12 genes (48–52%) [25], suggestive of acquisition by horizontal transfer. Further, genes from both these capsule groups are similar to capsule genes in *Neisseria meningitidis* and *Hemophilus influenzae* [44,45]. This observation raises the possibility that they may have shared a common ancestral origin. Given the high degree of homology of genes in group 2 capsule gene cluster regions 1 and 3, it has been postulated that capsular diversity is based on exchanges between the central, region 2 cassettes that encode for biosynthetic genes. Interestingly, the two group 3 capsule gene clusters analyzed to date (K54, K10) appear to have been acquired by an insertion sequence (IS)–mediated transposition event, with the site of insertion being the group 2 capsule gene *kpsM* [25,33]. Likewise, *E. coli* group 1 capsule shares similarities with those from *Klebsiella* and *Erwinia*, which also suggests a possible common origin [1].

Based on sequence data from *E. coli* and *Salmonella*, it has been argued that the diversity of O-antigens is also based on the transfer of *rfb* genes between species. Many *rfb* genes have low G + C content, and in some instances, IS-mediated transposition events have been implicated [46–48]. Data from my laboratory suggest that at least a portion of the O4-specific antigen *rfb* locus was derived from *Shigella flexneri* 2a or a common donor [49].

THE ROLE OF CAPSULE AND LPS IN THE PATHOGENESIS OF INFECTION

Capsule and the O-antigen are critical for protecting *E. coli* against the bactericidal activity of a variety of host defense components. Capsule and LPS also contribute to the pathogenesis of infection via their effects on the host's inflammatory and immunologic responses. These later aspects will be discussed in the last two sections of this chapter.

Intestinal Pathogenic Strains of *E. coli*

Capsule

If intestinal pathogenic isolates possess a capsule, it is typically a group 1 or 4 capsule or the group 1–like capsule colanic acid. Group 1 capsules may protect against desiccation outside the intestinal tract, but a role in pathogenesis has not been established. Some older studies, using nondefined ETEC mutants suggested that group 1 capsule may contribute to intestinal adherence [50], whereas others (also using nondefined mutants) had mixed [51] or contradictory results [52].

LPS

Although certain O-antigen serotypes are associated with various intestinal pathogenic strains (e.g., O157 and EHEC) [4], as with capsule, no defined role for LPS in the pathogenesis of intestinal pathogenic *E. coli* has been established. A potential role for O157 in adherence was evaluated [53,54], but the O-antigen diminished adhesion. Another study determined that O157 LPS augmented the cytotoxicity of the Stx toxin on endothelial cells *in vitro* [4]. As will be described later, the O-antigen is important for protecting extraintestinal pathogenic *E. coli* strains against the bactericidal effects of the cationic antimicrobial peptides. Whether this also occurs in intestinal pathogenic strains awaits evaluation.

Extraintestinal Pathogenic Strains of *E. coli* (ExPEC)

Group 1 Capsule Does Not Appear to Play a Role in Extraintestinal Infection

Some, if not all, ExPEC strains are capable of producing the group 1–like capsule colanic acid. However, my laboratory was unable to demonstrate a role for this capsule in the pathogenesis of extraintestinal *E. coli* infection. The inability to produce colanic acid had no effect on resistance to the bactericidal effects of serum or the cationic antimicrobial peptide bactericidal permeabilitiy increasing protein (BPI), nor on virulence *in vivo* in an abscess model or in a systemic infection model [55]. In contrast, the majority of ExPEC strains possess either a group 2 or 3 capsule [56], and these surface polysaccharides are critical for successful extraintestinal infection.

Group 2 and 3 Capsules and the O-Antigen Moiety of LPS are Epidemiologically Associated with ExPEC

Capsule

A number of studies have determined the prevalence of group 2 and 3 capsule in *E. coli* isolates responsible for a variety of extraintestinal infections. Most studies have evaluated strains causing urinary tract infection (UTI) or urosepsis [57–59], bacteremia [60], and meningitis [61,62], but *E. coli* isolates responsible for other infections such as spontaneous bacterial peritonitis (SBP) also have been assessed in one study [63]. Most of these strains possess genes for and/or express group 2 or 3 capsular polysaccharide, with the typical prevalence from highest to lowest being bacteremia/meningitis (80–85%), SBP (73%), pyelonephritis (55–65%), and cystitis (45–55%).

Although certain serotypes are more prevalent than others, more than 40 serotypes have been expressed in strains causing extraintestinal infection [12]. K1 is typically the most common serotype identified in most studies. The most striking example is the association of K1 strains and neonatal meningitis (54% [61] and 84% [62] of cerebrospinal fluid isolates in two studies). K1 strains are identified less frequently from various adult infections ranging from 11–39%

[57,60,62,64]. Other commonly identified serotypes include K5, K2, K12, K52, and K95 [11,60,65].

O-Antigen

A number of studies have demonstrated that the majority of *E. coli* responsible for extraintestinal infection also possess an O-antigen. Studies primarily have evaluated strains causing UTI or urosepsis [58,59,61,64,66], meningitis [61,62], SBP [63], and bacteremia [60,64,67]. Similar to group 2 or 3 capsules, most of these strains express a large variety of O-antigens. Prevalence typically ranges from 80 to 95% for bacteremia/meningitis/SBP/pyelonephritis and somewhat less for cystitis (50–62%). Certain serotypes were identified more frequently than others (O1, O2, O4, O6, O7, O16, O18, O75) [12,66].

This epidemiologic association of a number of group 2 and 3 capsules and O-antigen with ExPEC suggests, but does not establish, that they are an important virulence factors. Further, the more frequent identification of certain serotypes does not establish that they confer a greater degree of virulence than others. Those serotypes may be markers for strains that possess an overall greater repertoire of virulence factors [10,57]. However, the importance of group 2 and 3 capsules and O-antigen as virulence factors has been clearly supported a variety of additional *in vitro* and *in vivo* studies that will be described later.

Group 2 and 3 Capsules and O-Antigens Protect Against Phagocytosis and Complement Activity in Vitro

The ability to avoid clearance by the innate host response is a prerequisite for a successful extracellular pathogen such as ExPEC. Numerous *in vitro* studies have documented that some combination of group 2 and 3 capsules and O-antigen is capable of protecting ExPEC strains against phagocytosis and the bactericidal activity of complement.

Several studies have concluded that group 2 capsules protect against phagocytosis *in vitro*. Early studies demonstrated that K1 strains were resistant to phagocytosis *in vitro*, except in the presence of K1 antisera [60,68]. The resistance of non-K1 serotypes was variable [60]. Comparisons of nondefined capsule-positive and capsule-negative pairs [14,65,69,70] and isogenic pairs [71] further support the idea that at least some capsules (K1, K29, K5) protect against phagocytosis.

Even more work supports the concept that capsules are important in protecting against the bactericidal activity of complement *in vitro*. A large number of capsule-positive bacteremic isolates were serum-resistant *in vitro* [60]. Although comparisons were made using nondefined K1-deficient derivatives, K1-positive strains repeatedly have been shown to be complement-resistant [69,72–75], and a single study demonstrated the same findings using K23-positive and -negative strains [75]. Using isogenic derivatives, my laboratory has demonstrated that the group 3 K54 capsule is the major complement-resistance determinant [76]. The K5 capsule does not appear to be as important for complement resistance when nondefined [73,77,78] or isogenic capsule-positive and -negative pairs were compared [73,77,78].

Certain O-specific antigens also appear to afford protection against phagocytosis and the bactericidal activity of complement. Early *in vitro* evaluations have supported complete LPS as playing an important role in protection against complement-mediated lysis [79–81] and phagocytosis [82–84]. Unfortunately, some studies evaluated clinically inappropriate isolates or did not use proven isogenic derivatives lacking only the O-specific region of the LPS. More recently, studies from my laboratory using an O4-positive wild-type strain and an O4-deficient derivative demonstrated that the O4-antigen plays only a minor role in conferring complement resistance [49]. In contrast, using defined isogenic derivatives, the O75-antigen appears to contribute a minor degree of protection against phagocytosis and a greater degree of protection against the bactericidal activity of complement [71,85].

Taken together, the overall implications of this body of work demonstrate that protection against phagocytosis (in the absence of specific antisera) and complement-mediated bactericidal activity is conferred in large part by certain capsular and O-antigen serotypes (Table II). As a whole, the data to date suggest that capsule is more important than O-antigen. However, the overall relative importance of capsule versus O-antigen in protecting against each of these killing mechanisms awaits future studies in which additional serotypes are evaluated. Since most ExPEC strain possess both a group 2 or 3 capsule and an O-antigen, for a given strain one or another or both of these bacterial traits likely mediate this protection. As a result, the overwhelming majority of ExPEC are resistant to these critical host innate defense mechanisms. Of course, this is not surpris-

TABLE II The Ability of Capsules and O-Antigens to Confer Protection Against Various Host Defense Factors[a]

Bacterial virulence trait	Protects against complement	Protects against phagocytosis	Protects against cationic or cationic-like antimicrobial peptides
Group 1–like capsule (colanic acid)	No	Not tested	No
Group 2 or 3 capsules	Yes. K1, K54 major factors; K5 minor factor	Yes. K1, K5, K54 not tested	No. Only K54 tested
O-antigens	Yes. O75 major factor; O4 minor factor	Yes. O75, O4 not tested	Yes. O4 major factor, no others tested

[a]Based on *in vitro* studies with isogenic derivatives (except K1) deficient in capsule or O-antigen [51,57,73,78,85,89]. Extrapolation to other serotypes may be unreliable.

ing because it is necessary to avoid clearance by these host defense mechanisms to be an effective extracellular, extraintestinal pathogen, which is the case for ExPEC strains.

O-Antigen but Not Group 2 or 3 Capsules Confers Protection Against Cationic Antimicrobial Peptides

Cationic antimicrobial peptides and functionally equivalent molecules such as bactericidal permeability-increasing protein (BPI) have been recognized as other critical components of the host innate defense response. Typically, these are short peptides (12–50 amino acids) that possess a net positive charge and hydrophobic regions and facilitate interactions with membranes. One of their biologic properties is broad antimicrobial activity [86]. Cationic antimicrobial peptides are present in neutrophil granules and probably all mucosal surfaces, including the genitourinary tract, lungs, and intestine [87]. Clearly, resistance to this host defense component is required for a wide variety of pathogens.

Early studies established that resistance to cationic antimicrobial proteins was mediated by the O-antigen moiety of LPS [88]. Studies by my laboratory using defined isogenic derivatives of ExPEC strain CP9 demonstrated that the O4-antigen, but not the K54 group 3 capsule, conferred resistance to BPI [55]. Additional studies using the same set of isogenic derivatives determined that the O4-antigen also conferred protection against the cationic steroid squalamine, an antimicrobial compound isolated from shark tissue [89]. Interestingly, the presence of the group 3 K54 capsule increased the bactericidal activity of squalamine. It is tempting to speculate that certain cationic antimicrobial molecules may have developed increased activity in the presence of capsule in response to pathogens acquiring this virulence trait. Although studies are limited, it seems clear that the O-antigen is the critical defense determinant for protection against this important class of antimicrobial molecules (see Table II).

O-Antigen and Capsule Are Important in Conferring Virulence in Vivo, But Their Importance Varies in Different Infection Sites

Early studies using undefined and more recent studies using isogenic derivatives have assessed the role of group 2 or 3 capsules and O-antigen in various infection models.

Urinary Tract Infection Models

Early studies that compared capsule-positive and -negative strains found that the capsule-negative derivatives caused less bladder and renal colonization and lethality than their capsule-positive counterparts; however, isogenic strains were not used [78,90]. Likewise, *in vivo* studies, again using nonisogenic strains, also suggested that LPS may contribute directly to the development of UTI [78,91,92]. Studies from my laboratory using isogenic derivatives deficient in the group 3 K54 capsule, the O4-antigen or both have demonstrated that the O4-antigen but not the K54 capsule enhanced virulence in an ascending, unobstructed model of UTI [93]. In urine, the function of neutrophils may be depressed, complement levels are low, and anticomplement activity may be present [94–98]. Further, the efficacy of neutrophils within the urinary tract is unclear. Based on these observations, together with the importance of the O-antigen in our model system, I have postulated that cationic peptides may be a critical host defense component within the genitourinary tract. Further support of this hypothesis awaits future studies.

Through the use of signature-tagged mutagenesis and *in vivo* screening in a mouse UTI model, three independent K2 capsule-deficient mutant derivatives of CFT073 were identified recently (Michael Donnenberg, personal communication). Subsequent coinfection experiments confirmed that in the absence of the K2 capsule, bacterial survival is diminished, demonstrating a role for the K2 capsule in uropathogenesis. Although a contribution of the K54 capsule to survival within the urinary tract was not demonstrated [93], these findings are not necessarily contradictory. The K2 capsule may serve a different biologic function from the K54 capsule (e.g., confer resistance to cationic antimicrobial peptides). Alternatively, the K54 capsule may enhance survival within the urinary tract, but its contribution was missed because coinfection experiments were not performed. Future studies will be able to resolve these issues and further clarify the role of various group 2 and 3 capsules in UTI.

Systemic Infection Models

Using isogenic strains, my laboratory evaluated the contribution of the K54 capsule in systemic infection after intraperitoneal (IP) challenge and intravascular challenge in mice [99]. The K54 capsule enhanced virulence in each of these model systems. These results are consistent with findings from a study using a veterinary pathogen [100]. In contrast, loss of the O4-antigen increased virulence after IP challenge [49]. This finding was unexpected because some early studies using undefined strains suggested that LPS contributes to host death in systemic infection models [101]. Data will be discussed later that suggest that this finding may be due to the ability of the O4-antigen to modulate the host inflammatory response. In summary, these findings are consistent with the hypothesis that bacterial virulence traits that confer resistance to the bactericidal activity of complement (e.g., K54 capsule) and/or phagocytosis are important for systemic infection. Protection against cationic peptides (e.g., O-antigens) appears to be less important in this setting.

Meningitis Model

Most cases of neonatal meningitis are due to ExPEC strains that express a K1 capsule. The contribution of this capsular serotype has been evaluated in a neonatal rat model of meningitis. High levels of bacteremia are required for the development of meningitis in both humans and the rat model system. In rats, the group 2 K1 capsule but not the O18-specific antigen significantly contributes to high bacterial titers in blood and subsequent meningitis by conferring resistance to the bactericidal effects of complement and phagocytosis [102]. In subsequent studies, the K1 capsule did not appear to contribute to the penetration of the blood-brain barrier [103]. Therefore, the K1 capsule contributes to the pathogenesis of neonatal meningitis by enabling its survival within the host.

Abscess Model

Using isogenic derivatives of ExPEC strain CP9, my colleagues and I have shown that the K54 capsule enhances virulence in the rat granuloma pouch model that mimics an abscess [99]. O-antigen-deficient derivatives have not been evaluated in this model as of yet, but this finding is consistent with complement and/or phagocytosis as being important host defenses in this setting.

Pneumonia Model

Lastly, my laboratory also has used isogenic strains to evaluate the role of capsule and O-antigen in a rat pneumonitis model [104,105]. Some animals also were depleted of neutrophils and/or complement prior to bacterial challenge to assess the role of these host factors in infection. Both the K54 capsule and O4-antigen conferred a significant survival advantage to CP9 *in vivo*, as measured by bacterial growth/clearance. Both neutrophils and complement (likely mediated by nonbactericidal properties) were significantly important for the clearance of the wild-type pathogen; however, at high bacterial challenge inocula, both these defense components were not sufficiently efficient to prevent bacterial growth.

In summary, although both capsule and O-antigen are important virulence traits, their relative importance depends on the site of infection. Data from my laboratory suggest that this is due, in part, to variable activity or efficacy of various host defense factors at different sites. My laboratory's present conceptualization of this paradigm is based on the *in vitro* and *in vivo* evaluation of the wild-type ExPEC strain CP9 and its isogenic derivatives [49,55,93,99,104,105] (Table III). Data from the meningitis model also support this interpretation. Evaluation of additional strains and their respective capsule and O-antigen would be welcomed so that this model could be substantiated or refuted.

TABLE III The Importance of the Group 3 K54 Capsule and the O4-Antigen in Various Infection Models and the Defined or Predicted Critical Site-Dependent Host Defense Factors

Infection model	K54 capsule enhances virulence	O4-antigen enhances virulence	Critical site-dependent host factors
UTI	No	Yes	?Cationic antimicrobial peptides
Systemic infection	Yes	No	Complement/?neutrophils
Abscess	Yes	Not tested	Complement/?neutrophils/?cationic antimicrobial peptides
Pneumonia	Yes	Yes	Complement/neutrophils/?cationic antimicrobial peptides

However, the role of capsule and O-antigen in the pathogenesis of infection is not limited to its ability to protect *E. coli* against the bactericidal activity of various host defense components. The ability of these virulence traits to both stimulate and modulate the host inflammatory response is an additional layer of complexity that significantly contributes to the outcome of infection and is discussed in the next section.

SURFACE POLYSACCHARIDES STIMULATE AND ATTENUATE THE HOST INFLAMMATORY RESPONSE

Both Capsule and Lipid A Stimulate the Innate Host Proinflammatory Response

The manner in which the innate versus adaptive immune response recognizes infectious microbes is distinct [106]. The adaptive response uses T- and B-cell receptors that are generated somatically in a random manner during development. This results in an extensive repertoire capable of recognizing virtually any antigen. In contrast, the innate response uses a limited number of germ cell—encoded receptors. It appears that the evolutionary design of the innate response is to recognize a few highly conserved structures that are represented in a variety of pathogens. These structures have been termed *pathogen-associated molecular patterns* (PAMPs), and their cognate receptors are referred to as *pattern-recognition receptors* (PRRs). Examples of PAMPs include lipid A, peptidoglycan, lipoteichoic acid, mannans, bacterial DNA, glucans, and double-stranded RNA. Three classes of PRRs have been described: (1) secreted [e.g., mannose-binding lectin (MBL)], (2) endocytic (e.g., lectin-family macrophage mannose receptor), and (3) signaling (e.g., toll receptors). At least 10 mammalian toll-like receptors have been identified. Although we are beginning to understand the spectrum of PAMPs and their cognate PRRs that mediate pattern-recognition responses, more await identification. Further, even less is known about the roles of specific PRRs on cellular activation and the overall effect of different PAMPs on the innate immune response.

Lipid A

Bacterial lipid A has been the most extensively studied PAMP. It is a potent proinflammatory mediator. Effects of interest include recruitment and activation of pulmonary neutrophils and increased pulmonary vascular permeability with resulting noncardiogenic pulmonary edema and hypoxemia [107]. Although some of the observed effects have been directly attributable to endotoxin (e.g., endothelial damage), most are believed to be mediated through the activation of alveolar macrophages, the complement cascade, and subsequent recruitment and activation of neutrophils, monocytes, and macrophages. The signal-transduction pathway of the proinflammatory effects of lipid A has been clarified recently. LPS-binding protein binds lipid A, which facilitates transfer to CD14 on macrophages [108]. CD14–lipid A interacts with the MD-2 and TLR4 complex, resulting in nuclear translocation of NF-κB and subsequent activation of genes involved in the inflammatory response.

Group 3 K54 Capsule

My laboratory has postulated that at least certain group 2 or 3 capsules are PAMPs. Using the rat pneumonia model, my laboratory recently has demonstrated that the group 3 K54 capsule from the ExPEC strain CP9 is proinflammatory and stimulates neutrophil recruitment [104]. Unpublished data from my laboratory also has demonstrated that the K54 capsule induces pulmonary tumor necrosis factor α (TNF-α) and the chemokine MIP-2. Lastly, previously published data indirectly support the concept that the K54 capsule is proinflammatory. Mice were injected intraperitoneally with killed CP9 (wt) or two independent isogenic capsule-deficient derivatives. Since killed organisms were used, subsequent death was likely due to a bacterial product-mediated host response. At high challenge inocula, the mean survival of the mice given the capsule-deficient strains was significantly longer than that of the mice given the wild-type strain. Taken together, these findings suggest that although group 2 or 3 capsules confer resistance to a variety of host defenses, the host also recognizes capsule. This results in the induction of a protective proinflammatory response (or if unregulated, a detrimental response, e.g., septic shock). Therefore, I hypothesize that capsule, like lipid A, is an important mediator for the activation of the innate immune response.

O4-Antigen Attenuates the Innate Host Inflammatory Response

Data from my laboratory also support the concept that the overall effect of the O4-antigen is to attenuate the host's proinflammatory response. Using the rat pneumonitis model, my colleagues and I have demonstrated that the O4-antigen diminishes pulmonary neutrophil recruitment [104]. Further, LD_{50} values in mice after IP challenge with the wild-type ExPEC strain CP9 and isogenic O4-specific antigen—deficient derivatives are contrary to predictions. The LD_{50} valuess of the O4-specific antigen—deficient strains are significantly lower than those of CP9 [49]. These findings are consistent with the hypothesis that in the absence of the O4-specific antigen, an increased proinflammatory host response occurs that may be detrimental to the host. To date, the O-specific antigen usually is thought of as a bacterial component that protects against various host defense factors. However, in addition to this important role, another critical function of the O-specific antigen may be to attenuate the host inflammatory response. Potential benefits include a reduction in bactericidal activity generated by the host and/or a lower probability that the host response becomes unregulated (e.g., septic shock). Such a response may lead to premature death of the host and thereby decrease the likelihood of bacterial transmission to a new individual. It is interesting to note that most gram-negative *systemic* pathogens possess an O-specific antigen, whereas *mucosal* gram-negative pathogens generally do not. If these speculations are substantiated, an understanding of how the O-specific antigen attenuates host response could have significant implications both for the potential use of biologic modulators directed against the host response and for approaches based on inactivating bacterial components in attempts to modify sepsis syndromes.

CAPSULE AND O-ANTIGEN MAY SUBVERT THE HOST'S ACQUIRED IMMUNOLOGIC RESPONSE

Another important function of capsule and O-antigen is to deter the generation of a maximally efficacious acquired host immunologic response. This area has been relatively neglected but is probably as important as any aspect of bacterial virulence. At least three different mechanisms likely exist and contribute to the relative virulence of ExPEC strains.

The most obvious mechanism is the multiplicity of capsular and O-antigen serotypes possessed by ExPEC strains. Although serotype-specific antibodies have afforded protection in model systems against homologous strains [109,110], reinfection with a different serotype usually results in reinfection. A second mechanism has been attributed primarily to the K1 and K5 serotypes [14,111]. Some experimental data suggest that a poor to negligible antibody response occurs after exposure to these group 2 capsules. It has been postulated that this is due to structural similarities between these capsule and host structures. Lastly, O-antigen and capsule may shield antigenic epitopes on underlying structures such as lipid A and outer membrane proteins, thereby preventing an optimal immunologic response in natural infection [112,113]. They also prevent antibodies that are generated from reaching their targets deep to capsule and O-antigen [114]. These later functions may be the most critical. A number of outer membrane proteins and lipid A are conserved between *E. coli* strains. If natural bactericidal antibodies were able to be generated against and reach these targets, the host likely would be resistant to recurrent infection from various ExPEC strains. Future studies are warranted to determine the extent to which this occurs, its relative role in pathogenesis, and whether immunization with these targets (in contrast to natural infection) alters outcome.

SUMMARY

One needs to appreciate that the pathogenesis of infection and the nature of the responsible pathogen, including its surface polysaccharides, are dramatically different between intestinal and extraintestinal pathogenic *E. coli*. Further, when considering the role of capsule and LPS in each of these settings, three separate aspects should be considered. First is the role of capsule and O-antigen in protecting against various host defense components. Second is the ability of various surface polysaccharides to both stimulate (capsule, lipid A) and attenuate (O-antigen) the host inflammatory response. Lastly, these bacterial virulence traits likely play a critical role in diminishing the development of an optimal acquired host immunologic response against *E. coli*. A more complete understanding of these different layers in the infectious process is needed to realize improved treatment outcomes and/or the development of a successful vaccine.

REFERENCES

1. Whitfield, C., and Roberts, I. (1999). Structure, assembly and regulation of expression of capsules in *Escherichia coli*. *Mol. Microbiol.* 31:1307–1319.
2. Russo, T., and Singh, G. (1993). An extraintestinal, pathogenic isolate of *Escherichia coli* (O4/K54/H5) can produce a group 1 capsule which is divergently regulated from its constitutively produced group 2, K54 capsular polysaccharide. *J. Bacteriol.* 175:7617–7623.
3. Russo, T., and Johnson, J. (2000). A proposal for a new inclusive designation for extraintestinal pathogenic isolates of *Escherichia coli*: ExPEC. *J. Infect. Dis.* 181:1753–1754.

4. Nataro, J. P., and Kaper, J. B. (1998). Diarrheagenic *Escherichia coli* [published erratum appears in *Clin. Microbiol. Rev.* 1998;11(2):403]. *Clin. Microbiol. Rev.* 11(1):142–201.
5. Dho-Moulin, M., and Fairbrother, J. M. (1999). Avian pathogenic *Escherichia coli* (APEC). *Vet. Res.* 30(2–3):299–316.
6. De Rycke, J., Milon, A., and Oswald, E. (1999). Necrotoxic *Escherichia coli* (NTEC): Two emerging categories of human and animal pathogens. *Vet. Res.* 30(2–3):221–233.
7. Dozois, C. M., and Curtiss, R. (1999). Pathogenic diversity of *Escherichia coli* and the emergence of "exotic" islands in the gene stream. *Vet. Res.* 30(2–3):157–179.
8. Harel, J., and Martin, C. (1999). Virulence gene regulation in pathogenic *Escherichia coli*. *Vet. Res.* 30(2–3):131–155.
9. Yancey, R. J. (1999). Vaccines and diagnostic methods for bovine mastitis: fact and fiction. *Adv. Vet. Med.* 41(2–3):257–273.
10. Johnson, J. R., Delavari, P., Kuskowski, M., and Stell, A. L. (2001). Phylogenetic distribution of extraintestinal virulence-associated traits in *Escherichia coli*. *J. Infect. Dis.* 183(1):78–88.
11. Jann, K., and Jann, B. (1992). Capsules of *Escherichia coli:* Expression and biologic significance. *Can. J. Microbiol.* 38:705–710.
12. Orskov, F., Orskov, I., Jann, B., and Jann, K. (1977). Serology, chemistry, and genetics of O and K antigens of *Escherichia coli*. *Bacteriol. Rev.* 41:667–710.
13. Jann, B., and Jann, K. (1990). Structure and biosynthesis of the capsular antigens of *Escherichia coli*. *Curr. Top. Microbiol. Immunol.* 150:19–42.
14. Jann, K., and Jann, B. (1987). Polysaccharide antigens of *Escherichia coli*. *Rev. Infect. Dis.* 9(Suppl. 5):S517–S526.
15. Pearce, R., and Roberts, I. (1995). Cloning and analysis of gene clusters for production of the *Escherichia coli* K10 and K54 antigens: Identification of a new group of *serA*-linked capsule gene clusters. *J. Bacteriol.* 177:3992–3997.
16. MacLachlan, P. R., Keenleyside, W. J., Dodgson, C., and Whitfield, C. (1993). Formation of the K30 (group I) capsule in *Escherichia coli* O9:K30 does not require attachment to lipopolysaccharide lipid A-core. *J. Bacteriol.* 175(23):7515–7522.
17. Rahn, A., Drummelsmith, J., and Whitfield, C. (1999). Conserved organization in the *cps* gene clusters for expression of *Escherichia coli* group 1 K antigens: Relationship to the colanic acid biosynthesis locus and the *cps* genes from *Klebsiella pneumoniae*. *J. Bacteriol.* 181(7):2307–2313.
18. Gottesman, S., and Stout, V. (1991). Regulation of capsular polysaccharide synthesis in *Escherichia coli* K12. *Mol. Microbiol.* 5(7):1599–1606.
19. Jayaratne, P., Keenleyside, W. J., MacLachlan, P. R., *et al.* (1993). Characterization of *rcsB* and *rcsC* from *Escherichia coli* O9:K30:H12 and examination of the role of the *rcs* regulatory system in expression of group I capsular polysaccharides. *J. Bacteriol.* 175(17):5384–5394.
20. Keenleyside, W. J., Jayaratne, P., MacLachlan, P. R., and Whitfield, C. (1992). The *rcsA* gene of *Escherichia coli* O9:K30:H12 is involved in the expression of the serotype-specific group I K (capsular) antigen. *J. Bacteriol.* 174(1):8–16.
21. Hobbs, M., and Reeves, P. (1994). The JUMPstart sequence: A 39-bp element common to several polysaccharide gene clusters. *Mol. Microbiol.* 12:855–856.
22. Nieto, J., Bailey, M., Hughes, C., and Koronakis, V. (1996). Suppression of transcription polarity in the *Escherichia coli* hemolysin operon by a short upstream element shared by polysaccharide and transfer determinants. *Mol. Microbiol.* 19:705–713.
23. Stevens, M., Clarke, B., and Roberts, I. (1997). Regulation of the *Escherichia coli* K5 capsule gene cluster by transcription antitermination. *Mol. Microbiol.* 24:1001–1012.
24. Roberts, I. (1996). The biochemistry and genetics of capsular polysaccharide production in bacteria. *Annu. Rev. Microbiol.* 50:285–315.
25. Russo, T., Wenderoth, S., Carlino, U., *et al.* (1998). Identification, genomic organization and analysis of the group III capsular polysaccharide genes *kpsD*, *kpsM*, *kpsT*, and *kpsE* from an extraintestinal isolate of *Escherichia coli* (CP9, O4/K54/H5). *J. Bacteriol.* 180:338–349.

26. Orskov, I., and Nyman, K. (1974). Genetic mapping of the antigenic determinants of two polysaccharide K antigens, K10 and K54, in *Escherichia coli. J. Bacteriol.* 120:43–51.
27. Vimr, E. (1991). Map position and genome organization of the *kps* cluster for polysialic acid biosynthesis in *Escherichia coli* K1. *J. Bacteriol.* 173:1335–1338.
28. Roberts, I. S., Mountford, R., Hodge, R., *et al.* (1988). Common organization of gene clusters for production of different capsular polysaccharides (K antigens) in *Escherichia coli. J. Bacteriol.* 170(3):1305–1310.
29. Roberts, I., Mountford, R., High, N., *et al.* (1986). Molecular cloning and analysis of genes for production of K5, K7, K12, and K92 capsular polysaccharides in *Escherichia coli. J. Bacteriol.* 168(3):1228–1233.
30. Silver, R. P., Finn, C. W., Vann, W. F., *et al.* (1981). Molecular cloning of the K1 capsular polysaccharide genes of *E. coli. Nature* 289:696–698.
31. Bliss, J., and Silver, R. (1996). Coating the surface: A model for expression of capsular polysialic acid in *Escherichia coli* K1. *Mol. Microbiol.* 21:221–231.
32. Cieslewicz, M., and Vimr, E. (1996). Thermoregulation of *kpsF*, the first region 1 gene in the *kps* locus for polysialic acid biosynthesis in *Escherichia coli* K1. *J. Bacteriol.* 178(11):3212–3220.
33. Clarke, B. R., Pearce, R., and Roberts, I. S. (1999). Genetic organization of the *Escherichia coli* K10 capsule gene cluster: Identification and characterization of two conserved regions in group III capsule gene clusters encoding polysaccharide transport functions. *J. Bacteriol.* 181(7):2279–2285.
34. Finke, A., Jann, B., and Jann, K. (1990). CMP-KDO-synthetase activity in *Escherichia coli* expressing capsular polysaccharide. *FEMS Microbiol. Lett.* 69:129–134.
35. Heinrichs, D. E., Yethon, J. A., and Whitfield, C. (1998). Molecular basis for structural diversity in the core regions of the lipopolysaccharides of *Escherichia coli* and *Salmonella enterica. Mol. Microbiol.* 30(2):221–232.
36. Amor, K., Heinrichs, D. E., Frirdich, E., *et al.* (2000). Distribution of core oligosaccharide types in lipopolysaccharides from *Escherichia coli. Infect. Immun.* 68(3):1116–1124.
37. Gibb, A. P., Barclay, G. R., Poxton, I. R., and di Padova, F. (1992). Frequencies of lipopolysaccharide core types among clinical isolates of *Escherichia coli* defined with monoclonal antibodies. *J. Infect. Dis.* 166(5):1051–1057.
38. Whitfield, C., and Valvano, M. (1993). Biosynthesis and expression of cell-surface polysaccharides in gram-negative bacteria. *Adv. Microbiol. Physiol.* 35:135–246.
39. Whitfield, C., Amor, P., and Koplin, R. (1997). Modulation of the surface architecture of gram-negative bacteria by the action of surface polymer lipid A-core ligase and by determinants of polymer chain length. *Mol. Microbiol.* 23:629–638.
40. Schnaitman, C. A., and Klena, J. D. (1993). Genetics of lipopolysaccharide biosynthesis in enteric bacteria. *Microbiol. Rev.* 57(3):655–682.
41. Reeves, P. R., Hobbs, M., Valvano, M. A., *et al.* (1996). Bacterial polysaccharide synthesis and gene nomenclature. *Trends Microbiol.* 4(12):495–503.
42. Whitfield, C. (1995). Biosynthesis of lipopolysaccharide O antigens. *Trends Microbiol.* 3(5):178–185.
43. Finlay, B., and Falkow, S. (1997). Common themes in microbial pathogenicity revisited. *Microbiol. Mol. Biol. Rev.* 61:136–169.
44. Frosch, M., Edwards, U., Bousset, K., *et al.* (1991). Evidence for a common molecular origin of the capsule gene loci in gram-negative bacteria expressing group II capsular polysaccharides. *Mol. Microbiol.* 5(5):1251–1263.
45. Moxon, E., and Kroll, J. (1990). The role of bacterial polysaccharide capsules as virulence factors. *Curr. Top. Microbiol. Immunol.* 150:65–85.
46. Xiang, S. H., Hobbs, M., and Reeves, P. R. (1994). Molecular analysis of the *rfb* gene cluster of a group D2 *Salmonella enterica* strain: Evidence for its origin from an insertion sequence-mediated recombination event between group E and D1 strains. *J. Bacteriol.* 176(14):4357–4365.

47. Stevenson, G., Neal, B., Liu, D., *et al.* (1994). Structure of the O antigen of *Escherichia coli* K-12 and the sequence of its *rfb* gene cluster. *J. Bacteriol.* 176(13):4144–4156.
48. Reeves, P. (1993). Evolution of *Salmonella* O antigen variation by interspecific gene transfer on a large scale. *Trends Genet.* 9(1):17–22.
49. Russo, T., Sharma, G., Brown, C., and Campagnari, A. (1995). The loss of the O4 antigen moiety from the lipopolysaccharide of an extraintestinal isolate of *Escherichia coli* has only minor effects on serum sensitivity and virulence *in vivo. Infect. Immun.* 63:1263–1269.
50. Hadad, J., and Gyles, C. (1982). The role of K antigens of enteropathogenic *Escherichia coli* in colonization of the small intestine of calves. *Can. J. Comp. Med.* 46:21–26.
51. Nagy, B., Moon, H., and Isaacson, R. (1976). Colonization of porcine small intestine by *Escherichia coli*: Ileal colonization and adhesion by pig enteropathogens that lack K88 antigen and by some acapsular mutants. *Infect. Immun.* 13:1214–1220.
52. Runnels, P. L., and Moon, H. W. (1984). Capsule reduces adherence of enterotoxigenic *Escherichia coli* to isolated intestinal epithelial cells of pigs. *Infect. Immun.* 45(3):737–740.
53. Cockerill, F., Beebakhee, G., Soni, R., and Sherman, P. (1996). Polysaccharide side chains are not required for attaching and effacing adhesion of *Escherichia coli* O157:H7. *Infect. Immun.* 64(8):3196–3200.
54. Bilge, S. S., Vary, J. C., Dowell, S. F., and Tarr, P. I. (1996). Role of the *Escherichia coli* O157:H7 O side chain in adherence and analysis of an *rfb* locus. *Infect. Immun.* 64(11):4795–4801.
55. Russo, T., Sharma, G., Weiss, J., and Brown, C. (1995). The construction and characterization of colanic acid–deficient mutants in an extraintestinal isolate of *Escherichia coli* (O4/K54/H5). *Microb. Pathog.* 18:269–278.
56. Orskov, F., Orskov, I., Jann, B., and Jann, K. (1971). Immunoelectrophoretic patterns of extracts from all *Escherichia coli* O and K antigen test strains: correlation with pathogenicity. *Acta Pathol. Microbiol. Scand. [B]* 79(2):142–152.
57. Johnson, J. R., and Stell, A. L. (2000). Extended virulence genotypes of *Escherichia coli* strains from patients with urosepsis in relation to phylogeny and host compromise. *J. Infect. Dis.* 181(1):261–272.
58. Johnson, J. R. (1991). Virulence factors in *Escherichia coli* urinary tract infections. *Clin. Microbiol. Rev.* 4(1):80–128.
59. Orskov, I., Orskov, F., Birch-Andersen, A., *et al.* (1982). O, K, H and fimbrial antigens in *Escherichia coli* serotypes associated with pyelonephritis and cystitis. *Scand. J. Infect. Dis. Suppl.* 33(3):18–25.
60. Cross, A. S., Gemski, P., Sadoff, J. C., *et al.* (1984). The Importance of the K1 capsule in invasive infections caused by *Escherichia coli. J. Infect. Dis.* 149(2):184–193.
61. Korhonen, T. K., Valtonen, M. V., Parkkinen, J., *et al.* (1985). Serotypes, hemolysin production, and receptor recognition of *Escherichia coli* strains associated with neonatal sepsis and meningitis. *Infect. Immun.* 48(2):486–491.
62. Robbins, J. B., McCracken, G. H., Gotschlich, E. C., *et al.* (1974). *Escherichia coli* K1 capsular polysaccharide associated with neonatal meningitis. *New Engl. J. Med.* 290(22):1216–1220.
63. Soriano, G., Coll, P., Guarner, C., *et al.* (1995). *Escherichia coli* capsular polysaccharide and spontaneous bacterial peritonitis in cirrhosis. *Hepatology* 21(3):668–673.
64. Opal, S. M., Cross, A., Gemski, P., and Lyhte, L. W. (1988). Survey of purported virulence factors of *Escherichia coli* isolated from blood, urine and stool. *Eur. J. Clin. Microbiol. Infect. Dis.* 7(3):425–427.
65. Cross, A. (1990). The biologic significance of bacterial encapsulation. *Curr. Top. Microbiol. Immunol.* 150:87–95.
66. Orskov, I., and Orskov, F. (1985). *Escherichia coli* in extra-intestinal infections. *J. Hyg. (Lond.)* 95(3):551–575.
67. McCabe, W. R., Kaijser, B., Olling, S., *et al.* (1978). *Escherichia coli* in bacteremia: K and O antigens and serum sensitivity of strains from adults and neonates. *J. Infect. Dis.* 138:33–41.

68. Welch, W. D., Martin, W. J., Stevens, P., and Young, L. S. (1979). Relative opsonic and protective activities of antibodies against K1, O and lipid A antigens of *Escherichia coli*. *Scand. J. Infect. Dis.* 11(4):291–301.
69. Aguero, M. E., and Cabello, F. C. (1983). Relative contribution of ColV plasmid and K1 antigen to the pathogenicity of *Escherichia coli*. *Infect. Immun.* 40(1):359–368.
70. Horwitz, M. A., and Silverstein, S. C. (1980). Influence of the *Escherichia coli* capsule on complement fixation and on phagocytosis and killing by human phagocytes. *J. Clin. Invest.* 65(1):82–94.
71. Burns, S. M., and Hull, S. I. (1999). Loss of resistance to ingestion and phagocytic killing by O(–) and K(–) mutants of a uropathogenic *Escherichia coli* O75:K5 strain. *Infect. Immun.* 67(8):3757–3762.
72. Allen, P. M., Roberts, I., Boulnois, G. J., *et al.* (1987). Contribution of capsular polysaccharide and surface properties to virulence of *Escherichia coli* K1. *Infect. Immun.* 55(11):2662–2668.
73. Gemski, P., Cross, A. S., and Sadoff, J. C. (1980). K1 antigen–associated resistance to the bactericidal activity of serum. *FEMS Microbiol. Lett.* 9:193–197.
74. Leying, H., Suerbaum, S., Kroll, H. P., *et al.* (1990). The capsular polysaccharide is a major determinant of serum resistance in K-1 positive blood culture isolates of *Escherichia coli*. *Infect. Immun.* 58(1):222–227.
75. Vermeulen, C., Cross, A. C., Byrne, W. R., and Zollinger, W. (1988). Quantitative relationship between capsular content and killing of K1-encapsulated *Escherichia coli*. *Infect. Immun.* 56:2723–2730.
76. Russo, T., Moffitt, M., Hammer, C., and Frank, M. (1993). TnphoA-mediated disruption of K54 capsular polysaccharide genes in *Escherichia coli* confers serum sensitivity. *Infect. Immun.* 61:3578–3582.
77. Brooks, H. J. L., O'Grady, F., McSerry, M. A., and Cattell, W. R. (1980). Uropathogenic properties of *Escherichia coli* in recurrent urinary-tract infection. *J. Med. Microbiol.* 13:57–68.
78. Svanborg-Eden, C., Hagberg, L., Hull, R., *et al.* (1987). Bacterial virulence versus host resistance in the urinary tracts of mice. *Infect. Immun.* 55(5):1224–1232.
79. Pluschke, G., Mayden, J., Achtman, M., and Levine, R. P. (1983). Role of the capsule and the O antigen in resistance of O18:K1 *Escherichia coli* to complement-mediated killing. *Infect. Immun.* 42(3):907–913.
80. Porat, R., Johns, M. A., and McCabe, W. R. (1987). Selective pressures and lipopolysaccharide subunits as determinants of resistance of clinical isolates of gram-negative bacilli to human serum. *Infect. Immun.* 55(2):320–328.
81. Taylor, P. W. (1983). Bactericidal and bacteriolytic activity of serum against gram-negative bacteria. *Microbiol. Rev.* 47(1):46–83.
82. Stendahl, O., Normann, B., and Edebo, L. (1979). Influence of O and K antigens on the surface properties of *Escherichia coli* in relation to phagocytosis. *Acta Path. Microbiol. Scand.[B]* 87:85–91.
83. Medearis, D. N., Camitta, B. M., and Heath, E. C. (1968). Cell wall compostion and virulence in *Escherichia coli*. *J. Exp. Med.* 128:399–413.
84. Abe, C., Schmitz, S., Jann, B., and Jann, K. (1988). Monoclonal antibodies against O and K antigens of uropathogenic *Escherichia coli* O4:K12:H^- as opsonins. *FEMS Microbiol. Lett.* 51:153–158.
85. Burns, S. M., and Hull, S. I. (1998). Comparison of loss of serum resistance by defined lipopolysaccharide mutants and an acapsular mutant of uropathogenic *Escherichia coli* O75:K5. *Infect. Immun.* 66(9):4244–4253.
86. Hancock, R. E. and Scott, M. G. (2000). The role of antimicrobial peptides in animal defenses. *Proc. Natl. Acad. Sci. USA* 97(16):8856–8861.
87. Hancock, R. E., and Diamond, G. (2000). The role of cationic antimicrobial peptides in innate host defences. *Trends Microbiol.* 8(9):402–410.

88. Weiss, J., Beckerdite-Quagliata, S., and Elsbach, P. (1980). Resistance of gram-negative bacteria to purified bactericidal leukocyte proteins: relation to binding and bacterial lipopolysaccharide structure. *J. Clin. Invest.* 65(3):619–628.
89. Russo, T. A., and Mylotte, D. (1998). Expression of the K54 and O4 specific antigen has opposite effects on the bactericidal activity of squalamine against an extraintestinal isolate of *Escherichia coli. FEMS Microbiol. Lett.* 162(2):311–315.
90. Verweij-van Vught, A. M. J. J., van den Bosch, J. F., Namavar, F., *et al.* (1983). K Antigens of *Escherichia coli* and virulence in urinary-tract infection: Studies in a mouse model. *J. Med. Microbiol.* 16:147–155.
91. Taylor, P. W., and Koutsaimanis, K. G. (1975). Experimental *Escherichia coli* urinary infection in the rat. *Kidney Int.* 8:233–238.
92. Hagberg, L., Hull, R., Hull, S., *et al.* (1984). Difference in susceptibility to gram-negative urinary tract infection between C3H/HeJ and C3H/HeN mice. *Infect. Immun.* 46:839–844.
93. Russo, T., Brown, J., Jodush, S., and Johnson, J. (1996). The O4 specific antigen moiety of lipopolysaccharide but not the K54 group 2 capsule is important for urovirulence in an extraintestinal isolate of *Escherichia coli. Infect. Immun.* 64:2343–2348.
94. Acquatella, H., Little, P., deWardener, H., and Coleman, J. (1967). The effect of urine osmolality and pH on the bactericidal activity of plasma. *Clin. Sci.* 33:471–480.
95. Beeson, P., and Rowley, D. (1959). The anticomplementary effect of kidney tissue. *J. Exp. Med.* 110:685–697.
96. Chernew, I., and Braude, A. (1962). Depression of phagocytosis by solutes in concentrations found in kidney and urine. *J. Clin. Invest.* 41:1945–1953.
97. Norden, C., Green, G., and Kass, E. (1968). Antibacterial mechanisms of the urinary bladder. *J. Clin. Invest.* 47:2689–2700.
98. Suzuki, Y., Fukushi, F., Orikasa, S., and Kumagai, K. (1982). Opsonic effect of normal and infected human urine on phagocytosis of *Escherichia coli* and yeasts by neutrophils. *J. Urol.* 127:356–360.
99. Russo, T., Liang, Y., and Cross, A. (1994). The presence of K54 capsular polysaccharide increases the pathogenicity of *Escherichia coli in vivo. J. Infect. Dis.* 169:112–118.
100. Ngeleka, M., Harel, J., Jacques, M., and Fairbrother, J. M. (1992). Characterization of a polysaccharide capsular antigen of septicemic *Escherichia coli* O115:K"V165":F165 and evaluation of its role in pathogenicity. *Infect. Immun.* 60(12):5048–5056.
101. Smith, H. W., and Huggins, M. B. (1980). The association of the O18, K1 and H7 antigens and the ColV plasmid of a strain of *Escherichia coli* with its virulence and immunogenicity. *J. Gen. Microbiol.* 121:387–400.
102. Kim, K., Itabashi, H., Gemski, P., *et al.* (1992). The K1 capsule is the critical determinant in the development of *Escherichia coli* meningitis in the rat. *J. Clin. Invest.* 90:897–905.
103. Hoffman, J., Wass, C., Stins, W., and Kim, K. (1999). The capsule supports survival but not traversal of *Escherichia coli* K1 across the blood-brain barrier. *Infect. Immun.* 67:3566–3570.
104. Russo, T., Davidson, B., Priore, R., *et al.* (2000). Capsular polysaccharide and O-specific antigen divergently modulate pulmonary neutrophil influx in a rat *Escherichia coli* model of gram-negative pneumonitis. *Infect. Immun.* 68:2854–2862
105. Russo, T., Davidson, B., Priore, R., *et al.* (submitted).
106. Medzhitov, R., and Janeway, C. (2000). Innate immunity. *New Engl. J. Med.* 343:338–344.
107. Brigham, K., and Meyrick, B. (1986). Endotoxin and lung injury. *Am. Rev. Respir. Dis.* 133:913–927.
108. Raetz, C. (1993). Bacterial endotoxins: Extraordinary lipids that activate eucaryotic signal transduction. *J. Bacteriol.* 175:5745–5753.
109. Kaijser, B., Larsson, P., Nimmich, W., and Soderstrom, T. (1983). Antibodies to *Escherichia coli* K and O antigens in protection against acute pyelonephritis. *Prog. Allergy* 33:275–288.

110. Salles, M.-F., Mandine, E., Zalisz, R., *et al.* (1989). Protective effects of murine monoclonal antibodies in experimental septicemia: *E. coli* antibodies protect against different serotypes of *E. coli*. *J. Infect. Dis.* 159:641–647.
111. Kasper, D. L. (1986). Bacterial capsule: Old dogmas and new tricks. *J. Infect. Dis.* 153:407–415.
112. van der Ley, P., de Graaff, P., and Tommassen, J. (1986). Shielding of *Escherichia coli* outer membrane proteins as receptors for bacteriophages and colicins by O-antigenic chains of lipopolysaccharide. *J. Bacteriol.* 168:449–451.
113. Bradley, D., Howard, S., and Lior, H. (1991). Colicinogeny of O157:H7 enterohemorrhagic *Escherichia coli* and the shielding of colicin and phage receptors by their O-antigenic side chains. *Can. J. Microbiol.* 37:97–104.
114. van der Ley, P., Kuipers, O., Tommassen, J., and Lugtenberg, B. (1986). O-antigenic side chains of lipopolysaccharide prevent binding of antibody molecules to an outer membrane pore protein in Enterobacteriaceae. *Microbiol. Path.* 1:43–49.

Index